OUR SOILS

AND THEIR MANAGEMENT

SIXTH EDITION

ROY L. DONAHUE, Ph.D.

Professor Emeritus
Soil Science
Michigan State University
East Lansing

ROY HUNTER FOLLETT, Ph.D.

Professor
Extension Agronomy
Department of Agronomy
Colorado State University
Fort Collins

Agronomist
American Registry of Certified Professionals in Agronomy, Crops, and Soils

RODNEY W. TULLOCH, Ph.D.

Associate Professor
Agricultural Education
University of Kentucky
Lexington

OUR SOILS

AND THEIR MANAGEMENT

Increasing Production Through
Environmental Soil and Water Conservation
and Fertility Management

INTERSTATE PUBLISHERS, INC.

Danville, Illinois

OUR SOILS AND THEIR MANAGEMENT

Sixth Edition

Library of Congress Catalog Card No. 88-82225

4 5 6
7 8 9

ISBN 0-8134-2848-3

This Sixth Edition is professionally dedicated to a North Carolina farm boy, **Dr. Hugh Hammond Bennett** (1881 - 1960), who became the organizer and first chief of the USDA - Soil Conservation Service. He earned the affectionate title "Father of Soil Conservation." His name was entered into the Agricultural Hall of Fame at Bonner Springs, Kansas, in April 1985. The home of his birth in Anson County, North Carolina, was donated to the North Carolina Soil Conservation Society to be used as a conservation education center. (In his right hand is a probe used for determining soil surface crusting.)

(Photo: Courtesy, USDA - Soil Conservation Service)

PREFACE TO THE SIXTH EDITION

Soils are the basic building blocks for life. We are dependent on soils for food for ourselves, feed for animals, and fiber for all types of purposes. A knowledge of soils is useful for anyone and is essential for persons engaged in most agriculturally related jobs. A good understanding of soils can also lead to a number of interesting and challenging careers.

The sixth edition of *Our Soils and Their Management* consists of 21 chapters. Eleven chapters provide a basic understanding of soils, plant nutrition and fertilization, soil and water management, and soils judging. The information in these chapters provides the scientific basis for improved land use and management. Since different types of plants, as well as various methods of plant production, require different practices, nine separate chapters for plant groups or enterprise areas have been included. These nine chapters relate to rangelands, pastures, field crops, vegetable gardens, turf and ornamental plants, greenhouses and nurseries, orchards, forests, and vegetating of disturbed areas. The last chapter deals with the training and qualifications necessary for and the job opportunities available in various careers in soil and crop management. Updated and extensive conversion tables follow in the Appendix.

The wide range of subject matter makes this edition useful to those engaged in farming, other agricultural occupations, or occupations which relate to soils, land use, conservation, natural resources, the environment, ecology, geography, and many more areas.

A unique feature of this sixth edition is a section at the end of Chapter 1 entitled "Profiles of Major Soils in the 50 States."

ACKNOWLEDGMENTS

For their patience, encouragement, and understanding, the authors express appreciation to their wives, Lola, Barbara, and Charlotte, respectively.

Secretaries deserving special mention for timeliness and skill include: Theora Correll, Forsyth, Missouri; Doreen Jordan, Extension Agronomy Secretary, Colorado State University; and Phyliss Dedman, Department of Vocational Education, University of Kentucky.

The authors wish to acknowledge Dr. Roy Roath, Range Specialist, Colorado State University, for his extensive contributions to Chapter 12, "Rangeland Management."

For help in obtaining the photographs included in Section 1:16, we thank Dr. Ellis Knox, National Leader for Soil Survey Investigations, Soil Conservation Service, Washington, D.C. The authors also appreciate the information on "Andisols," a new soil order approved in mid-1989, furnished by Dr. John E. Witty, National Leader, Soil Classification, Soil Conservation Service, Washington, D.C.

We also thank each state director of agricultural education for helping with a survey of teacher-reviewers to improve the sixth edition of *Our Soils and Their Management.* Several agricultural education teachers also made valuable contributions to this edition.

Additional thanks go to Alan Heard, Assistant State Conservationist, the U.S. Department of Agriculture – Soil Conservation Service, Lexington, Kentucky; Dr. Richard Stinson, Department of Horticulture and Department of Agricultural Education, and Dr. John W. White, Department of Horticulture, The Pennsylvania State University, University Park; and Ms. Vickie M. Tarvin, Communications Director, Conservation Technology Information Center, West Lafayette, Indiana. A special thanks goes to Dr. Charlotte Tulloch, University of Kentucky, who provided editorial assistance on the manuscript.

The authors appreciate the superior professionalism of Interstate's Editorial Department, especially that of Ronald L. McDaniel, Vice-President–Editorial; Pat Ward, Production Editor; and Greg Ottarski, former Editorial Assistant.

The following reviewers from the University of Kentucky receive a special thank-you from the authors: Dr. Dean Knavel, Department of Horticulture, for reviewing Chapter 17 and providing numerous photographs; Dr. A. J. Powell, Department of Agronomy, for reviewing Chapter 16; Dr. John G. Strang, Department of Horticulture, for reviewing Chapter 18; Dr. Kenneth L. Wells, Department of Agronomy, for reviewing Chapters 6 and 7; and Dr. Richard Barnhisel, Department of Agronomy, for reviewing Chapter 20.

A special thanks also goes to Dr. James E. Christiansen, Professor of Agricultural Education, Texas A&M University, College Station, who reviewed Chapter 21, "Careers in Soil and Crop Management and Allied Subjects."

CONTENTS

TEACHER-REVIEWERS

The following teachers of agricultural education, who were identified by their state directors of agricultural education, responded to a survey on how to improve *Our Soils and Their Management.* The authors wish to thank them as well as other teachers who provided many excellent suggestions.

Teacher	***School***	***State***
Billy Bryan	Falkville	Alabama
Jasper P. Wagner, Jr.	Ider	Alabama
Nathan L. Moore	Westwood	Alabama
Dick Allen	San Luis Obispo	California
David Whitehead	Madison County	Georgia
William C. Wilder	Columbia City Joint	Indiana
Byron Callahan	Rensselaer Central	Indiana
Ken Stringer	Oldham County	Kentucky
Bradley A. Leger	Midland	Louisiana
Charles B. Cramer	Walkersville	Maryland
Jimmy C. Weeks	Shannon	Mississippi
Daniel R. Watts	Fairview	Montana
Dennis Mottl	Fullerton	Nebraska
Wilbur Palmer	Alvirne	New Hampshire
Eugene Walker	Ronda	North Carolina
Gerald Wettlaufer	Bottineau	North Dakota
Tim Reichert	Johnstown-Monroe	Ohio
Larry Lokai	Northwestern Clark	Ohio
William D. Waidelich	Southeastern Ross	Ohio
Roger A. Wehde	Dell Rapids	South Dakota
Barney McClure	Cleburne	Texas
Harvey Johnson	Hutto	Texas
Randy Mosely	Irion County	Texas
Danny Beck	Lamesa	Texas
Ivan P. Davis, Jr.	Buckingham Voc. Cen.	Virginia
W. Roger Mitchell	Lee Davis	Virginia

Teacher	*School*	*State*
Virgil Wilkins	Hundred	West Virginia
Howard Henderson	Valley	West Virginia
Don C. Leibelt	Green Bay East	Wisconsin
Bill Stewart	Douglas	Wyoming
Thomas M. Hale	Caribou Vo Tech Center	Maine
Ralph Gallagher	Fort Fairfield	Maine
Paul R. Lynch	Presque Isle	Maine
Frederick S. Warren	Wachusett Regional	Massachusetts
Charles Daigle	Enosburg Falls Area	Vermont
Randy Anderson	Pierson Taylor	Florida
Robert G. Brown	Eastmont	Washington

The authors welcome your comments on this edition of *Our Soils and Their Management* and encourage you to make suggestions for improvements that might be incorporated into the next revision. Write to them in care of the publisher.

CHAPTER 1

Soils and Soil Management

"Great opportunities exist for teaching and learning about new perspectives in soil science, soil survey interpretations, and land use planning." — Gerald W. Olson, Cornell University, 1981

OUTLINE

□ □ □

1:1 □ □ OVERVIEW

Soil is a natural, three-dimensional body at the earth's surface. It is capable of supporting plants and has properties resulting from the integrated effect of climate and living matter acting on earthy parent material, as conditioned by topography over periods of time.

Soils develop from geologic material as a result of five factors. If there is significant variation in any one factor, a different soil will be formed. The five factors are (1) parent material, (2) climate, (3) life, (4) topography, and (5) time.

Physical characteristics of soil include texture, structure, aeration, tilth, and moisture. Soil chemical and colloidal characteristics determine the abundance and availabilities of nutrient elements and cation exchange. Biological (life) character of soil includes all plants and animals of all sizes.

Drainage in soil means internal drainage and not direct surface runoff. Seven soil drainage classes are defined.

Soil profiles are described in relation to soil classification (taxonomy), diagnostic horizons, and the six soil categories: order, suborder, great group, subgroup, family, and series. A soil map of the United States is presented, based on soil orders and suborders.

Prime and unique lands, the 78 Benchmark soils, and soil management for pollution control complete the text for Chapter 1, along with a section on the unusual uses of clay soils. Clay soils are sometimes eaten by people and cattle and plastered on the bodies of people as mud to beautify.

1:2 □ SOIL FORMATION

The following five factors are responsible for soil formation (soil genesis): (1) parent material, (2) climate, (3) life, (4) topography, and (5) time.

1:2.1 □ Parent Material

Nature deposited sandstone in some places and shale or limestone in other places. At still different locations, molten rock (**magma**) from the interior of the earth was slowly pushed upward, forming and exposing granite.

But whatever bedrock nature left, weathering by rainwater, heating and cooling, freezing and thawing, and the action of plants and animals have been working for centuries to wear the rock away. At the same time that the rocks have been rotting, other forces in nature have been building a soil from the decaying rock products.

New chemicals have been formed in the soil from the decay products of the rocks and plants. Humus has been built from plant and animal residues. Clay minerals have, at the same time, been synthesized into something both physically and chemically different from the minerals in the parent rocks. Examples of synthesized clay soil minerals not present in parent rocks are **montmorillonite** and **kaolinite**.

Thus, we can observe that soils developed from sandstone are coarse-textured and sandy. Soils from shale are clay in texture but are not very fertile. Soils that develop from granite bedrock are usually sandy loams of low fertility. In contrast, the soils developed from limestone or basalt bedrock are usually dark in color, fine in texture, and very fertile (Note 1.1).

Note 1.1 — Minerals and Rocks as Parent Materials

Most soils develop from parent materials that came from minerals and rocks.

Minerals are naturally occurring inorganic elements or compounds with definite crystal structure, physical properties, and chemical composition. The most abundant mineral in nature is quartz. Quartz has a rhombohedral (6-sided) crystal structure, a hardness of 7 (diamond has a hardness of 10), and a chemical composition of silicon dioxide (SiO_2).

Rocks are mixtures of minerals. For example, sandstone, a rock, is impure quartz.

1:2.2 ▫ Climate

Imagine that a large limestone rock as big as a house suddenly appeared above the waves in the Gulf of Mexico near New Orleans, in the Atlantic Ocean near New York City, in Lake Michigan near Chicago, or in the Pacific Ocean near San Francisco. Soon afterward, exposure to the rain and sun would result in the surface of the rock becoming soft. Changes in temperature from day to night and from season to season would break the rock into scale-like pieces. Rainwater would dissolve some of the limestone and soften other parts of the rock. Wetting and drying and heating and cooling would cause expansion and contraction of the loosened rock particles. In some of the tiny cracks and depressions there would now be enough water and available plant nutrients to support lower forms of plant life. A soil is being born.

Starting from limestone bedrock, the number of years of weathering and synthesis of minerals required to make soil depends mostly upon the climate. For example, in the Upper Peninsula of Michigan it has taken 9,000 years to develop soil 3 to 6 inches (7.6 – 15.2 cm) deep. This is in comparison to a soil 12 to 16 inches (30.5 – 40.6 cm) deep that has developed from limestone bedrock near Richmond, Indiana, during a period of 15,000 years. Thus, in northern Michigan it has taken approximately 2,000 years for 1 inch (2.5 cm) of soil to develop from limestone, whereas in the warmer climate of Indiana it has taken approximately 1,000 years to develop 1 inch (2.5 cm) of soil from limestone.

If the bedrock in northern Michigan and Indiana had been granite (which weathers more slowly) instead of limestone, the number of years required to produce 1 inch (2.5 cm) of soil from granite would have been perhaps 10 times longer. In central Canada, less than 1 inch (2.5 cm) of soil has developed from granite rock in 10,000 years.

In the humid lowland tropical climates of Hawaii and Puerto Rico, where it never freezes, rock weathering continues every day in the year. For this reason, soils develop faster and deeper than in temperate regions. By contrast, permanently frozen rocks and glacial debris in northern Alaska may *never* develop a soil profile (Note 1.2).

Note 1.2 — How Soils Develop

Soils are synthesized from weathered rock products (lithosphere) on specific slopes (toposphere) while being acted upon by living organisms (biosphere) in distinct climates (atmosphere) over characteristic periods of time in years (chronosphere).

Weathering is defined as "all physical and chemical changes produced by atmospheric agents in rocks or other deposits at or near the earth's surface." These changes result in disintegration and decomposition of the material.

1:2.3 □ Life

Lower forms of life, such as **lichens** (a group of plants composed of a fungus and an alga in a symbiotic relationship), are among the first to grow under the harsh environments that exist on slightly weathered limestone bedrock. After many years of growth and death of these tiny plants, the environment is thereby improved to such an extent that other plants of a higher form, such as the mosses, become established. Generation after generation of lichens and mosses grow, reproduce, and die until the organic matter from their decaying tissues helps to make a better place for seed-bearing higher plants to start growing. Many annual weeds will migrate into the area, followed in a few years by some grasses and shrubs; then finally, if the rainfall is adequate, a hardwood forest will gradually dominate the vegetation on what was once bare limestone rock. With less rainfall, grasses will be the dominant vegetation.

Bacteria and fungi and many birds and other animals are constantly a part of the environment during soil formation. Animals and plants help in the process of disintegration and decomposition of rocks into soil (Figure 1.1). For example, of the animal life that we can see, earthworms, ants, and many rodents are always manipulating the soil material and thus modifying the continuous formation of soils from rocks.

Fig. 1.1 — In a few thousand years these granite rocks will become soil. *Top*: Heating and cooling by the sun has caused a scaling off (exfoliation) of the rock surface. *Bottom*: Tree roots are forcing the rocks apart to hasten exfoliation and soil formation. (Arrows point to roots.) (Courtesy, Roy L. Donahue)

There are about 1,800 species of earthworms throughout the world, the most common of which in the United States belong to the family Lumbricidae (the night crawlers). All species ingest (eat) coarse organic matter and burrow into the soil. The processes of soil profile formation are therefore delayed; but with earthworms, the soil is both chemically and physically a better medium for plant growth. Old earthworm holes are ideal channels for plant roots.

1:2.4 □ Topography

If this same limestone boulder (discussed in Section 1:2.2) were tilted slightly in order to permit rainwater to drain more rapidly, the time required to develop a hardwood forest would be increased. With less water staying on the boulder after each rain, plants would grow more slowly. This would result in less organic matter being returned to the rock to help the next generation of plants to grow.

On the other hand, if the large boulder were to have a slight depression in the center in order to hold more rainwater, the process of soil formation, plant growth, and plant succession would be faster. Thus, when the whole countryside is hilly, the steep slopes develop soil slowly, while the more level areas make soil faster. Soil development can be retarded by too much water, however.

1:2.5 □ Time

A soil forms from parent material by continuous weathering and the accumulation of organic matter on the surface. Just when parent material changes to soil is always difficult to determine. But as lime and other salts, clay, iron, and humus compounds move downward as parent material weathers, and as organic matter accumulates on the surface, we can say that a soil is being formed. In fact, this can be said when *any* of these activities can be seen or measured.

The age of a soil may be considered from two aspects: absolute and relative. **Absolute age** refers to the number of years the soil has been forming (developing), without interruption, from parent material. **Relative age** refers to the stage of maturity of the soil, as indicated by the soil profile. For example, the absolute age of a particular silt loam soil in Missouri that has been developing from windblown loess parent material on a steep slope may be 500,000 years, but the relative age may still be young because of constant surface erosion that washes away the upper A-horizon almost as fast as it is being formed.

By contrast, in Alaska, the absolute age of certain glacial material may be only 30 years, yet the relative age of the soil that has developed a surface horizon is also young. In North Carolina where weathering is more intense than in Alaska, a recognizable surface soil horizon has developed from parent material after only 50 years.

The kind of soil that has developed from a particular parent material in a designated climate with certain plants and animals under a given topography over a designated time will have a physical, chemical, and biological character that will be discussed in the next three sections.

1:3 □ PHYSICAL CHARACTER OF SOIL

The physical character of a soil and its suitability for plant growth depends on six factors: (1) texture, (2) structure, (3) consistence, (4) aeration, (5) tilth, and (6) moisture. These six factors are explained as follows.

Soil texture refers to the relative proportions of sand, silt, and clay particles in a mass of soil. The basic **textural classes**, in order of increasing proportion of fine particles, are: sand, loamy sand, sandy loam, loam, silt loam, silt, sandy clay loam, clay loam, silty clay loam, sandy clay, silty clay, and clay (Figure 1.2). (See also Figure 1.9.)

Figure 1.2 is a graphic guide for naming the textural class of a soil after the percentage of sand, silt, and clay has been determined in a laboratory.

By definition, sand particles vary in diameter from 2.0 to 0.05 millimeters, silt particles average from 0.05 to 0.002 millimeters in diameter, and clay particles are less than 0.002 millimeters in diameter. (Individual particles of silt can be seen only under a microscope, but clay particles cannot be seen.)

In nature, a soil is almost always a mixture (textural class) of sand, silt, and clay particles. To give names to natural soils, the triangle in Figure 1.2 was developed. For example, clay as a textural class consists of soil particles below 0.002

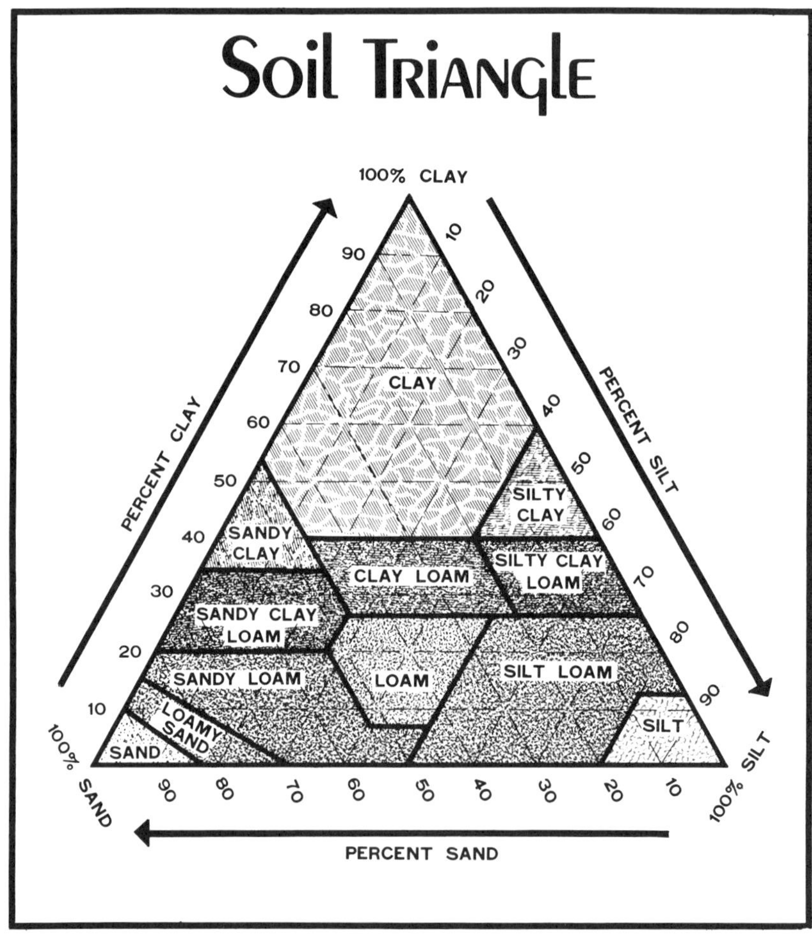

Fig. 1.2 — Graphic guide for soil textural classification (see Figure 1.10). (Courtesy, USDA - Soil Conservation Service) (Redrawn by Instructional Materials Services, Texas A&M University)

millimeters in diameter, but clay as a textural class may contain less than 40 percent silt, less than 20 percent sand, and over 40 percent clay.

Soil structure means the arrangement of primary soil particles (sand, silt, clay) into compound particles or aggregates. The principal forms of soil structure are ***platy*** (laminated), ***prismatic*** (vertical axis of aggregates longer than horizontal), ***columnar*** (prisms with rounded tops), ***blocky*** (angular or subangular), and ***granular***. ***Structureless*** soils are either ***single-grained*** (each grain by itself, as in dune sand) or ***massive*** (the particles adhering without any regular cleavage, as in many hardpans) (Figure 1.3).

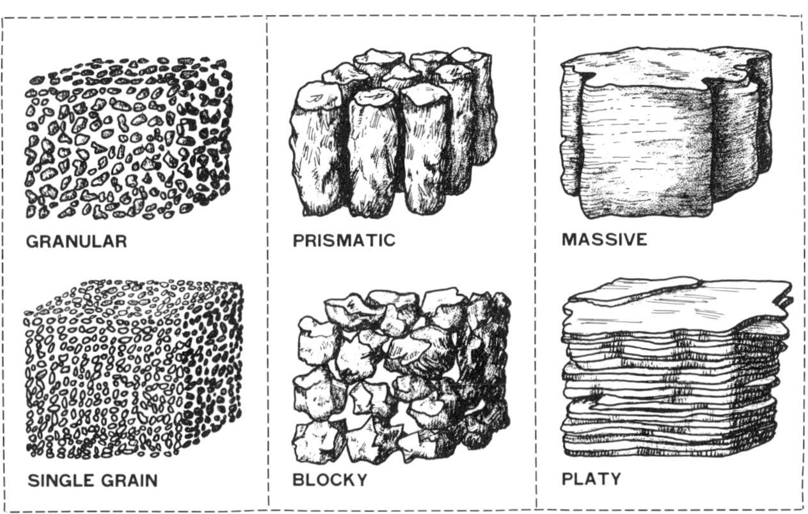

Fig. 1.3 — Principal soil structural classes. In addition to these shown are *columnar* (similar to *prismatic*) and *crumb* (porous granular). *Single grain* is not always classified as soil structure. (Courtesy, Instructional Materials Services, Texas A&M University)

Soil consistence is the feel of the soil and the ease with which a lump can be crushed by the fingers. Terms commonly used to describe consistence are:

- ***Loose*** — Noncoherent when dry or moist; does not hold together in a mass.
- ***Friable*** — When moist, crushes easily under gentle pressure between the thumb and forefinger and can be pressed together into a lump.
- ***Firm*** — When moist, crushes under moderate pressure between the thumb and forefinger, but resistance is distinctly noticeable.
- ***Plastic*** — When wet, readily deformed by moderate pressure but can be pressed into a lump; will form a "wire" when rolled between the thumb and forefinger.

- ***Sticky*** — When wet, adheres to other material and tends to stretch somewhat and pull apart rather than to pull free from other material.
- ***Hard*** — When dry, moderately resistant to pressure; can be broken with difficulty between the thumb and forefinger.
- ***Soft*** — When dry, breaks into powder or individual grains under very slight pressure.

Soil aeration relates to the exchange of air in soil with air from the atmosphere. The air in a well-aerated soil is similar to that in the atmosphere (21 percent oxygen), while the air in a poorly aerated soil is several thousand times higher in carbon dioxide (up to 20 percent) and lower in oxygen because of excess water.

Soil tilth is the physical condition of the soil as related to tillage, seedbed preparation, seedling emergence, and root penetration.

Soil moisture (soil water) is defined as the water held in soil pores (voids, capillaries) between soil particles and within soil granules. Soil water and soil air both occupy the same soil pores. The relative amount of each often fluctuates hour by hour. **Available soil moisture** is soil water held against the pull of gravity, between the **field capacity** and the **wilting point** (Figures 1.4 and 1.5).

Note: All soil survey reports have tables listing the physical and chemical properties of each soil map unit.

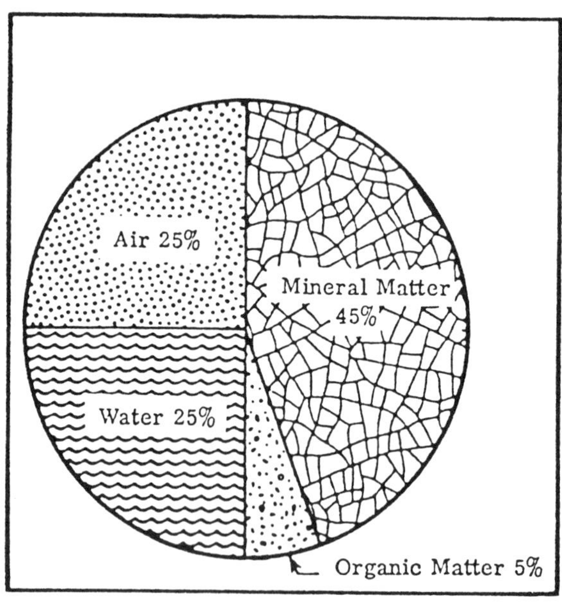

Fig. 1.4 — Volume composition of loam soil in proper physical condition (tilth) for normal plant growth. (Courtesy, Instructional Materials Services, Texas A&M University)

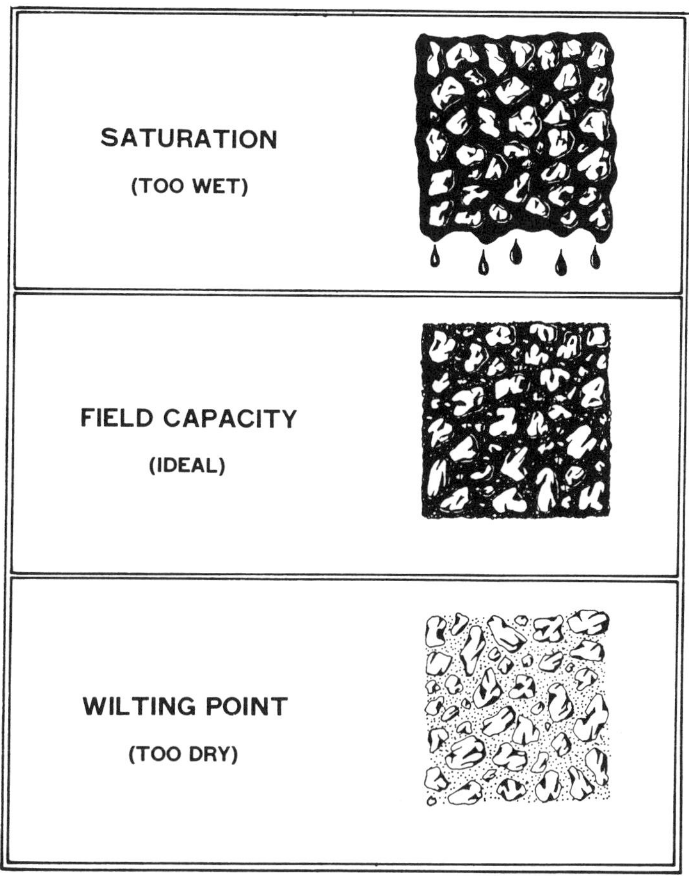

Fig. 1.5 — Soil moisture conditions of soils in relation to plant growth. (Courtesy, Instructional Materials Services, Texas A&M University)

1:4 □ CHEMICAL AND COLLOIDAL CHARACTER OF SOIL

Soils vary widely in their chemical and colloidal make-up. This variation is due to the chemical composition of the parent materials and to the climate, plant, animal life, and age under which the soil developed.

Nitrogen in representative surface soils varies from 0.02 percent in the sandy soils of Florida and Virginia to approximately 0.15 percent in a limestone soil in the Great Lakes states. The nitrogen in the soil is stored almost entirely as a part of the organic matter, and when the organic matter is decomposed by chemical and biological processes, nitrogen is released for plant growth as ammonium or nitrate.

Generally the soils in the United States are extremely low in phosphorus, varying from a trace to 0.3 percent in surface soils. The phosphorus content is highest in the soils of the central basin in Tennessee which were developed from a limestone parent material high in phosphorus.

The percentage of potassium is low in sandy soils but fairly high in silt and clay soils. Potassium is especially plentiful in soils in the regions of low rainfall because it has not leached out.

Of the 16 essential nutrients for plants, oxygen is the most abundant element in the earth's crust, with an average of 47 percent. Iron is the second most abundant element in the earth's crust; the total iron content in most soils ranges from 0.5 to 5.0 percent.

There is very little lime in most soils of the eastern United States and humid areas of the Northwest, Hawaii, and Alaska. In high rainfall areas, lime leaches out rapidly from the surface of sandy soils. In these regions of high rainfall, the lime content of most clays is higher than that of the sands because there is less leaching through the clay soils. Many soils which have developed from limestone are still high in lime in the surface soil. Soils that have developed in drier areas are rich in lime because there has been very little loss of lime by leaching.

Colloidal particles are clays and organic matter (humus) smaller than about 0.000,008 inch (0.002 mm) in diameter. Such particles are fine enough to carry a negative charge and to stay in suspension to make "muddy" water (Note 1.3).

All soil survey reports have tables listing the physical and chemical properties of each map unit.

1:5 ▫ BIOLOGICAL CHARACTER OF SOIL

The average soil supports a large population of tiny plants and animals, many of which can be seen only with the aid of a microscope. Among the important organisms are single-celled bacteria and fungi, which vary from simple yeast cells to complex mushrooms, actinomycetes, thread bacteria, and algae. Certain fungi and actinomycetes are the sources of **antibiotics** (germ killers) (Figure 1.6). In addition, in soils there are many small animals such as single-celled protozoa, nematodes (tiny worms), as well as earthworms and certain insects.

After 20 years of no-till corn on 10 to 15 percent slopes in Ohio, earthworms had made an average of nine vertical holes per square inch of soil surface (1.41/cm^2). Infiltration was increased by 2.6 inches (6.6 cm) per year, and erosion was decreased to a trace.

The majority of these microorganisms obtain their food and energy by breaking down complex organic substances manufactured by higher plants. Without microorganisms, humans probably could not survive, since dead plants and animals would soon accumulate and foul the earth's surface. Also, the nutrients in dead plants and animals are needed to recycle and feed other plants.

The various groups of organisms grow best under quite different environments. Most bacteria prefer well-drained, slightly acid to alkaline soils with moderate temperatures and good aeration. Fungi prefer acid soils, and algae are

Note 1.3 — Cation Exchange Capacity of Soil

Both fine clays (mineral particles) and small organic particles (humus) react in the soil somewhat like negative acid chemical salts. Both clay and humus have the characteristic of adsorbing (physically holding) positive ions (cations) such as calcium (Ca^{2+}), magnesium (Mg^{2+}), potassium (K^+), ammonium (NH_4^+), sodium (Na^+), hydrogen (H^+), and aluminum (Al^{3+}). The cations so held cannot be leached with pure rainwater but plant roots absorb the first four for growth and reproduction. The total amount of exchangeable cations that can be held by the soil, expressed in terms of milliequivalents ($\frac{1}{1,000}$ of equivalent weight) per 100 grams of soil at neutrality (pH 7.0) or at some other stated pH value, is known as the **cation exchange capacity.**

So that you can better understand chemical reactions, a few chemical terms are defined here:

- **Element** — A substance that cannot be further subdivided. Examples: phosphorus (P), potassium (K), and calcium (Ca).
- **Compound** — A substance comprising two or more elements in fixed proportions. Example: glucose (a simple sugar) — $C_6H_{12}O_6$, the first product of photosynthesis.
- **Atom** — The smallest particle (an element) with no charge. Examples: phosphorus (P), potassium (K), and calcium (Ca).
- **Ion** — One or more atoms that have an electrical charge. Examples: calcium (Ca^{2+}), ammonium (NH_4^+), and nitrate (NO_3^-).
- **Cation** — A positively charged ion. Examples: calcium (Ca^{2+}) and ammonium (NH_4^+).
- **Anion** — A negatively charged particle. Examples: nitrate (NO_3^-) and sulfate (SO_4^{2-}).
- **Atomic weight** — The relative weights of the elements. Examples: hydrogen (H) = 1; oxygen (O) = 16; and calcium (Ca) = 40.
- **Valence** — The number of charges on an ion. Examples: hydrogen (H^+) and oxygen (O^{2+}).
- **Molecular weight** — The sum of the atomic weights of the elements in a compound. Example: calcium carbonate ($CaCO_3$) — calcium = 40, carbon = 12, 3 oxygens (3 × 16) = 48. Total = 100 (molecular weight).
- **Equivalent weight** — The weight associated with one unit of valence or the atomic weight divided by the valence. Chemicals react in equivalent weights. One equivalent weight of a substance reacts completely with one equivalent weight of another substance. Example:

NaOH (sodium hydroxide)	+	HCl (hydrochloric acid)	=	NaCl (sodium chloride)	+	H_2O (water)

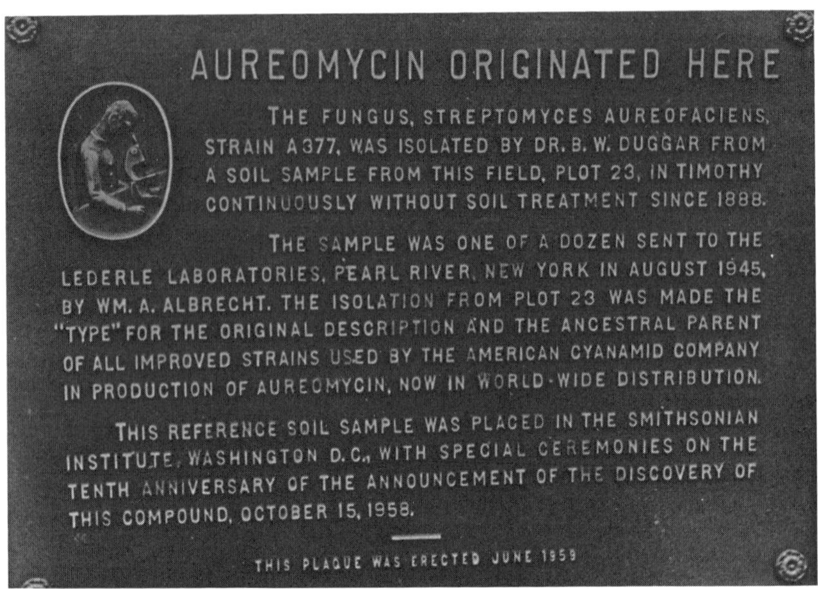

Fig. 1.6 — The fungus *aureomycin*, the source of the *antibiotic* by the same name, was first discovered and identified from a *soil* on a research plot of very low fertility at the University of Missouri at Columbia. It is used in many livestock feeds. (Courtesy, Dept. of Soil Science, University of Missouri)

found near the surface of the soil where light and water are available. While most groups depend on organic matter for food and energy, several types obtain their energy by oxidizing inorganic elements such as sulfur and nitrogen. They get their carbon from the carbon dioxide of the soil atmosphere. A few organisms can live in poorly drained soils and in the absence of free oxygen.

The "nitrogen-fixers" are important to farmers. Some groups are growing free in the soil and some are attached to the roots of certain plants called "legumes"; both groups are able to take inert nitrogen gas from the air and convert it into a form which plants can use. Some nonlegumes can also fix atmospheric nitrogen (Note 2.2).

The decomposition of organic matter in the soil by microorganisms completes a vital cycle in the ecosystem. The plant grows and absorbs nutrient elements from the soil, matures, and dies. By microbial breakdown of the dead plant, these plant food elements are released and again become available for the use of a new generation of plants. Without this decay process, most of the carbon in our universe would become locked up in complex plant and animal tissue, since the amount of available carbon dioxide in the atmosphere is quite limited. When organic matter decays, various acids are formed in the soil. These may react with mineral compounds to make vital plant food elements such as phosphorus, calcium, magnesium, and potassium more soluble and available to growing plants.

During the process of organic matter decay, an organic colloidal material which helps to cement the finer soil particles into larger, water-stable granules is produced. A soil which is granulated in this fashion is said to have good structure (tilth) and is often described as being "mellow" or easy to work. When it rains, the water soaks into a well-granulated soil instead of running off the surface.

It should be emphasized that organic matter decay by soil microbes is a continuous process, and fresh organic residues must be continually added to the soil if the maximum benefits of this complex process are to be realized from year to year.

Several species of **mycorrhizae** (fungi) have a beneficial (**symbiotic**) relationship with many plants, especially trees and shrubs. Mycorrhizal inocula are now available on the U.S. market.

Not all the activities of soil organisms are beneficial to plants and humans. Some groups produce diseases in crops, such as wilt in tomatoes or alfalfa and soft rots in vegetables. Some of the actinomycetes cause scab of potatoes. Certain organisms such as nematodes attack the roots of plants, causing serious damage. Microorganisms may also compete with crop plants for nutrients, especially for nitrogen.

1.6 ▫ DRAINAGE CHARACTER OF SOIL

The word **drainage** is commonly defined to explain how fast water moves off the **surface** of a soil (runoff). **Internal drainage** means the frequency and duration of periods of saturation or partial saturation during soil formation, as opposed to **altered drainage,** which is commonly the result of artificial drainage or irrigation but may be caused by the sudden deepening of channels or the blocking of drainage outlets. Seven classes of natural soil drainage are recognized by the U.S. Department of Agriculture – Soil Conservation Service in describing and mapping soils for the National Cooperative Soil Survey:

- ***Excessively drained*** — Water is removed from the soil very rapidly. Excessively drained soils are commonly very coarse-textured, rocky, or shallow. Some are steep. All are free of the mottling related to wetness.
- ***Somewhat excessively drained*** — Water is removed from the soil rapidly. Many somewhat excessively drained soils are sandy and rapidly pervious. Some are shallow. Some are so steep that much of the water they receive is lost as runoff. All are free of the mottling related to wetness.
- ***Well-drained*** — Water is removed from the soil readily, but not rapidly. It is available to plants throughout most of the growing season, and wetness does not inhibit growth of roots for significant periods during most growing seasons. Well-drained soils are commonly medium-textured. They are mainly free of mottling.

- ***Moderately well-drained*** — Water is removed from the soil somewhat slowly during some periods. Moderately well-drained soils are wet for only a short time during the growing season, but periodically they are wet long enough that most mesophytic crops are affected. They commonly have a slowly pervious layer within or directly below the two surface horizons, or periodically receive high rainfall, or both.
- ***Somewhat poorly drained*** — Water is removed slowly enough that the soil is wet for significant periods during the growing season. Wetness markedly restricts the growth of mesophytic crops such as corn unless artificial drainage is provided. Somewhat poorly drained soils commonly have a slowly pervious layer, a high water table, additional water from seepage, nearly continuous rainfall, or a combination of these.
- ***Poorly drained*** — Water is removed so slowly that the soil is saturated periodically during the growing season or remains wet for long periods. Free water is commonly at or near the surface for long enough during the growing season that most mesophytic crops cannot be grown unless the soil is artificially drained. The soil is not continuously saturated in layers directly below plow depth. Poor drainage results from a high water table, a slowly pervious layer within the profile, seepage, nearly continuous rainfall, or a combination of these.
- ***Very poorly drained*** — Water is removed from the soil so slowly that free water remains at or on the surface during most of the growing season. Unless the soil is artificially drained, most mesophytic crops cannot be grown. Very poorly drained soils are commonly level or depressed and are frequently ponded and may be mapped as wetlands.

All Histosols (organic soils) are *very poorly drained* and must be artificially drained before growing upland crops. However, when Histosols are drained, the organic matter decomposes more rapidly, and the soil surface subsides as much as 1 inch (2.5 cm) per year. For example, Histosols in the Everglades of Florida are shallow over limestone bedrock and are subsiding so rapidly that by the year 2,000 they will be too thin to use for agricultural crops. There is no practical solution to this problem. (See Chapter 8.)

1:7 □ THE RESULTING SOIL PROFILE

A vertical section (**profile**) of a soil usually shows three natural layers known as horizons. The **horizons** can be identified by color changes, by texture, by structure, or by differences in chemical composition. The topsoil or surface mineral layer is the **A-horizon,** the subsoil is the **B-horizon,** the parent material is the **C-horizon,** and underlying bedrock is designated as the **R-horizon** (Figure 1.7). (A soil profile with a third dimension is known as a **pedon**.)

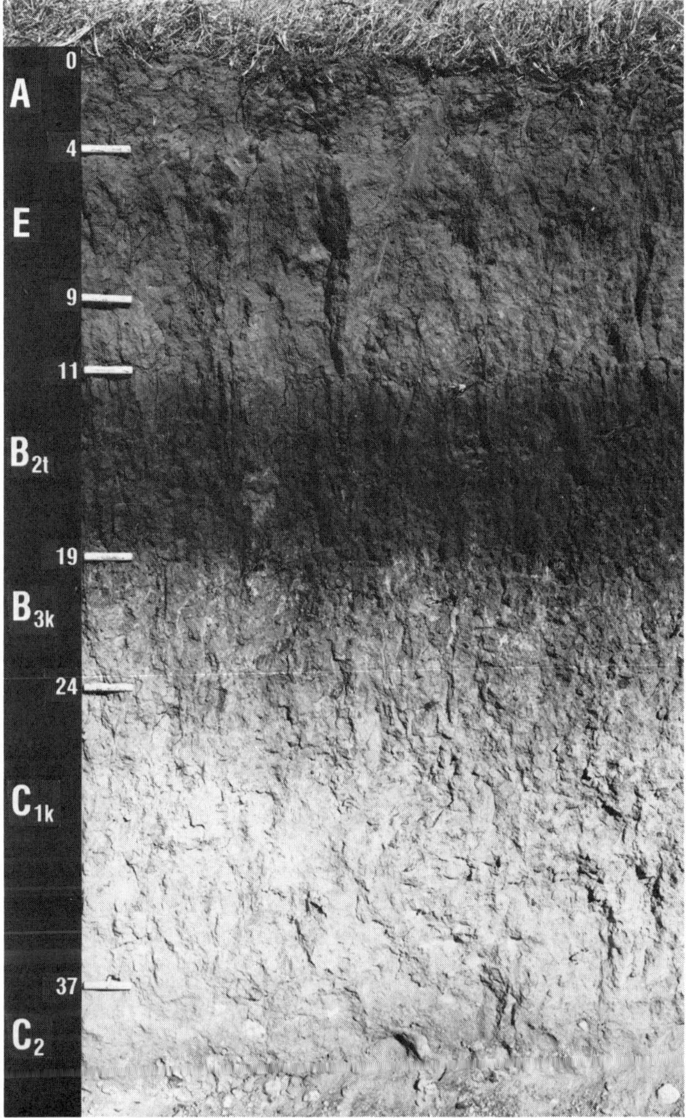

Fig. 1.7 — Vertical soil differences are the result of unequal rates of addition and removal of materials from the various horizons. The soil series shown above is a Dawes silt loam with a distinct B_{2t}-horizon. (The numbers represent the depth in inches.) This soil series has developed in the northern Great Plains under native grass. (Courtesy, USDA - Soil Conservation Service)

Note: (1) The "t" in "B_{2t}" indicates an increase in clay content (from the German word *ton*, meaning "clay"). The "k" (formerly Ca) stands for an accumulation of calcium carbonate. (2) The R (rock)-horizon is at an indefinite depth below the C-horizon. (3) Scale is in inches (inches × 2.54 = cm). (4) See Section 1:16.

The major horizons of mineral soils are as follows, but few soils have all of these horizons:

- ***O-horizon*** — An organic layer of fresh and decaying plant residue at the surface of a mineral soil.
- ***A-horizon*** — The mineral horizon at or near the surface in which an accumulation of humified organic matter is mixed with the mineral material. Also, a plowed surface horizon, some of which was originally part of a B-horizon.
- ***E-horizon*** — The mineral horizon in which the main feature is loss by leaching of silicate clay, humus, iron, or aluminum, or some combination of these.
- ***B-horizon*** — The mineral horizon below an O- , A- , or E-horizon. The B-horizon is, in part, a layer of transition from the overlying horizon to the underlying C-horizon. The B-horizon also has distinctive characteristics, such as accumulation of clay, sesquioxides, humus, or a combination of these; prismatic or blocky structure; redder or browner colors than those in the A-horizon; or a combination of these. The combined A- and B-horizons are generally called the **solum,** or true soil. If a soil does not have a B-horizon, the A-horizon alone is the solum.
- ***C-horizon*** — The mineral horizon or layer, excluding indurated bedrock, that is little affected by soil-forming processes and does not have the properties typical of the A- or B-horizon. The material of a C-horizon may be either like or unlike that in which the solum formed. If the material is known to differ from that in the solum, the Arabic numeral 2 precedes the letter C.
- ***R-horizon*** — Consolidated rock (unweathered bedrock) beneath the soil. The rock commonly underlies a C-horizon, but it can be directly below an A- or B-horizon if there is no C-horizon.

Soil maps are accompanied by soil profile descriptions, which are the records of the observations the soil scientist makes when holes are dug in soils and the various layers or horizons are examined. Each soil survey area or published soil survey report has numerous soil profile descriptions, which give the properties of the soils that are significant for their use. Figure 1.8 gives a schematic picture of the pit and soil profile that a soil scientist describes for each soil series when making a map.

1:8 □ SOIL CLASSIFICATION (SOIL TAXONOMY)

A system of soil classification, called **soil taxonomy**, has been developed in

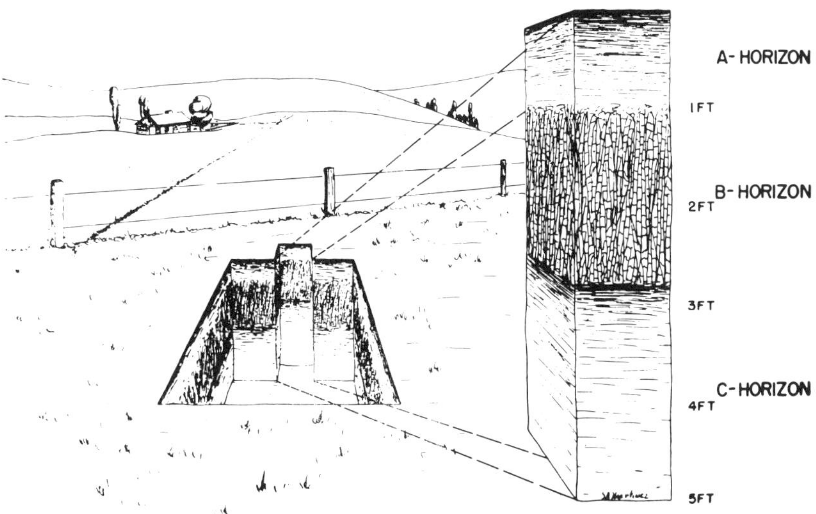

Fig. 1.8 — Schematic drawing of pit and soil profile as described by a soil scientist making a profile description during a soil survey. (Courtesy, Iowa State University)

the United States and was officially adopted in 1965. It is intended for worldwide use and therefore must provide for classifying all of the soils in the world. Furthermore, it must be as logical and usable as possible. Such an undertaking necessarily involves the viewpoints of the people who will be using the system; the more viewpoints considered, the more usable the system is likely to be for the greatest number of people. The system is dynamic and subject to change with new technology. (Note: Because of newer technology, a new soil order was added in 1989.)

The categories of the 1965 soil classification system include **order, suborder, great group, subgroup, family**, and **series** (Figure 1.9). The nomenclature of the system above the level of the soil series is new. The names have been systematically coined mostly from Latin and Greek roots.

The 1965 soil classification system is based largely upon several kinds of diagnostic horizons that may or may not be present in any particular soil. The most important of these horizons are given brief definitions below:

- **Mollic epipedon** — The thick, dark-colored upper horizon typical of grassland soils such as those in central United States. It cannot be strongly acid nor can it have very wide carbon to nitrogen (C:N) ratios (Figure 1.7).

- **Histic epipedon** — An organic surface horizon.

- **Umbric epipedon** — A thick, dark-colored surface horizon that is too acid and / or has too wide a C:N ratio to be a mollic epipedon.

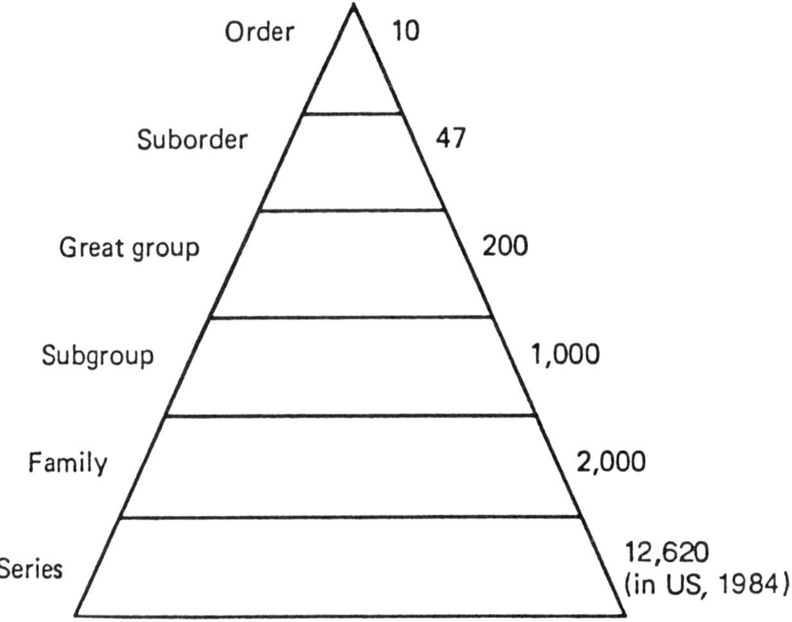

Fig. 1.9 — The six categories in the U.S. system of soil taxonomy and the number of *taxa* (units) in each category. *Note:* In 1989 an eleventh soil order was officially recognized. It is predominantly found in New Zealand. (Courtesy, USDA - Soil Conservation Service)

- **Ochric epipedon** — A surface horizon that is too light-colored or too thin to be a mollic epipedon.
- **Argillic horizon** — A soil horizon with a definite increase in clay from the horizon above. Usually it contains about 20 percent more clay than overlying soil. Clay films are evident on soil peds, in pores, and/or on sand grains. (The B_{2t}-horizon of Figure 1.7 is an example.)
- **Calcic horizon** — A soil horizon with more than 15 percent calcium carbonate and at least 5 percent more calcium carbonate than the horizon below. (The C_{1k}-horizon of Figure 1.7 is an example.)
- **Natric horizon** — An argillic horizon that has a high content (over 15 percent) of exchangeable sodium and forms in arid environments.
- **Spodic horizon** — A soil horizon with an illuvial (moved in from above) concentration of iron and aluminum oxides and humus but little or no clay accumulation and which forms in cool, humid environments.
- **Cambric horizon** — A soil horizon that has developed the color and/or structure of a subsoil but does not qualify as an argillic or spodic horizon.
- **Oxic horizon** — A highly weathered soil material high in iron and aluminum oxides which forms in old tropical, humid environments.

- **Melanic epipedon** — A black surface soil horizon high in organic matter from grass residues.

There are 10 soil orders in the 1965 system. Important characteristics of these orders are as follows (the italicized syllable of each is used for naming the classes at lower categorical levels) (note: A new soil order, Andisols, was added in 1989):

- **_Ent_isols** (from rec*ent* = new) — Very young soils with little or no horizon development (8 percent each of U.S. and world).
- **_Vert_isols** (from *verto* = turn, as in in*vert*) — "Self-swallowing" soils that have large cracks during the dry season into which surface soil falls. They contain much clay that swells when wet and causes a "churning" process (1 percent of U.S.; 2 percent of world).
- **In_cept_isols** (from *inceptum* = beginning) — Young soils that have had time to develop only the more rapidly formed horizons such as histic, umbric, or ochric epipedons (surface layers) or cambric horizons (18 percent of U.S.; 9 percent of world).
- **_Arid_isols** (from *arid* = dry) — Soils of dry regions (12 percent of U.S.; 19 percent of world).
- **_Moll_isols** (from *mollis* = soft) — Soils with mollic epipedons. Wet soils and soils with argillic horizons are included if they have mollic epipedons (25 percent of U.S.; 9 percent of world).
- **_Spod_osols** (from *spodos* = wood ashes) — Refers to light gray (A_2-horizon) — soils with spodic horizons; they are highly leached soils of cool, humid forested areas (5 percent of U.S.; 4 percent of world).
- **_Alf_isols** (from *Al, Fe* = symbols for aluminum and iron) — Soils that have argillic horizons and are not too highly leached. The epipedon may be ochric or umbric but not mollic (13 percent each of U.S. and world).
- **_Ult_isols** (from *ult*imate) — Similar to Alfisols but more leached so that they have more red and yellow (13 percent of U.S.; 6 percent of world).
- **_Ox_isols** (from *oxides*) — Soils with an oxic horizon which form in an old, humid, tropical environment (<1 percent of U.S.; 8 percent of world).
- **_Hist_osols** (from *histos* = tissue) — Organic soils (<1 percent of U.S.; 1 percent of world).
- **_And_isols** (from *melas-anos* = black) — Black mineral soils with a melanic epipedon and 60 percent of entire soil volcanic ejecta, including 5 percent volcanic glass.

The 1965 soil taxonomy system has six categories (levels of generalization). The 11 orders defined above constitute the highest category (the one with the broadest generalization and the fewest classes). The lowest category is composed of 12,620 soil series in the United States. A system of nomenclature has been developed to give meaning and distinctiveness to the names of the classes of the four highest categories. Each syllable in these names tells something about the soil. The six categories for the Dawes soil series, the method of naming classes in each, and an example of each are listed below (see Figure 1.7).

Category	*Method of Naming*	*Example*
Order	Names end with "-sol"	Molli*sol*
Suborder	Names end with symbol (formative element) from the name of the order	Ust*oll*
Great group	Prefix added to suborder name	*Pale*ustoll
Subgroup	Adjective added to great group name	*Typic* Paleustoll
Family	Divisions of subgroup based on texture, mineralogy, and temperature, respectively	Fine clayey, mixed mesic family
Series	Geographic names	Dawes (see Figure 1.7)

A soil series such as Dawes consists of a group of soils that have profiles that are almost alike, except for differences in texture of the surface layer or of the underlying material. All the soils of a series have horizons that are similar in composition, thickness, and arrangement.

Analyzing the above example from the top down reveals that the Dawes series has a mollic epipedon, occurs in subhumid to semiarid climates (**Ustoll** comes from *ustus* meaning "burnt"), and has a relatively complex profile suggesting a number of alternating cycles of humidity and aridity. The adjective **Typic** in the subgroup name indicates that they are typical Paleustolls. The **fine clayey** family designation indicates clay content between 35 and 59 percent in the subsoil (Figure 1.10), with no one kind of clay mineralogy predominating, and with average annual temperatures between 47° and 59° F (8° and 15° C). The series name "Dawes" came from Dawes County in northwestern Nebraska where the soil was first recognized. The soil mapping unit on all soil maps consists of the series name "Dawes," plus the textural class of the A-horizon (silt loam), plus the slope class and the erosion class.

Note: Before 1965 the term *Dawes silt loam* was designated *soil type*. The term *soil type* is now obsolete. The textural class of the surface soil horizon, the slope class, and the erosion class are now properly designated as **phases of soil series**.

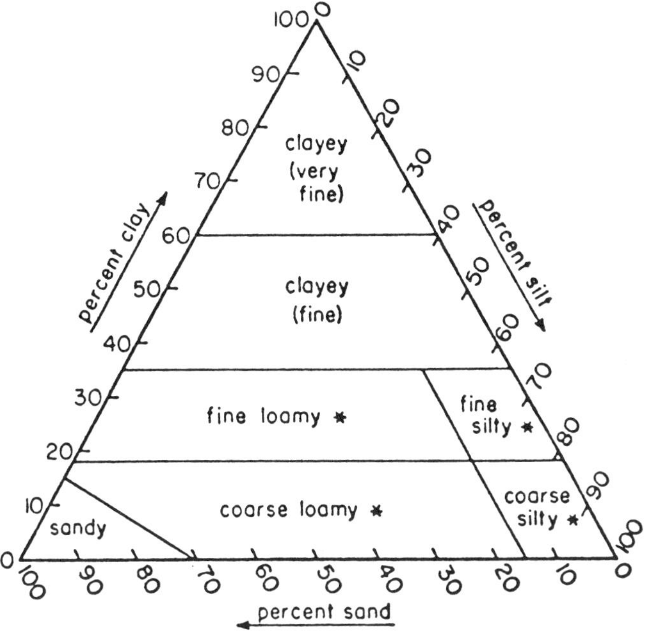

Fig. 1.10 — These textural classes for soil *family groupings* are broadened from those in Figure 1.2 to more generally correspond to soil - plant growth relationships. (Courtesy, USDA - Soil Conservation Service)

1:9 □ SOIL SURVEY REPORTS

The boundaries and properties of soil mapping units in counties throughout the United States can be obtained from the detailed soil survey reports published by the Soil Conservation Service of the U.S. Department of Agriculture.

Soil surveys are fundamental to sound soil and water management, crop production, homesite location, and land use. Soil surveys describe properties of soils that affect farm management decisions.

1:9.1 □ Table of Contents

The "contents" of most soil survey reports are as follows:

Index to map units
Summary of tables
Foreword
    General nature of the survey area
    How this survey was made

General soil map units
Detailed soil map units
Prime farmland
Use and management of the soils
    Crops and pasture
    Woodland management and productivity
    Recreation
    Wildlife habitat
    Engineering
Soil properties
    Engineering index properties
    Physical and chemical properties
    Soil and water features
    Physical and chemical analyses of selected soils
    Engineering index test data
Classification of the soils (soil series and their morphology)
Formation of the soils (factors of soil formation)
References
Glossary
Tables
General soil map (scale: 1 inch = 3 miles; 1:190,080).
Detailed soil maps (scale: 4 inches = 1 mile; 1:15,840).
(*Note*: Scales will vary, depending on the intended uses of the maps.)

1:9.2 □ Yield Goals

Yield goals is a common term used when a farmer submits a soil sample for chemical testing to obtain fertilizer / lime recommendations. To keep yield goals within practical limits, consult a recent soil survey report under this heading: "Summary of Tables: Land Capability Classes and Yields per Acre of Crops and Pasture." Here all soil map units are listed alphabetically and expected yields are given for all adapted crops grown under good management. Also listed for pastures are predicted animal unit months (AUMs) of grazing per year.

1:9.3 □ Water Management

Water management is detailed in the "tables" of soil survey reports by soil name and map symbol in relation to limitations for pond construction, drainage, terraces, and grassed waterways.

1:10 □ GENERAL SOIL MAP OF THE UNITED STATES

The most recent system of soil classification and soil taxonomy has been used to make a general soil map of the United States at the level of soil orders and suborders (Figure 1.11).

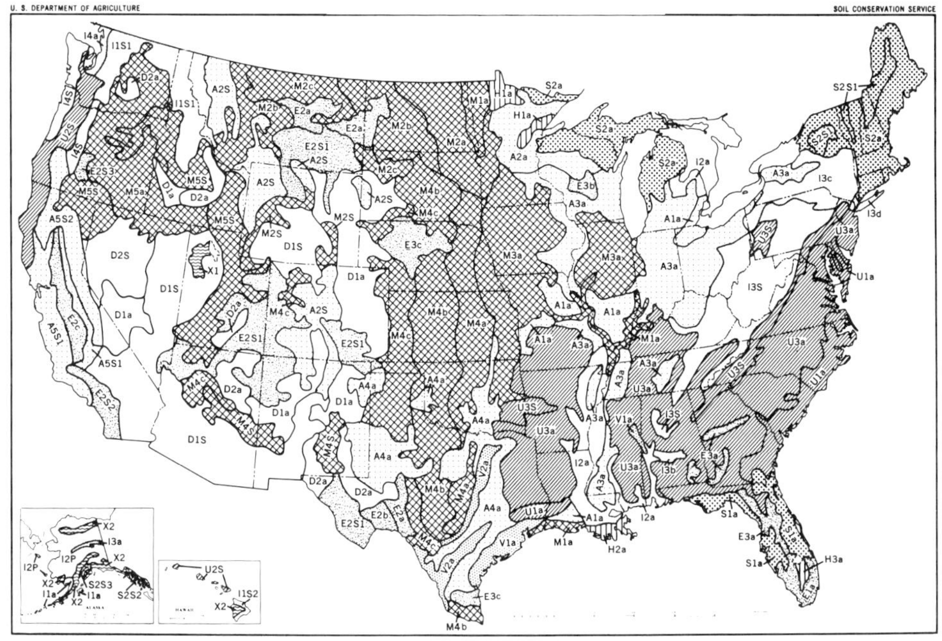

Fig. 1.11 — General soil map of the United States. (See legend below.)

ALFISOLS

AQUALFS

A1a Aqualfs with Udalfs, Haplaquepts, Udolls; gently sloping.

BORALFS

A2a Boralfs with Udipsamments and Histosols; gently and moderately sloping.

A2S Cryoboralfs with Borolls, Cryochrepts, Cryothods, and Rock outcrops; steep.

UDALFS

A3a Udalfs with Aqualfs, Aquolls, Rendolls, Udolls, and Udults; gently or moderately sloping.

USTALFS

A4a Ustalfs with Ustochrepts, Ustolls, Usterts, Ustipsamments, and Ustorthents; gently or moderately sloping.

XERALFS

A5S1 Xeralfs with Xerolls, Xerorthents, and Xererts; moderately sloping to steep.

A5S2 Ultic and lithic subgroups of Haploxeralfs with Andepts, Xerults, Xerolls, and Zerochrepts; steep.

ARIDISOLS

ARGIDS

D1a Argids with Orthids, Orthents, Psamments, and Ustolls; gently and moderately sloping.

D1S Argids with Orthids, gently sloping; and Torriorthents, gently sloping to steep.

ORTHIDS

D2a Orthids with Argids, Orthents, and Xerolls; gently or moderately sloping.

D2S Orthids, gently sloping to steep, with Argids, gently sloping; lithic subgroups of Torriorthents and Xerorthents, both steep.

ENTISOLS

AQUENTS

E1a Aquents with Quartzipsamments, Aquepts, Aquolls, and Aquods; gently sloping.

ORTHENTS

E2a Torriorthents, steep, with borollic subgroups of Aridisols; Usterts and aridic and vertic subgroups of Borolls; gently or moderately sloping.

E2b Torriorthents with Torrerts; gently or moderately sloping.

E2c Xerorthents with Xeralfs, Orthids, and Argids; gently sloping.

E2S1 Torriorthents; steep, and Argids, Torrifluvents, Ustolls, and Borolls; gently sloping.

E2S2 Xerorthents with Xeralfs and Xerolls; steep.

E2S3 Cryorthents with Cryopsamments and Cryandepts; gently sloping to steep.

PSAMMENTS

E3a Quartzipsamments with Aquults and Udults; gently or moderately sloping.

E3b Udipsamments with Aquolls and Udalfs; gently or moderately sloping.

E3c Ustipsamments with Ustalfs and Aquolls; gently or moderately sloping.

HISTOSOLS

HISTOSOLS

H1a Hemists with Psammaquents and Udipsamments; gently sloping.

H2a Hemists and Saprists with Fluvaquents and Haplaquepts; gently sloping.

H3a Fibrists, Hemists, and Saprists with Psammaquents; gently sloping.

INCEPTISOLS

ANDEPTS

I1a Cryandepts with Cryaquepts, Histosols, and Rock land; gently or moderately sloping.

I1S1 Cryandepts with Cryochrepts, Cryumbrepts, and Cryorthods; steep.

I1S2 Andepts with Tropepts, Ustolls, and Tropofolists; moderately sloping to steep.

AQUEPTS

I2a Haplaquepts with Aqualfs, Aquolls, Udalfs, and Fluvaquents; gently sloping.

I2P Cryaquepts with cryic great groups of Orthents, Histosols, and Ochrepts; gently sloping to steep.

OCHREPTS

I3A Cryochrepts with cryic great groups of Aquepts, Histosols, and Orthods; gently or moderately sloping.

I3b Eutrochrepts with Uderts; gently sloping.

I3c Fragiochrepts with Fragiaquepts, gently or moderately sloping; and Dystrochrepts, steep.

I3d Dystrochrepts with Udipsamments and Haplorthods; gently sloping.

I3S Dystrochrepts, steep, with Udalfs and Udults; gently or moderately sloping.

UMBREPTS

I4a Haplumbrepts with Aquepts and Orthods; gently or moderately sloping.

I4S Haplumbrepts and Orthods; steep, with Xerolls and Andepts; gently sloping.

MOLLISOLS

AQUOLLS

M1a Aquolls with Udalfs, Fluvents, Udipsamments, Ustipsamments, Aquepts, Eutrochrepts, and Borolls; gently sloping.

BOROLLS

M2a Udic subgroups of Borolls with Aquolls and Ustorthents; gently sloping.

M2b Typic subgroups of Borolls with Ustipsamments, Ustorthents, and Boralfs; gently sloping.

M2c Aridic subgroups of Borolls with Borollic subgroups of Argids and Orthids, and Torriorthents; gently sloping.

M2S Borolls with Boralfs, Argids, Torriorthents, and Ustolls; moderately sloping or steep.

UDOLLS

M3a Udolls, with Aquolls, Udalfs, Aqualfs, Fluvents, Psamments, Ustorthents, Aquepts, and Albolls; gently or moderately sloping.

USTOLLS

M4a Udic subgroups of Ustolls with Orthents, Ustochrepts, Usterts, Aquents, Fluvents, and Udolls; gently or moderately sloping.

M4b Typic subgroups of Ustolls with Ustalfs, Ustipsamments, Ustorthents, Ustochrepts, Aquolls, and Usterts; gently or moderately sloping.

M4c Aridic subgroups of Ustolls with Ustalfs, Orthids, Ustipsamments, Ustorthents, Ustochrepts, Torriorthents, Borolls, Ustolls, and Usterts; gently or moderately sloping.

M4S Ustolls with Argids and Torriorthents; moderately sloping or steep.

XEROLLS

M5a Xerolls with Argids, Orthids, Fluvents, Cryoboralfs, Cryoborolls, and Xerorthents; gently or moderately sloping.

M5S Xeralls with Cryoboralfs, Xerolfs, Xerorthents, and Xererts; moderately sloping or steep.

SPODOSOLS

AQUODS

S1a Aquods with Psammaquents, Aquolls, Humods, and Aquults; gently sloping.

ORTHODS

S2a Orthods with Boralfs, Aquents, Orthents, Psamments, Histosols, Aquepts, Fragiochrepts, and Dystrochrepts; gently or moderately sloping.

S2S1 Orthods with Histosols, Aquents, and Aquepts; moderately sloping or steep.

S2S2 Cryorthods with Histosols; moderately sloping or steep.

S2S3 Cryorthods with Histosols, Andepts, and Aquepts; gently sloping to steep.

ULTISOLS

AQUULTS

U1a Aquults with Aquents, Histosols, Quartzipsamments, and Udults; gently sloping.

HUMULTS

U2S Humults with Andepts, Tropepts, Xerolls, Ustolls, Orthox, Torrox, and Rock land; gently sloping to steep.

UDULTS

U3a Udults with Udalfs, Fluvents, Aquents, Quartzipsamments, Aquepts, Dystrochrepts, and Aquults; gently or moderately sloping.

U3S Udults with Dystrochrepts; moderately sloping or steep.

VERTISOLS

UDERTS

V1a Uderts with Aqualfs, Eutrochrepts, Aquolls, and Ustolls; gently sloping.

USTERTS

V2a Usterts with Aqualfs, Orthids, Udifluvents, Aquolls, Ustolls, and Torrerts; gently sloping.

AREAS WITH LITTLE SOIL

X1 Salt flats.

X2 Rock land (plus permanent snow fields and glaciers).

SLOPE CLASSES

Gently sloping — Slopes mainly less than 10 percent, including nearly level.
Moderately sloping — Slopes mainly between 10 and 25 percent.
Steep — Slopes mainly steeper than 25 percent.

1:11 □ PRIME AND UNIQUE AGRICULTURAL LANDS

To increase efficiency of food production and in anticipation of federal and state land use legislation, the U.S. Department of Agriculture – Soil Conservation Service started a system of using soil maps to locate prime and unique agricultural lands. There are 345 million acres (140 million ha) in the United States classified as prime farmland of which 230 million acres (93 million ha) are used in growing crops. The remaining acres of prime farmland are in pasture, range, forest, and other land uses with about a 50 percent potential for conversion to cropland.

Prime agricultural lands are the most productive lands for raising the common food and fiber crops; whereas unique agricultural lands are those most productive for the less common but high-value-per-acre crops such as rice, cranberries, pineapples, and orchard products, as well as for cut flowers and nursery products. A table on prime farmland is given for map units in soil survey reports.

To be classified as prime agricultural lands, all areas on a soil map must meet these eight minimum requirements:

1. The mean annual soil temperature must be greater than 32° F (0° C) and the mean summer soil temperature greater than 46° F (8° C).
2. Adequate soil moisture must be available during the growing season either from rainfall and stored soil moisture or from irrigation, or both. If irrigated, the water must be of suitable quality so as not to cause plant toxicity or soil deterioration. (See Chapters 9 and 10.)
3. There must not be excessive moisture; that is, the soil should not be flooded more often than once in two years, and the water table must be maintained at a depth below the normal rooting zone of the crops expected to be grown.
4. The soil must be deep enough over hardpan, gravel, or bedrock to permit adequate soil moisture storage and unhampered root development.
5. The surface soil must not be so gravelly, cobbly, or stony as to interfere seriously with power machinery.
6. The permeability must be at least 0.06 inch (0.15 cm) per hour in the upper 19.7 inches (50 cm) of soil.
7. For the crops to be grown, the soil must not be excessively acid, alkaline, saline, or sodic.
8. The soil must be on gentle slopes (<6 percent) and must not be excessively erodible.

Unique agricultural lands are more difficult to characterize because of the variation in soil suitability of the crops to be grown. However, in general, unique agricultural lands must possess these qualities:

1. A location that has a unique combination of soil qualities, temperature, humidity, air drainage, slope, and nearness to market that favors the growth and / or distribution of the specific food or fiber crop.
2. A growing season that is long enough and a soil temperature that is high enough to produce a good harvest of the specific crop.
3. Available soil moisture from precipitation, stored soil moisture, or irrigation that is adequate for the specific crop.

1:12 □ Benchmark Soils

Benchmark soils were first proposed by Dr. L. D. Swindale at the University of Hawaii for soil families of the tropics. Later, 78 soil series in Benchmark soil families were designated for all 50 U.S. states.

Soil families are groups of soils having similar percentages of sand, silt, and clay; similar kinds of minerals and temperature regimes; and similar other soil properties important to the production of principal crop plants (Table 1.1).

Table 1.1 — The 78 Soil Series Classified in Benchmark Soil Families in the United States[1]

Soil Series Classified in Benchmark Soil Families in the United States				
Aiken	Canfield	Houghton	Minto	Sassafras
Alderwood	Cecil	Houston	Mohave	Scobey
Altamont	Chilcott	Houston Black	Molokai	Sharkey
Amarillo	Clarion	Hoytville	Montoya	Sharpsburg
Amery	Crider		Mutnala	Showlow
Arvada	Drummer	Indio	Myakka	Simas
Aschoff	Fayette	Keith	Norfolk	Tetonka
Barbary	Fort Collins	Kirkland	Palmarejo	Travessilla
Barnes	Frederick	Lucky Star	Palouse	Valentine
Bayamon	Fullerton	Maile	Parleys	Vergennes
Becket	Gilman	Malbis	Plainfield	Wahiawa
Beltsville	Grenada	Mardin	Porters	Walla Walla
Bladen	Hagerstown	Marshall	Pullman	
Blazon	Halii	Melbourne	Rains	
Bonner	Hiko Peak	Mexico	Rhoades	
Bridget	Hilo	Miami	Rifle	
	Holloway	Mimbres	San Joaquin	

[1]Source: Dr. Ellis Knox, National Leader for Soil Survey, USDA - Soil Conservation Service. Sent by letter dated March 12, 1986, to Roy L. Donahue.

1:13 □ SOIL MANAGEMENT FOR POLLUTION CONTROL

Pollution of surface waters and groundwaters from agriculture can be reduced by wise soil management practices that include:

1. Reducing erosion sediment transport. This restricts the movement of or-

ganic and mineral soil particles along with manures, phosphorus fertilizer, and pesticides. Terracing, contour tillage, and residue mulch tillage are three common soil management practices to reduce sediment transport. (See Chapter 6.)

2. Applying nitrogen fertilizers in small applications at times when plants are growing rapidly. This soil management practice will reduce deep percolation losses of nitrates which could pollute groundwaters. (See Chapters 4 and 5.)

1:14 ▫ UNUSUAL USES OF CLAY SOILS

In two southern states, one of the authors of this book observed some people and cattle eating clay. In Africa, cattle eat certain kinds of clay. Furthermore, some tribal groups plaster their bodies with mud, a practice that has now spread to parts of the United States.

1:15 ▫ REFERENCES

Davidson, Donald A. *Soils and Land Use Planning.* New York: Longman, Inc., 1980, 364 pp.

Edwards, C. A., and J. R. Lofty. *Biology of Earthworms.* Ontario, California: Bookworm Publishing Co., 1972, 283 pp. (*Note*: This is one of the few books on earthworms that is 100 percent scientific; most others promote the sale of earthworms or the cult of organic gardening.)

Edwards, W. M., and L. D. Norton. "Effect of Macropores upon Infiltration into Non-tilled Soil." Transactions of 13th Congress of the International Society of Soil Science, Vol. 2, 1986, pp. 47, 48.

El-Ashry, Mohamed T., and Diana C. Gibbons. "Managing the West's Water." *Journal of Soil and Water Conservation*, Vol. 42, No. 1, January – February 1987, pp. 8 – 13.

Fact Book of Agriculture, 1988. Misc. Pub. 1063, U.S. Department of Agriculture, Washington, D.C.

Festervant, D. F., Gerald Manis, and Rich Mayer. "Soil Survey of Perry County, Missouri." Washington, D.C.: U.S. Dept. of Agriculture – Soil Conservation Service, September 1986.

Gutzi, S. C., I. Hague, J. McIntire, and J. E. S. Stares, eds. "Management of Vertisols in Sub-Saharan Africa." International Livestock Centre for Africa, Addis Ababa, Ethiopia, 1988, 435 pp.

Kladivko, Eileen J., Alec D. Mackay, and Joe M. Bradford. "Earthworms as a Factor in the Reduction of Soil Crusting." *Soil Science Society of America Journal*, Vol. 50, 1986, pp. 191 – 196.

Knox, Ellis G., National Leader, Soil Survey Investigations, U.S. Dept. of Agriculture – Soil Conservation Service, Washington, D.C. Personal correspondence with Roy L. Donahue regarding the new soils order Andisols.

Marx, D. H., *et al.* "Commercial Vegetative Inoculum of *Pisolithus tinctorius* and Inoculation Techniques for Development of Ectomycorrizae on Bare-Root Tree Seedlings." *Forest Science, Monograph 25*, 1984.

Robbins, C. W. "Reclamation and Reuse of Irrigation Sediments." *Journal of Soil and Water Conservation*, Vol. 42, No. 1, January – February 1987, pp. 24 – 26.

Sanitary Landfill Design and Operation. U.S. Environmental Protection Agency, Washington, D.C.: Supt. of Documents, 1980, 58 pp.

Smith, Darrell. "Have You Hugged Your Nightcrawlers Today?" *Farm Journal*, December 1986, pp. A – 4, A – 6.

Soil Survey Staff. "Soil Taxonomy: A Basic System of Soil Classification for Making and Interpreting Soil Surveys." *Agr. Handbook No. 436*. Washington, D.C.: U.S. Dept. of Agriculture – Soil Conservation Service, December 1975, 754 pp.

Swindale, L. D. "A Soil Research Network Through Tropical Soil Families." In L. D. Swindale, ed., *Soil-Resource Data for Agricultural Development*. University of Hawaii, Honolulu, 1978, pp. 210 – 218.

U.S. Dept. of Agriculture. "Classification of Soil Series." Soil Conservation Service, Washington, D.C., 1988, 282 pp.

U.S. Dept. of Agriculture. "Hydric Soils of the United States, 1985." Soil Conservation Service, October 1985, unpaged.

U.S. Dept. of Agriculture. "Perspective on Prime Lands." Seminar on the Retention of Prime Lands. Soil Conservation Service, Washington, D.C., July 16 – 17, 1975.

Witty, John E., National Leader, Soil Classification, National Soil Survey Center, U.S. Dept. of Agriculture – Soil Conservation Service, Washington, D.C. Personal correspondence with Roy L. Donahue, December 21, 1988, 51 pp.

1:16 ▫ PROFILES OF MAJOR SOILS IN THE 50 STATES[1]

Here is a unique collection of photographs never before assembled in a textbook. They are soils declared of major importance by the soil scientists of the U.S. Department of Agriculture – Soil Conservation Service in cooperation with those of the respective land-grant state universities. Some have been officially declared "state soils"; others are under active consideration for such recognition. Also, a few of the soil series whose profiles are shown here are listed in Table 1.1 in Section 1:12.

[1]U.S. Dept. of Agriculture – Soil Conservation Service. *Agr. Handbook No. 296*, 1981; and "Soil Taxonomy by Soil Series," January 12, 1988, unnumbered.

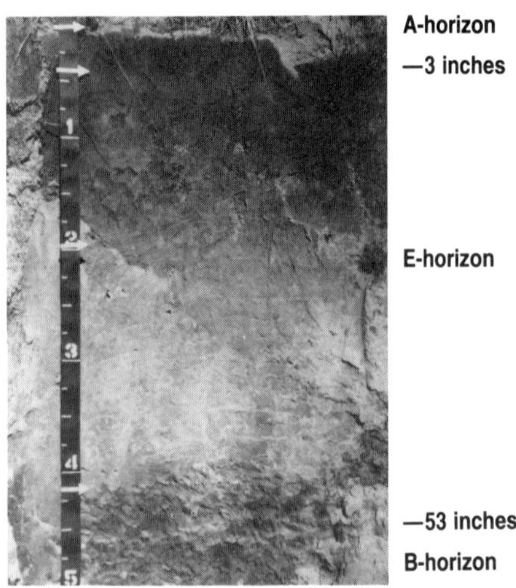

Alabama—Troup soil series. Soil order of Ultisols. Soil family—loamy, siliceous, thermic Grossarenic Paleudults. The soil order of Ultisols comprises 12.8 percent of the soils of the United States. The soil is acid throughout the profile. Mean annual precipitation is 60 inches (152 cm). Native vegetation is southern pine with mixed hardwoods. Cleared areas are used for growing peanuts, watermelons, and garden vegetables. (Courtesy, USDA–Soil Conservation Service)

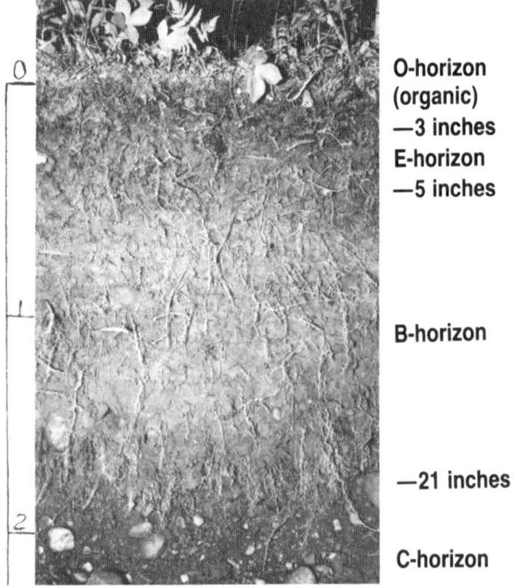

Alaska—Chulitna soil series. Soil order of Spodosols. Soil family—medial over sandy-skeletal, mixed Typic Cryorthods. The soil order of Spodosols comprises 4.8 percent of the soils of the United States. The average annual precipitation is about 29 inches (73 cm). The mean annual temperature is 33°F (1°C) (1 degree above freezing). All horizons are very strongly acid. Original vegetation was paper birch, white spruce, and aspen. A few cleared areas are planted to bromegrass, timothy, barley, Irish potatoes, cabbage, lettuce, and peas. (Note the extensive roots in all horizons.) (Scale is in feet; feet x 30.5 = cm) (Courtesy, USDA–Soil Conservation Service)

Arizona—Eagar soil series. Soil order of Mollisols. Soil family—loamy-skeletal, mixed Typic Calciborolls. The soil order of Mollisols comprises 25.1 percent of U.S. soils. This alkaline soil receives about 15 inches (37.5 cm) of annual precipitation. The short grass plains is used mostly for grazing. (Scale is in feet; feet x 30.5 = cm) (Courtesy, USDA–Soil Conservation Service)

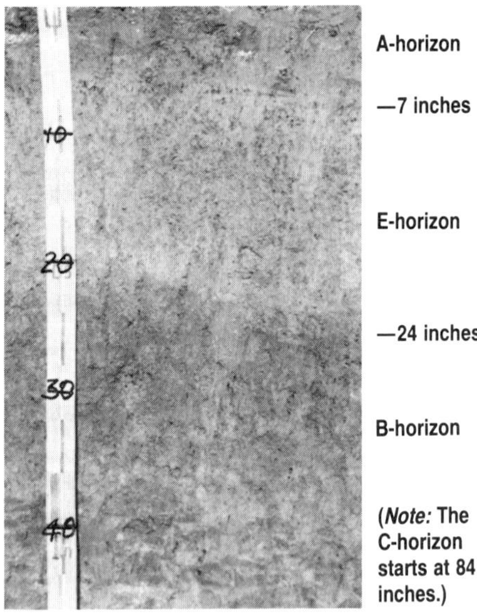

Arkansas—Stuttgart soil series. Soil order of Alfisols. Soil family—fine, montmorillonitic, thermic Typic Natrudalfs. The soil order of Alfisols comprises 13.5 percent of all soils in the United States. All horizons are acid. Average annual precipitation is about 50 inches (127 cm). Soils are nearly level. Most areas are planted to rice, soybeans, and wheat. (Scale is in inches; inches x 2.54 = cm) (Courtesy, USDA–Soil Conservation Service)

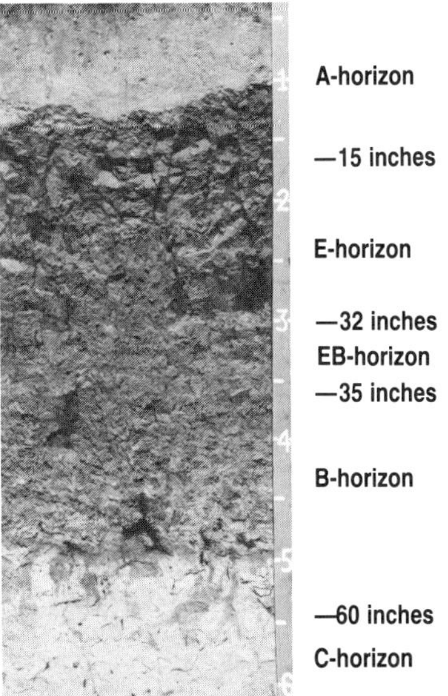

California—Narlon soil series. Soil order of Ultisols. Soil family—clayey, mixed, thermic Typic Albaquults. The soil order of Ultisols comprises 12.8 percent of all U.S. soils. All horizons are acid. Mean annual precipitation is 16 inches (41 cm). Original vegetation was short annual grasses and brush. Present use is for range grazing. (Scale is in feet; feet x 30.5 = cm) (Courtesy, USDA–Soil Conservation Service)

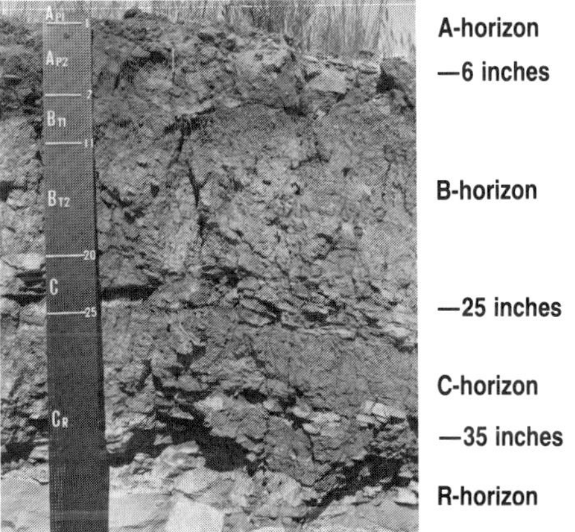

Colorado—Sharps soil series. Soil order of Aridisols. Soil family—fine silty, mixed, mesic Ustollic Haplargids. The soil order of Aridisols comprises 11.6 percent of the soils of the United States. Mean annual precipitation is about 15 inches (38 cm). All horizons are neutral to alkaline. Elevations average 6,200 feet (1,891 m). Native vegetation is sagebrush, cactus, pinyon pine, juniper, western wheatgrass, and Indian ricegrass. Used for range grazing, dryland farming, and irrigated crops. (Scale is in inches; inches x 2.54 = cm) (Courtesy, USDA–Soil Conservation Service)

Note: The C-horizon consists of shale and soft sandstone. It is a mixture of parent material (C) and bedrock (R).

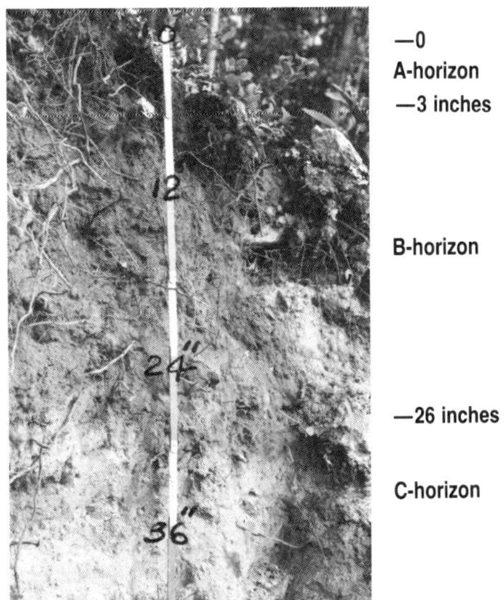

Connecticut—Charlton soil series. Soil order of Inceptisols. Soil family—coarse loamy, mixed, mesic Typic Dystrochrepts. The soil order of Inceptisols comprises 18.2 percent of the soils of the United States. This well-drained, strongly acid soil occurs in New England and New York. It is used primarily for fruits, vegetables, hay, and pasture. (Scale is in inches; inches x 2.54 = cm) (Courtesy, USDA–Soil Conservation Service)

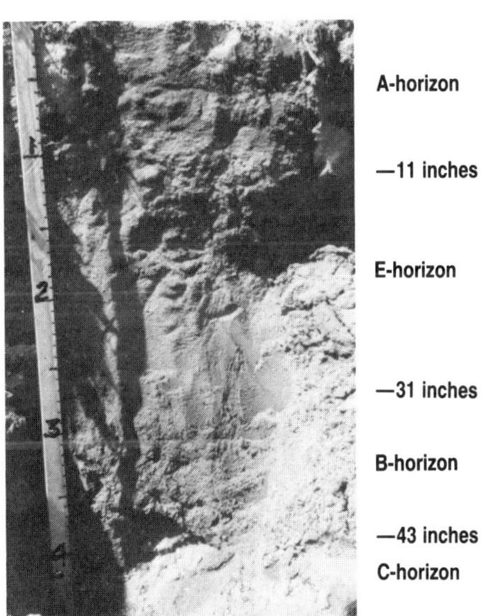

Delaware—Evesboro soil series. Soil order of Entisols. Soil family—mesic, coated Typic Quartzipsamments. The soil order of Entisols comprises 8 percent of the soils of the United States. All horizons are strongly acid. The A- and E-horizons are loamy, the B-horizon is a sandy loam, and the C-horizon varies from a sand to a clay loam. (Scale is in feet; feet x 30.5 = cm) (Courtesy, USDA–Soil Conservation Service)

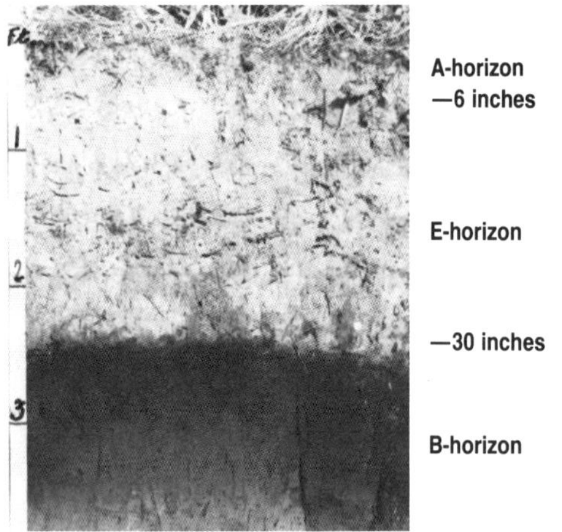

Florida—Myakka soil series. This is a Benchmark soil (see Table 1.1). Soil order of Spodosols. Soil family—sandy, siliceous, hyperthermic Aeric Haplaquods. This soil family occurs only in Florida and in 1988 was being proposed as the state soil. The soil order of Spodosols comprises 4.8 percent of the soils of the United States. The water table varies from 10 to 40 inches (25–102 cm) in depth. The soils are acid in all horizons. Mean annual precipitation is 50 inches (127 cm). Native vegetation was longleaf pine and slash pine with an undergrowth of sawpalmetto. Cleared areas are used for citrus, truck crops, and pasture grasses and legumes. (Courtesy, USDA–Soil Conservation Service)

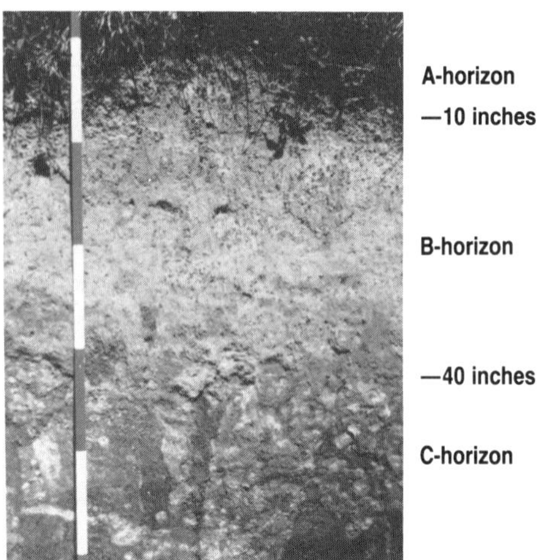

Georgia—Tifton soil series. Soil order of Ultisols. Soil family—fine loamy, siliceous, thermic Plinthic Paleudults. All soil horizons are acid. Ironstone nodules are in all horizons. The soil order of Ultisols comprises 12.8 percent of all soils in the United States. Mean annual precipitation is 46 inches (117 cm). Native vegetation was longleaf pine, slash pine, and mixed southern hardwoods. Most of the Tifton soils are cultivated to cotton, corn, peanuts, tobacco, vegetable crops, and small grains. (The range pole is marked in feet; feet x 30.5 = cm) (Courtesy, USDA–Soil Conservation Service)

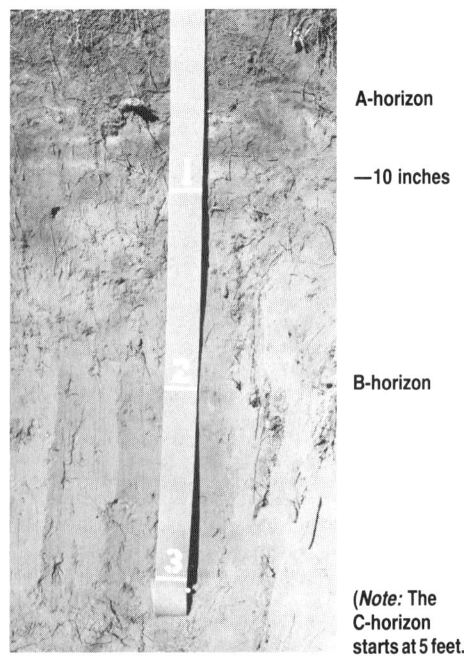

Hawaii—Hanipoe soil series. Soil order of Inceptisols. Soil family—medial, isomesic Typic Dystrandepts. The soil order of Inceptisols comprises 18.2 percent of all soils in the United States. This slightly acid to neutral soil has a mean annual precipitation of 40 inches (102 cm). Elevation is 5,000 to 6,500 feet (1,525–1,982 m). Native vegetation was tall grasses and shrubs. Present use is mostly for pasture. (Courtesy, USDA–Soil Conservation Service)

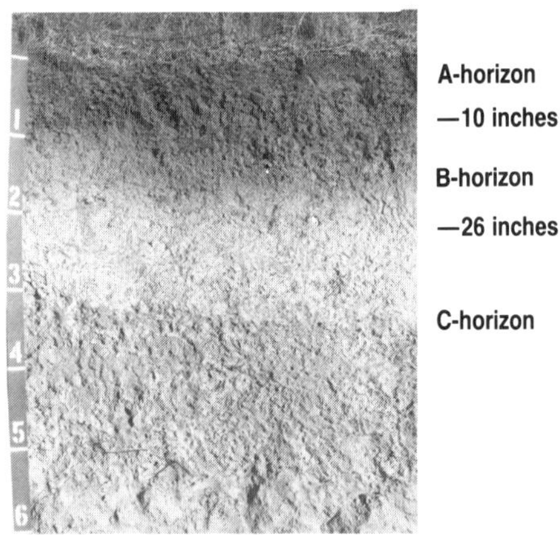

Idaho—Rexburg soil series. Soil order of Mollisols. Soil family—coarse silty, mixed, frigid Calcic Haploxerolls. The soil order of Mollisols comprises 25.1 percent of all U.S. soils. This deep, well-drained, alkaline soil receives 15 inches (37.5 cm) of annual precipitation. It is noted for the production of Irish potatoes, wheat, barley, and range grasses. (Scale is in feet; feet x 30.5 = cm) (Courtesy, USDA–Soil Conservation Service)

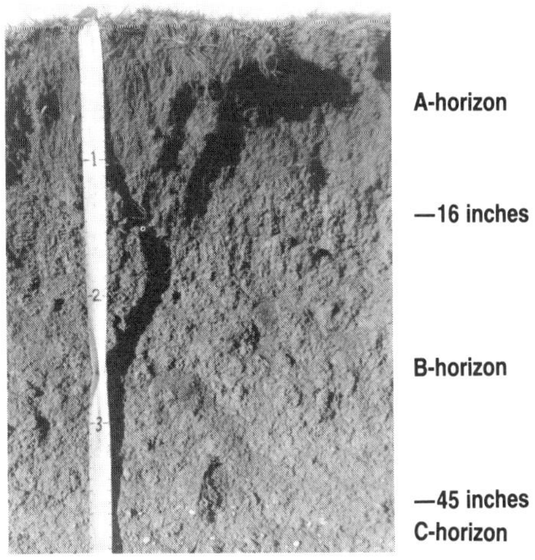

Illinois—Flanagan soil series. Soil order of Mollisols. Soil family—fine, montmorillonitic, mesic Aquic Argiudolls. The soil order of Mollisols comprises 25.1 percent of all U.S. soils. These somewhat poorly drained, slightly acid soils were originally in tall grass prairie and are now planted mostly to corn and soybeans. (Scale is in feet; feet x 30.5 = cm) (Courtesy, USDA–Soil Conservation Service)

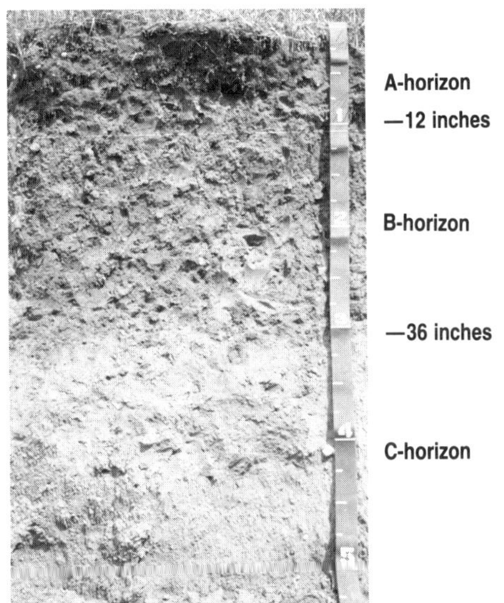

Indiana—Miami soil series. This is a Benchmark soil (see Table 1.1). Soil order of Alfisols. Soil family—fine loamy, mixed, mesic Typic Hapludalfs. The soil order of Alfisols comprises 13.5 percent of all soils in the United States. The pH is slightly acid to alkaline. Mean annual precipitation is 36 inches (91 cm). The original vegetation was an oak-hickory forest. Most areas are cleared and used for corn, soybeans, small grains, hay, and pasture. (Scale is in feet; feet x 30.5 = cm) (Courtesy, USDA–Soil Conservation Service)

Note: In 1988, the Miami soil series was proposed as the state soil of Indiana.

Iowa—Tama soil series. Soil order of Mollisols. Soil family—fine silty, mixed, mesic Typic Argiudolls. The soil order of Mollisols comprises 25.1 percent of all U.S. soils. These soils occur in subhumid and semiarid parts of the United States and the world. Mollisols are neutral to alkaline, have a dark and thick A-horizon, and are fertile. (Inches x 2.54 = cm) (Courtesy, USDA–Soil Conservation Service)

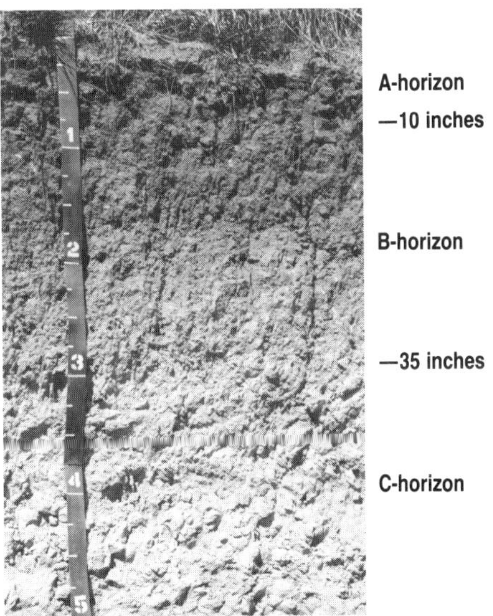

Kansas—Harney soil series. Soil order of Mollisols. Soil family—fine, montmorillonitic, mesic Typic Argiustolls. The soil order of Mollisols comprises 25.1 percent of the soils of the United States. Mean annual precipitation is 23 inches (58 cm). Original vegetation was grasses; presently, most of these soils are planted to wheat and grain sorghum. (Scale is in feet; feet x 30.5 = cm) (Courtesy, USDA–Soil Conservation Service)

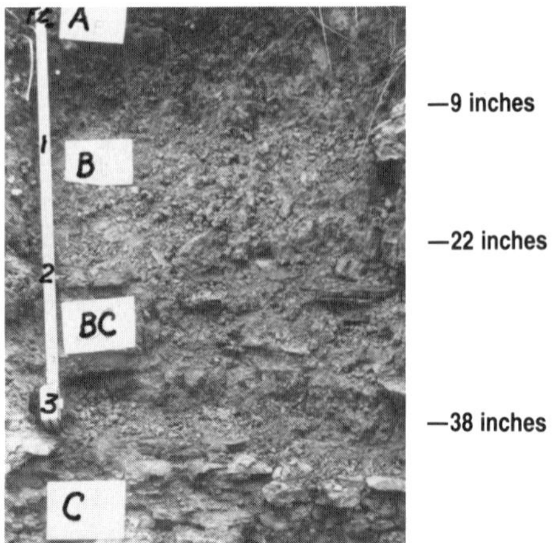

Kentucky—Eden soil series. Soil order of Alfisols. Soil family—fine, mixed, mesic Typic Hapludalfs. The soil order of Alfisols comprises 13.5 percent of the soils of the United States. (Inches x 2.54 = cm; feet x 30.5 = cm) (Courtesy, USDA–Soil Conservation Service)

Note: (1) The BC-horizon has natural characteristics of both B- and C-horizons. (2) Mostly fertile agricultural soils in humid, temperate regions. In Kentucky they are planted to corn, soybeans, small grains, and pastures.

Louisiana—Crowley soil series. Soil order of Alfisols. Soil family—fine, montmorillonitic, thermic Typic Albaqualfs. The soil order of Alfisols comprises 13.5 percent of all soils in the United States. The A- and E-horizons are acid; the B-horizon is neutral to alkaline. The parent materials are alluvial (water-deposited) sediments. The B-horizon is an example of prismatic soil structure created by montmorillonite clay, which expands and contracts greatly when wetted and dried. (See Figure 1.3.) Mean annual precipitation is 59 inches (150 cm). Topography is level. The water table, from December to April, is at 1 foot (30.5 cm). The native vegetation was tall wetland grass prairie. Most of the Crowley soil series is used for the production of rice in rotation with soybeans, pasture grasses/legumes, corn, and cotton. (Courtesy, USDA–Soil Conservation Service)

Maine—Plaisted soil series. Soil order of Spodosols. Soil family—coarse loamy, mixed, frigid Typic Haplorthods. The soil order of Spodosols comprises 4.8 percent of all soils in the United States. All horizons are strongly acid. Mean annual precipitation is 40 inches (102 cm). Native vegetation was birch, beech, maple, hemlock, white pine, spruce, and balsam fir. Cleared areas are used for growing Irish potatoes, small grains, and vegetables. (Scale is in inches; inches x 2.54 = cm) (Courtesy, USDA–Soil Conservation Service)

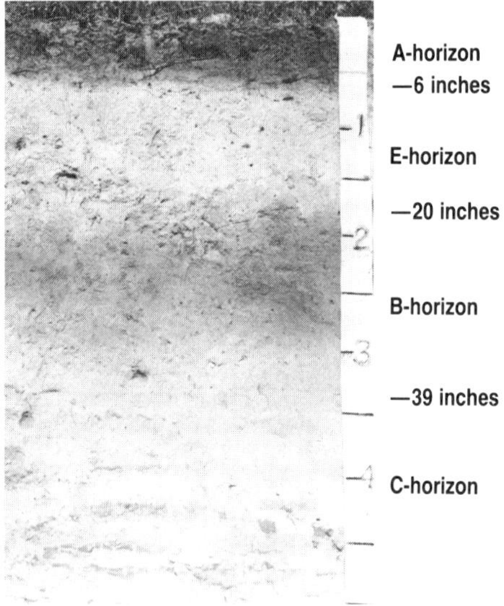

Maryland—Fort Mott soil series. Soil order of Ultisols. Soil family—loamy, siliceous, mesic Arenic Hapludults. The soil order of Ultisols comprises 12.8 percent of all soils in the United States. All soil horizons are acid. Mean annual precipitation is 48 inches (122 cm). Native vegetation was hardwoods, mostly oak and hickory. Present use is primarily for commercial and garden vegetables. (Scale is in feet; feet x 30.5 = cm) (Courtesy, USDA–Soil Conservation Service)

Massachusetts—Paxton soil series. In 1988 this series was being proposed as the state soil. Soil order of Inceptisols. Soil family—coarse loamy, mixed, mesic Typic Dystrochrepts. The soil order of Inceptisols comprises 18.2 percent of all soils in the United States. All horizons are strongly acid. The C-horizon is very compact. The large surface stone and the nearly spherical stone between 2 and 3 feet are granite. Smaller stones are schists or gneiss. Mean annual precipitation is 43 inches (109 cm). Original vegetation was hardwoods. Present use is mostly for hay and pasture crops. (Scale is in feet; feet x 30.5 = cm) (Courtesy, USDA–Soil Conservation Service)

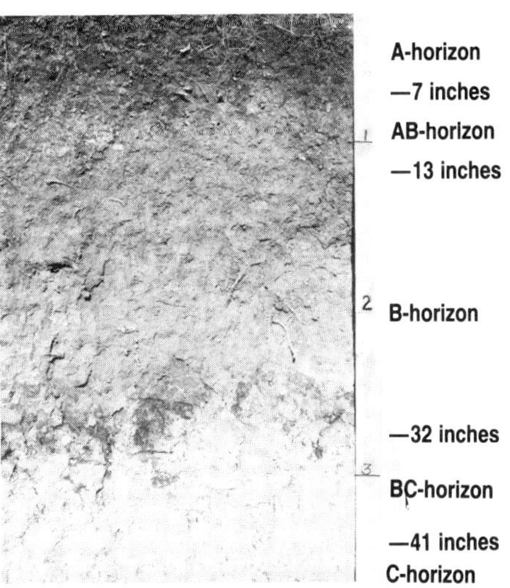

Minnesota—LeSueur soil series. Soil order of Mollisols. Soil family—fine loamy, mixed, mesic Aquic Argiudolls. The soil order of Mollisols comprises 25.1 percent of all soils in the United States. Mean annual precipitation is 30 inches (76 cm). The A-horizon is neutral, the B-horizon is acid, and the C-horizon is alkaline. Native vegetation was elm, basswood, and sugar maple. Most of the soil series is used for growing corn, soybeans, small grains, and grass-legume hay. (Scale is in feet; feet x 30.5 = cm) (Courtesy, USDA–Soil Conservation Service)

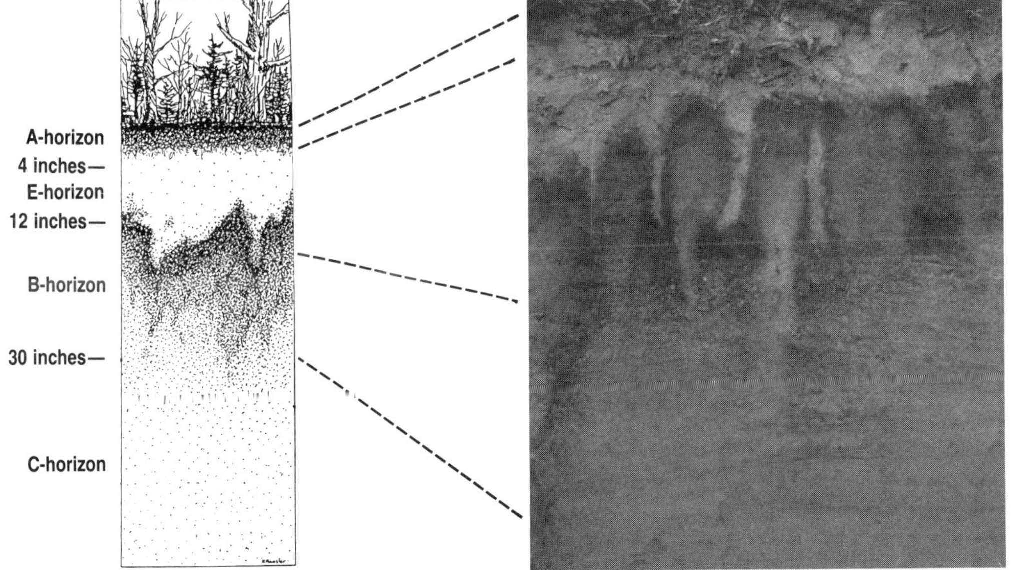

Michigan—Kalkaska soil series. The state soil. Soil order of Spodosols. Soil family—sandy, mixed, frigid Typic Haplorthods. The soil order of Spodosols comprises 4.8 percent of the soils of the United States. Most Kalkaska soils are in hardwood and pine forests. Cropped areas are planted to small grains, Irish potatoes, and hay-pasture crops. Annual precipitation is 30 inches (75 cm). (Inches x 2.54 = cm) (Courtesy, USDA–Soil Conservation Service)

Note: The tongues of the E-horizon appear to be old tree tap root channels.

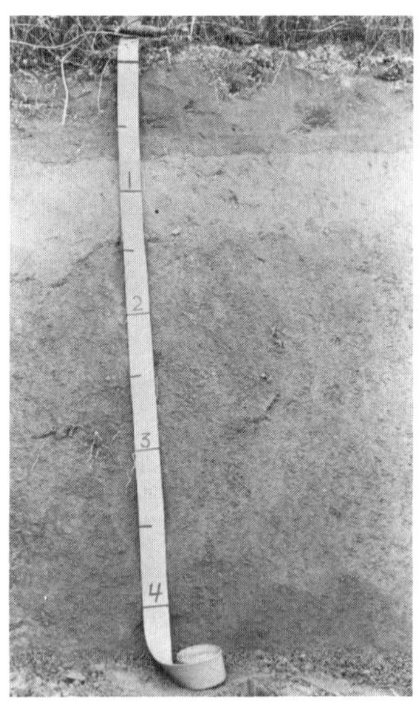

Mississippi—Smithdale soil series. Soil order of Ultisols. Soil family—fine loamy, siliceous, thermic Typic Hapludults. The soil order of Ultisols comprises 12.8 percent of all U.S. soils. All soil horizons are strongly acid. Mean annual precipitation is 52 inches (132 cm). Original vegetation was shortleaf pine, loblolly pine, and slash pine with a mixture of hardwoods, mostly oaks. Small clearings are used for gardens, small grains, and pastures. (Scale is in feet; feet x 30.5 = cm) (Courtesy, USDA–Soil Conservation Service)

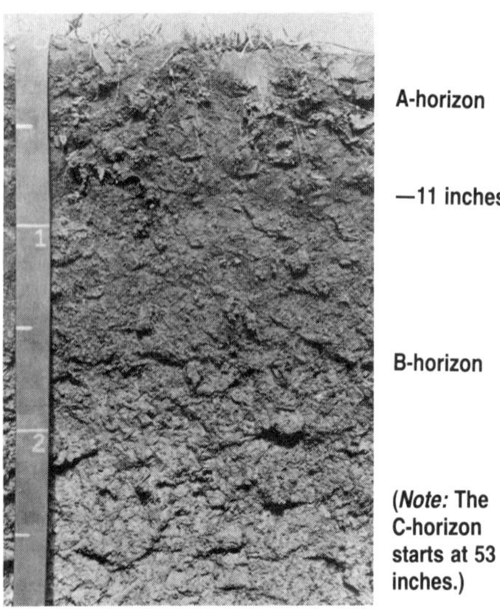

Missouri—Grundy soil series. Soil order of Mollisols. Soil family—fine, montmorillonitic, mesic Aquic Argiudolls. The soil order of Mollisols comprises 25.1 percent of all soils in the United States. Mean annual precipitation is 35 inches (89 cm). Native vegetation was tall grass prairie. Most of these soils are planted to corn, soybeans, and grain sorghum. (Scale is in feet; feet x 30.5 = cm) (Courtesy, USDA–Soil Conservation Service)

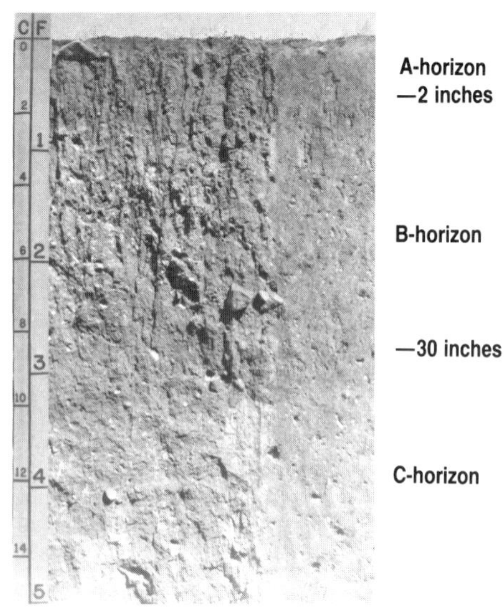

Montana—Scobey soil series. This is a Benchmark soil (see Table 1.1). Soil order of Mollisols. Soil family—fine, montmorillonitic Aridic Argiborolls. The soil order of Mollisols comprises 25.1 percent of all soils in the United States. Native vegetation was short grasses such as western wheatgrass and blue gramagrass. Mean annual precipitation is about 12 inches (30.5 cm). (Courtesy, USDA–Soil Conservation Service)

Nebraska—Holdrege soil series. The state soil. Soil order of Mollisols. Soil family—fine silty, mixed, mesic Typic Argiustolls. The soil order of Mollisols comprises 25.1 percent of all U.S. soils. Average annual precipitation is about 20 inches (51 cm). Range grasses used for grazing are the land use on about half of the area; the other half is devoted to wheat, alfalfa, and grain sorghum. (On the scale, C x 10 = cm and F = feet; feet x 30.5 = cm) (Courtesy, USDA–Soil Conservation Service)

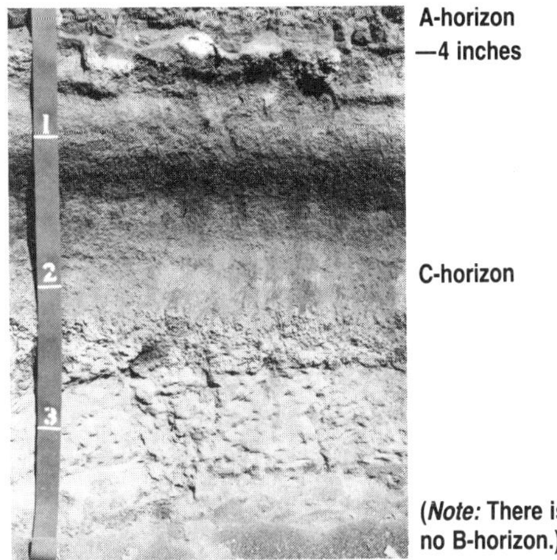

Nevada—Land soil series. Soil order of Aridisols. Soil family—fine silty, mixed, thermic Typic Salorthids. The soil order of Aridisols comprises 11.6 percent of all U.S. soils. Elevation is about 1,800 feet (549 m). Mean annual precipitation is about 5 inches (12.7 cm). Original vegetation was short grasses, mesquite bushes, and big sagebrush. Concrete driveways are often destroyed by the formation and growth of sodium sulfate crystals. (Courtesy, USDA–Soil Conservation Service)

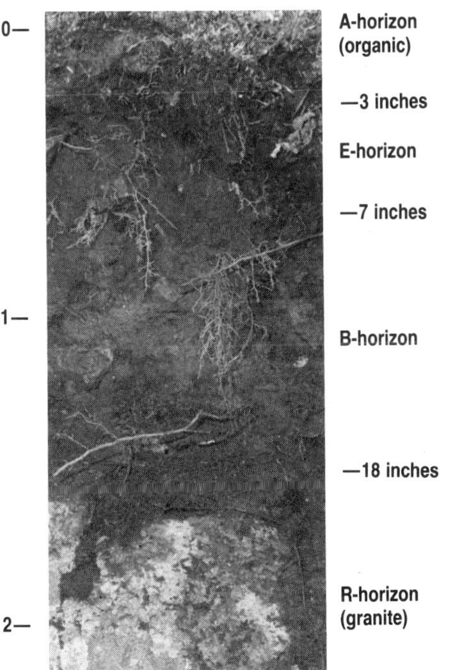

New Hampshire—Lyman soil series. The state soil. Soil order of Spodosols. Soil family—loamy, mixed, frigid Lithic Haplorthods. The soil order of Spodosols comprises 4.8 percent of all U.S. soils. (Scale is in feet; feet x 30.5 =cm) (Courtesy, USDA–Soil Conservation Service)

Note: There is no C-horizon.

New Jersey—Pascack soil series. Soil order of Ultisols. Soil family—coarse loamy, mixed, mesic Aquic Hapludults. The soil order of Ultisols comprises 12.8 percent of all U.S. soils. Acid is in all horizons. Mean annual precipitation is 44 inches (112 cm). Native vegetation was oak and hickory. Present use is mostly for corn, soybeans, truck crops, and nurseries. (Inches x 2.54 = cm) (Courtesy, USDA–Soil Conservation Service)

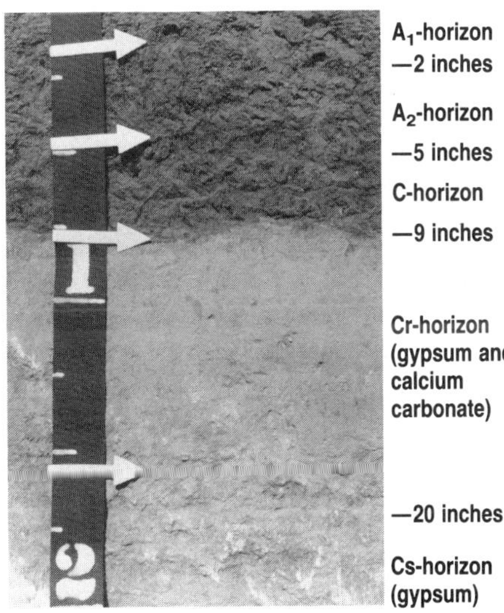

New Mexico—Holloman soil series. Soil order of Entisols. Soil family—loamy, gypsic, thermic, shallow Typic Torriorthents. The soil order of Entisols comprises 8 percent of all soils in the United States. Mean annual precipitation is 11 inches (28 cm). Original vegetation was short grasses, now used for range grazing of cattle. (Scale is in feet; feet x 30.5 = cm) (Courtesy, USDA–Soil Conservation Service)

Note: (1) There is no B-horizon. (2) This is the soil on the U.S. White Sands Missile Range.

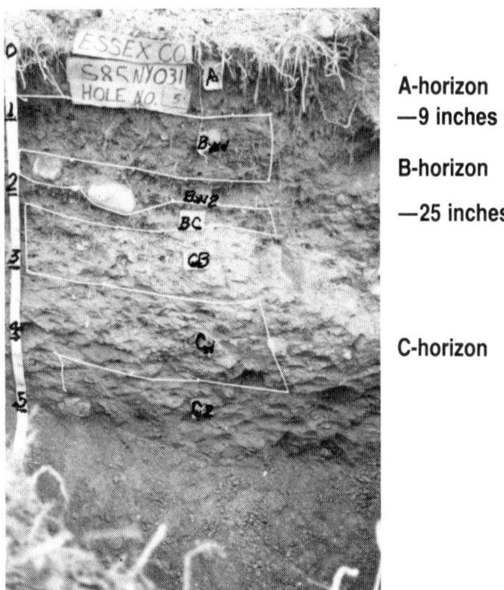

New York—Nelli soil series. Soil order of Inceptisols. Soil family—coarse loamy, mixed, mesic Typic Utrochrepts. The soil order of Inceptisols comprises 18.2 percent of all soils in the United States. Mean annual precipitation is 32 inches (81 cm). Original vegetation was sugar maple, basswood, white ash, and other hardwoods. Cleared areas are used for grass-legume hay, corn, small grains, vegetables, and fruit. Parent materials (C_1- and C_2-horizons) are high in calcic limestone. The A-horizon is slightly acid; other horizons are neutral (pH 7). (Scale is in feet; feet x 30.5 = cm) (Courtesy, USDA–Soil Conservation Service)

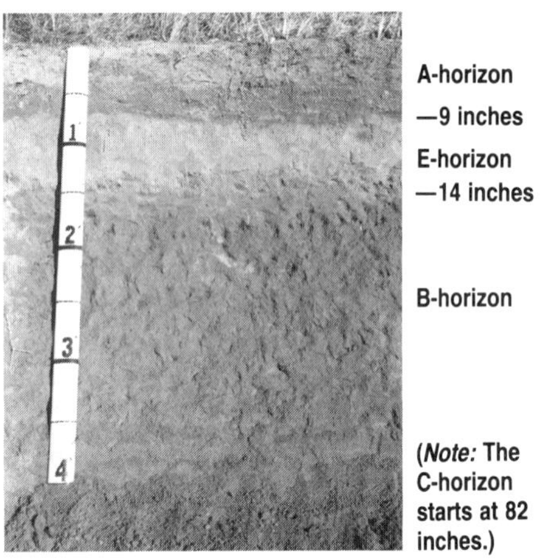

North Carolina—Norfolk soil series. This is a Benchmark soil (see Table 1.1). Soil order of Ultisols. Soil family—fine loamy, siliceous, thermic Typic Paleudults. The soil order of Ultisols is very extensive in the lower South and comprises 12.8 percent of all soils in the United States. Mean annual precipitation is 50 inches (127 cm). All horizons are strongly acid. Original vegetation was southern pines and mixed southern hardwoods, mostly oaks. Present use is for peanuts, tobacco, corn, and cotton. (Scale is in feet; feet x 30.5 = cm) (Courtesy, USDA–Soil Conservation Service)

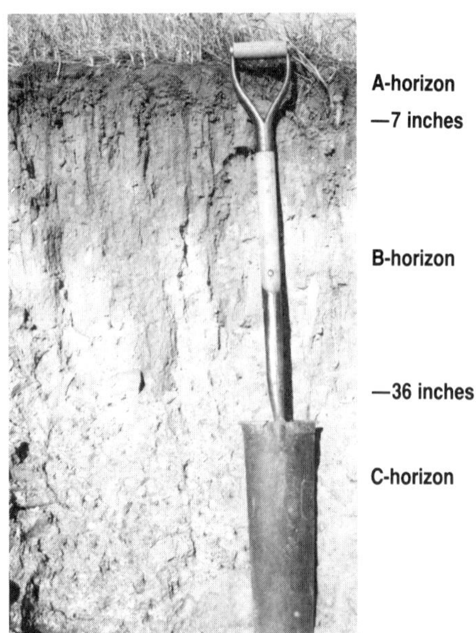

North Dakota—Barnes soil series. This is a Benchmark soil (see Table 1.1). Soil order of Mollisols. Soil family—fine loamy, mixed Udic Haploborolls. The soil order of Mollisols comprises 25.1 percent of all U.S. soils. Annual precipitation is 17 inches (42.5 cm). (Courtesy, USDA–Soil Conservation Service)

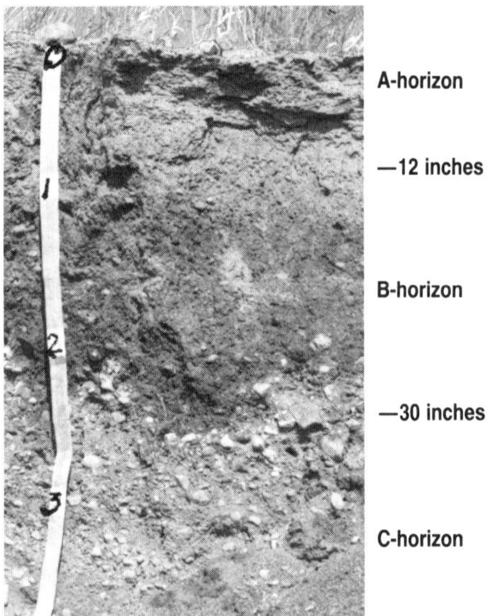

Ohio—Eldean soil series. Soil order of Alfisols. Soil family—fine, mixed, mesic Typic Hapludalfs. The soil order of Alfisols comprises 13.5 percent of the soils of the United States. This well-drained soil is used for corn, soybeans, small grains, hay, and pasture. Annual precipitation is 38 inches (95 cm). (Scale is in feet; feet x 30.5 = cm) (Courtesy, USDA–Soil Conservation Service)

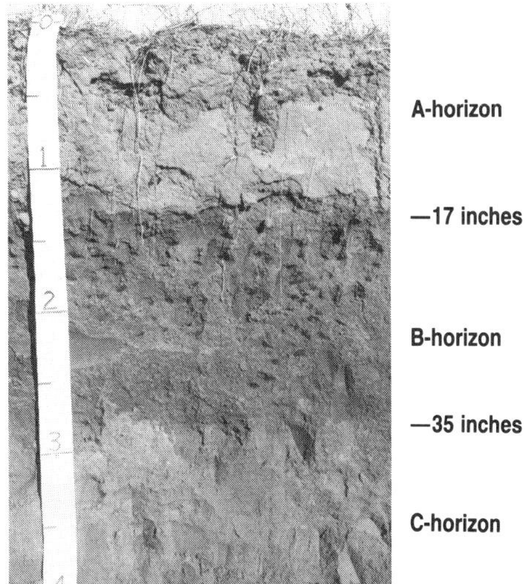

Oklahoma—Port soil series. Soil order of Mollisols. Soil family—fine silty, mixed, thermic Cumulic Haplustolls. The soil order of Mollisols comprises 25.1 percent of all U.S. soils. Mean annual precipitation is 30 inches (76 cm). The soil is alkaline in all horizons. Native vegetation was tall grass prairie with a scattering of pecan, bur oak, and cottonwood. Present use is mainly for alfalfa, small grains, grain sorghum, and cotton. (Scale is in feet; feet x 30.5 = cm) (Courtesy, USDA–Soil Conservation Service)

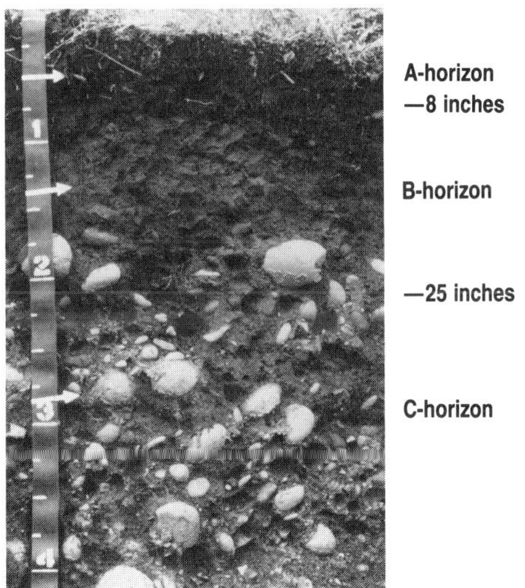

Oregon—Multnomah soil series. Soil order of Inceptisols. Soil family—fine loamy over sandy or sandy-skeletal, mixed, mesic Dystric Xerochrepts. The soil order of Inceptisols comprises 18.2 percent of all U.S. soils. Average annual precipitation is 50 inches (125 cm). Original vegetation was Douglas fir, Oregon white oak, bigleaf maple, and western redcedar. Cleared areas are planted to truck crops, nurseries, and pasture grasses. (Scale is in feet; feet x 30.5 = cm) (Courtesy, USDA–Soil Conservation Service)

Pennsylvania—Hagerstown soil series. This is a Benchmark soil (see Table 1.1). Soil order of Alfisols. Soil family—fine, mixed, mesic Typic Hapludalfs. The soil order of Alfisols comprises 13.5 percent of all soils in the United States. The A-horizon is slightly acid; other horizons are neutral. Mean annual precipitation is 38 inches (94 cm). Native vegetation was mixed hardwoods. Present use is for corn, small grains, and grass-legume pastures. (Scale is in feet; feet x 30.5 = cm) (Courtesy, Roger Pennock, Jr., Pennsylvania State University, and USDA–Soil Conservation Service)

Note: An extensive and very productive soil.

Rhode Island—Hinkley soil series. Soil order of Entisols. Soil family—sandy-skeletal, mixed, mesic Typic Udorthents. The soil order of Entisols comprises 8 percent of the soils of the United States. These excessively drained, droughty, acid soils receive 40 to 50 inches (102–127 cm) of annual precipitation. Forest trees, fruits, and vegetables are the principal crops grown. (Courtesy, USDA–Soil Conservation Service)

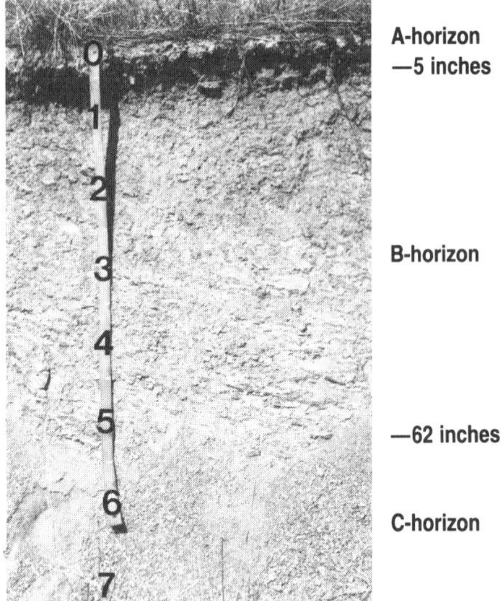

A-horizon
—5 inches

B-horizon

—62 inches

C-horizon

South Carolina—Cataula soil series. Soil order of Ultisols. Soil family—clayey, kaolinitic, thermic Hapludults. The soil order of Ultisols comprises 12.8 percent of all U.S. soils. These soils receive about 50 inches (125 cm) of average annual precipitation. Shortleaf pine and loblolly pine are growing on most of these soils. Cleared areas are planted to cotton, corn, small grains, and pastures. (Scale is in feet; feet x 30.5 = cm) (Courtesy, USDA–Soil Conservation Service)

A-horizon
—9 inches

B-horizon

—50 inches

C-horizon

Tennessee—Memphis soil series. Soil order of Alfisols. Soil family—fine silty, mixed, thermic Typic Hapludalfs. The soil order of Alfisols comprises 13.5 percent of all U.S. soils. The most level of these strongly acid, well-drained soils are used for cotton, small grains, soybeans, hay, and pasture. Steeper areas are used for timber production. (Scale is in feet; feet x 30.5 = cm) (Courtesy, USDA–Soil Conservation Service)

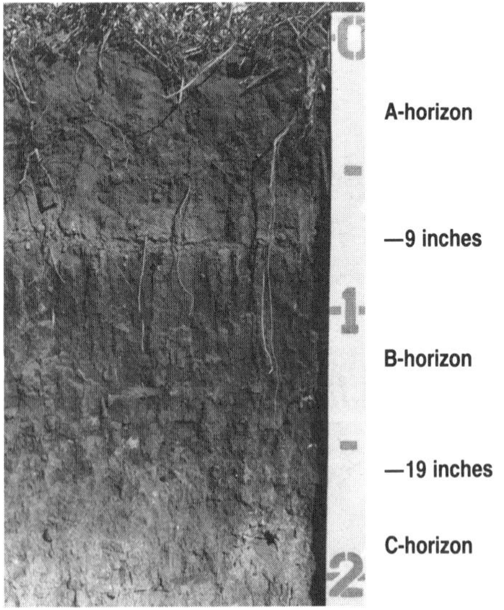

A-horizon

—9 inches

B-horizon

—19 inches

C-horizon

South Dakota—Nahon soil series. Soil order of Mollisols. Soil family—fine, montmorillonitic , frigid Udic Natriborolls. The soil order of Mollisols comprises 25.1 percent of all U.S. soils. The Nahon soil series is high in sodium, is alkaline, and is moderately well-drained. It receives about 20 inches (50 cm) of annual precipitation. Principal crops grown are small grains, alfalfa, and pasture and range grasses. (Scale is in feet; feet x 30.5 = cm) (Courtesy, USDA–Soil Conservation Service)

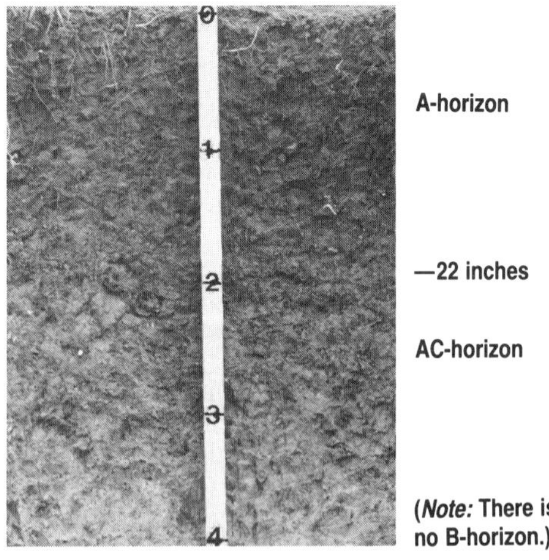

A-horizon

—22 inches

AC-horizon

(*Note:* There is no B-horizon.)

Texas—Branyon soil series. Soil order of Vertisols. Soil family—fine, montmorillonitic, thermic Udic Pellusterts. The soil order of Vertisols comprises 1 percent of all U.S. soils. Original vegetation was tall grass prairie. Cleared areas are used extensively for grain sorghum, small grains, and hay. Average annual precipitation is 40 inches (100 cm). (Scale is in feet; feet x 30.5 = cm) (Courtesy, USDA–Soil Conservation Service)

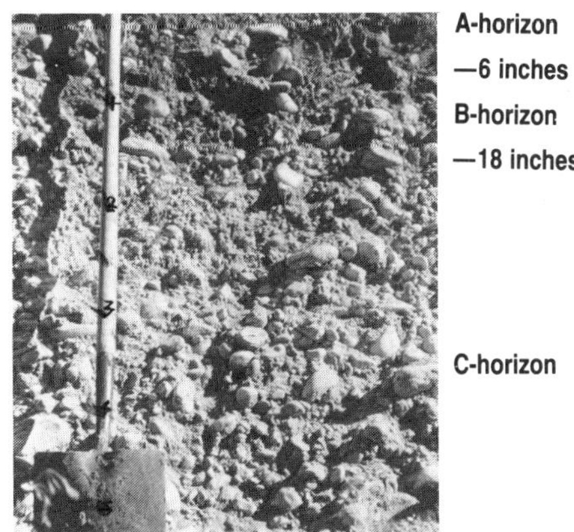

Utah—Bingham soil series. Soil order of Mollisols. Soil family—fine loamy, over sandy or sandy-skeletal, mixed, mesic Calcic Argixerolls. The soil order of Mollisols comprises 25.1 percent of all soils in the United States. Mean annual precipitation is 16 inches (41 cm). All soil profiles are neutral to alkaline. Native vegetation was bluebunch wheatgrass, big sagebrush, and rubber rabbitbrush. Irrigated areas are in orchards, small grains, peas, tomatoes, and alfalfa. (Scale is in feet; feet x 30.5 = cm) (Courtesy, USDA–Soil Conservation Service)

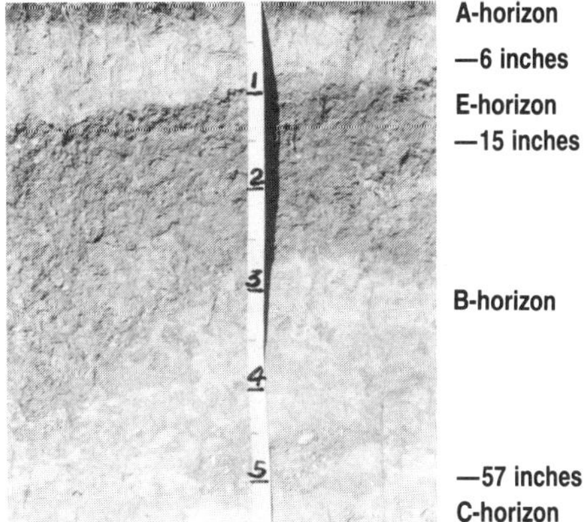

Virginia—Emporia soil series. Soil order of Ultisols. Soil family—fine loamy, siliceous, thermic Typic Hapludults. The soil order of Ultisols comprises 12.8 percent of all U.S. soils. All horizons are very strongly acid. Mean annual precipitation is 48 inches (122 cm). Original vegetation was loblolly pine, oaks, hickories, sweet gum, and maples. Cleared areas are used primarily for peanuts, soybeans, cotton, corn, and tobacco. (Scale is in feet; feet x 30.5 = cm) (Courtesy, USDA–Soil Conservation Service)

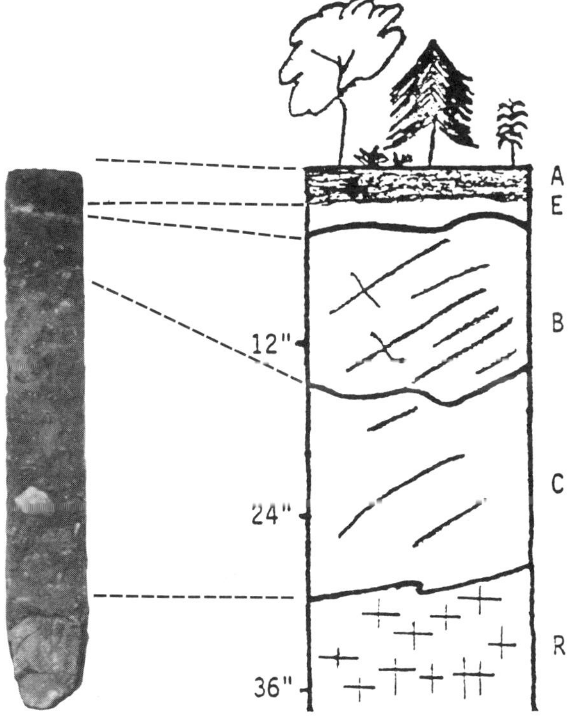

Vermont—Tunbridge soil series. The official state soil. Soil order of Spodosols. Soil family—coarse loamy, mixed, frigid Typic Haplorthods. The soil order of Spodosols comprises 4.8 percent of all soils of the United States. (Courtesy, USDA–Soil Conservation Service)

Note: The E-horizon is whitish; the "E" stands for eluviation (moving out of clay, humus, and nutrients). The "R" stands for rock; in this instance, granite rock. Soil is mostly in hardwood-conifer forests.

Washington—Athena soil series. Soil order of Mollisols. Soil family—fine silty, mixed, mesic Pachic Haploxerolls. The soil order of Mollisols comprises 25.1 percent of all soils in the United States. Mean annual precipitation is 17 inches (43 cm). The A-horizon is neutral, and the B-horizon is alkaline. Natural vegetation is tall grass prairie. On cropland, wheat is common. (Scale is in feet; feet x 30.5 = cm) (Courtesy, USDA–Soil Conservation Service)

West Virginia—Potomac soil series. Soil order of Entisols. Soil family—sandy-skeletal, mixed, mesic Typic Udifluvents. The soil order of Entisols comprises 8 percent of all soils in the United States. A soil formed along the edges of streams, used primarily for hay and pasture crops. Average annual precipitation is 42 inches (105 cm). (Courtesy, USDA–Soil Conservation Service)

Note: There is no B-horizon because the soil material is young, sandy, and cobbly and has not had time for one to develop.

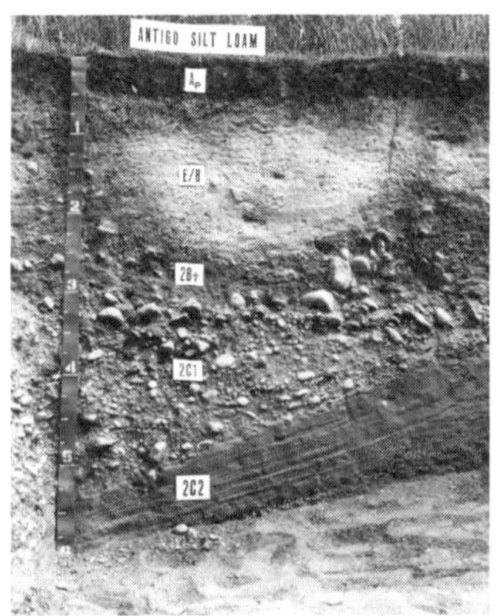

Wisconsin—Antigo silt loam. The state soil. Soil order of Alfisols. Soil family—fine silty over sandy or sandy-skeletal, mixed Typic Glossoboralfs. The soil order of Alfisols comprises 13.5 percent of all U.S. soils. (Scale is in feet; feet x 30.5 = cm) (Courtesy, USDA–Soil Conservation Service)

Note: (1) "Ap" means the plowed layer; "E/B" means a natural mixture of the E (eluvial)-horizon and the B-horizon; "2B" means the second B-horizon; the "t" in "2Bt" stands for *ton,* German for *clay;* "2C1" is the first horizon of the second parent material layer; and "2C2" is the second horizon of the second parent material. (2) The soil is acid to 6 feet (1.8 m). (3) The parent materials were deposited by glaciers on outwash plains between 10,000 and 11,000 years ago. (4) Mean annual precipitation is 28 inches (70 cm). (5) Most of these soils have been cleared from native hardwood and pine and are planted in corn, small grains, forage, potatoes, and green beans.

Wyoming—Farson soil series. Soil order of Aridisols. Soil family—coarse loamy, mixed Borollic Haplargids. The soil order of Aridisols comprises 11.6 percent of all soils of the United States. Elevation is about 6,560 feet (2,000 m), and annual precipitation is about 10 inches (25 cm). Original vegetation was short grasses, big sagebrush, and rabbitbrush. Irrigated areas are used to produce small grains, hay, and pasture grasses. (Scale is in inches; inches x 2.54 = cm) (Courtesy, USDA–Soil Conservation Service)

CHAPTER 2

Soils, Nutrition, and Plant Growth

"The yield of a crop is limited by the deficiency of one necessary element even though all others are present in adequate amounts." — Justus von Liebig, 1840

OUTLINE

This farm is successful basically because of a very fertile soil that supplies adequate nutrients for maximum economic plant growth and yield. (West Virginia) (Courtesy, USDA - Soil Conservation Service)

□ □ □

2:1 □ OVERVIEW

A practical example is given of a farmer who applies all of the correct soil, water, and fertility management to produce a world's record corn yield. Returning to the soil only those plant nutrients taken from the soil will not maintain productivity.

Soil scientists, plant scientists, and their collaborators have been very successful in conducting research, extension, and education in soils, nutrition, and plant growth. All 50 states in the United States have field research stations and laboratories where this work is done. Furthermore, money and personnel have been supplied by the U.S. Department of Agriculture. Also, 12 international agricultural research stations have been established around the world. This state, national, and international network has made remarkable progress in soils, nutrition, and plant growth on which all primary food, feed, and fiber production depends.

From the soil, plants receive water, oxygen, and 13 of the 16 nutrients essential for growth. Of these 16 elements, 15 are also essential for animals. One, boron, is not essential for animals. However, people and animals require 10 more elements which plants do not need: sodium, iodine, selenium, vanadium, chromium, cobalt, nickel, fluorine, silicon, and arsenic.

The law of limiting factors states that at any one time, one nutrient limits plant growth and yield. Each of the 16 elements is equally essential, although plants absorb them in unequal amounts. A soil test can be relied upon to discover this one-at-a-time limiting factor.

Plant deficiency symptoms vary from nutrient-to-nutrient and often from plant-to-plant. Also symptoms of deficiency of any one nutrient are sometimes masked by pesticides, insects, diseases, and a deficiency of another nutrient. By the time a deficiency can be observed, it is often too late to add that nutrient to the soil to benefit the current crop. Some nutrient deficiency symptoms can be corrected on the current crop by foliar application. (See Chapter 4.)

2:2 □ A PRACTICAL APPLICATION

One of the best examples of the successful application of the principles of soil management, nutrition, and plant growth is as follows:

In 1985 Herman Warsaw, a farmer in McLean County, Illinois, raised 370 bushels of shelled corn per acre (23,227 kg / ha). This was a world record. Such a yield was almost ½ pound per square foot — enough shelled corn to cover the blackness of the productive Corn Belt soil.

How did he do it?

1. With continuous corn.
2. With a chisel plow — never a turning plow.
3. With 16,000 pounds per acre (17,936 kg/ha) of corn stalk residue chopped into the soil.
4. With fertilizer and lime applications based on a soil test.
5. With supplemental irrigation.
6. With excellent management.

When Herman Warsaw harvested his world record yield of 370 bushels of corn grain per acre (23,227 kg/ha), he removed from the field only 479.71 pounds per acre (537.6 kg/ha) of plant nutrients (Table 2.1).

Table 2.1 — Corn Grain Chemical Composition and Total Nutrients in 370 Bushels of Corn per Acre

Nutrient	Composition[1] (%)	Total Nutrients in 370 Bushels of Corn Grain[2] (lbs/acre)
Nitrogen (N)	1.4	290.08
Phosphorus (P)	0.3	62.16
Potassium (K)	0.3	62.16
Calcium (Ca)	0.01	2.07
Magnesium (Mg)	0.2	41.44
Sulfur (S)	0.1	20.72
	(parts per million)	
Iron (Fe)	35.0	0.73
Zinc (Zn)	10.0	0.21
Manganese (Mn)	4.0	0.08
Copper (Cu)	3.0	0.06
		Total = 479.71

[1] Douglas M. Considine and Glenn D. Considine. *Foods and Food Production Encyclopedia.* New York: Van Nostrand Reinhold Co., 1982, p. 627.

[2] *Note:* (a) 370 bu. of corn = 20,720 lbs. = 9,407 kg; pounds × 0.454 = kg.
(b) Smaller quantities of chlorine, boron, and molybdenum were also removed from the soil in the grain.

Why not apply to the soil only the pounds (kg) of nutrients removed in the grain? Wouldn't this maintain the productive capacity of the soil? This idea was researched for the first time around 1840 and was declared inadequate.

2:3 ▫ RESEARCH, EXTENSION, AND TEACHING

Justus von Liebig (1803 – 1873), a German *laboratory* chemist, believed that by analyzing plants and putting back into the soil all the elements the plants removed, soil productivity could be maintained.[1]

To test this proposition, in 1843 the Rothamsted Experiment Station near London was established by J. B. Lawes and J. H. Gilbert. After several years of exhaustive *field research,* they concluded that the chemical composition of plants is no indication of the amounts or kinds of nutrients required by the plants. Then, how can the nutrient needs of plants be determined accurately? There is no *one* answer because of extreme variability in soils, plants, climates, diseases, and insect infestations. For these reasons, all the countries of the world and all the states in the United States have agricultural research stations to solve many problems, including those of soils, nutrition, and plant growth.

The principal research, extension, and teaching successes of soil and crop scientists have included:

1. Establishing field research and education plots to push back the frontiers of soil science, plant nutrition, and plant growth.
2. Establishing soil-testing services to extend the data from the field plot research to farmers, students in high schools and colleges, and the concerned public. Soil tests are used to determine economical lime, fertilizer, and organic residue recommendations and to assess the reclamation potential of saline and sodic soils and the quality of irrigation water.
3. Working with other environmental scientists, the lay public, and politicians to design and operate sanitary landfills and to use sewage effluents safely as irrigation water and as organic fertilizer.
4. Cooperating with the Tennessee Valley Authority and private fertilizer companies in field testing new fertilizer formulations (Figure 2.1).
5. Cooperating with crop scientists, horticulturists, livestock specialists, farmers, ranchers, and municipalities in finding the most efficient use of new fertilizer formulations, lime, animal manures, and sewage sludges on specific soil series in all climatic regions of the United States.
6. Cooperating with the U.S. Agency for International Development in the State Department, the Consultative Group on International Agriculture, the 12 International Agricultural Research Centers, the United Nations Development Program, the Food and Agriculture Organization of the United Nations, the International Fertilizer Development Center, and the International Bank for Reconstruction and Development (World Bank) in research and extension development of food, feed, and fiber throughout the world (Figures 2.2, 2.3, and 2.4).
7. Breeding and selecting crop plants with hybrid vigor and greater yield potential and better nutritional quality through biogenetic research.

---

[1]This may be called the "Bank Account Hypothesis," which has been proven false. However, von Liebig did make two major contributions to soils, nutrition, and plant growth: (a) He proved that plants do not absorb humus as humus particles and (b) he developed the law of limiting factors (see Section 2:8).

Fig. 2.1 — Mr. and Mrs. Schroeder and their family in Whiteside County, Illinois, were proud to see their name stand out in this field. The letters of the name are greener grass than the rest of the field, the result of an experimental soil test which limed and fertilized with nitrogen, phosphorus, and potassium to supplement the nutrient-supplying capacity of the soil. (Courtesy, University of Illinois)

Fig. 2.2 — Research in the United States has been extended by the international research organizations to this scene in India where modern corn hybrids are now grown successfully. (Courtesy, U.S. Agency for International Development)

Fig. 2.3 — An agricultural education teacher helps to educate, train, and guide students in solving problems related to all aspects of agricultural resource management, including soils, nutrition, and plant growth. (Courtesy, Rodney W. Tulloch)

Fig. 2.4 — Field days are just one of the many ways by which experiment station and extension workers communicate their findings to producers. (Photo by Roy Hunter Follett)

An example of research successes is given here for rices of the world. From 1972 to 1989, the International Rice Research Institute (IRRI) collected seed with resistance to specific stresses from more than 10,000 selections around the world. The most important of these include: **resistance** to drought, insects, diseases, nematodes, and rodents and **tolerance** of deep water, cold weather, high salinity, highly acid soils, highly alkaline soils, low phosphorus, low iron, high iron, and **allic** (high aluminum) soils, a capability that can be destroyed by steaming.

Soil and crop scientists and their collaborators have been so successful that a world-renowned agricultural administrator has written that about 50 percent of world food, feed, and fiber production is due to the use of fertilizers.

2:4 □ PRODUCTIVE SOILS

Soils are productive when they contain adequate amounts of all 13 essential soil elements in forms readily available to plants every day, are in a good physical condition to support plants, and contain just the right amount of water and air for desirable top and root growth. Toxic substances such as total soluble salts and sodium must be in low concentrations. Some soils are capable of suppressing certain plant diseases.

Too little nitrogen or other essential nutrients or soil water, even for a day, may reduce crop yields. If the soil has a tillage pan or a surface crust so that it is too wet after a rain and too dry a few days later, plant growth is stunted. To be more specific, all crop plants need the same kinds of elements, as well as water and air, but plants differ in the relative amounts of their requirements of these essentials. For example, blueberries, alfalfa, and rice all require the same elements, as well as air and water, but blueberries grow on soils very low in both available nutrients and water, while alfalfa requires a very fertile soil which is constantly moist. Rice and alfalfa both have a high water requirement, but alfalfa must have plenty of air mixed with the water such as occurs in well-drained soils, whereas rice does better when the soil is flooded.

Some soils are in such physical and chemical condition as to encourage plant roots to grow deeply and to extend long distances laterally. These soils are ideal because the plants growing in them will be windfirm, drought-resistant, and capable of absorbing water and nutrients from a large volume of soil. By contrast, plant roots may be restricted by naturally or artificially compacted layers, infertile horizons, too much or too little soil moisture, or soluble salts in toxic quantities.

Water and air occupy pore spaces in the soil. Following a heavy and prolonged rain, the soil pores may be almost completely filled with water for a few hours. After several hours, some water will have moved downward in response to gravity, and the larger pores will be emptied of water and filled with air. With a further loss of water by evaporation or transpiration, air will replace more of the water. The next soaking rain will repeat this cycle.

Adequate amounts of nutrients for desirable plant growth may be "available" according to any chemical test but actually deficient because of soil physical conditions. Tillage pans, cloddy surface soils, surface crusts, or lack of soil aggregation may reduce nutrient availability and plant growth for one of the following reasons:

1. By restricting the exchange of oxygen and carbon dioxide in the soil, the ability of plant roots to translocate nutrients to the leaves is reduced.
2. By reducing water infiltration from rain or irrigation, water may become the first limiting factor in plant growth although all nutrients are present in adequate amounts.
3. By retarding the growth of nitrifying bacteria, the soil nitrogen may remain as unavailable protein instead of breaking down to release available ammonium and nitrate ions.
4. By physically restricting root elongation, especially deep root penetration or lateral root extension, the volume of soil in contact with plant roots and the total nutrients absorbed will be limited and plant growth restricted.

2:5 □ ELEMENTS ESSENTIAL FOR PLANTS, PEOPLE, AND ANIMALS

Carbon, hydrogen, and oxygen form the plant structure. They are readily obtained from air and water. Normally, the plant obtains its remaining nutrients from the soil solution. The primary nutrients, nitrogen, phosphorus, and potassium, and the secondary nutrients, calcium, magnesium, and sulfur, are referred to as ***macronutrients*** — *macro-*, meaning "large" — because they are required by growing plants in relatively large amounts. The ***micronutrients,*** boron, chlorine, cobalt, copper, iron, manganese, molybdenum, and zinc, are so-called because — *micro-*, meaning "small" — they are required by the growing plant in relatively small amounts.

Since farm livestock are mostly forage-eating animals, it is of vital concern to know whether the elements essential for plants are also essential for animals. Elements known to be essential for people and animals are carbon, hydrogen, oxygen, phosphorus, potassium, nitrogen, sulfur, calcium, iron, magnesium, manganese, copper, zinc, sodium, iodine, selenium, molybdenum, chromium, chlorine, and cobalt. Boron[2] is essential for plants but not for people and animals. Sodium, iodine, selenium, chromium, fluorine, silicon, and cobalt are essential for people and animals but not for plants. If raised on dirt (soil), pigs will get enough iron in their diets because of the large amount of iron in the soil.

---

[2]A research report on boron describes it as an essential element for women. Small amounts of boron prevented **osteoporosis** (brittle bones). Boron supplements also increased the growth of poultry. (U.S. Dept. of Agriculture, *Agricultural Research,* Vol. 35, No. 10, November – December 1987, p. 12)

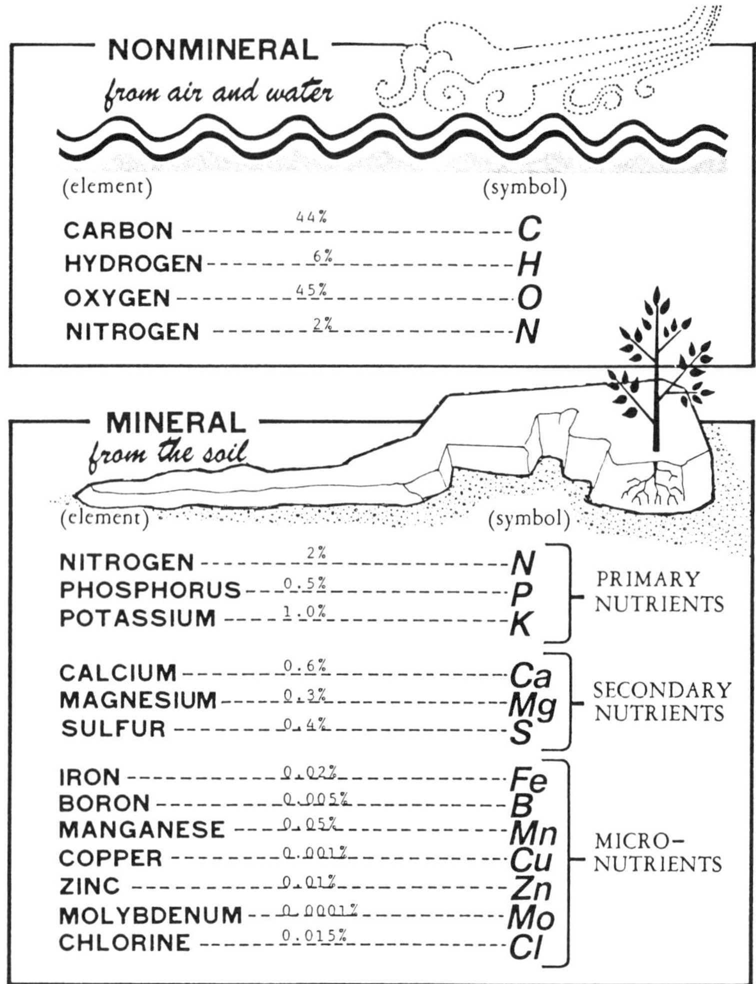

Fig. 2.5 — Elements essential for the growth of seed-bearing plants and their average percentage composition in plants. Smaller quantities of copper, chlorine, zinc, and molybdenum are also present. Note: Nitrogen is a nonmineral (gas) in the atmosphere, but it must be changed to a mineral before plants can absorb it. Example: ammonium nitrate, a mineral, available to plants in the ionic forms, NH_4^+ and NO_3^+. (Courtesy, Instructional Materials Services, Texas A&M University, and the American Assn. of Plant Food Control Officials)

2:6 □ HOW PLANTS ABSORB NUTRIENTS

All crop plants feed mostly by root uptake of nutrients from the soil. Small amounts of nutrients can be absorbed through the leaves. Carbon, as carbon dioxide, enters the plant almost entirely through the stomata of the leaves. Water also is absorbed through the leaves, but the relative amount is small compared with that which enters the roots.

Besides obtaining some nutrients through their leaves and stems, plants obtain most of their nutrients through their roots from: (1) the soil solution, (2) exchangeable ions on the surface of clay and humus particles, (3) readily decomposable minerals and decomposable organic matter, and (4) applied lime and fertilizers (Note 2.1).

Clay and soil organic matter serve as the principal storehouses for plant nutrients. Through the decomposition of organic matter by bacteria, fungi, and actinomycetes, plant nutrients become available to growing plants.

Almost all nitrogen is held in the soil in the organic matter. Organic matter also accounts for 10 to 50 percent of the total soil phosphorus and up to 80 percent of total soil sulfur. Boron and molybdenum reserves seem to be stored both in organic matter and in clay.

From the soil, plants accumulate nitrogen, phosphorus, and sulfur; as a result, plants nearly always contain a higher percentage of these elements than the soil in which the plants are growing. Conversely, soils almost always contain a higher percentage of iron, calcium, potassium, magnesium, and manganese than do the plants growing in them.

Note 2.1 — How Plants Absorb Nutrients

Plants absorb nutrients primarily by root hairs absorbing elements from the soil solution. Root hairs and all surfaces of clay and humus particles are surrounded by *electronegative* particles. In the soil solution this results in the attraction of masses of *electropositive* ions to the root and soil surface. These electropositive ions include **aluminum** (Al^{3+}), **calcium** (Ca^{2+}), **magnesium** (Mg^{2+}), **potassium** (K^{+}), **ammonium** (NH_4^{+}), **sodium** (Na^{+}), and **hydrogen** (H^{+}). The ions so attracted (adsorbed) are not leachable with deionized (distilled) water, but plant roots can absorb them by a process known as **contact exchange.** For example, a root hair may exchange two hydrogen ions for one calcium ion. Plant roots are capable of absorbing any nutrient, independently from any other, in greater concentration than that in the soil solution.

The total amount of exchangeable cations that can be held by a given mass of soil is known as **cation exchange capacity.** It is expressed in milliequivalents (meq) per 100 grams of soil at neutrality (pH 7) or at some other stated pH value. The exchange capacity of mineral soils usually ranges between 2 and 50 meq per 100 grams of soil. One meq per 100 grams is equivalent to 400 pounds of calcium (Ca) per acre — the amount of calcium in a half-ton of pure limestone.

In a well-limed agricultural soil in a humid region, about 80 to 90 percent of the exchange capacity is ordinarily occupied with calcium plus magnesium; there are about 5 to 10 times as many calcium ions as magnesium ions. Potassium may be expected to occupy 2 to 5 percent of the capacity, and hydrogen (acid) the remainder. In arid regions, sodium may constitute a large percentage of the exchange complex.

The average cation exchange capacity of sandy soils is about 10 meq per 100 grams of soil (dry weight basis). For loams it may be about 20, for clays around 50, and for Histosols (peats and mucks) about 200.

To a much lesser extent, plants are also capable of absorbing anions by a similar mechanism. These include nitrates (NO_3^{-}), chlorides (Cl^{-}), phosphates ($H_2PO_4^{-}$), and sulfates (SO_4^{2-}) (especially the latter two).

Plant roots also absorb from the soil solution and by contact exchange many cations and anions which they apparently do not need. Plant leaves and stems can also absorb many essential nutrients; in addition, they absorb toxic chemicals, such as herbicides, and then die.

2:7 □ PLANT GROWTH AND NUTRIENT UPTAKE

The crop plants vary considerably in the nutrients which they absorb from the soil (Table 2.2). Note that the relative amount of the nutrient content of most crops is in this order: N > K > P > Mg > S. When crops are harvested and removed from the soil where they were grown, the soil loses the nutrients and gradually becomes depleted. To maintain the productivity of the soil, the loss of nutrients by plant removal must be replaced either by the slow decomposition of soil minerals or by manure, sludge, compost, lime, and fertilizer.

Table 2.2 — Plant Nutrient Content of Selected Crops with Specified Yields per Acre Indicated[1]

lb. / A	Corn			Soybeans[2]			Wheat		
	120	160	200 bu.	40	60	80 bu.	40	80	100 bu.
N	160	213	266	224	315	416	75	166	188
P_2O_5	68	91	114	38	58	78	27	54	68
K_2O	160	213	266	144	205	250	81	184	203
Mg	39	52	65	15	24	32	12	24	30
S	20	26	33	14	20	26	10	20	25
	Cotton (Lint)			Grain Sorghum			Potatoes		
	750	1,125	1,500 lb.	6,000	8,000	10,000 lb.	300	600	900 cwt.
N	105	143	180	178	238	297	150	300	450
P_2O_5	45	54	63	63	84	105	48	96	144
K_2O	65	96	126	180	240	300	270	540	810
Mg	17	26	35	30	40	50	24	40	72
S	15	23	30	28	38	47	14	27	41
	Alfalfa[2]			Clovergrass			Hyb. Bermudagrass		
	4	6	8 tons	3	6	7 tons	6	8	10 tons
N	225	338	450	150	300	350	258	368	460
P_2O_5	60	90	120	45	90	105	60	96	120
K_2O	240	360	480	180	360	420	288	400	500
Mg	20	30	40	15	30	35	18	26	34
S	20	30	40	15	30	35	30	44	58

[1]Courtesy, Potash and Phosphate Institute, "Better Crops," Fall 1988, p. 38.
[2]Legumes get most of their nitrogen from the air.

Note: The figures given are the total amounts taken up by the crop in both the harvested and the aboveground unharvested portions. These numbers are estimates for indicated yield levels, taken from research studies, and should be used only as general guidelines.

Nutrient uptake by the plant through the season follows a pattern similar to dry matter accumulation (Figure 2.6). However, nutrient uptake is somewhat faster early in the season and precedes growth throughout the season. The amounts of nutrients taken up early in the growing season are relatively small, but it is important to have adequate amounts present during the early period of leaf growth so that the growth of leaves will not limit yields.

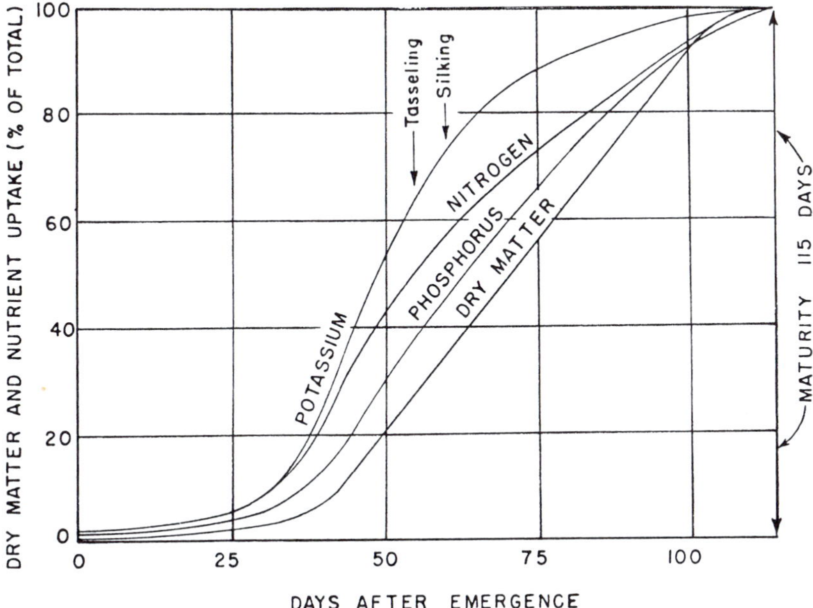

Fig. 2.6 — The uptake of nutrients by a corn plant and the increase in dry matter in relation to the number of days after emergence. (Courtesy, Iowa State University)

Larger amounts of nitrogen, phosphorus, and potassium are taken up by the plant during the early part of the growing season. The supply of these nutrients to the plants must be adequate during this time, or translocation of these elements from the leaves to other parts of the plant will result in premature death of some of the leaves. This, of course, reduces leaf area for **photosynthesis** (food manufacture) and reduces the growth and yield.

Through the process of photosynthesis, plants use energy from sunlight to manufacture sugar from carbon dioxide in the air and water and nutrients in the soil. Some of this sugar is then converted to starches, proteins, and other compounds that make up the dry weight of the plant.

Essentially, all photosynthesis occurs in the leaves. Generally, the more adequate the essential elements, the more leaf area a plant has and the more dry weight it will produce. Corn leaves are produced early in the growing season — largely before mid-July — then the stalk and tassel form, then the husks, cob, and finally the grain. Assuming no other limiting factors, the potential amount of grain that can be produced at the end of the season depends upon the leaf area produced early in the season. This has been determined by exhaustive research.

2:8 ▫ LAW OF LIMITING FACTORS

One of the most important principles of plant nutrition is the **law of limiting**

factors. This law states that yield or plant growth is limited by the essential factor which is in the shortest relative supply. Figure 2.7 illustrates this concept. The water level in the barrel is limited by the lowest stave. Yield is also limited by the nutrient or other growth factor, such as water, which may be the most limiting.

Fig. 2.7 — An illustration of the law of limiting factors. The level of production can be no higher than the lowest stave. The lowest stave shown here is nitrogen. After adequate nitrogen is supplied by fertilizer, phosphorus would limit the yield of crop; then potassium would limit crop yield after phosphorus was supplied. What are the limiting factors in a particular soil with which you are familiar? Only a soil test can answer this question satisfactorily.

This law applies not only to nutrients and other growth factors but also to management variables. It is a waste of money to apply high rates of fertilizers to crops unless there is adequate water, the proper varieties are used, correct plant populations are obtained, and weeds, insects, and diseases are controlled.

State experiment station researchers and USDA scientists conduct field experiments in each state to study special problems caused by variations in plants, soils, and climate. This type of research is necessary to gain new information and knowledge about factors that may limit or may increase plant production. Researchers publish many technical reports in scientific and professional journals to keep other scientists informed of new findings. Besides these formal publications, the results of research are communicated to the public in several ways. Scientists continually supply new information to extension personnel, who spread it through popular bulletins and pamphlets, radio and TV broadcasts, news releases, meetings, and field days. The progressive farmer takes advantage of this type of information so that he or she can eliminate some of the factors that may be limiting the yields by having the soil tested.

Each of the 13 soil-related essential nutrients that *could* be a growth- and yield-limiting factor will be discussed.

2:9 □ NITROGEN (N) AND PLANT GROWTH

The atmosphere contains about 78 percent nitrogen gas (N_2), yet nitrogen is one of our most critical elements for plant growth. The reason is that most plants cannot utilize nitrogen as a gas; it must first be combined into some stable form such as urea or ammonium nitrate fertilizer. Legumes are the principal exception (Note 2.2).

Note 2.2 — Legume Inoculant Groups

Legumes inoculated with the proper group of *Rhizobia* bacteria are capable of fixing enough atmospheric nitrogen (N_2) for themselves and a following crop. However, legumes need a higher level of soil calcium, magnesium, phosphorus, and potassium than nonlegumes.

Rhizobia bacteria inoculant groups for the most common legumes are as follows (all legumes in each group can be inoculated with the same strain of bacteria):

- □ ***Alfalfa group*** — Black medic clover, bur clover, sour clover, and button clover.
- □ ***Clover group*** — Alsike, crimson, red, white, ladino, and subterranean clovers.
- □ ***Pea and vetch group*** — Garden, sweet, and Austrian peas and purple vetch.
- □ ***Bean group*** — Garden, navy, and kidney beans.
- □ ***Soybean group*** — All soybean varieties.
- □ ***Cowpea group*** — All cowpeas, all peanuts, lima beans, all lespedezas, and kudzu vine.
- □ ***Lupine group*** — All lupines.
- □ ***Birdsfoot trefoil group*** — All birdsfoot trefoil varieties.
- □ ***Crownvetch group*** — All crownvetches.
- □ ***Black locust group*** — All black locusts, including rose acacia (bristly locust).

Plants absorb nitrogen either as the ammonium (NH_4^+) or the nitrate ion (NO_3^-). The ammonium ions can be held in an exchangeable and available form on the surfaces of clay and humus, but bacteria soon transform the ammonium to nitrates, which are readily leachable. There is no good storehouse for available forms of nitrogen. The only storehouse of any kind for nitrogen is soil organic matter, which must decompose before the nitrogen can be absorbed by plants. Furthermore, nearly 4 percent of the available soil nitrogen is lost each year by wind and water erosion and by leaching.

Nitrogen is a constituent of all living plant cells, and each molecule of **chlorophyll** contains four atoms of nitrogen. Nitrogen makes plants darker green and more succulent; it also makes larger cells with thinner cell walls. In addition, nitrogen increases the proportion of water and decreases the percentage of calcium in plant tissues.

Nitrogen influences plant growth in these ways, summarized as follows:

1. It gives a dark green color to all plants.
2. It increases leaf and stem growth.
3. It aids in seed production of many grasses, especially grasses under semiarid range conditions.
4. It increases the protein content of some food and feed crops.
5. It improves the quality of bread made from wheat.
6. It encourages the growth of grasses in a legume-and-grass mixture.
7. On some plants in certain soils, *too much* nitrogen decreases winter-hardiness, delays maturity, decreases resistance to some insects and diseases, weakens straw, increases water content, and lowers the quality of fruits and vegetables.
8. In deep sandy soils, nitrogen in excess of plant uptake is readily leached in nitrate form to contaminate the groundwater. (In 1987, this was confirmed in northeastern Iowa.)

2:10 ▫ PHOSPHORUS (P) AND PLANT GROWTH

Phosphorus nutrition is doubly critical; the total supply of phosphorus in most soils is usually low, and its relative availability is also low. Nearly 10 percent of the available soil phosphorus is lost each year by wind and water erosion.

The total phosphorus in an average arable soil is approximately 0.1 percent, only a small part of which is available to the plant at any one time. Under ideal conditions, as plants take in phosphorus from the soil solution, other phosphorus ions replace it from slowly soluble compounds in the soil. There is no efficient mechanism on clay or humus for holding exchangeable and available phosphorus, although there is some exchangeable phosphorus held on the surface of humus particles. Upon decomposition, humus releases available phosphorus.

Phosphorus availability is low in strongly acid soils because of the formation of iron and aluminum phosphates, from which phosphorus is very slowly available. In alkaline soils, tricalcium phosphate forms readily to reduce the availability of soil phosphorus. (See Chapter 3.)

The nucleus of each plant cell contains phosphorus; for that reason, cell division and growth are not possible without adequate phosphorus. Phosphorus is concentrated in cells near the most actively growing part of both roots and shoots, where cells are dividing rapidly.

Phosphorus reacts on plant growth in these ways:

1. Stimulates root formation and growth.
2. Hastens maturity.
3. Aids in cell division and reproduction.
4. Encourages flower development, pollination, and seed formation.
5. Increases legume growth.
6. Increases the protein and mineral content of grasses and legumes.
7. Makes plants more winter-hardy.
8. Enables legumes to compete more favorably with grasses (Figure 2.8).
9. Increases the percentage of phosphorus and calcium in food and feed crops.
10. Aids in legume nodule formation (Figure 2.9).

On some soils too much may cause zinc deficiency, especially in beans. On soils low in phosphorus, most trees and shrubs and some field crops may form a mutual association with a fungus such as genus *Glomus*. This fungus assists plant roots to absorb more nutrients and to grow larger.

Fig. 2.8 — Phosphate enables legumes to compete more favorably with grasses. The same mixture of sweetclover and oats was planted, but the phosphate on the left stimulated the sweetclover more than it did the oats. (Texas Blacklands) (Courtesy, Texas A&M University)

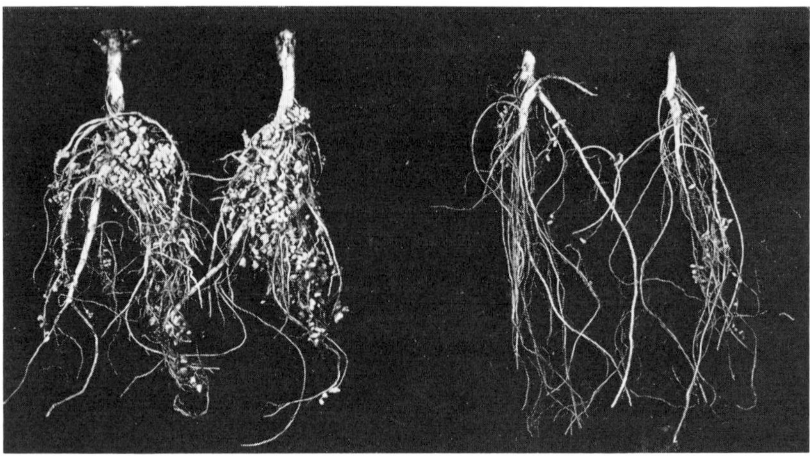

Fig. 2.9 — Phosphate aided in the formation of legume nodules on these sweetclover roots. *Left:* With phosphate. *Right:* Without phosphate (Courtesy, Texas A&M University)

2:11 □ POTASSIUM (K) AND PLANT GROWTH

The amount of total potassium in some soils is sufficient to last forever, yet the money spent for potassium fertilizers continues to increase. An explanation of this apparent contradiction lies in the fact that much of the *total* potassium is in soil minerals that are only very slowly available to plants. Soils may contain 2 percent total potassium, only $\frac{1}{50}$ of which is in a readily available form at any one time. However, during the growing season, approximately half of the potassium absorbed by the plant may come from the exchangeable form and the other half from relatively insoluble minerals which decompose and thereby release their potassium.

Until a few years ago it was thought that all potassium in the plant stayed in soluble form. Radioactive potassium techniques have demonstrated that as much as one-third of plant potassium is fixed as a part of plant proteins. Potassium also helps to maintain cell permeability, aids in the movement of carbohydrates, keeps iron more mobile in the plant, and increases the resistance of plants to certain diseases.

Potash affects legumes, grasses, and most other plants in these ways:

1. Imparts plant vigor and resistance to certain diseases.
2. Aids in **translocation** (moving foods) from the leaves to the roots.
3. Favors the growth of legumes in competition with other plants.
4. Produces stiff stems and thus reduces **lodging** (plant stems breaking).
5. Increases grain plumpness.
6. Imparts winter-hardiness.
7. Aids in transformation and translocation of sugars, starches, celluloses, and other carbohydrates.

In some soils too much potassium or ammonium in forage can cause low magnesium in livestock blood, a disease known as **grass tetany (hypomagnesemia).**

With intensive corn production practices on some Ultisols, high potassium fertilization may also require additional boron fertilizer.

2:12 □ CALCIUM (Ca) AND PLANT GROWTH

Calcium in the soil in the United States may average 1 percent, but variations are great. Calcium minerals are fairly soluble and leach out, causing low-calcium soils in humid regions. Almost one-fourth of soil calcium loss each year is due to wind and water erosion. Most of the reserve calcium in the soil is in the form of calcium carbonate (limestone) or, in the arid West, a mixture of calcium carbonate and calcium sulfate (gypsum). Plants obtain calcium from the soil solution and from exchangeable calcium ions (Ca^{2+}) on the surface of clay and humus particles.

Calcium tends to make cells more selective in their absorption of nutrients, since it is a constituent of the middle layer of each cell wall. Rapidly growing root tips are especially high in calcium, indicating that calcium is needed in large quantities for cell division and growth.

Calcic limestone ($CaCO_3$) added to the soil supplies calcium which serves these functions in the plant:

1. Promotes early root growth.
2. Improves general plant vigor and growth.
3. Encourages grain and seed production.
4. Increases stiffness of the straw in grains.
5. Maintains strength and selective permeability of cell walls.
6. Neutralizes acids produced in plants.
7. Increases the calcium content of food and feed crops.
8. Encourages the nodule formation of legumes.
9. Reduces plant uptake of radioactive elements harmful to humans.
10. Regulates intake of other elements.
11. Reduces toxic aluminum in acid soils. (See Chapter 3.)

2:13 □ MAGNESIUM (Mg) AND PLANT GROWTH

Chlorophyll a, b, c, and d each contain one atom of magnesium in each molecule; therefore, there could be no green plants without magnesium. There is almost as much magnesium as calcium in average soils, yet there are many soils along the Atlantic coast that are extremely deficient in magnesium.

Reserve magnesium occurs mostly in dolomitic limestone, a rock which consists of a mixture of calcium and magnesium carbonate. Dolomitic limestone is not so readily decomposed as is calcic limestone; for that reason, the amount of exchangeable magnesium in soils is usually less than that of exchangeable calcium.

It has been demonstrated that magnesium aids in the uptake of phosphorus. This fact is of particular importance to the livestock industry, because the phosphorus content most desirable for the growth of forage is inadequate for satisfactory animal nutrition. Low magnesium in forage can cause grass tetany in livestock.

Magnesium serves these purposes in plant nutrition:

1. Aids in maintaining a dark green leaf color (Figure 2.10).

Fig. 2.10 — Cucumbers and snap beans need extra magnesium when growing in low-magnesium soils. *Top:* Cucumbers — (left) no magnesium; (right) 160 pounds per acre of magnesium (MgO). *Bottom:* Snap beans — (left) no magnesium; (right) 160 pounds per acre (179 kg / ha) of magnesium (MgO). (Courtesy, New Jersey Agr. Exp. Sta., Rutgers University)

2. Regulates the uptake of other plant nutrients.
3. Acts as a carrier of phosphorus in the plant.
4. Promotes the formation of oils and fats.
5. Plays a part in the movement of starch in the plant.

2:14 ▫ SULFUR (S) AND PLANT GROWTH

Sulfur occurs in small amounts in the soil, averaging perhaps 0.15 percent in a typical soil. A large part of the sulfur which plants use comes either from sulfates, which are a by-product of ordinary superphosphate fertilizer (14 percent S), or from gypsum; or it is released by the biological decomposition of organic matter. The burning of fossil fuels also releases sulfur into the atmosphere.

Many plant proteins contain sulfur, as does an oil produced by members of the cabbage family.

Sulfur functions in plant growth in these ways:

1. Gives increased root growth.
2. Helps maintain dark green color.
3. Promotes nodule formation on legumes.
4. Stimulates seed production.
5. Encourages more vigorous plant growth.

2:15 ▫ MICRONUTRIENTS AND PLANT GROWTH

Iron, manganese, zinc, copper, boron, molybdenum, and chlorine are listed as micronutrients because they are used by plants in such small amounts. These elements may limit plant growth either because there may not be a sufficient amount of them in the soil, or, as is more often the case, because some condition in the soil reduces their availability. Micronutrients are abundant in all soil organic matter.

All of the micronutrients except molybdenum are more soluble and available in an acid soil; molybdenum solubility increases with liming. (See Chapter 3.)

The micronutrients perform many important functions in plants. Let's look at them one by one.

Iron (Fe)

1. Is most important in chlorophyll formation.
2. Is active in oxidation-reduction reactions.
3. Plays a vital role in the activity of enzyme systems.

Manganese (Mn)

1. Together with iron, assists in the formation of chlorophyll.

2. Is active in carbohydrate formation.
3. Accelerates the germination of seeds and maturity of plants.
4. Is active in oxidation-reduction reactions.
5. Affects vitamin content of plants.

Zinc (Zn)

1. Is essential to the formation of chlorophyll.
2. Influences seed production and grain yield.
3. Is essential in the formation of growth hormones.

Copper (Cu)

1. Acts as a regulator of several biochemical processes that occur in plants.
2. Acts as a catalyst to help route the various nutrient ions into their proper growth functions.
3. Is active in oxidation-reduction reactions.
4. In small excess, is toxic to plants.

Boron (B)

1. Is essential for pollination and reproduction.
2. Influences flower and seed formation.
3. Influences oxygen supply to plant tissues and roots.
4. Is closely related to calcium performance in plants.
5. In small excess, is toxic to plants.

Molybdenum (Mo)

1. Is essential for the *Rhizobia* bacteria that live in nodules of legume roots.
2. Influences the reduction of nitrates in protein synthesis.
3. Is instrumental in starch formation, amino acid formation, and vitamin formation.
4. In small excess, is toxic to plants and animals.

Chlorine (Cl)

The specific function of chlorine in plants is not known.

2:16 □ PLANT NUTRIENT DEFICIENCY SYMPTOMS

Fields of healthy, vigorous plants indicate a fertile soil. However, when the supply of any *one* of the essential elements is low, growth is retarded. (See Section 2:8.) Plant nutrient deficiency symptoms may be regarded as the language plants use to indicate the nature of their distress. Certain types of deficiency symptoms at times can provide a better indication of the nutrient relationship between the plant and soil than can be obtained from detailed chemical analysis of the plant or soil.

Deficiency symptoms of plants are sometimes confused or masked by insect damage, disease, pesticide injury, or air pollution. Many visual symptoms of nutrient deficiencies are difficult to interpret under field conditions, because factors other than plant nutrient supply constantly affect the plant. For example, prolonged periods of cold and wet weather may cause typical phosphorus or zinc deficiency on corn, but as the weather conditions improve, the deficiency symptom may disappear. Furthermore, plants already weakened by nutrient deficiency are often more severely affected by diseases and insects or damaged by unfavorable weather, making positive identification of symptoms more difficult.

Some plants are better indicators of nutrient deficiencies than others. Plants with a wide expanse of broad leaves, such as corn and sorghum, are better indicators of the supply of available nutrients than are small grain plants with their narrow leaves.

The best background training for the diagnosis of deficiency symptoms is practical experience in observing plants grown under known field conditions. Such experience, along with a good knowledge of the growth characteristic of the crop, usually results in an accurate diagnosis of the nutrient deficiency. Also, the use of a simple semiquantitative chemical test for the detection of nitrogen, phosphorus, or potassium compounds in plant tissue can be a valuable aid in diagnosing nutrient deficiencies. (See Chapter 14.)

Not all plant food element shortages display themselves by visual symptoms. Greater losses are sustained by farmers from **hidden hunger** than from acute shortages because it goes unnoticed and uncorrected. Plant tissue tests or plant analysis can help determine if hidden hunger exists. With acute deficiencies, by the time the shortages are discovered, it is often too late to do anything for that crop. This information will help to plan for the prevention of deficiencies on the next crop.

Demonstration plots and check strips on suspected areas of nutrient deficiencies are helpful for discovering hidden hunger. It is a good practice for farmers to use check strips every year to warn of hidden hunger and to help determine the most profitable levels of fertilization.

The various plant nutrient deficiencies can be described by certain characteristic symptoms as follows:

Nitrogen (N) Deficiency Symptoms

1. Sickly, yellowish-green foliage.
2. Slow, dwarfed growth.
3. Drying up (**firing**) of the lower or bottommost leaves of the plant, proceeding upward as the season advances. In plants such as corn, small grains, beets, beans, and grasses, the firing starts at the tips of the bottom

leaves and proceeds along the **midribs** (centers) of the leaves toward the stems.

Phosphorus (P) Deficiency Symptoms

1. Slow-growing, dwarfed plant.
2. Lower leaves are purple along the margins, beginning at the tips. Margins and tips of leaves may eventually die, leaving signs of "firing."
3. In the case of corn, sorghums, and small grains, the remainder of the plant has a deep bluish-green or purple cast.
4. Small, slender stalks in the case of corn. Small grains fail to **stool** (new plants arising from the base of a single plant).
5. Low grain yields, slow to mature.

Potassium (K) Deficiency Symptoms

1. Lower leaves "scorched" or "burned" on margins and tips. These dead areas may fall out, leaving ragged edges. In corn, small grains, and grasses, "firing" starts at the tips of the leaves and proceeds down from the edges or margins, usually leaving the midribs green.
2. Weak stalks, poor root development. Causes plants such as corn to **lodge** (break over) (Figure 2.11).

Fig. 2.11 — Strong stalks and root systems are a major factor responsible for today's high corn yields. Research has shown low potassium as one of the main causes of poor root development and stalk lodging. (Courtesy, Potash and Phosphate Institute)

Calcium (Ca) Deficiency Symptoms

1. Young leaves at the growing points (tops of the plants) become "hooked" in appearance and die back at the tips and along the margins.
2. Leaves have wrinkled appearance.
3. Young leaves remain folded in some cases.
4. Short and many-branched roots.

Magnesium (Mg) Deficiency Symptoms

1. A general loss of green color which starts in the bottom leaves and later moves up the stalks. The veins of the leaves remain green, thus causing a striped appearance.
2. Weak stalks with long, branched roots.
3. In corn, definite and sharply defined series of yellowish-green, light yellow, or even white streaks on all leaves throughout entire plant (Figure 2.12).
4. Leaves curve upward along the margins.

Sulfur (S) Deficiency Symptoms

1. Young leaves (uppermost leaves) light green, having even lighter veins. Entire plant shows a general pale yellow color (Figure 2.13).

Fig. 2.12 — Magnesium deficiency in corn appears as yellow and green stripes parallel to the veins of the leaves. (Photo by Roy Hunter Follett)

Fig. 2.13 — Sulfur deficiency on peach leaves. *Left:* Normal dark green leaf. *Center:* Leaf showing moderate sulfur deficiency. *Right:* Leaf showing severe sulfur deficiency. (Courtesy, Washington State University)

2. Short, slender plants.
3. Slow, stunted growth.

Boron (B) Deficiency Symptoms

1. Younger or bud leaves affected.
2. Young leaves light green.
3. Terminal bud or growing point usually dies after distortion at tips and bases of young leaves.
4. Fails to set seed. High degree of barren stalks in corn.
5. In alfalfa is known as **white top** (Figure 2.14).
6. On pears results in small, corky fruit that contains cracks (Figure 2.15).
7. In cauliflower causes **hollow heart.** The hollow stem rots, and the entire plant may rot if severely deficient.

Zinc (Zn) Deficiency Symptoms

1. Older or lower leaves mostly affected.

Fig. 2.14 — The alfalfa plant at the extreme left is normal; the four to the right of it show successive stages of boron deficiency. The plant at the extreme right is dwarfed and its top has turned white; hence, the name "white top" for this deficiency of alfalfa. (Courtesy, Pacific Coast Borax Company)

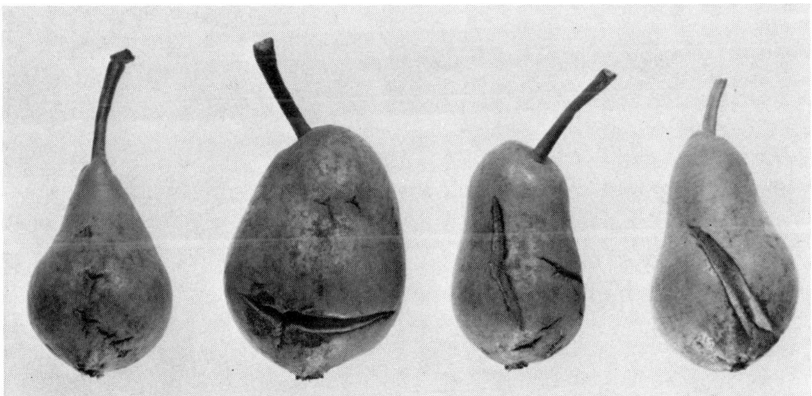

Fig. 2.15 — Boron deficiency of pears results in small corky fruit that contains cracks. (Courtesy, Washington State University)

2. In corn, **chlorotic** (yellow to white) stripes on each side of the midribs, especially on the lower and middle leaves of the plant (Figure 2.16). Some leaves show distinct light and dark green stripes.
3. Slow, stunted growth.
4. In beans, slow, stunted growth. General yellowing of the upper foliage with a browning or bronzing of the older or lower leaves.

Fig. 2.16 — Zinc deficiency on corn leaves with chlorotic (yellow to white) stripes on each side of midribs. (Photo by Roy Hunter Follett)

Iron (Fe) Deficiency Symptoms

1. Younger or bud leaves affected.
2. Pale to bright yellow color of entire plant, particularly noticeable on new leaves. Leaf margins may **fire** (turn brown).
3. Lower leaves of corn may be striped if deficiency is mild.
4. Leaf veins remain green (Figure 2.17).
5. Stalks are short and slender.

Copper (Cu) Deficiency Symptoms

1. In corn, the entire plant pales and the youngest leaves turn yellow.
2. Younger parts of plants are affected first and dieback is typical.
3. Alfalfa plants fade and take on a grayish cast. There is sharply reduced growth; plants take on bushy appearance.

Fig. 2.17 — Iron-deficient corn showing chlorosis (yellowing) between veins. (Photo by Roy Hunter Follett)

Manganese (Mn) Deficiency Symptoms

1. Younger or bud leaves affected.
2. Spots of dead tissue on leaves.
3. Veins remain green; tissue between veins chlorotic, similar to iron deficiency, but the green veins are wider.
4. Deficiency symptoms in small grain appear as grayish areas near the base of small plants.

Molybdenum (Mo) Deficiency Symptoms

1. In oats, bluish coloration of outer seed covering (**glumes**).
2. In legumes, older leaves pale greenish-yellow to yellow.
3. Plants dwarfed.

Chlorine (Cl) Deficiency Symptoms

Deficiency symptoms have not been recognized.

2:17 □ MINERAL DEFICIENCY AND TOXICITY FOR LIVESTOCK

Plants are capable of taking from the soil only that which is "available." It is often true that nutrients may be adequate for normal plant growth, but when the plants are eaten by livestock, the nutrients may not be in sufficient concentration for adequate livestock nutrition. Mineral supplements are then required. Furthermore, a plant can absorb certain elements which it may or may not need which are harmless to the plant but toxic to livestock.

Many animals that eat only forage that has been grown on soils deficient in one or more essential elements chew on wood, bones, mud, or manure. Such a depraved appetite is known as **pica.** When this occurs, the cure seems to be to lime and fertilize the pastures, feed a mineral supplement, or both.

Cobalt has not yet been added to the list of elements essential for plants, but animals require it in their nutrition. Cobalt deficiency has been reported in livestock in the following states because the soils and forage plants do not contain sufficient cobalt: Oregon, Michigan, Louisiana, Arkansas, Florida, South Carolina, North Carolina, Virginia, Delaware, New Jersey, Massachusetts, New York, and New Hampshire. A deficiency of cobalt in soils and forage may cause **bush sickness** in sheep and cattle.

Copper deficiency in livestock due to inadequate copper in the soil has been reported in Oregon, Florida, and Virginia. Too little copper in forage may result in **hypocupremia.**

In certain parts of California, Nevada, Utah, Colorado, Texas, and Florida, molybdenum may be accumulated by certain plants, and while it is nontoxic to the plants, it may be toxic to animals eating the plants. Excess molybdenum in animals is called **molybdenosis.**

Selenium toxicity **(selenosis)** in livestock has been known in several western states since 1857, but the cause was not understood until 1928. Although selenium is essential for animals, it is not necessary for plants. Soil containing as little as one part of selenium per million may produce vegetation that concentrates selenium, which is poisonous to livestock. In the following states many plants accumulate enough selenium to be toxic when eaten by livestock: Nevada, Arizona, Utah, Colorado, Kansas, Wyoming, South Dakota, Nebraska, Montana, and North Dakota. **Lathyrism** is a disease of domestic animals caused usually by eating roughpea (genus *Lathyrus)* high in selenium.

Although iodine has not been proved essential for plants, humans and animals require it. The "goiter belt" is an area where both animals and humans often develop a **goiter** (swollen neck) due to a deficiency of iodine. Goiter is common in Washington, Oregon, northern California, Nevada, Utah, Idaho, Montana, Colorado, Wyoming, North Dakota, South Dakota, Minnesota, Iowa, Wisconsin, northern Illinois, Michigan, Indiana, Ohio, northern Kentucky, West Virginia, and Pennsylvania. Too much iodine in animals may cause **iodism.**

Grass tetany is a magnesium deficiency occurring in cattle on lush, green pastures low in magnesium. It is characterized by **hypomagnesemia** (too little magnesium), extreme **hyperirritability** (nervousness), and convulsions. Plants may absorb too little magnesium when there is a deficiency of soil magnesium or an excess of available ammonium, potassium, or exchangeable aluminum, which decreases absorption of magnesium by plants.

The tetany that occurs in magnesium-deficient animals is very similar to the **hypocalcemia** (too little calcium) tetany in calcium-deficient animals. The blood plasma levels of these two elements are depressed markedly in animals eating forage deficient in magnesium or calcium, respectively.

Small amounts of fluorine are essential for animals and assist in reducing tooth decay and in strengthening bones in both animals and humans; a slight excess will cause a pitting of the teeth. Excess fluoride in animals, known as **fluorosis,** can cause an animal to lose its teeth and, in more extreme cases, cause bone damage so that the animal cannot bear the pain of standing. Excessive amounts of fluorine may accumulate in soils and plants near certain phosphate and aluminum factories and injure or kill all vegetation.

When nitrate nitrogen is at a high level in the soil and plant growth is retarded, the plants may accumulate sufficient nitrates to be toxic when fed to livestock. This is true whether the plant growth is retarded by drought, shading, low temperature, or an unbalanced nutrient status of the soil.

Boron and lithium toxicity to plants may be caused by an excess of boron or lithium in certain irrigation waters in arid regions.

Strontium 90 comes from radioactive fallout resulting from the explosion of certain kinds of atomic bombs. It is harmful to all people and animals that eat plants or animal products that have accumulated strontium 90. However, plants growing on soils high in lime accumulate less strontium 90. (See Chapter 3.)

2:18 □ REFERENCES

Animal Health: Livestock and Pets, The 1984 Yearbook of Agriculture. U.S. Department of Agriculture, Washington, D.C., 684 pp.

Brady, Nyle C. *The Nature and Properties of Soils,* 9th ed. New York: Macmillan Publishing Co., 1984, 750 pp.

Cook, Ray C., and Boyd G. Ellis. *Soil Management: A World View of Conservation and Production.* New York: John Wiley & Sons, 1987, 414 pp.

Crowder, Bradley M., Marc O. Ribando, and C. Edwin Young. "Agriculture and Water Quality." Agr. Inform. Bul. No. 548, U.S. Dept. of Agriculture–Economic Research Service, 1988, 6 pp.

Donahue, Roy L., Raymond W. Miller, and John C. Shickluna. *Soils: An Introduction to Soils and Plant Growth,* 5th ed. Englewood Cliffs, New Jersey: Prentice-Hall, Inc., 1983, 668 pp.

Fact Book of Agriculture, 1988. Misc. Pub. 1063, U.S. Department of Agriculture, Washington, D.C., 168 pp.

Follett, Roy H., Larry S. Murphy, and Roy L. Donahue. *Fertilizers and Soil Amendments.* Englewood Cliffs, New Jersey: Prentice-Hall, Inc., 1981, 558 pp.

Hausenbuiller, R. L. *Soil Science: Principles and Practices,* 3rd ed. Dubuque, Iowa: Wm. C. Brown Group, 1985, 610 pp.

Official Publication No. 40. West Lafayette, Indiana: Association of American Plant Food Control Officials, 1987, 56 pp.

Schneider, R. W., ed. *Suppressive Soils and Plant Disease.* St. Paul: The American Phytopathological Society, 1982, 88 pp.

Soils Improvement Committee, California Fertilizer Association. *Western Fertilizer Handbook,* 7th ed. Danville, Illinois: The Interstate Printers & Publishers, Inc., 1985, 288 pp.

Tisdale, Samuel L., Werner L. Nelson, and James D. Beaton. *Soil Fertility and Fertilizers,* 4th ed. New York: Macmillan Publishing Co., 1985, 754 pp.

Woodruff, J. R., F. W. Moore, and H. L. Musen. "Potassium, Boron, Nitrogen, and Lime Effects on Corn Yield and Earleaf Nutrient Concentrations." *Agronomy Journal,* Vol. 79, 1987, pp. 520 – 524.

CHAPTER 3

Soil Acidity and Liming

"Soils are becoming more acid in the U.S. because farmers are applying only 17% as much lime as needed." — Based on 1988 estimates by the authors

OUTLINE

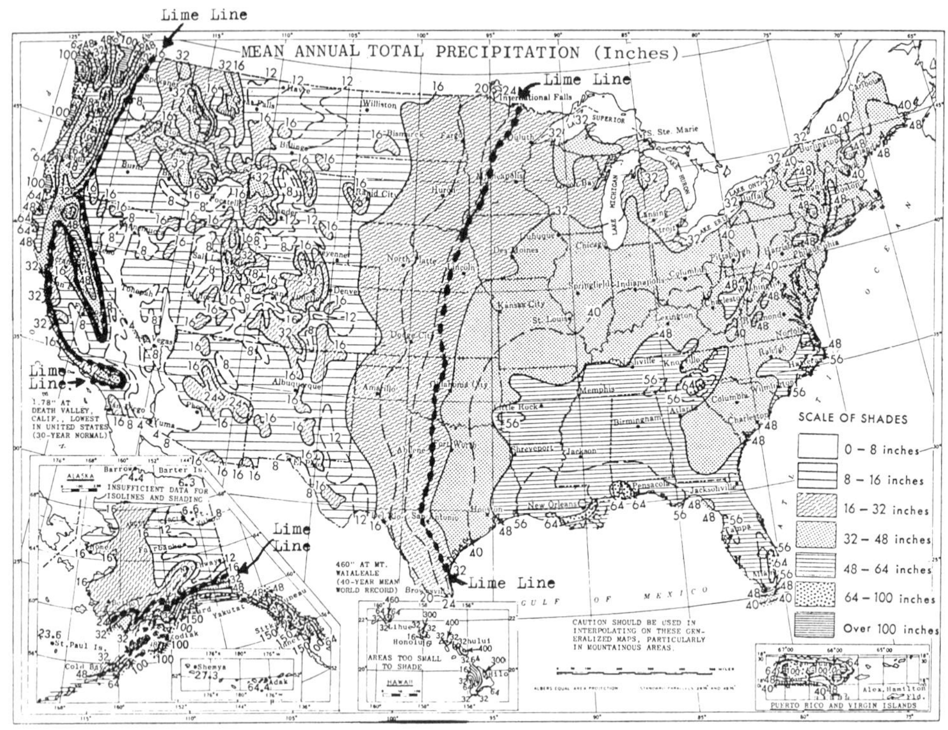

The three lime lines indicate that lime may be needed on high – lime-demanding crops such as alfalfa in areas receiving more precipitation than that along the lime lines. (Courtesy, National Oceanic and Atmospheric Administration)

□ □ □

3:1 □ OVERVIEW

To understand why soils become acid, observe the frontispiece to this chapter. There are three **lime lines**. Along these lime lines, precipitation is approximately equal to evaporation of water from an open water surface. With less rainfall, lime tends to accumulate in the surface horizons of the soil. Conversely, with more rainfall, lime tends to be leached from the root zone of the soil, thus making the soil continuously more acid. Therefore, lime may be needed on the well-drained, readily leached loams and sandy loams in the midwestern, eastern, and southern states, as well as on soils in the high rainfall areas of Washington, Oregon, and California and on the coastal soils of southern Alaska. Plants on some humid area soils on old landscapes in Hawaii respond to lime, but the areas are too small to be shown on a map of this scale.

In the humid United States, acid soils are becoming more acid because farmers are applying only 17 percent as much lime as needed. Principal causes of soil acidification are acid precipitation, removal of calcium and magnesium in harvested crops, use of acid-forming fertilizers, and erosion / leaching.

The conventional way to express soil acidity is by pH. These numbers are negative logarithms of the hydrogen ion concentration of the soil solution.

Most crop plants on well-drained mineral soil in temperate regions grow best at a soil pH of 6.5 to 7.0, and on organic soils and well-drained soils in tropical and subtropical regions, they grow best at a pH of 5.0 to 5.5. Legumes in general respond better than grasses to lime applications on acid soils.

Limes applied to agricultural soils are mostly calcic ground limestone ($CaCO_3$) and dolomitic ground limestone [$(CaMgCO_3)_2$]. Lime requirements can be determined readily by a soil test.

3:2 □ WHY SOILS BECOME ACID

In humid regions, soils either are acid or are becoming acid when the content of calcium and magnesium is naturally low, is being removed, or is being neutralized. Humid region soils are becoming more acid for these reasons:

1. Acid parent materials are naturally low in calcium and magnesium.
2. Acid rainfall, moving through the soil, leaches calcium and magnesium out of the root zone. Furthermore, rainfall is becoming more acid because of nitrogen and sulfur oxides that are created from the burning of coal and from the exhausts of motorized vehicles, respectively.
3. Harvested crops deplete the soil of calcium and magnesium.

4. Most fertilizers are acid-forming, and the most popular ones are the most acid.
5. Erosion and leaching remove calcium and magnesium.

Acid parent materials are those containing mostly silicon dioxide (SiO_2) found predominantly in light-colored igneous rocks such as granite and sandstone and in sandy coastal areas. However, there is not enough information available to use in estimating how much acidity is contributed to soils by acid parent materials.

In the United States, the amount of soil acidity generated by acid precipitation, harvested crops, acid fertilizers, and erosion and leaching is estimated to be equivalent to 176 million tons (160 million mt) per year of agricultural limestone. This compares with the 30 million tons (27 million mt) of limestone used by farmers — only 17 percent of the amount needed just to *balance* the annual generation of soil acids.

For example, in Wisconsin the percentage of soils tested in 1968 – 1973 that needed lime was 28; in 1982 – 1985 lime was needed on 51 percent of the soils.

Agricultural limestone is assumed to be 85 percent pure calcium carbonate, which is approximately the purity required by most state laws. The four principal causes of acid soil formation will be discussed separately (Table 3.1).

Table 3.1—Origin of Acid Formation in Soils of the United States[1]

Origin of Soil Acid	Agricultural Limestone ($CaCO_3$) Equivalent Required to Neutralize Soil Acidity Generated (million tons / year)
Acid precipitation (rain)	5
Harvested crops	17
Acid-forming fertilizers	37
Accelerated erosion and leaching	117
TOTAL	176

[1] Roy Hunter Follett, Larry S. Murphy, and Roy L. Donahue. *Fertilizers and Soil Amendments.* Englewood Cliffs, New Jersey: Prentice-Hall, Inc., 1981, p. 404.

Note: Tons × 0.907 = metric tons.

3:2.1 □ Acid Rain

Acid rain in the United States is estimated to neutralize acidity generated by 5 million tons (4.5 million mt) of 85 percent pure calcium carbonate. Stated in another way, if all this acid rain were to fall on crop and pastureland, it would require the application of 5 million tons (4.5 million mt) just to *maintain* existing soil acidity levels.

Rain is made more acid by the oxides of sulfur and nitrogen fed into the atmosphere from internal combustion engines and from the burning of coal. These oxides combine with rain to make sulfuric acid and nitric acid, respectively (Figure 3.1).

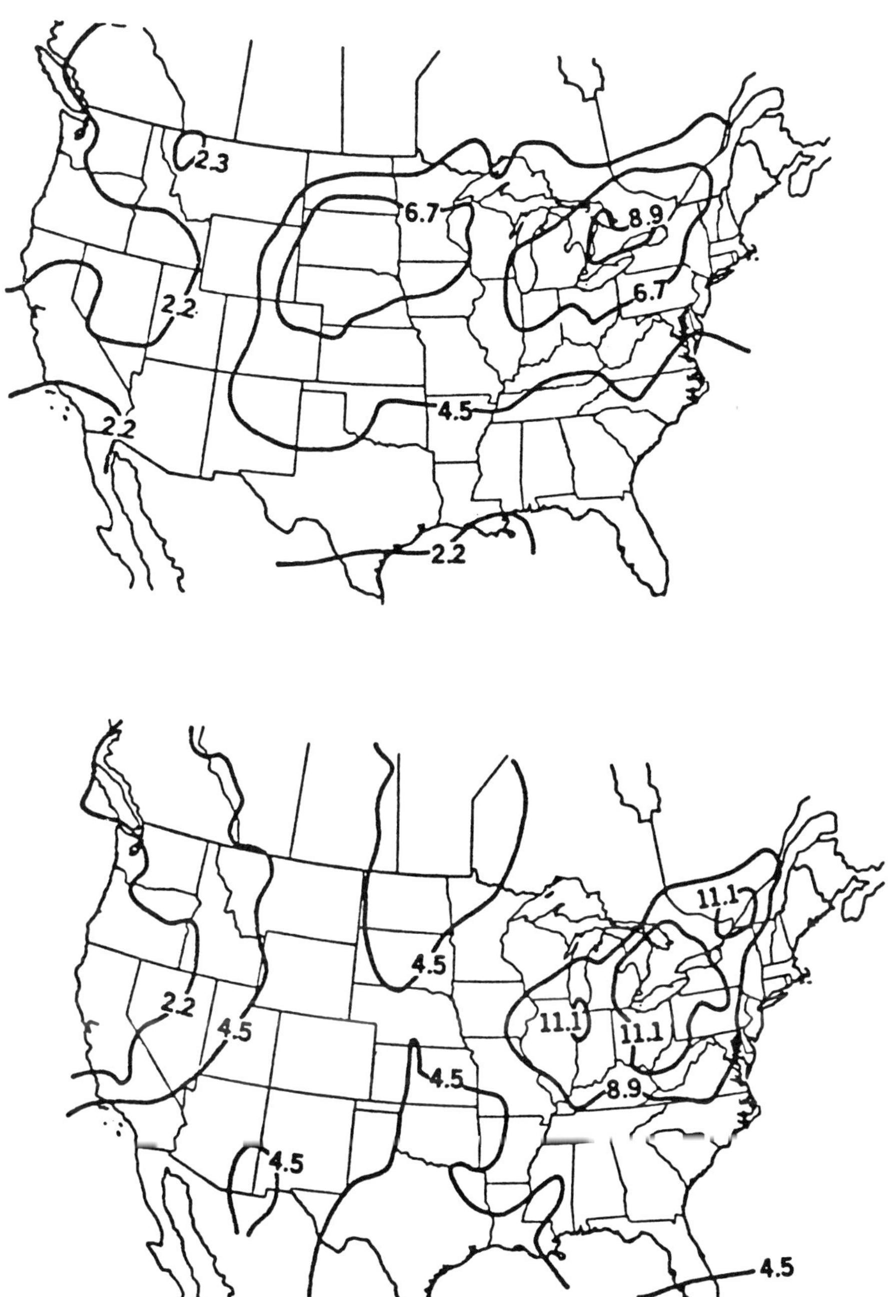

Fig. 3.1 — Acid precipitation (acid rain) is caused by oxides of nitrogen and sulfur from fossil fuels. *Top:* Ammonium nitrogen (NH_4^+-N) plus nitrate nitrogen (NO_3^--N) in acid rain (pounds of N/acre in 1981). *Bottom*: Oxides of sulfur in acid rain (pounds of S / acre in 1981). (Courtesy, USDA – Soil Conservation Service, 1987)

The amount of sulfur dioxide (a gas) spewed into the atmosphere each year is about 33 million tons (30 million mt). Also each year 22 million tons (20 million mt) of nitrogen oxides are released in the United States. The burning of coal in the United States increased by 25 percent from 1975 to 1980, but since then coal consumption has been decreasing because oil and gas have been cheaper. Nitrogen oxides originate from motor vehicle exhausts and from refineries. Catalytic converters on all new motor vehicles have reduced the *rate* of increase of nitrogen oxide emissions, but oxides from other sources continue to increase faster.

However, not all nitrogen (N) and sulfur (S) from acid rain can be considered as detrimental. Both N and S are essential plant nutrients, and on least sensitive clay soils high in calcium and magnesium, acid rain may increase plant growth. For example, in Wisconsin each acre each year receives from the atmosphere 12 pounds of N (13.4 kg / ha) and 15 pounds (16.8 kg / ha) of S, totaling about $6 per acre per year at fertilizer prices. A comparable annual value in the Tennessee Valley is $1.44 per acre ($1.61 / ha).

3:2.2 □ Harvested Crops

Calcium and magnesium in crops harvested and removed from the field represent a net loss of the principal acid-neutralizing elements. These losses are estimated to equal 17 million tons (15 million mt) of agricultural grade limestone each year. An estimate of this origin of soil acid production can be made in this way: There are about 400 million acres (162 million ha) of crops harvested in the United States. Assume the calcium and magnesium content of all harvested crops averaged 12 pounds per acre (13 kg / ha) of calcium (Ca) and 10 pounds per acre (11 kg / ha) of magnesium (Mg). The total amount of calcium removed in harvested crops each year would be 12 × 400 million, or 4.8 billion pounds, and the total amount of magnesium removed would be 10 × 400, or 4 billion pounds. To change calcium (Ca) to calcium carbonate equivalent, multiply the amount of pure Ca by 2.5 (see Table 3.7), that is, 4.8 billion pounds × 2.5 = 12 billion pounds. To convert pure calcium carbonate to the average purity of agricultural lime, divide by 85 percent as a decimal (0.85), that is, 12 billion ÷ 0.85 = 14 billion pounds. To change 4 billion pounds of magnesium (Mg) to calcium carbonate equivalent, multiply 4 × 4.12. The result is 16.5 billion pounds. Then, to convert to agricultural lime equivalent, divide 16.5 by 0.85. The result is 19.4 billion pounds.

In summary:

1. Agricultural lime equivalent from calcium in harvested crops = 14 billion pounds (6.4 billion kg).
2. Agricultural lime equivalent from magnesium in harvested crops = 19.4 billion pounds (8.8 billion kg).
3. 14 + 19.4 = 33.4 billion pounds. Converted to tons = 33.4 billion pounds

÷ 2,000 = 16.7 million tons of lime of agricultural quality. This is rounded to 17 million tons (15.4 million mt) (tons × 0.907 = metric tons; pounds / acre × 1.12 = kg / ha).

3:2.3 ▫ Acid-forming Fertilizers

The most popular nitrogen and phosphorus fertilizers are acid-forming. Only potassium fertilizers do not make soils more acid. The greatest tonnages of nitrogen fertilizers used in the United States are, in decreasing order: anhydrous ammonia, urea, nitrogen solutions, and ammonium nitrate.

Per pound of nitrogen (N), all of these nitrogen fertilizers will create the same soil acidity. For example, for each pound of nitrogen (N) used, 1.8 pounds of pure limestone must be applied to correct the soil acidity created. Diammonium phosphate is twice as acid, and monoammonium phosphate is almost three times as acid per pound of nitrogen (N) as the straight nitrogen fertilizers.

3:2.4 ▫ Accelerated Erosion and Leaching

The U.S. Department of Agriculture – Soil Conservation Service estimated from detailed surveys in the 50 states that erosion sediments in 1977 were 60 percent greater than the previous estimate made in the 1930's. The actual estimated amount of sediments eroded from all sources — cropland, pastureland, forestland, rangeland, construction sites, surface mining, and streambanks — was 6.4 billion tons per year (5.8 billion mt / yr), one-third of which was in the Corn Belt.

From cropland plus pastureland the erosion rate is 3.5 billion tons per year (3.2 billion mt / yr). The authors estimate that about 1.5 billion tons (1.4 billion mt) of this sediment originates on humid-region cropland where lime may be a limiting factor in crop production. From the soil science literature, the authors estimate that these 1.5 billion tons (1.4 billion mt) of erosion sediment contain about 1.5 percent calcium (Ca) and 0.7 percent magnesium (Mg). Using these estimates, the authors conclude that each year in the United States, about 22.5 million tons (20.4 million mt) of calcium (Ca) and 10.5 million tons (9.5 million mt) of magnesium (Mg) are eroded from acid soils on croplands and pasturelands. The technique for converting these values to agricultural limestone ($CaCO_3$) equivalent is as follows:

$$\text{Agricultural limestone equivalent of Ca} = \frac{22.5 \text{ million tons of Ca} \times 2.5}{0.85} = 66{,}176{,}000 \text{ (rounded to 66 million tons)}$$

$$\text{Agricultural limestone equivalent of Mg} = \frac{10.5 \text{ million tons of Mg} \times 4.12}{0.85} = 50{,}894{,}000 \text{ (rounded to 51 million tons)}$$

or a total of 117 million tons (106 million mt) of agricultural limestone ($CaCO_3$) equivalent eroded from acid soils used for cropland plus pastureland. The assumption is that farmers must apply 117 million tons (106 million mt) of lime each year just to replace that lost by erosion and leaching.

3:3 □ SOIL SENSITIVITY TO ACIDIFICATION

Although all humid region soils are *becoming* acid (lowering of pH because of acid rain, harvested crops, acid fertilizers, and accelerated erosion and leaching), soils vary in the *rate* of acid formation. Within the same environment, sensitive soils become acid faster because they are coarse-textured (sandy) and low in calcium and magnesium. Least sensitive to becoming acid (maintaining soil pH) are clay soils high in calcium and magnesium. (See Note 3.1.)

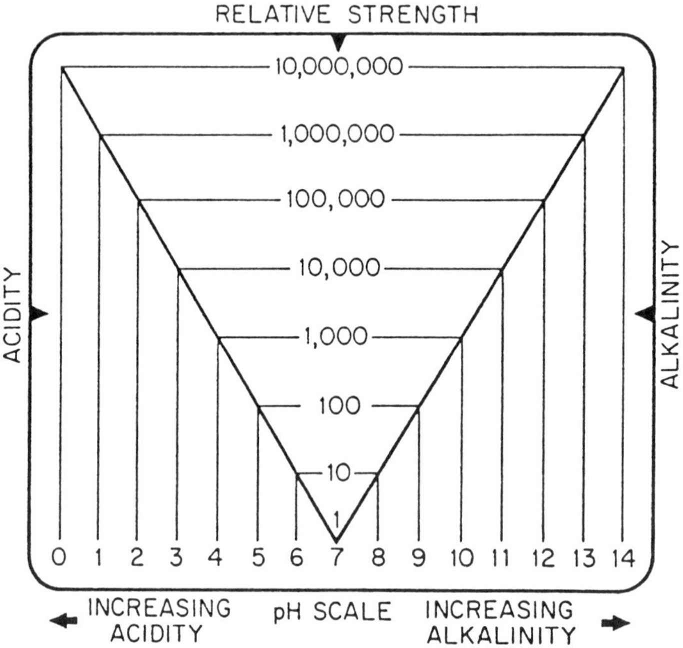

Fig. 3.2 — The pH scale depicted graphically. (Courtesy, Beckman Instrument Company)

3:4 □ SOIL ACIDITY AND SOIL PRODUCTIVITY

There is usually a fairly narrow range of soil pH in which plants grow best; there is a wider pH range which they tolerate. Note in Figure 3.3 that most plants are benefitted by lime at a pH range of 6.0 to 6.5.

At a soil pH below about 5.0, most crop plants will not produce at their maxiumum because of one or more of these reasons:

Note 3.1 — The Meaning of Soil pH

The term ***pH*** was proposed for use by Sören Peer Lauritz Sörensen (1868 – 1939), a Danish chemist. The letters are always written with a lower case "p" and a capital "H". pH was proposed for use as a convenient way to write very small amounts of hydrogen ion concentrations $[H^+]$. The "p" probably originated from the word "power," negative logarithmic power. The numbers used to express pH are negative logarithmic numbers (to the base 10). Therefore,

$$pH = \text{minus} \log_{10} [H^+]$$

$$= \log_{10} \frac{1}{[H^+]} \text{ in grams per liter of solution}$$

The **logarithm** of a number is the power to which the number 10 must be raised to give that number. For example, the logarithm of 10 is 1 (10^1); of 100 is 2 (10^2); of 0.1 is minus 1 (10^{-1}); of 0.01 is minus 2 (10^{-2}); and so on to the logarithm of 0.000 000 1 gram per liter of $[H^+]$ which is minus 7 (10^{-7}) = pH of 7 (neutral).

At a pH of 7, the concentration of hydrogen ions $[H^+]$ per liter of solution is 0.000 000 1 gram, written in the pH scale as 7. A pH of 6 has 0.000 001 gram of ionized hydrogen per liter. Note that 0.000 001 is 10 times greater than 0.000 000 1, meaning that a soil with a pH of 6 is 10 times more acid than a soil with a pH of 7. In like manner, each decreasing whole pH number is 10 times more acid than the preceding number. A pH of 5 is 10 times more acid than a pH of 6 and 100 times more acid than a pH of 7. All pH numbers above 7 indicate **alkalinity** and below 7, **acidity**. (See Figure 3.2.) The entire **pH scale** is from 0 to 14. The concentration of hydrogen ions for any fractional pH, such as 6.5, can be obtained by consulting a table of logarithms usually found in introductory mathematics, chemistry, or physics textbooks.

The pH scale of primary concern to soil science, soil management, plant nutrition, and crop production is shown in Figure 3.3. Note that the pH scale is displayed from 5.0 to 7.5, the pH range of most cropland soils.

Note: The pH of familiar substances may be of interest:

Lye — 13.0
Household ammonia — 12.0
Limewater — 11.0
Milk of magnesia — 10.0
Sodic soils — 9.0 - 11.0
Baking soda — 8.5
Ground limestone — 8.3
Soils saturated
with calcium — 8.3
Fresh eggs — 7.6 - 8.0

Blood — 7.5
Milk — 6.3 - 6.6
Black coffee — 5.0
Beer — 4.0 - 5.0
Tomatoes — 4.0
Most acid soil — 3.0
Carbonated drinks — 2.0 - 4.0
Vinegar — 2.4 - 3.4
Lemon juice — 2.0
Gastric fluid — 1.3

1. Toxicity of hydrogen, iron, manganese, and/or aluminum.
2. Deficiency of nitrogen, phosphorus, calcium, magnesium, sulfur, and/or molybdenum.
3. Slower decomposition of organic matter because of reduced microbial activity.
4. Lower cation exchange capacity.

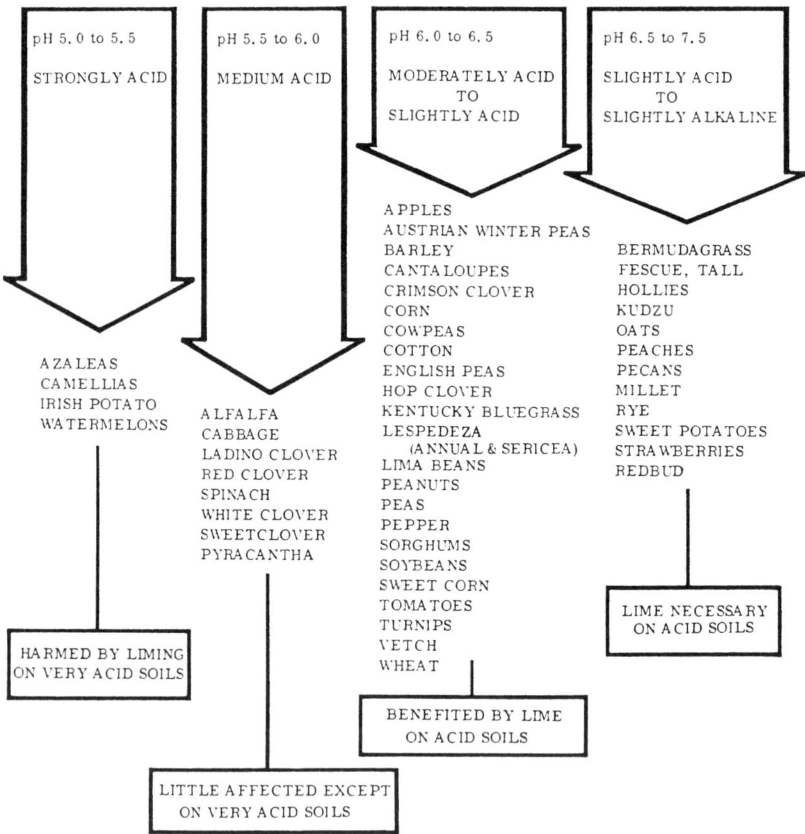

Fig. 3.3 — Soil pH ranges suitable for 48 common plants. (Courtesy, Instructional Materials Services, Texas A&M University)

Relative availability of 12 of the 13 essential elements plants absorb from the soil is displayed by world soil groups. Figure 3.4 shows the relative availability of soil nutrients in well-drained mineral soils of temperate regions, and Figure 3.5 portrays comparable data for all organic soils (peats and mucks) of the world. This figure also applies to well-drained soils on old land surfaces in tropical and subtropical regions.

Plants vary in pH tolerance among varieties but more so among species. For example, there is a variation in pH tolerance of varieties of barley, wheat, alfalfa,

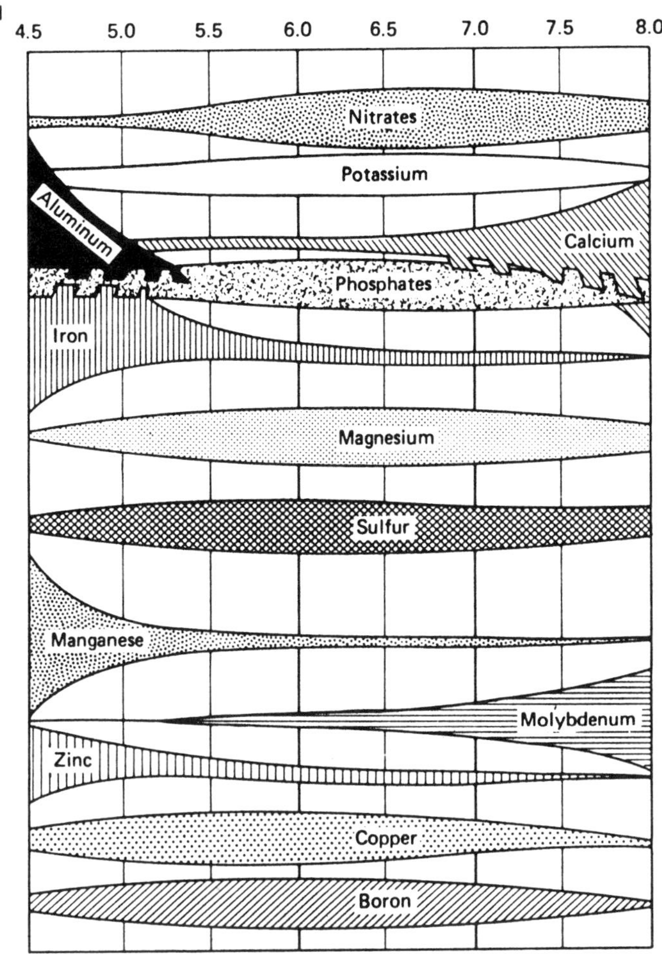

Fig 3.4 — The relative availability of 12 essential plant nutrients in well-drained *mineral* soils in temperate regions in relation to soil pH. A pH range between 6.0 and 7.0 (between heavy lines) is considered ideal for most plants. The thirteenth essential plant nutrient from the soil, chlorine, is not shown because its availability is not pH-dependent. Aluminum is not an essential nutrient for plants but is shown because it may be toxic below a soil pH of about 5.2. Above a pH of 6.8 calcium may tie up some available phosphorus. (See Figure 3.5.) (Courtesy, University of Kentucky)

soybeans, ryegrass, rice, Irish potatoes, tomatoes, peanuts, snapbeans, and sunflowers.

3:5 ▫ CROP RESPONSE TO LIME

Based on many field research and demonstration plots in the temperate region, estimates have been made of crop yield increases resulting from liming soils to specific pH ranges. These data are given in Table 3.2. Note in the table

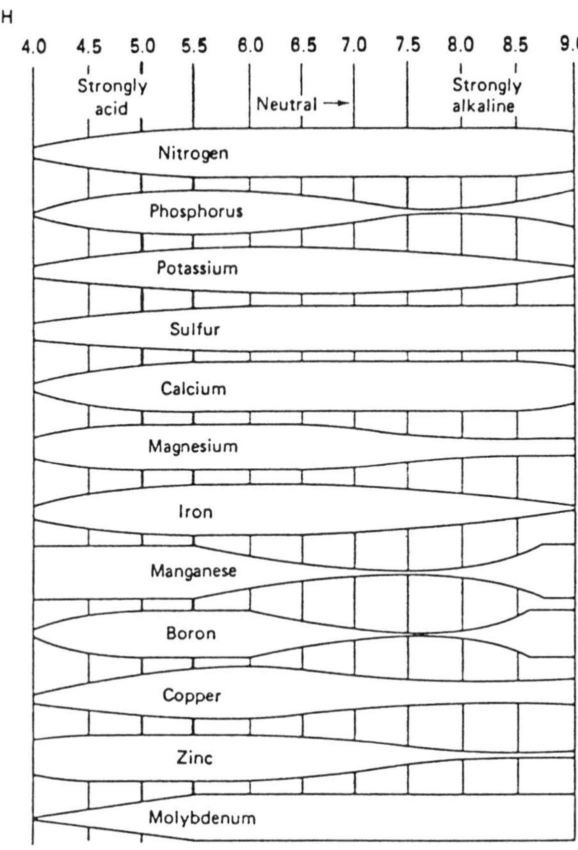

Fig. 3.5 — The relative availability of essential plant nutrients in *organic* soils and in well-drained mineral soils on old land surfaces in tropical and subtropical regions. A pH range of 5.0 to 5.5 is considered ideal for the optimum growth of most plants. Deduction: The more the organic matter in a *mineral* soil, the lower the "ideal" soil pH. (See Figure 3.4.) (Courtesy, Michigan State University)

Table 3.2—Estimated Yield Increases Resulting from Liming Soils to Specific pH Ranges[1]

	Percent Yield Increase Predicted from the Harvested Crop		
	Soil pH Increased by Lime Applications from		
Crop	5.0 to 5.5	5.5 to 6.0	6.0 to 6.5
Alfalfa	75	50	25
Lespedeza, annual	50	25	1
Barley	20	10	1
Cotton	20	10	1
Soybeans	20	10	1
Wheat	20	10	1
Corn	15	5	1
Grain sorghum	15	5	1
Tobacco	15	5	1

[1]Primary source: Tennessee Valley Authority.

that for alfalfa, 75 percent, 50 percent, and 25 percent yield increases, respectively, can be expected as the soil is limed from pH 5.0 to 5.5, 5.5 to 6.0, and 6.0 to 6.5, respectively. In Wisconsin the percentage of protein in alfalfa was increased from 12 to 16 percent when soil pH was increased from 5.0 to 7.0 by liming. From other data it has been confirmed that on some soils alfalfa responds economically to a soil that has been limed to a pH of 7.5. From these data in Table 3.2 the more responsive crops after alfalfa are annual lespedezas, barley, cotton, soybeans, and wheat. Least responsive at these pH's are corn, grain sorghum, and tobacco.

On Savanna fine sandy loam in Arkansas with a pH of 5.4, 2 tons of lime per acre (4½ mt / ha) eliminated blossom-end rot on "Traveler 76" variety of tomato.

Limestone application at the rate of 1 ton per acre (2 mt / ha) on a Norfolk fine sandy loam in North Carolina reduced toxic aluminum by 85 percent, increased calcium plus magnesium by 93 percent, and increased the yield of soybeans from 11 to 35 bushels per acre (740 – 2,354 kg / ha), a 218 percent increase. The pH was increased from 4.9 to 5.9, a tenfold (1,000 percent) decrease in the intensity of acidity.

Several lime experiments on Ultisols in Alabama resulted in increases in soybean yields because of a reduction in aluminum toxicity and more molybdenum availability. Alfalfa in a soil in Idaho with a pH of 5.4 was limed according to a soil test with 1.5 tons per acre (3.36 mt / ha) of agricultural limestone. Note the response of alfalfa in Figure 3.6.

Fig. 3.6 — The soil in the pot on the right has a pH of 5.4. The same soil in the pot on the left received 1.5 tons per acre (3.36 mt / ha) of agricultural limestone, according to a soil test, to pH 6.5. The crop is alfalfa. (Courtesy, Extension Service, University of Idaho)

In the Southeast, lime on soils with a pH of 4.7 to 5.5, applied according to a soil test, increased corn yields from 16 to 58 percent and alfalfa yields from 69 to 270 percent.

Cotton responded to the application of lime more than seven times greater in dollar value at a pH of 5.0 than at a pH of 5.9 (Figure 3.7). Lime applied according to a soil test in Maryland increased soybean yields 21 percent and alfalfa yields from 14 to 22 percent. Returns per dollar spent on lime were from $6 to $7 for soybeans and from $5 to $12 for alfalfa.

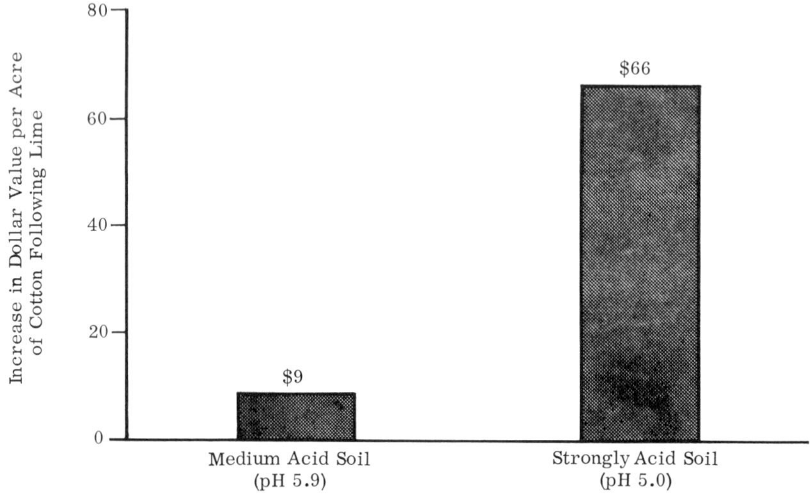

Fig. 3.7 — One ton (0.9 mt) of lime on a medium acid soil produced increased cotton yields valued at $9 per acre ($22 / ha). The same amount of lime on a strongly acid soil produced increased yields valued at $66 per acre ($163 / ha). (Courtesy, North Carolina State University)

Liming research from southwestern Canada, just north of western Montana, Idaho, and Washington, on mineral soils with a pH between 4.5 and 5.0, indicates that (Figure 3.8):

1. Legumes responded more to lime than grasses did; and of the legumes, alfalfa gave the greatest increase in yield.
2. Bromegrass and timothy increased in yield between 20 and 30 percent when limed; whereas, creeping red fescuegrass responded the least, about 10 percent.

Research in Wisconsin resulted in these conclusions:

1. Alfalfa yield nearly doubled when soil was limed from pH 5.0 to 6.0 and increased another 15 percent when limed to 6.6.
2. Corn yields increased 20 percent when soil was limed to pH 6.8 from 5.7.
3. Soybean yields increased 40 percent when soil pH was increased by lime from 5.2 to 6.0 but only 7.5 percent more from 6.0 to 6.6 (Figure 3.9).

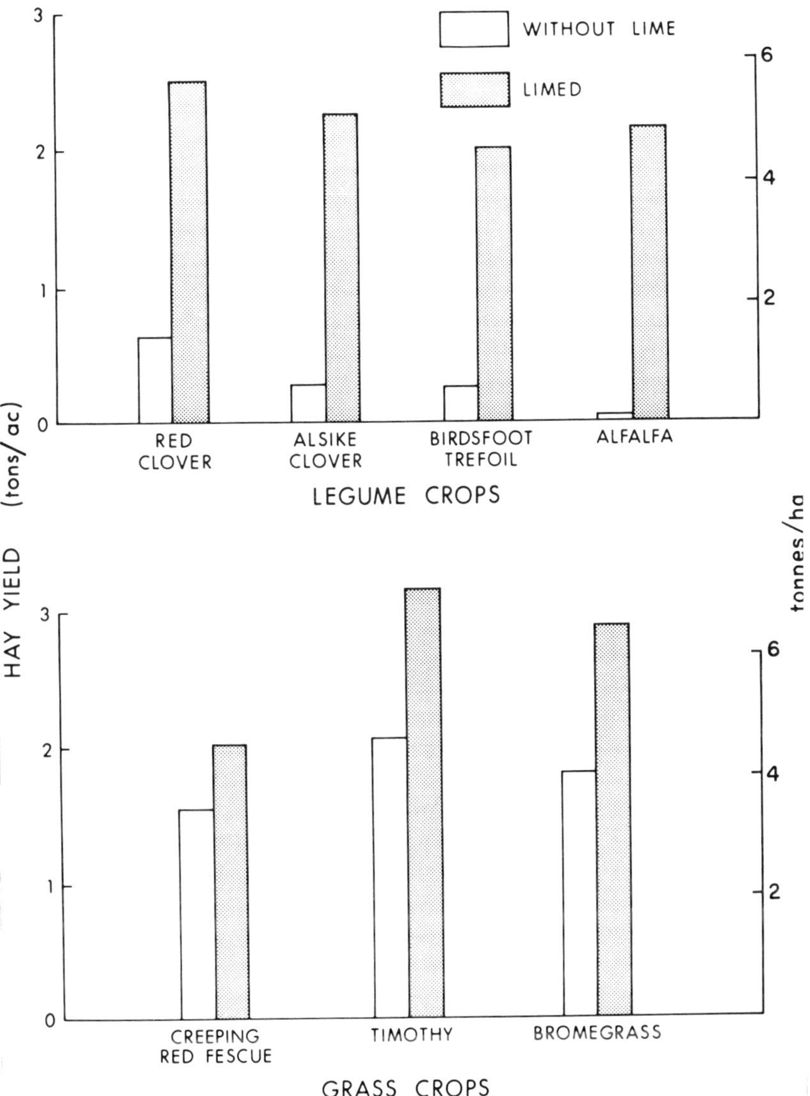

Fig. 3.8 — The comparative response of four legumes and three grasses to lime applied according to a soil test on soils with a pH between 4.5 and 5.0 (very strongly acid). In general, legumes respond more to lime than grasses do. (Alberta and British Columbia, just north of northwestern U.S.) (Courtesy, Canada Department of Agriculture)

Note: "Tonnes / ha" means metric tons per hectare.

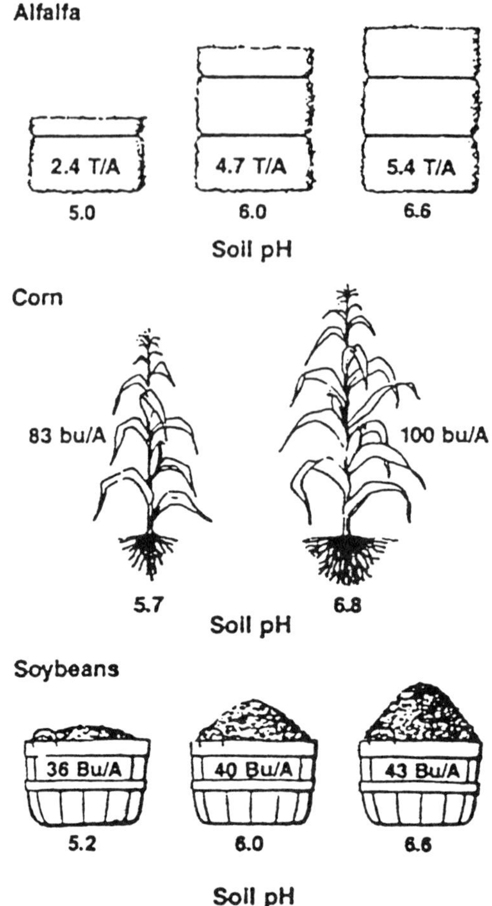

Fig. 3.9 — In Wisconsin, alfalfa and soybean yields increased by liming soil to a pH of 6.6 and corn to a pH of 6.8. (Source: University of Wisconsin Extension Circular A2240, 1985)

Ryegrass on strongly acid soils (pH 4.7) in Louisiana *did not* yield more when the soil pH was maintained by liming to pH 5.7.

Although not usually considered a liming material, on a Florida Histosol (organic soil) calcium silicate increased the yield of rice by 50 percent, and the yields of sugarcane and sugar that followed in rotation increased by 23 percent and 25 percent, respectively.

3:6 □ KINDS OF LIME

Almost all of the lime used to increase the productivity of acid soils consists of finely ground limestone. There are many forms of limestone, however, and other minor sources of agricultural lime that are not called limestone.

Limestone occurs in nature as a sedimentary rock that is composed mostly of calcium carbonate ($CaCO_3$). Its origin is in the *oceans* by chemical and biological precipitation of calcium carbonate from the **ambient** (surrounding) waters over the centuries. Oceans receded, and the deposits were solidified into limestone, as we see it today in a weathered form (Figure 3.10). Such deposits that were not solidified are called **chalk**. *Freshwater* deposits of sediments that are high in calcium carbonate are known as **marl**. In a few places, oyster shells and clam shells are ground and used as a lime to enhance the productivity of acid soils.

Some solidified limestone may have been formed near ocean waters high in magnesium; in this instance, the limestone may be rich in magnesium as well as calcium and is called **dolomite** or **dolomitic limestone**.

Pure calcium carbonate ($CaCO_3$) occurs to a limited extent in nature as crystalline **calcite**. It has an oxide composition of 56 percent CaO and 44 percent CO_2. Pure dolomite also occurs as crystals, with a molecular chemical composition of $CaMg(CO_3)_2$ and an oxide composition of CaO — 30 percent; MgO — 22 percent; and CO_2 — 48 percent.

Limestone in nature is always impure calcite, impure dolomite, or an impure mixture of the two. Impurities are usually sand, silt, clay, and / or iron.

For all practical purposes, unless naturally occurring limestone, marl, ground shells, and chalk are about 85 percent pure calcium carbonate or calcium and magnesium carbonate, it is not economical to develop them as a source of agricultural lime. However, state—not federal—laws regulate the use of lime.

Fig. 3.10 — No one has the ability to look at a limestone quarry like this and determine whether the lime is primarily calcic (calcium carbonate) or dolomitic (calcium and magnesium carbonate), as shown here; neither can anyone determine the percentage of purity without a chemical test. (Courtesy, Lee Lime Company, Lee, Massachusetts)

The liming materials that are officially recognized by the Association of American Plant Food Control Officials for use in agriculture are as follows:

1. **Agricultural liming materials** are those whose calcium and magnesium content are in forms that are capable of reducing soil acidity.
2. **Agricultural slag** is a fused silicate of calcium, magnesium, or both, which is capable of neutralizing soil acidity and is ground fine enough to react readily in the soil.
3. **Air-slaked lime** is a product derived from exposing calcium and/or magnesium oxide to the CO_2 and moisture of the air and consists usually of a mixture of oxides, hydroxides, and carbonates of calcium or calcium and magnesium.
4. **Ground limestone** is calcitic or dolomitic limestone that has been ground fine enough to pass the respective state law standards.
5. **Ground shells** are a product prepared by grinding mollusk (such as oyster) shells to such fineness that 50 percent of the product will go through a 100-mesh (150-micron) sieve.
6. **Ground shell marl** is a marl containing shells that have been ground so that at least 75 percent of the marl will pass through a 100-mesh (150-micron) sieve.
7. **High calcic liming materials** are products containing 25 percent or more of calcium (Ca), and at least 91 percent of the Ca+Mg is Ca.
8. **High magnesic liming** materials are products containing 6 percent or more of magnesium (Mg).
9. **Hydrated lime** consists chiefly of calcium hydroxide [$Ca(OH)_2$] and magnesium hydroxide [$Mg(OH)_2$].
10. **Marl** is an earthy material containing varying amounts of calcium carbonate that have been deposited over the centuries at the bottom of a freshwater pond. Impurities are usually clay and organic matter.
11. **Pulverized limestone** is calcitic or dolomitic limestone that has been ground fine enough so that 100 percent of it will go through a 20-mesh (850-micron) sieve and at least 75 percent of it will pass through a 100-mesh (150-micron) sieve.
12. **Quicklime, burned lime, caustic lime, lump lime,** and **unslaked lime** are calcined limestone that has been burned at temperatures between 662° F (350° C) and 1,517° F (825° C) to drive off the CO_2. The products are mostly CaO, MgO, or a mixture of the two. In all chemistry textbooks, "lime" is CaO and "magnesia" is MgO.
13. **Waste lime,** or **by-product lime**, is any industrial waste, or by-product, that contains calcium or magnesium or both, in forms that will neutralize soil acidity. Common names used in its designation are **acetylene lime, calcium silicate, gashouse lime, lime-kiln ashes,** or **tanners' lime**. Composition is so variable that each source should be tested before use.
14. **Chalk** is not officially recognized as a liming material but is sometimes so used.
15. **Gypsum, landplaster,** or **crude calcium sulfate**, is a product consisting

chiefly of calcium sulfate with combined water ($CaSO_4 \cdot 2H_2O$) and is incapable of neutralizing soil acidity. It shall contain not less than 70 percent $CaSO_4 \cdot 2H_2O$. However, gypsum reduces aluminum toxicity more readily than calcium or magnesium carbonate because it is more soluble.

16. **Magnesium sulfate** is a product consisting chiefly of that material with or without combined water: Epsom salts ($MgSO_4 \cdot 7H_2O$), kieserite ($MgSO_4 \cdot H_2O$), and calcined kieserite ($MgSO_4$).
17. **Calcined brucite** is a magnesium product concentrated from brucite limestone. It consists chiefly of magnesium oxide with lesser amounts of calcium hydroxide, silicates, and sesquioxides.

3:7 □ LIME REQUIREMENT

The **lime requirement** of a soil is the amount of lime of stated purity and stated fineness that is required to raise the pH of an acid soil to a predetermined level.

> *Example:* Soil from field A is a loam that needs 5 tons per acre (11 mt/ha) of pulverized dolomitic limestone (35 percent CaO equivalent plus 10 percent MgO equivalent) with a fineness that permits all of it to pass through a 20-mesh (850-micron) sieve and 75 percent of it to pass through a 100-mesh (150-micron) sieve. This amount of lime is predicted to raise the pH from 5.0 to 6.5.

The lime requirement has been determined traditionally by an integration of soil pH and assumed buffer capacity of the soil based upon its clay content. For instance, Table 3.3 details the tons of lime recommended at four pH ranges and

Table 3.3—Approximate Tons of Agricultural Limestone Required to Raise the pH of the 7-Inch Plow Layer of Five Contrasting Soil Textural Classes with 5 Percent Organic Matter Under Four pH Ranges

Texture of 7-Inch (18-cm) Plow Layer	pH Range			
	4.5 to 4.9	5.0 to 5.4	5.5 to 5.9	6.0 to 6.4
	... *(tons of lime recommended per acre*[1]*)* ...			
Sands	2½	2	1½	½[2]
Loamy sands	3	2½	2	1
Sandy loams	4	3	2½	1½
Clay loams and loams	5	4	3	2
Clays and silty clays	6	5	4	2

[1] Lime recommendations based on a ground limestone material having a neutralizing value of 90 percent with 100 percent of it passing through a 20-mesh (850-micron) sieve and 75 percent passing through a 100-mesh (150-micron) sieve.

[2] It is preferable to recommend a minimum of 1 ton per acre (2.24 mt/ha) so as to obtain uniform application and to justify the expense of application.

Notes:
1. For each inch of depth of plowing below 7 inches (18 cm), increase the rate of lime applied by 15 percent.
2. To convert from tons per acre to metric tons per hectare, multiply by 2.24.
3. For each 1 percent increase in soil organic matter above 5 percent, *reduce* the rate of lime by ½ ton per acre. This seems to be a contradiction because humus buffers the soil; however, the more the humus the lower the pH requirement of plants. (See Figure 3.5.)

for five ranges in soil texture. Note that to achieve a given increase in soil pH, the greater the clay percentage the greater the amount of lime; and the less acid the pH range within the same soil texture the less lime is required.

If you live in an area shown in Figure 3.11 as "very low" or"low" in magnesium, a soil test for magnesium should be made. If the soil test indicates a magnesium content of medium or less, then dolomitic limestone is the only lime to use, even when it costs more. There are also other advantages in using a dolomitic limestone besides that of furnishing magnesium. Pound for pound, dolomitic limestone neutralizes more acidity than does calcic limestone. This is due to the fact that magnesium is lighter; 1 pound (0.45 kg) of lime containing magnesium has 19 percent more ions to neutralize acidity than a pound of calcic limestone.

Another advantage of dolomitic limestone is that it lasts longer in the soil because it is less soluble. Since it lasts longer, larger amounts can be spread less frequently, thus saving time and money.

If a quick change in pH is wanted, burned lime (CaO) or hydrated lime [$Ca(OH)_2$] can be used. In some areas, wood ashes are available and will react quickly in the soil. Also, the limestone can be ground finer to obtain results more quickly.

Liming materials are usually so plentiful that a choice exists between two or more types of limes and also among various degrees of fineness.

Any liming material used should be fine enough to react in time for the current crop, yet coarse enough to last about three to five years. The finer the lime the more quickly it is available to plants. Coarser limestone lasts longer. The most desirable fineness and purity for most limestones consist of this simplified specification: All through a 10-mesh (2-mm) sieve, one-fourth through a 100-mesh (150-micron) sieve with all **fines** (finely ground lime) left in, and a minimum calcium carbonate equivalent rating of 0.85 (100 percent pure calcium carbonate has a rating of 1.0).

Fluid lime (liquid lime, lime suspension) is made from fine lime [usually 60-mesh (250-micron)] as a dispersing agent, **attapulgite clay** (stabilizing agent) as a suspending agent, and either water or fluid fertilizer as a suspending medium. Commonly, the mixtures used contain 50 to 75 percent lime, 0.5 to 5.0 percent attapulgite clay, a small amount of a **dispersing agent** (substance to reduce settling) such as tetrasodium pyrophosphate or lignin sulfonate, with the remainder of the suspension either fluid fertilizer or water or both.

Fluid lime can be combined and spread with fluid fertilizers such as nitrogen, potassium, and sulfur but *not with phosphorus* because of reduced solubility.

There seem to be soils and situations where suspension lime has a place, both agronomically and economically. The answers, however, are not all available at this time. Further research and development by both institutional researchers and commercial firms will undoubtedly define the place of fluid lime over the next several years. Because of its quick reaction but short-term effects, a one-year land lease is a prime situation for the use of fluid lime.

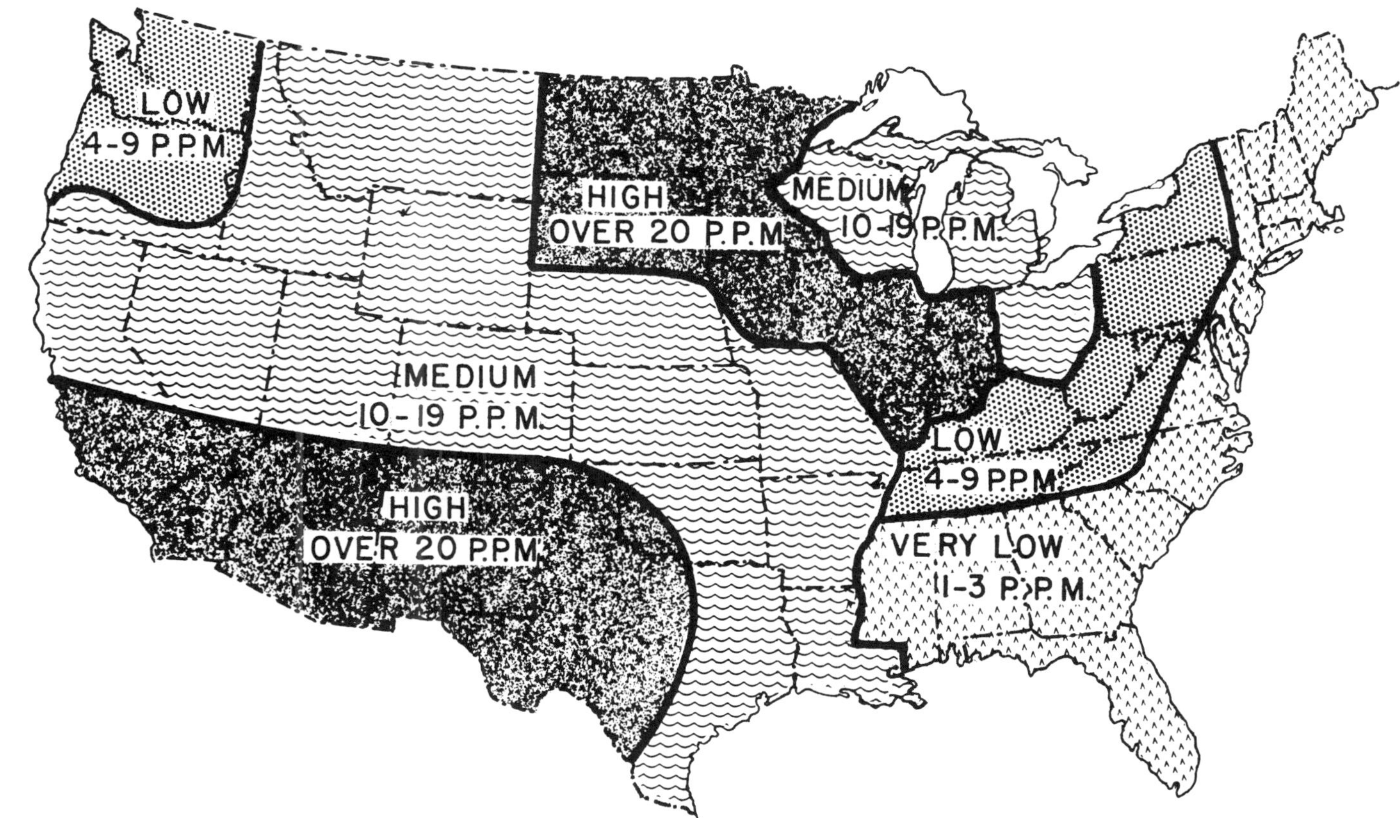

Fig. 3.11 — Magnesium content of drainage waters of the United States, in parts per million (ppm). The magnesium content of the surface soils usually corresponds to that of drainage waters, but the only certain way to find out is to have the soil tested. (Courtesy, USDA – Soil Conservation Service)

3:8 □ APPLYING LIME

The question of *when* to apply lime can be answered very simply and directly: *As soon as possible.* Reasons for such an answer include the slow-acting and long-lasting nature of lime to correct soil acidity, the resulting greater availability of native soil nutrients, and the greater efficiency of applied fertilizers.

How to apply lime also brings a direct reply: With whatever kind of spreader is available. In most communities in humid regions, there are usually one or more commercial companies from which lime can be ordered; most of them have lime-spreading trucks that will spread the amounts of dry or fluid lime on each field or pasture that are specified from a soil test (Figure 3.12).

Limestone may be spread at any time of the year when the soil is firm enough to support the spreading equipment and when crops do not interfere. It is usually desirable to apply lime at least three months in advance of the time it is actually needed by a crop such as clover or alfalfa. This gives sufficient time for some of the lime to dissolve and to react with the soil so that it will be available for the following crop. Fluid lime may become available more quickly.

When large applications of lime are needed, say around 10 tons per acre (22 mt / ha), it is best to plow down half of it and then disc in the other half.

Once the soil is adequately limed, it is usually desirable to test the soil every three years to determine when to lime again. An experiment in Ohio on Wooster silt loam found that once the soil had been properly limed, it took 300 pounds per acre (336 kg / ha) per year to maintain the desired lime level. This amount was applied as 1 ton per acre (2 mt / ha) every six or seven years.

Such conclusions, however, always need reassessment because more pounds of more acid fertilizer are currently being applied, and rainwaters in urbanizing and industrializing areas are becoming more acid.

The traditional method of applying lime on fields has always included the recommendation to incorporate soil and lime with a disc or harrow or plow. The reason given was that lime is slowly soluble and moves down into the root zone very slowly, an average of 1 to 2 inches (2.5 – 5 cm) a year. New tillage systems, however, upset convention. How does a person apply lime on no-tillage corn?

Research in Virginia has found an answer. Over an eight-year period surface-applied lime on no-tillage corn has more than doubled the average corn yield and increased water use efficiency nearly three times as compared with the same amount of lime where the corn received conventional tillage.

The newest technique of applying lime is to apply it as a finely divided suspension fluid. With the increasing popularity of fluid fertilizers, it is logical also to spread lime as a fluid with the same equipment. Furthermore, the fluid lime can be mixed and spread at the same time with fluid nitrogen, potassium, or sulfur fertilizer, but not with fertilizer containing phosphorus, because lime reduces the solubility of phosphorus by the formation of tricalcium phosphate [$Ca_3(PO_4)_2$].

Fig. 3.12 — *Top:* The most popular way to have lime spread on a field is to hire a contract spreader with a V-shaped bed and a spinner wheel, as illustrated here in Virginia. Suitable overlapping will reduce the unevenness of application. *Bottom:* "Big Wheels" for spreading *fluid lime* and fluid fertilzers. (*Top:* Courtesy, Virginia Polytechnic Institute and State University. *Bottom:* Courtesy, Big Wheels, Inc.)

3:9 □ TOO MUCH LIME?

Even though most limes applied on fields are slow-acting, there is a serious hazard of spreading too much on deep, coarse, sandy soils, soils low in organic matter, and highly weathered soils occurring on old land surfaces in tropical and subtropical humid climates such as those in the Southern Atlantic, in the Gulf Coast states, in Puerto Rico, and in Hawaii.

Too much lime on any soil may cause these hazards to efficient plant production:

1. Plants may be **chlorotic** (yellow or white) due to iron deficiency.
2. Zinc, copper, manganese, and/or boron may be rendered less soluble and less available to plants.
3. Phosphorus may become less available because of the formation of slowly soluble tricalcium phosphate [$Ca_3(PO_4)_2$].

One apparent cure for overliming is the application of large amounts of organic matter such as animal manures or sewage sludges. Coarse limestone, dolomitic limestone, minor elements, an organic mulch, and the turning under or incorporation of a cover crop or crop residues have all been used to prevent or correct overliming injury. (See Chapter 5.)

3:10 □ LIME CONVERSION FACTORS

Assuming 100 percent purity and equal fineness for all liming materials, the relative acid-neutralizing value can be converted from one form of lime to another (Table 3.4). To understand how to use this table, observe the entry on the

Table 3.4—Lime Conversion Factors[1]

To Convert from (Column A)	To (Column B)	Multiply Column A by
Calcium (Ca)	Calcium oxide (CaO)	1.40
Calcium (Ca)	Calcium carbonate ($CaCO_3$)	2.50
Calcium (Ca)	Magnesium (Mg)	0.61
Calcium (Ca)	Magnesium oxide (MgO)	1.01
Calcium (Ca)	Magnesium carbonate ($MgCO_3$)	2.10
Calcium oxide (CaO)	Calcium (Ca)	0.71
Calcium oxide (CaO)	Calcium carbonate ($CaCO_3$)	1.78
Calcium oxide (CaO)	Magnesium (Mg)	0.43
Calcium oxide (CaO)	Magnesium oxide (MgO)	0.72
Calcium oxide (CaO)	Magnesium carbonate ($MgCO_3$)	1.50
Calcium carbonate ($CaCO_3$)	Calcium (Ca)	0.40
Calcium carbonate ($CaCO_3$)	Calcium oxide (CaO)	0.56
Calcium carbonate ($CaCO_3$)	Magnesium (Mg)	0.24
Calcium carbonate ($CaCO_3$)	Magnesium oxide (MgO)	0.40
Calcium carbonate ($CaCO_3$)	Magnesium carbonate ($MgCO_3$)	0.84

(Continued)

Table 3.4 (Continued)

To Convert from (Column A)	To (Column B)	Multiply Column A by
Magnesium (Mg)	Magnesium oxide (MgO)	1.66
Magnesium (Mg)	Magnesium carbonate ($MgCO_3$)	3.50
Magnesium (Mg)	Calcium (Ca)	1.67
Magnesium (Mg)	Calcium oxide (CaO)	2.33
Magnesium (Mg)	Calcium carbonate ($CaCO_3$)	4.17
Magnesium oxide (MgO)	Magnesium (Mg)	0.60
Magnesium oxide (MgO)	Magnesium carbonate ($MgCO_3$)	2.10
Magnesium oxide (MgO)	Calcium (Ca)	1.00
Magnesium oxide (MgO)	Calcium oxide (CaO)	1.40
Magnesium oxide (MgO)	Calcium carbonate ($CaCO_3$)	2.50
Magnesium carbonate ($MgCO_3$)	Magnesium (Mg)	0.29
Magnesium carbonate ($MgCO_3$)	Magnesium oxide (MgO)	0.48
Magnesium carbonate ($MgCO_3$)	Calcium (Ca)	0.48
Magnesium carbonate ($MgCO_3$)	Calcium oxide (CaO)	0.67
Magnesium carbonate ($MgCO_3$)	Calcium carbonate ($CaCO_3$)	1.19

[1] Calculated, using the following atomic weights: calcium, 40; magnesium, 24; oxygen, 16; carbon, 12; and hydrogen, 1. For example, in line 1 above, to convert from Ca to CaO equivalent, divide the molecular weight of CaO by the atomic weight of Ca = 56 ÷ 40 = 1.40. Therefore, to convert Ca to CaO, multiply the lb. (kg) of Ca by 1.40 to obtain the equivalent lb. (kg) of CaO.

last line. To convert magnesium carbonate ($MgCO_3$) to calcium carbonate ($CaCO_3$) equivalent, multiply magnesium carbonate by 1.19. This means that the same weight of pure magnesium carbonate has 1.19 times (119 percent) more acid-neutralizing value than pure calcium carbonate. This value also can be calculated by using atomic and molecular weights. Magnesium has an atomic weight of 24; carbon, 12; and oxygen, 16. Therefore, the molecular weight of $MgCO_3 = 24 + 12 + 48 = 84$. The atomic weight of $CaCO_3 = 40 + 12 + 48 = 100$. The relative acid-neutralizing value of $MgCO_3$ would be $100 \div 84$, or 1.19.

3:11 ▫ REFERENCES

Adams, Donald D., and Walter P. Page, eds. *Acid Deposition: Environmental, Economic, and Policy Issues*. New York: Plenum Publishing Corp., 1985, 522 pp.

Adams, Fred, ed. *Soil Acidity and Liming*. American Society of Agronomy, Crop Science Society of America, and Soil Science Society of America, 1984, 380 pp.

Ahmad, F., and K. H. Tan. "Effect of Lime and Organic Matter on Soybean Seedlings Grown in Aluminum-toxic Soils." *Soil Science Society of America Journal*, Vol. 50, No. 3, May – June 1986, pp. 656 – 661.

Anderson, D. L., D. B. Jones, and G. H. Snyder. "Response of a Rice-Sugarcane Rotation to Calcium Silicate Slag on Everglade Histosols." *Agronomy Journal*, Vol. 79, 1987, pp. 531 – 535.

Association of American Plant Food Control Officials. *Official Publication No. 40*. West Lafayette, Indiana: Purdue University, 1987, 63 pp.

Association of American Plant Food Control Officials. *Official Publication No. 42*. Virginia Dept. of Agriculture, Richmond, 1989.

Bates, Robert L., and Julia A. Jackson, eds. *Glossary of Geology*, 2nd ed. Falls Church, Virginia: American Geological Institute, 1980, 751 pp.

Burmester, C. H., J. F. Adams, and J. W. Adom. "Response of Soybean to Lime and Molybdenum on Ultisols in Northern Alabama." *Soil Science Society of America Journal*, Vol. 52, 1988, pp. 1,391–1,394.

"Canada's Environment: An Overview." Ontario, Canada: Supply and Services, May 1986.

"Commercial Fertilizers." U.S. Dept. of Agriculture – Statistical Reporting Service, Crop Reporting Board, November 1985.

Donahue, Roy L., Raymond W. Miller, and John C. Shickluna. *Soils: An Introduction to Soils and Plant Growth*, 5th ed. Englewood Cliffs, New Jersey: Prentice-Hall, Inc., 1983.

Farm Chemicals Handbook. Willoughby, Ohio: Meister Publishing Co., 1988 (published annually).

Follett, Roy H., Larry S. Murphy, and Roy L. Donahue. *Fertilizers and Soil Amendments*. Englewood Cliffs, New Jersey: Prentice-Hall, Inc., 1981.

Follett, Roy H., and Ronald F. Follett. "Soil and Lime Requirement Tests for the 50 States and Puerto Rico." *Journal of Agronomic Education*, Vol. 12, 1983, pp. 9 – 17.

Kamrath, Eugene J., and Charles D. Foy. "Lime-Fertilizer-Plant Interactions in Acid Soils." In O. P. Englestad, ed., *Fertilizer Technology and Use*, 3rd ed. Soil Science Society of America, 1985, pp. 91 – 151.

McLean, E. O. "Recommended pH and Lime Requirement Tests." In W. C. Dahnke, ed., *Recommended Chemical Soil Test Procedures for the North Central Region*. North Central Regional Pub. No. 221, 1980, pp. 8 – 24.

Morris, D. R., *et al.* "Soil and Annual Forage Crops Response to Liming." *Agronomy Abstracts*. American Society of Agronomy, 1987.

Noble, A. D., M. E. Sumner, and A. K. Alva. "The pH Dependency of Aluminum Phytotoxicity Alleviation by Calcium Sulfate." *Soil Science Society of America Journal*, Vol. 52, 1988, pp. 1,398–1,402.

NSA Aglime Fact Book. Washington, D.C.: National Stone Foundation, 1986, 65 pp.

Radcliffe, D. E., R. L. Clark, and M. E. Summer. "Effect of Gypsum and Deep-rooted Perennials on Subsoil Mechanical Impedance." *Soil Science Society of America Journal*, Vol. 50, 1986, pp. 1566 – 1570.

Rechcigl, J. E., D. D. Wolf, R. B. Reneau, Jr., and W. Kroontje. "Surface Lime Influence on No-Till Alfalfa Grown in an Acid Soil." In W. L. Hargrove and F. C. Boswell, "Proceedings of the 1985 Southern Region No-Till Conference," Griffin, Georgia, July 16, 17, pp. 112 – 116.

Schulte, E. E., and K. A. Kelling. "Aglime — Key to Increased Yield and Profits." University of Wisconsin, Ext. Cir. A2240, March 1985, 4 pp.

Yost, Russell, Goro Uehara, Michael Wade, M. Sudjadi, I.P.G. Widjaja, and Zhi-Cheng Li. "Expert Systems in Agriculture: Determining Lime Recommendations for Soils of the Humid Tropics." Research Extension Series 089.8, University of Hawaii, March 1988.

CHAPTER 4

Fertilizers

"The whole of agriculture centers on one point: The nourishing of plants."—Francis Home, 1757

OUTLINE

This helicopter has been adapted for spraying or dusting insecticides, fungicides, herbicides, defoliants, and dry or water-soluble fertilizers. The downwash of the rotor blades assists in uniform coverage of the foliage. It is anticipated that an increasing acreage of crops will be fertilized with urea, water-soluble phosphorus, and micronutrients at the same time that they are sprayed with insecticides or fungicides. (Courtesy, Bell Helicopter Co.)

□ □ □

4:1 □ OVERVIEW

To **fertilize** a soil means to enrich it, to make the soil more productive. It also means to add to the soil one or more plant nutrient elements so that the lack of these elements will not limit plant growth. The continuous use of soil without replacement of the essential nutrients removed by crops leads inevitably to lower productive capacity. Modern agriculture, with its high crop yields, is dependent upon an adequate supply of plant nutrients. Materials containing these plant nutrients—falsely called plant foods—are referred to as fertilizers. Fertilizers supply essential plant nutrients which are used by plants to manufacture foods such as proteins, carbohydrates, and fats.

Nearly all soils in their native state contain adequate quantities of plant nutrients to support some sort of plant life. However, the plant nutrient levels are inadequate for modern agriculture where high crops yields are essential for profitable crop production. Fertilizers applied to soils correct these inadequate nutrient levels. They allow the farmer to eliminate plant nutrient deficiencies as factors limiting plant growth.

Fertilizers increase soil fertility and provide a means of maintaining high soil fertility levels. They replace nutrients removed from soil by harvested crops and those lost by erosion, leaching, and denitrification, and those nutrients fixed in unavailable forms.

In the early days, fertilizers were made largely from animal and plant processing by-products. These materials were usually low in plant nutrient content. As fertilizers improved, manufacturers have been able to develop new techniques to manufacture concentrated high analysis fertilizers which provide farmers with more plant nutrients per ton of fertilizer. This reduces the manufacturing, distribution, and application costs per unit of plant nutrient for both the fertilizer manufacturer and the farmer and lowers the cost of food and fiber to the consumer.

4:2 □ WHAT'S IN THAT FERTILIZER BAG?

Each bag or package of fertilizer includes chemical compounds that contain one or more of the 16 **essential plant nutrients**. These are carbon (C), hydrogen (H), oxygen (O), nitrogen (N), phosphorus (P), potassium (K), calcium (Ca), magnesium (Mg), sulfur (S), iron (Fe), zinc (Zn), copper (Cu), manganese (Mn), boron (B), molybdenum (Mo), and chlorine (Cl).

Carbon, hydrogen, and oxygen, comprising more than 90 percent of plant tissue, are obtained from air and water. Their presence in fertilizer should not be considered when determining the supply of plant nutrients contained in a fertilizer. The other nutrients are usually obtained from the soil, but when they are deficient in the soil, they must be added for good plant growth. Materials containing these 13 plant nutrient elements are known as **fertilizers**.

Nitrogen, phosphorus, and potassium are known as the **macro (major** or **primary) plant nutrients**. Calcium, magnesium, and sulfur are known as the **secondary plant nutrients**. The **micro (minor** or **trace) elements** include boron, copper, iron, manganese, molybdenum, zinc, and chlorine. This classification is based upon the proportion of these essential elements normally found in plants. (See Chapter 2.)

4:2.1 □ Fertilizer Grade/Analysis/Brand

Fertilizer grade is expressed as a set of three numbers such as 10-10-10, 16-8-8, 10-10-5, and 5-20-20. Fertilizer grade is the minimum guaranteed percentage of plant nutrients in a fertilizer. The numbers "5-20-20" on a fertilizer bag mean that the manufacturer guarantees that it contains (always in this order) 5 percent total nitrogen (N), 20 percent available phosphate (P_2O_5), and 20 percent water-soluble potash (K_2O). The remaining 55 percent of the product consists of other elements necessary to stabilize the chemical compounds such as calcium, chlorine, and oxygen. If a nutrient is missing, it is represented by a zero; thus, 45-0-0 for urea, 0-45-0 for triple superphosphate, 0-0-60 for potassium chloride (muriate of potash), and 18-46-0 for diammonium phosphate. With secondary and micronutrients, there is no legal order of arrangement of the elements after the N, P_2O_5, and K_2O values for the statement of a grade which guarantees these additional nutrients. The order of arrangement varies with the need of the additional plant nutrients and by state regulations. Generally, the percentages claimed for the major and secondary plant nutrients are expressed in whole numbers, although this regulation varies in the different states. The minimum percentages of micronutrients, which are claimed and guaranteed, may be stated in less than whole numbers, usually as decimal percentages. Many states have minimum percentage values, which must be claimed, before any micronutrient may be guaranteed. State laws penalize fertilizer manufacturers when the **actual analysis** of a fertilizer, as determined in a laboratory, varies more than a specified amount from the grade (guaranteed analysis).

The term ***brand*** refers to the word, design, and/or trademark used by companies to describe one or several grades of commercial fertilizers which they sell. Brand names are usually registered in the state in which they are marketed. Also, brand trademarks are almost always federally copyrighted.

4:2.2 □ Oxide Versus Elemental Basis

Fertilizer terminology has been in confusion: the elemental terms have been used for designating nitrogen and the oxide terms for phosphorus and potassium. Nitrogen (N) and most other plant nutrients have customarily been described according to the actual amount of the element present. But phosphorus (P) is traditionally listed according to the chemical equivalence of phosphorus pentoxide (P_2O_5), even though P_2O_5 contains only 44 percent phosphorus (P). Potassium (K) is commonly described in terms of the chemical equivalence of potassium oxide (K_2O) which contains only 83 percent potassium (K).

Most scientific journals now require that the amounts of plant nutrients used in test plots be expressed on the elemental basis when the data are summarized. Figure 4.1 shows how to make conversions from one system to the other. A 6-24-24 fertilizer under the oxide system becomes a 6-10.5-20 under the elemental system (Figure 4.2).

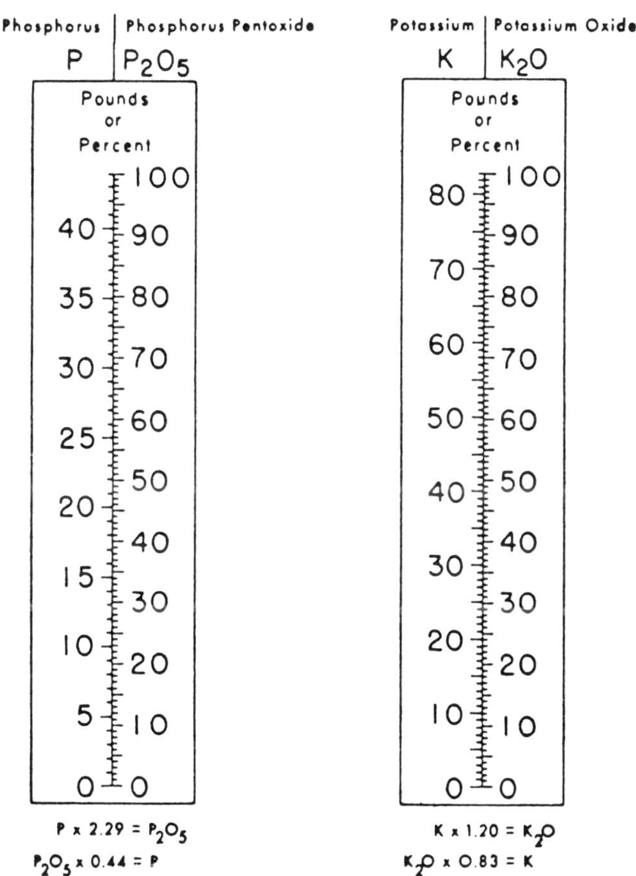

Fig. 4.1—Charts to convert phosphorus (P) to phosphorus pentoxide and potassium (K) to the oxide (K_2O), and vice versa.

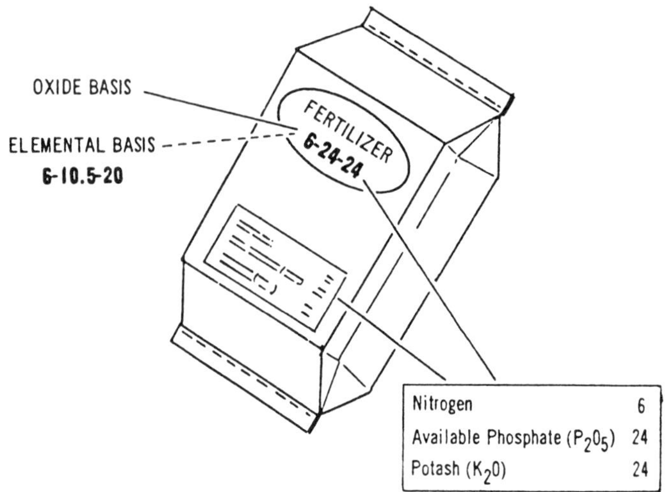

Fig. 4.2—What's in the fertilizer bag? A 6-24-24 fertilizer grade under the oxide system becomes a 6-10.5-20 under the elemental system.

4:2.3 □ Fertilizer Ratio

The term ***ratio of a fertilizer*** denotes the proportion of one plant nutrient percentage to another. For example, a fertilizer which has equal percentages of N, P_2O_5, and K_2O, such as a 10-10-10, has a 1-1-1 ratio. Likewise, a fertilizer such as a 5-10-20, which has twice as much P_2O_5 as it has N and four times as much K_2O as it has N, has a 1-2-4 ratio.

4:2.4 □ Formulation

In order to achieve a certain grade and ratio, the manufacturer has to determine the **formulation**—the fertilizer materials to use and the amounts of each. The choice depends upon the available materials, costs, ease of handling, compatibility of materials in the mixture, and the physical condition of the final product. Formulation can become quite complicated, and computers are often used to make the calculations rapidly.

A formulation for a ton of 5-20-20 fertilizer might be:

125	pounds ammonia (82-0-0)
412	pounds phosphoric acid (0-52-0)
902	pounds normal superphosphate (0-20-0)
667	pounds muriate of potash (0-0-62)
2,106	pounds
– 106	pounds (water loss in process)
2,000	pounds of 5-20-20 fertilizer

4:3 □ NITROGEN (N)

Nitrogen in its pure state is a colorless, tasteless, odorless, inert gas. The atmosphere contains about 78 percent nitrogen gas (N_2). Higher plants other than legumes and a few nonlegumes cannot use pure nitrogen; nor can the nitrogen be put into fertilizers in this form. Pure nitrogen must be combined with other elements before it can be used as a fertilizer.

Most of the nitrogen in soils is found in the organic matter, which in turn is found largely in the topsoil, or plow layer. The soils of the earth vary in their organic matter content; they also vary in their nitrogen content. In the United States the total nitrogen in the topsoil varies from about 1,500 pounds per acre (1,680 kg/ha) to more than 7,000 pounds per acre (7,850 kg/ha).

Fortunately, we have unlimited nitrogen. Above every acre of the earth's surface there are about 35,000 tons (78,470 mt/ha) of nitrogen that can be utilized either by factories that fix atmospheric nitrogen or by properly inoculated leguminous plants, such as clovers and alfalfa. Legumes, when inoculated and supplied with proper nutrients, have the ability to take nitrogen from the air and convert it into plant nutrients. (See Note 2.2.) A few nonleguminous plants also have the ability to support bacteria that fix atmospheric nitrogen. Likewise, atmospheric nitrogen is fixed by some free-living bacteria and algae. In addition, rain and snow wash 5 to 10 pounds of nitrogen per acre (5.6–11.2 kg/ha) per year from the air into the soil. This has been fixed by lightning and released by fossil fuels. (See Section 3:2.1.)

Nitrogen is quickly exhausted from our soils by erosion, leaching, denitrification, and the growing of crops. Therefore, nitrogen must be replaced frequently to maintain soil productivity. The key to maintaining soil nitrogen is to maintain organic matter and to apply nitrogen fertilizer. Also, the value of a legume for adding nitrogen to a cropping system has long been recognized.

Plants can use nitrogen only when it is properly combined with oxygen or hydrogen. In these combined ionic forms, nitrogen is known as **nitrate (NO_3^-)** or **ammonium (NH_4^+) nitrogen**. Nitrogen in organic matter is changed by biological action into the ammonium and nitrate forms.

Nitrogen gas of the atmosphere is the starting material of fertilizer nitrogen. Nitrogen is reacted with hydrogen under great pressure and heat to produce gaseous ammonia (82 percent N), commonly called **anhydrous ammonia**. In the United States the hydrogen gas is obtained almost entirely from the reforming of natural gas, which is mostly methane (CH_4). A fertilizer plant requires about 38,130 cubic feet (1,080 cu m) of natural gas to produce 1 ton of ammonia. The amount of natural gas used to produce 5 tons (4.5 metric tons) of ammonia would heat an average home in Iowa for one year. This same amount of natural gas, converted to nitrogen fertilizer, could result in enough extra corn production to satisfy the minimum protein and caloric requirements of 275 persons for one year.

More than 90 percent of all nitrogen fertilizers consist of ammonia or fertilizers made from ammonia (Figure 4.3). Anhydrous ammonia, liquid ammonia,

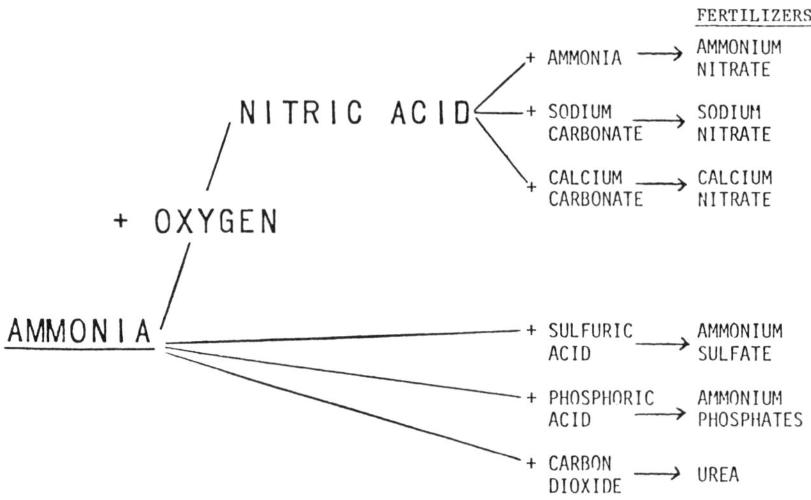

Fig. 4.3—More than 90 percent of all nitrogen fertilizers consist of ammonia or fertilizers made from ammonia.

ammonium nitrate, urea, ammonium sulfate, synthetic sodium nitrate, and ammonium phosphate are made with ammonia as the source of nitrogen. Only Chilean nitrates of soda and cyanamide do not use ammonia in their manufacture, but these fertilizers have almost disappeared from the U.S. market.

4:3.1 ▫ Nitrogen Fertilizers

Many different chemical and physical forms of nitrogen fertilizer are available. Under certain conditions one form may have an advantage over another. However, if properly applied, the various forms of nitrogen fertilizer are generally equally effective. The grade, chemical form, physical form, and recommended method of applying the important sources of nitrogen fertilizer are presented in Table 4.1.

Table 4.1—Sources of Nitrogen Fertilizer, Grade, Physical Form, and Recommended Method of Application[1]

Fertilizer	Grade (N-P_2O_5-K_2O)	Physical Form	Recommended Method of Application
Anhydrous ammonia (NH_3)	82-0-0	High-pressure liquid	Must be injected 6–8 in. (15–20 cm) deep in friable, moist soil.
Aqua ammonia ($NH_3 + H_2O$)	20-0-0 to 24-0-0	Low-pressure liquids	Must be injected 2–3 in. (5–8 cm) deep in friable, moist soil.
Low-pressure N solutions ($NH_4NO_3 + NH_3 +$ (H_2O))	37-0-0 to 41-0-0	Low-pressure liquids	Must be injected 2–3 in. (5–8 cm) deep in friable, moist soil.

(Continued)

Table 4.1 (Continued)

Fertilizer	Grade ($N-P_2O_5-K_2O$)	Physical Form	Recommended Method of Application
Pressureless N solutions (NH_4NO_3 + urea + H_2O)	28-0-0 to 32-0-0	Pressureless liquids	Spray on surface or side-dress. Incorporate surface application to prevent volatilization loss of NH_3 from the urea.
Ammonium nitrate (NH_4NO_3)	33.5-0-0 to 34-0-0	Dry prills, granules	Broadcast or side-dress. Can be left on the soil surface.
Ammonium sulfate $(NH_4)_2SO_4$	20-0-0	Dry granules	Broadcast or side-dress. Can be left on the soil surface.
Urea ($NH_2 \cdot CO \cdot NH_2$)	45-0-0	Dry prills, granules	Broadcast or side-dress. Incorporate surface application to prevent volatilization loss of NH_3 from the urea.
Sodium nitrate ($NaNO_3$)	16-0-0	Dry granules	Broadcast or side-dress. Can be left on the soil surface.
Calcium nitrate $Ca(NO_3)_2$	15.5-0-0	Dry granules	Broadcast or side-dress. Can be left on the soil surface.
Ammonium nitrate of lime ($NH_4NO_3 + CaCO_3$)	20.5-0-0	Dry granules	Broadcast or side-dress. Can be left on the soil surface.
Diammonium phosphate $(NH_4)_2HPO_4$	18-46-0	Dry granules	Broadcast or apply in the row. Can be left on the soil surface.
Potassium nitrate (KNO_3)	13-0-44	Dry granules	Broadcast or apply in the row. Can be left on the soil surface.
Ammonium phosphate	10-34-0 to 11-37-0	Pressureless liquids	Spray on surface or side-dress.

[1]Source: L. M. Walsh. "Soil and Applied Nitrogen." University of Wisconsin, Ext. Fact Sheet A 2519, 1973.

4:3.2 ▫ Slow-Release Nitrogen Fertilizers

The nitrogen in most chemical fertilizers is readily soluble and available for use by plants. However, plants need nitrogen every day in their life rather than a surplus one day and a deficiency the next. During the last few years several slowly available forms of nitrogen fertilizer have been developed. Among the products developed are urea-formaldehyde, metal ammonium phosphates, and sulfur-coated urea. Greenhouse and field tests have shown these materials to be a slowly available source of nitrogen especially for use on sandy soils or for turfgrasses and specialty crops. However, they cost more per unit of nitrogen.

4:4 ▫ PHOSPHORUS (P)

Phosphorus is one of the major nutrients that plants, people, and animals must have in order to survive. It is a major ingredient of fertilizers produced in various compounds and grades. Plants get their phosphorus from the soil in which they grow; however, most soils do not contain enough available phosphorus to produce maximum plant growth without adding a phosphorus fertilizer.

Phosphate is never plentiful. The phosphate content of the earth's surface averages only 0.1 percent. Necessary in all plant growth, phosphate assists in cell division and in the formation of fats and proteins. Called by some a "life-generating element," phosphate is concentrated in the seeds and fruits of plants.

In its pure form, phosphorus is a very active substance that soon bursts into flame when exposed to the air. Phosphorus must, therefore, be combined with another element, usually calcium, before it can be used as a fertilizer.

The native phosphate in soils is often bound or fixed in some unavailable form. In acid soil, iron and aluminum react with phosphate to tie it up and to keep plants from getting it. On the other hand, in alkaline soils, calcium and magnesium often make some phosphate unavailable. Add these two hazards to the fact that nearly all soils in the world are low in total phosphate, and immediately this element takes on new and critical significance for us.

Phosphorus fertilizers are derived from the mining and treatment of **phosphate rock**, which occurs in sedimentary deposits, usually from marine origins. In the past, unprocessed phosphate rock was applied directly as a fertilizer material. Today, however, very little is used as a direct-application source of phosphorus. Studies over the years have shown that the phosphorus available to growing plants from phosphate rock is quite low; thus, phosphate rock must be processed to increase the phosphorus availability (Figures 4.4 and 4.5).

Rock phosphate is the source of nearly all phosphorus fertilizer sold in the United States. Insoluble rock phosphate is treated with sulfuric or nitric acid to convert it to more available superphosphate or ammonium phosphate. The acid used in the process of making ordinary and triple superphosphate is neutralized so that the use of a phosphate fertilizer results in very little residual acidity when it is applied to the soil. Monoammonium and diammonium phosphates, however, are acid-forming. Some of the common phosphate fertilizers are listed in Table 4.2.

4:5 □ POTASSIUM (K)

Potassium, averaging about 2.5 percent in our surface soils, is in adequate supply almost everywhere. That is, the total amount is sufficient. *Available* supplies are often low, however, in the more humid areas of the United States. A soil may contain as much as 20 tons of total potash (K_2O) per acre (45 mt/ha), yet need potash fertilizer to supply readily available potash for proper plant growth.

In a pure form, potassium is a highly active and caustic element. Before it can be used by the plant or handled as a fertilizer, potassium must be combined with some other element. Usually chlorine, sulfur, and oxygen, or nitrogen and oxygen are the elements found in combination with potassium in a fertilizer.

Fig. 4.4—Here a giant bucket is moving rock phosphate ore into a shallow pit where water under high pressure "liquefies" it for transmission by pipe line to the processing plant where the ore is concentrated (beneficiated) (see Figure 4.5). (Courtesy, International Minerals and Chemical Corporation)

Fig. 4.5—Phosphate rock ore is pumped from the mine to this processing plant where the ore is concentrated and stored outside, awaiting drying and shipping. (Courtesy, International Minerals and Chemical Corporation)

Table 4.2—Fertilizer Sources of Phosphorus (P)

Name of Fertilizer	Approximate Chemical Formula	Grade: Oxide Basis (N-P_2O_5-K_2O)	Grade: Elemental Basis (N-P-K)
Ordinary superphosphate	$Ca(H_2PO_4)_2 + CaSO_4$	0-20-0	0-8.8-0
Triple superphosphate	$Ca(H_2PO_4)_2$	0-45-0	0-19.8-0
Monoammonium phosphate	$NH_4H_2PO_4$	11-48-0	11-21.1-0
Diammonium phosphate	$(NH_4)_2HPO_4$	18-46-0	18-20.2-0
Ammonium polyphosphate:	$NH_4H_2PO_4 + (NH_4)_3HP_2O_7$		
Liquid form		10-34-0	10-15-0
Dry form		15-62-0	15-27.3-0

Originally this country depended largely on European mines for its potash supplies. The production of potash started in the United States in 1916 using the brines of Searles Lake near Trona, California. Later potash discoveries were made near Carlsbad, New Mexico; Salduro Marsh in Utah; and in brines in eastern Michigan.

In Saskatchewan, Canada, the largest potash reserve in the world was discovered, and the first deep mine started production in 1962 (Figure 4.6). Other large producers of potash fertilizer are West Germany, the U.S.S.R., and France. Minor producers are Spain, Israel, and Italy.

Fig. 4.6—Canadian potash is exported from this mine and refinery to help feed the world's exploding population. Demand for North American potash is great and rising at an average annual rate of about 6 percent—the fastest of any plant nutrient. This plant provides 12 percent of the current world supply of potash. The project is the world's largest potash mine and refinery. (Courtesy, International Minerals and Chemical Corporation)

The principal potassium fertilizers are presented in Table 4.3. The most common potassium fertilizer for use on field crops is potassium chloride (muriate of potash). This is the least expensive source of potassium, and it is just as effective as the other sources. For that reason it is usually recommended, except when the crop also needs sulfur or magnesium. Also, some specialty crops require the use of the sulfate form of potassium to maintain crop quality. For ex-

Table 4.3—Fertilizer Sources of Potassium (K)

Name of Fertilizer	Chemical Formula	Grade	
		Oxide Basis $N\text{-}P_2O_5\text{-}K_2O$	Elemental Basis N-P-K
Potassium chloride (muriate of potash)	KCl	0-0-60	0-0-49.8
Potassium sulfate	K_2SO_4	0-0-50	0-0-41.5 + 17.6% sulfur
Potassium-magnesium sulfate	$K_2SO_4 \cdot 2MgSO_4$	0-0-22	0-0-18.3 + 22.7% sulfur and 11.2% magnesium
Potassium nitrate	KNO_3	13-0-44	13-0-36.5

ample, tobacco will not burn properly when chloride is added to the soil, so it should be fertilized with a sulfate form of potassium. Also potassium nitrate seems to have special beneficial effects when it is used on tobacco plants.

4:6 □ SECONDARY PLANT NUTRIENTS

The secondary plant nutrients are calcium, magnesium, and sulfur.

4:6.1 □ Calcium (Ca)

On the average, approximately 3.5 percent of the earth's crust is composed of calcium, but since plants cannot move about as animals do, the calcium within reach of every growing plant must be adequate. When calcium is not adequate (and this is the case in most humid region soils) it is added to the soil, usually as ground limestone. The most productive soils in the world are abundantly supplied with calcium.

Calcium is found in the middle layer of the cell walls in all green plants. There it acts as a guard to the cells, permitting only those nutrients that are listed in the "social register" to pass.

In addition to its direct use by plant cells, calcium, when added to an acid soil, makes many other nutrients more available to the growing plant. Three of the nutrients thus mobilized are phosphorus, nitrogen, and molybdenum. While these indirect effects are highly important, of almost equal importance is the fact that lime stimulates all biological activity, and this results in greater soil granulation, better aeration, increased root growth, and higher yields. Furthermore, calcium hastens microbial breakdown of organic residues and thus releases some of all essential nutrients. (See Chapter 5.)

Limestone used to correct soil acidity is the predominant source of applied calcium. The amount of calcium added in limestone plus the relatively large amounts of exchangeable calcium in the soil far exceed the 50 to 100 pounds per acre (56–112 kg/ha) of calcium commonly removed by crops. The amount of calcium in several liming and fertilizing materials is listed in Table 4.4. (See Chapter 3.)

Table 4.4—Fertilizer / Lime Sources of Calcium (Ca)

Carrier	Percent Calcium (Ca)
Slaked lime, $Ca(OH)_2$	54
Calcite, $CaCO_3$	40
Ground calcic limestone, 90% pure	36
Dolomite ($CaCO_3 + MgCO_3$)	22
Gypsum, $CaSO_4$	22
Ordinary superphosphate, 0-20-0	20
Triple superphosphate, 0-46-0	14

4:6.2 □ Magnesium (Mg)

The most economical way of correcting magnesium deficiency on an acid soil is to apply dolomitic limestone. However, if the pH is already high or if the crop being grown, such as potatoes, requires an acid soil, then other carriers of magnesium, such as Epsom salt ($MgSO_4$) and potassium magnesium sulfate ($K_2SO_4 \bullet 2MgSO_4$), must be used. If excessive potassium fertilization caused the magnesium deficiency, Epsom salt is preferred. Correction of magnesium deficiency with Epsom salt or potassium magnesium sulfate requires 50 to 100 pounds per acre (56–112 kg/ha) of magnesium when broadcast, or 10 to 20 pounds per acre (11–22 kg/ha) when applied in the row.

The magnesium content of common magnesium carriers is given in Table 4.5.

Table 4.5—Fertilizer / Lime Sources of Magnesium (Mg)

Carrier	Percent Magnesium (Mg)
Dolomitic limestone	Variable
Dolomite ($CaCO_3 + MgCO_3$)	8 to 20
Epsom salt ($MgSO_4 \bullet 7H_2O$)	10
Kieserite ($MgSO_4 \bullet H_2O$)	18
Potassium magnesium sulfate ($K_2SO_4 \bullet 2MgSO_4$)	11

The importance of adequate magnesium for both plant growth and nutrition of animals being fed grass forages is now recognized. Animals grazed on pasture low in magnesium sometimes develop hypomagnesemia (grass tetany). Grass tetany is a nutrient deficiency disease of beef cattle, dairy cattle, and sheep associated with low magnesium levels in the blood serum. This is sometimes a problem where calcite (calcic) limestone ($CaCO_3$) has been used as a liming material. Additions of excessive ammonium nitrogen and potassium fertilizers have also increased the incidence of grass tetany.

4:6.3 □ Sulfur (S)

Soils commonly contain 100 to 600 pounds per acre (112–672 kg/ha) of total sulfur. Nearly all of this sulfur is in an unavailable organic form. As the organic matter decomposes, a small portion of this sulfur is converted into the sulfate form (SO_4^{2-}). Like nitrate (NO_3^-) nitrogen, this form is soluble and moves downward in the soil when heavy rains occur.

Burning coal, and to a lesser extent, burning oil and gas, results in contamination of the atmosphere with sulfur. This atmospheric sulfur is taken from the air and deposited on the land in rainwater. In 1980 the amount of sulfur returned to the soil from the atmosphere in the states that are not highly industrialized was 1 pound per acre (1.1 kg/ha) and 100 pounds per acre (112 kg/ha) in industrialized states.

A number of different sources of sulfur fertilizer and their relative solubilities are presented in Table 4.6. All sulfate forms of sulfur can be surface applied or incorporated with equal effectiveness. Elemental or pure sulfur (S) is insoluble and must be transformed into sulfate (SO_4^{2-}) sulfur by soil bacteria before it can be used by the plant. Therefore, to be effective, elemental sulfur should be worked into the soil several months in advance of the time the crop needs it. Top-dressed elemental sulfur is not very effective because the transformation of the unavailable sulfur to available sulfate is often too slow to benefit the current crop.

Table 4.6—Fertilizer Sources of Sulfur (S)

Name of Fertilizer	Chemical Formula	Grade N-P_2O_5-K_2O	Percent Sulfur (S)
Very soluble			
Ammonium sulfate	$(NH_4)_2SO_4$	21-0-0	24
Potassium sulfate	K_2SO_4	0-0-50	18
Potassium magnesium sulfate	$K_2SO_4 \cdot 2MgSO_4$	0-0-22	23
Magnesium sulfate (Epsom salt)	$MgSO_4 \cdot 7H_2O$	0-0-0	14
Ordinary superphosphate	$Ca(H_2PO_4)_2 + CaSO_4$	0-20-0	14
Slightly soluble			
Calcium sulfate (Gypsum)	$CaSO_4 \cdot 2H_2O$	0-0-0	17
Insoluble			
Elemental sulfur	S	0-0-0	88–100

4:7 □ MICRONUTRIENTS

Micronutrients are so called because plants use them in only minute quantities. Only traces of some forms of the micronutrients essential for plants are found in the soil. These elements are boron, copper, zinc, and molybdenum. Two micronutrients occur abundantly in the earth's crust as follows: iron, 4 percent, and manganese, 0.1 percent. Chlorine is also abundant, especially in arid regions.

Micronutrients, especially copper, zinc, manganese, and molybdenum, act mostly as **catalysts**. Catalysts are highly important agents for speeding up reactions. They are not used up in the process, but they must be present, otherwise the chemical reaction will not take place. In this role they are vital to the growth and health of all living things. A deficiency or one of more micronutrients may occur under these conditions:

1. The soil may be naturally low in a given element. This is generally true of boron in certain soils in humid regions.
2. The solubility of the soil's supply of some element may be reduced to the point of deficiency by overliming. This sometimes occurs with manganese, iron, boron, zinc, copper, and phosphorus.

3. The nature of the crop may be such that it has extra need for a particular element. This seems to be true of pecan trees, for example, which need relatively large amounts of zinc.
4. The character of the soil may be such that plants growing on it require some special elements, such as in peat soils, on which more manganese, zinc, molybdenum, and copper are usually needed.
5. The acidity of the soil may be so high that it interferes with the solubility of an element. This applies to molybdenum.
6. The soil may be so alkaline that it makes certain elements almost insoluble. Plants growing on such soil are commonly lacking in available iron.
7. The soil may contain only mere traces of all of the micronutrient elements. This tends to be true of the sandier soils, notably in Florida.

Unsatisfactory plant growth in many areas is traceable to the lack of one or more of these micronutrients. Lack of boron, for example, may lower the yields of alfalfa and other crops. Deficiencies of boron can be conveniently corrected by adding it to commercial fertilizers. In some cases, micronutrients are also effectively used as separate materials applied directly to the soil and as a spray or dust applied to the growing crop. With some of these elements, the range between beneficial and detrimental amounts is very narrow, so it is very important that they are used according to directions given by an expert.

Other elements, such as sodium, cobalt, vanadium, and silicon, affect plant growth, although they are not now classified by most scientists as essential plant nutrients. This is especially true of sodium which, for some crops and some conditions, seems to serve a specific function of its own in promoting plant growth. For other crops and other conditions sodium has the ability to substitute for a portion of the potassium requirement.

4:7.1 □ Boron (B)

The soluble form of boron in soils, an anion, is easily leached. It is more easily leached from sandy soils than from fine-textured clay soils. Low moisture levels reduce boron availability. For this reason, boron deficiency, especially in alfalfa, peanuts, and cotton, is most frequent under drought conditions.

Application rates are usually in the range of 0.2 to 3.0 pounds of boron (B) per acre (0.22–3.36 kg/ha) with the lower rates for the most sensitive crops such as beans, corn, and cotton. Sugar beets and alfalfa require higher rates. Soil application is preferred to foliar sprays except for crops which have routine spray programs.

Toxicity can be a serious problem when boron is applied excessively. For example, application rates adequate for sugar beets can cause the following crop of corn or beans to fail because of boron toxicity. Analyses of plants, soils, and irrigation water are the best means of determining boron levels and needs.

The major forms of boron in fertilizers are shown in Table 4.7.

Table 4.7—Fertilizer Sources of Boron (B)

Fertilizer	Formula (Form)	Percent Boron (B)
Boron frits	Variable	10–17
Borax	$Na_2B_4O_7 \cdot 10H_2O$	11
Boric acid	H_3BO_3	17
Sodium pentaborate	$Na_2B_{10}O_{16} \cdot 10H_2O$	18
Sodium tetraborate		
Fertilizer borate—46	$Na_2B_4O_7 \cdot 5H_2O$	14
Fertilizer borate—65	$Na_2B_4O_7$	20
Solubor	$Na_2B_{10}O_{16} \cdot 10H_2O$	20

4:7.2 □ Copper (Cu)

Copper may occur in soils as metallic copper (Cu^{2+}), insoluble salts, water-soluble copper compounds, adsorbed by clay colloids, and as copper organic compounds. Total amounts of copper in soils vary widely. Organic soils generally contain relatively high amounts of copper. However, it is held so tightly by the organic fraction that relatively small amounts of copper are available for plant growth.

Highly organic soils, such as peat and muck soils (Histosols), are often deficient in available copper. Sandy soils, which are low in organic matter content, do not hold copper against leaching loss. Therefore, these soils are generally more deficient in copper than are fine-textured clay soils.

The copper requirements of crops are quite small, and the most common fertilizer source is copper sulfate. Copper fertilizers are available in inorganic or organic forms. Copper can be broadcast or banded in soils or applied as a foliar spray. Broadcasting is the principal method of application. Applications of recommended amounts usually last for five to eight years depending on the crop and soil. The copper content of several fertilizer sources is given in Table 4.8.

Copper toxicities can build up on clay soils as a result of the use of copper fertilizer or copper-containing fungicides. Copper does not leach in fine-textured soils, and the crop requirement is small; consequently, excess levels of copper can build up in clay soils and remain for long periods of time. Toxic levels of cop-

Table 4.8—Fertilizer Sources of Copper (Cu)

Source	Formula (Form)	Percent Copper (Cu)
Copper sulfate	$CuSO_4 \cdot 3Cu(OH)_2$	13–53
Cuprous oxide	Cu_2O	89
Cupric oxide	CuO	75
Copper acetate	$Cu(C_2H_3O_2)_2 \cdot H_2O$	32
Copper oxalate	$CuC_2O_4 \cdot \frac{1}{2}H_2O$	40
Copper ammonium phosphate	$Cu(NH_4)PO_4 \cdot H_2O$	32
Copper chelate	$Na_2CuEDTA$	13

per reduce or eliminate seed germination and possibly reduce the availability of zinc, phosphorus, and iron.

Recommended rates of copper fertilization should be followed closely. When a cumulative total of 30 pounds per acre (34 kg/ha) of copper has been applied, applications should be discontinued. Further yield increases due to additional copper are unlikely.

4:7.3 ◻ Iron (Fe)

Total iron in the soil is high but often it is not available, especially in alkaline soils. Iron deficiency generally appears as irregular areas in the field associated with areas of high soil pH, free calcium carbonate, and low organic matter. The iron-deficient areas frequently have had the topsoil removed by erosion or leveled for irrigation.

Severe iron deficiency in a soil may require switching from a sensitive crop such as sorghum or beans to a more tolerant crop such as wheat or alfalfa.

Soil application of iron-containing compounds for the elimination of iron deficiency has not been economically possible on a field scale. Iron **chelates** that will correct iron deficiency are available, but they are usually too expensive to use on a field crop. (Chelates are strongly bonded metals such as copper, iron, manganese, and zinc with organic compounds, resulting in availability over a wide pH range.) Correcting iron deficiency by applying a soluble iron salt, such as iron sulfate, is not very effective on high pH soils. Soluble iron salts are rapidly converted to insoluble iron compounds; thus, they are unavailable to plants.

Applying iron as a foliar spray (Figure 4.7) is very effective in restoring the green color to the plants, but it may not restore top yields. Iron sprays are most ef-

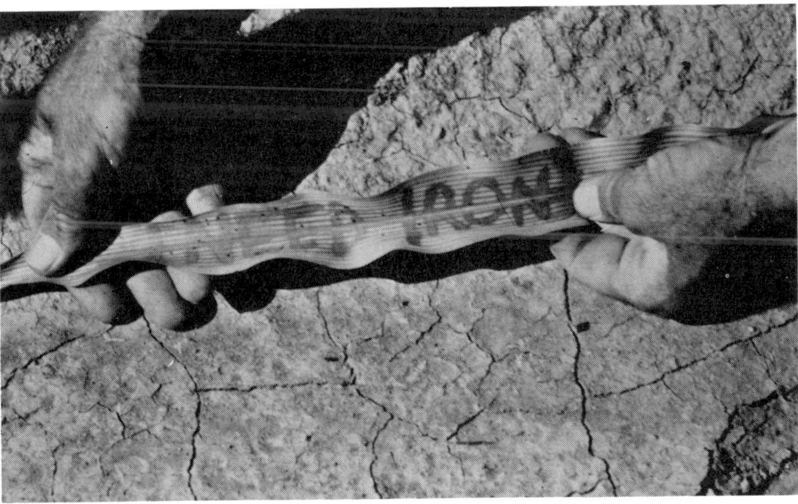

Fig. 4.7—Foliar applications of iron give quick response. The brush marks show where a solution of iron was painted to illustrate immobility of iron in iron-deficient plants. (Courtesy, W. W. Harris, Fort Hays State University)

fective when applied to young plants and when repeated at 10-day to 2-week intervals. Iron sulfate or iron chelates can be used effectively as a spray.

Manure and sewage sludge applications are quite effective in eliminating iron-deficiency problems when applied at the rate of about 20 tons per acre (44.8 mt/ha). (See Chapter 5.)

4:7.4 ▫ Manganese (Mn)

Manganese reactions in soils are quite complex. The organic matter content and the soil pH as well as moisture content, temperature, and aeration have considerable influence on manganese availability to plants. Deficiencies are observed in the eastern half of the United States on sandy, gravelly, highly calcareous, or overlimed soils. Manganese deficiencies usually occur on well-drained soils. Drainage of a previously wet area may result in a decrease in available soil manganese levels.

Certain crops such as small grains, soybeans, and sweet potatoes appear to be more sensitive to manganese requirements.

Manganese sulfate is the most common carrier of manganese. It is very soluble and can be used as a soil or foliar treatment. Chelated forms are effective when foliarly applied but not as effective as manganese sulfate when soil-applied. The manganese concentration of most common carriers is presented in Table 4.9.

Table 4.9—Fertilizer Sources of Manganese (Mn)

Source	Formula (Form)	Percent Manganese (Mn)
Manganese sulfate	$MnSO_4 \cdot 3H_2O$	26–28
Manganese oxide	MnO	41–68
Manganese carbonate	$MnCO_3$	31
Manganese chloride	$MnCl_2$	17
Manganese oxide	MnO_2	63
Manganese chelate	MnEDTA	12
Manganese frits	Variable	10–25

Applying 5 pounds per acre (5.6 kg/ha) of manganese in the row can correct most deficiencies. One pound per acre (1.1 kg/ha) of elemental manganese (Mn) (4 pounds per acre [4.5 kg/ha] of manganese sulfate) or 0.15 pound per acre (0.17 kg/ha) of elemental manganese in the chelate form in 20 gallons (76 liters) of water is recommended for foliar applications. For severe deficiencies a second spray may be needed. Foliar applications should be considered as emergency measures.

Row applications of manganese are more effective. The lower pH near the band as a result of the acidifying effects of the banded fertilizer increases effectiveness of row-applied manganese. For broadcast application to be as effective as a row application, 5 to 10 times as much manganese must be applied.

4:7.5 □ Molybdenum (Mo)

Molybdenum is one of the newest micronutrients added to the list required for plant growth. Nodule bacteria on roots of legumes also need this element in the fixation of atmospheric nitrogen.

Molybdenum deficiency occurs most frequently in leached, acid soils. Therefore, liming an acid soil is one way to reduce molybdenum deficiency, provided the soil contains sufficient total quantities of the element.

Molybdenum may be applied as a seed treatment, soil application, or foliar spray. Row application with fertilizers is also satisfactory. Molybdenum applications of as little as 2 ounces (57 gm) per acre of Mo equivalent may be adequate to supply sufficient molybdenum for normal plant growth. Some molybdenum sources are given in Table 4.10.

Excess molybdenum can be toxic to seedlings and livestock and can adversely affect uptake and movement of iron in the plant.

Table 4.10—Fertilizer Sources of Molybdenum (Mo)

Source	Formula (Form)	Percent Molybdenum (Mo)
Ammonium molybdate	$(NH_4)_6Mo_7O_{24} \cdot 2H_2O$	54
Molybdenum trioxide	MoO_3	66
Molybdenum frits	Variable	30
Sodium molybdate	$Na_2MoO_4 \cdot 2H_2O$	39

4:7.6 □ Zinc (Zn)

Zinc deficiencies have been noted in nearly all states and are frequently associated with areas where the topsoil has been removed by leveling for irrigation, by erosion, or by terrace channel construction. Soils that are deficient in zinc are frequently low in organic matter, are sandy, and have an alkaline soil pH. Very high available soil phosphorus levels produced by overfertilization or too much native phosphorus in the soil may induce a severe zinc deficiency on soils low in available zinc (Figure 4.8).

Crop sequence can also bring on zinc deficiency. Corn and beans are often zinc-deficient following a crop of sugar beets unless the corn and beans are fertilized with zinc. As is the case with most micronutrients, zinc-deficient areas are usually irregular and spotty in fields.

Zinc deficiency can be readily corrected by the application of zinc fertilizer. Zinc fertilizer applications can be either banded at planting or broadcast preplant, with little difference in response when applied at an adequate rate. Foliar applications are effective in making foliage a darker green but are not as effective in producing yield response and should be considered as a salvage measure. Manure and sewage sludge applications are quite effective in eliminating zinc deficiency when applied at the rate of about 20 tons per acre (44.8 mt/ha).

Fig. 4.8—On a soil with borderline zinc levels, zinc deficiency was induced by phosphorus applied *in the row* (left two rows). The same amount of phosphorus *broadcast* (right two rows) did not cause zinc deficiency. (Courtesy, Kansas State University)

The most common inorganic source is zinc sulfate, which is considered an excellent source of zinc whether applied as a granular material in a bulk blend or incorporated into solid or fluid fertilizers. Zinc oxide is another inorganic source of zinc that is effective when it is incorporated into solid or fluid fertilizers. Application of granular zinc oxide on calcareous soils should be avoided due to its low solubility. Organic zinc sources (i.e., chelates) also are available and are generally more effective per pound of zinc than the inorganic sources. The most common sources of zinc fertilizers are listed in Table 4.11.

Table 4.11—Fertilizer Sources of Zinc (Zn)

Source	Formula (Form)	Percent Zinc (Zn)
Zinc sulfate	$ZnSO_4 \cdot H_2O$	36
Zinc oxide	ZnO	78
Zinc carbonate	$ZnCO_3$	56
Zinc chelate	$Na_2ZnEDTA$	14

4:7.7 □ Chlorine (Cl)

Chlorine is found almost universally in soils and plants. It is considered to be a minor constituent of the **lithosphere** (earth's crust), as the earth's crust contains an average of only 0.05 percent chlorine. The majority of chlorine in soils is believed to originate from salts trapped in soil parent material, from volcanic emissions, and from marine **aerosols** (airborne particles). Most of the soil chlorine commonly occurs as soluble salts, such as NaCl, $CaCl_2$, and $MgCl_2$. Chlorine is sometimes the principal anion in extracts of saline or sodic soils. The quantity of chlorine in soil solutions may range from less than 0.5 to over 6,000 ppm.

Since chlorine ions are easily leached out, the chlorine content of well-drained soils is usually very low. Chlorine is one of the first elements to be removed in the process of weathering, and this is the reason why most of the world's chlorine is either in salt deposits or in the oceans.

It was not until the mid-1950's that this element was shown to be essential for plant growth. Chlorine can be readily transported in plant tissues. A useful function for chlorine is to act as the counterion during rapid potassium **fluxes** (changes), thus contributing to the **turgor** (normal swelling) of leaves and other plant parts.

The following is a list of materials that contain chlorine:

Ammonium chloride	66% Cl
Calcium chloride (deicing salt)	65% Cl
Magnesium chloride	74% Cl
Potassium chloride (muriate of potash)	47% Cl
Sodium chloride (common table salt)	60% Cl

Another source of chlorine is rainwater. Estimated levels of chlorine in inland areas are about 10 pounds per acre (11 kg/ha) per year. Coastal areas can receive hundreds of pounds of chlorine per acre per year. Chlorine in the air may come from industrial sources or sea sprays. Irrigation water from wells, streams, or rivers may also contain chlorine.

4:8 □ MULTINUTRIENT FERTILIZERS

Multinutrient fertilizers refer to any fertilizer containing two or more plant nutrients. These fertilizers may be primary chemical compounds, physical mixtures of chemical compounds, or multinutrient slurry formulations. Primary chemical compounds are generally referred to as **materials** and physical mixes as **blends**. Multinutrient fertilizers may be solids or fluids. The fluids consist of true solutions or suspensions.

Present technology allows the manufacture of multinutrient fertilizers to fit almost any desired combination and ratio of nutrients, as determined by a soil test.

The most common classes of multinutrient fertilizers are:

- **Granulated formulations**—Granulated formulations are solid materials in which nutrients are fairly evenly distributed in each pellet. These materials are presently being formulated into uniform granules with good physical properties. They have the advantage of being rather homogeneous and having desirable physical properties.

- **Bulk-blends**—Bulk-blends are solid materials containing varying proportions of at least two of the major nutrients. They are physical mixes of dry fertilizers and are not as homogeneous as granulated formulations. Their primary advantage is that they are cheaper to produce and can be custom-mixed at any mixing plant to a wide variety of grades.
- **Fluid formulations**—There are two classes of fluid fertilizers: **clear liquids** and **suspensions**. The clear liquids are *true solutions* that are lower in analysis and easier to apply than suspensions. Suspensions are *supersaturated solutions* that settle out if they are not stirred constantly. For this reason, suspensions usually have fine clay added to stabilize them, but they must still be stirred constantly even while being applied, to reduce settling out.

4:9 □ SOIL TESTING

A soil test is a measure of the nutrient-supplying power of the soil. If the soil tests "very high for all nutrients," the soil is capable of supplying all nutrients except a small amount as a **pop-up** (starter). When the soil tests "very low," almost all of the nutrients must be supplied as fertilizer. (See Section 4:10.5.)

High crop yields are possible only when all essential elements are present in the soil in the right proportions and in adequate and available amounts at all times. Such a balance between the nutrient needs of the crop and the ability of the soil to supply the nutrients at the right time is usually very difficult to maintain. Soil testing and plant nutrient deficiency symptoms aid in diagnosing the cause of low crop yields and in maintaining the proper plant nutrient balance. Soil tests help to estimate the supply of available nutrients, while plant deficiency symptoms indicate extreme shortages of certain essential elements.

Valuable information can be obtained from testing soil and from observing plant deficiency symptoms. But these techniques cannot supply answers to unsatisfactory plant growth when the cause is the wrong variety, wrong planting time, dry weather, wet weather, compacted soils, low temperatures, high temperatures, diseases, insect damage, air pollution, or excessive amounts of herbicides.

4:9.1 □ When to Test

For general rotation crops, soils should be tested at least once every three years. There is a distinct advantage in testing the last sod year of the rotation because lime, if needed, can be applied before **breaking** (turning under) sod. This not only makes for easier spreading and less soil compaction but also provides ample time for the lime to become mixed in the surface soil before acid-sensitive legumes are seeded.

Soils used intensively, such as in gardens and greenhouses or used for high-value field crops, should be tested prior to planting each crop.

4:9.2 □ Sample Carefully

Lime and fertilizer recommendations can be no better than the samples tested. Poor samples result in fertilizer recommendations that are misleading and costly. Good samples result in reliable recommendations which can save money.

Since the small amount of soil sent to the laboratory for testing must indicate the needs of 2 to 3 million pounds (0.9–1.4 million kg) of soil in each acre that will eventually be treated with fertilizer, it is obvious that the sample of soil must be carefully taken and must be representative of the soil in that field. In order for the sample to be representative of the area that is tested, a number of cores—one from each of at least 10 to 15 sites, and preferably from at least 20 to 30 sites within the areas—should be collected (Figure 4.9). As these cores or slices from

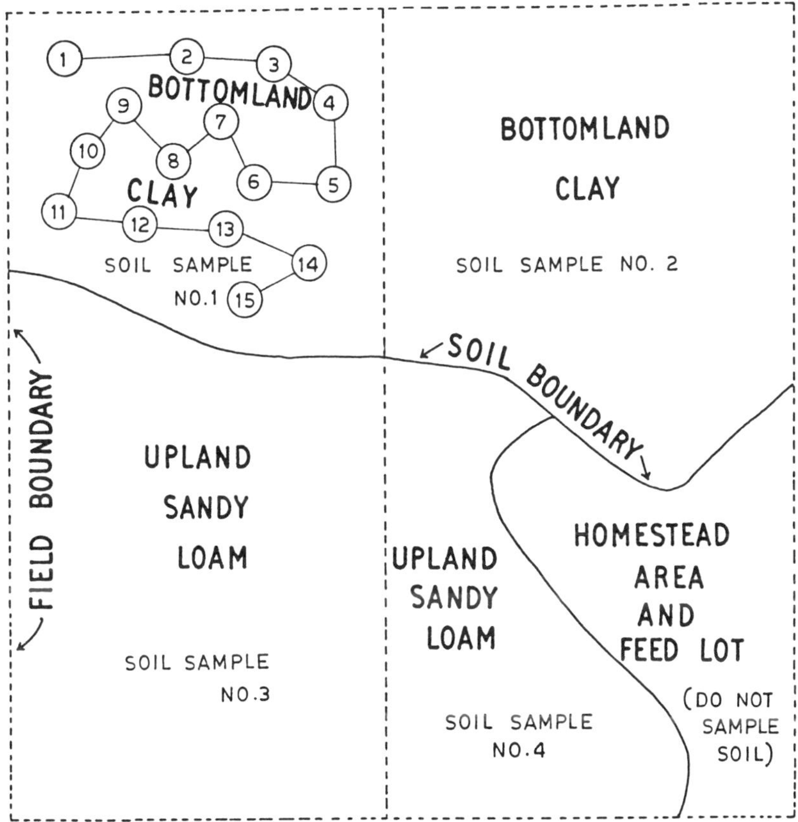

Fig. 4.9—An example of how a field may be divided for the collection of soil samples for chemical testing. *Top left:* Bottomland clay soil, showing 15 spots from which a subsample is collected, mixed, and a 1-pound sample is put in a box or bag, labeled "No. 1," and sent to the soil testing laboratory. *Top right:* Another bottomland clay soil, but the field has been farmed separately. This will be sampled the same way as the first field and the sample labeled "No. 2." *Bottom left:* Upland sandy loam field will be sampled in the same way and labeled "No. 3." *Bottom right:* Upland sandy loan is the same soil as "No. 3" but the cropping and fertilizing have been different. It is sampled in a similar manner to all other fields and labeled "No. 4." Homestead areas, feedlots, and unusual areas that do not represent an entire field are not to be sampled. (Courtesy, Michigan State University)

the area are obtained, they should be placed into a clean, plastic bucket and thoroughly mixed. This mixing of the individual cores taken from within each sampling area is necessary to prepare a composite sample which is the average of the entire area. When the sample is thoroughly mixed it should be placed in a bag or box which should be labeled. (Special containers may be obtained from your state soil testing laboratory.) After the sample is ready and the information sheet (furnished by the local county extension office) is completed, they should be taken or sent to the local county extension office for further processing (Figure 4.10). Any agricultural education teacher or county agricultural agent will be able to supply any further information which may be needed.

Fig. 4.10—*Top left:* Tools needed for soil sampling. *Top right:* Using a shovel to cut a soil slice. *Center left:* Reducing the sample. *Center right:* Using a soil probe. *Bottom left:* Mixing the samplings with the hands in a plastic (not metal) bucket. *Bottom right:* Information sheets (an essential item in the soil testing program) are available at county extension offices. Each laboratory has its own container or has instructions for the proper type to be used. (Courtesy, Roy Hunter Follett)

If special containers are unavailable, any clean pint plastic or ice cream container that can be tightly closed should prove satisfactory. Rusty or otherwise contaminated metal cans should not be used—any foreign material may affect the soil test.

Changes in nutrient availability in the soil samples, particularly phosphorus, have resulted from the soil samples being artificially dried prior to the analysis. Samples should not be forced to dry by being placed in the sun, on radiators, or inside conventional or microwave ovens before being submitted to the laboratory. Wet soil samples should be spread out and allowed to dry in the shade or in a room at normal room temperature. Forced drying may change the test results.

4:9.3 □ Provide Complete Information

The more complete the information provided, the better the fertilizer recommendation will be.

The information sheet must be completely filled in and must accompany the soil sample. The package should be labeled with the owner's name and address and the field number. Because ink will smear if it gets wet, the package should be marked with either a lead pencil or a wax marker.

Information usually requested on the information sheet follows.

1. Previous crops grown during one cycle of the rotation.
2. Crop or crops to be grown.
3. When the field was last limed and fertilized and the rate of application of each.
4. Whether the field will receive manure or sewage sludge for the crop being grown, and the amount.
5. Depth of plowing.
6. Soil series or soil management group.
7. Whether drainage is good, poor, or intermediate.
8. Yield goal.
9. Whether irrigation is to be used.
10. Special problems or conditions such as deficiency symptoms noticed.

Most of the 50 states in the nation have a state-supported soil testing service (Figure 4.11). In addition, many states have county laboratories which operate as branch services for the central laboratory. There are also many private soil testing services (Figure 4.12).

Based upon the soil tests, lime and fertilizer recommendations are made in a variety of ways from state to state. In some states, one person at the central laboratory makes all of the recommendations; in others, the county extension agents make them.

4:9.4 □ Be Sure to Have Soil Tested

Amounts of previously applied soil phosphorus and potassium, as deter-

Fig. 4.11—Soil testing is a science that utilizes some of the most modern and sophisticated analytical equipment available on the market today. For example, it is necessary to use an atomic absorption spectrophotometer to analyze for zinc and copper contents in soils. (Courtesy, The Ohio State University)

Fig. 4.12—The soil sampler in this picture is quite useful for fields that can be driven across by a vehicle. A few commercial soil testing laboratories offer a soil sampling service to producers. (Courtesy, Harris Laboratories, Inc., Lincoln, Nebraska)

mined by soil tests, are important sources of these elements to plants. The residual supply of available soil phosphorus and potassium will govern, in part, how much of these elements to apply as fertilizer to the current crop to maintain adequate fertility and proper nutrient balance.

Certainly, the safest way to avoid nutrient deficiencies is to have the soil tested. This is as important a practice as ensuring adequate drainage or weed control. Haphazard selection and application of a fertilizer will lead to unbalanced fertility conditions which will promote excessive uptake of some elements and too little of others, resulting in low yields and, consequently, low farm income.

4:10 □ FUNDAMENTALS OF FERTILIZER APPLICATION

Soil tests indicate the relative fertility level of the soil. Where soil fertility levels are high, only small amounts of commercial fertilizers may be needed for maximum crop yields. Where the fertility level of the soil is low, fertilizers must furnish a major portion of the nutrients needed by the crops, and large amounts of fertilizers must be applied if maximum yields are to be attained.

Maintaining soil fertility at medium or high levels is a desirable long-term soil fertility program goal. Continued high crop yields are most certain when soils contain abundant phosphorus and potassium along with well-balanced supplies of other nutrients.

Fluid fertilizers containing N or combinations of N, P, and K have the same effect on plant growth as equivalent amounts of ***dry fertilizers*** when applied at equivalent rates and when similar placement is used. However, dry fertilizers are more economical (Figure 4.13).

Fig. 4.13—Fluid fertilizer is ideal for top-dressing many types of crops. Shown here is a flotation-type truck that can be used for applying fluid (liquid) fertilizer to grain sorghum residue prior to plowing or discing. The large tires reduce the hazard of compacting the soil. (Courtesy, Ag-Chem Equipment Co., Minneapolis)

4:10.1 □ Methods of Fertilizer Application

Fertilizer application methods include broadcasting, top-dressing, side-dressing, row placement, "pop-up" (at time of planting), foliar, and soil injection.

Fertilizers should be applied so that: (1) growing plants can use the fertilizer efficiently, (2) there is little or no injury to plants from fertilizer, and (3) the application can be accomplished as quickly and easily as is economically possible.

The greatest hazard in fertilizer placement is the danger of injury to the germinating seedlings. Nitrogen and potassium salts in fertilizers are mainly responsible for fertilizer injury. Phosphorus is not as "salty" and moves very slowly from the point of placement. Even though it may be applied in the water-soluble form, phosphorus is rapidly immobilized by soil conditions once it is applied to the soil.

Nitrogen salts move up and down in the soil, depending on the direction of water movement. Nitrate nitrogen moves more readily in the soil than does ammonium nitrogen because nitrates are not held (adsorbed) on clay or humus surfaces as are ammonium ions.

The rate of potassium salt movement in the soil is somewhere between that of phosphorus and ammonium nitrogen.

Soil water movement is largely vertical. If the soil is wet (above field capacity) the movement will be largely downward. As the soil dries, water moves upward by capillary action and carries with it the dissolved fertilizer salts, which may be deposited on the surface just above the fertilizer band. This may appear as a white or brown deposit on the soil surface. A rain immediately after planting, followed by a long dry period, promotes fertilizer injury. The dissolved fertilizer salts move upward near the roots of the germinating seedlings. If the concentration of dissolved fertilizer salts is high enough, water will be pulled out of the plant cells through **exosmosis**. This causes a "firing" at the tips of the plant leaves. The plant may die if the damage is severe enough and if the weather remains dry. Death is due to a loss of water from plant cells. The plant dries out, and it appears as if it had been placed in an oven.

4:10.2 □ Broadcast Applications Before Planting

Broadcasting fertilizer refers to the spreading of fertilizer by use of spreader trucks (Figure 4.14) or tractor-drawn farm equipment.

Advantages of broadcast applications include:

1. Large amounts of fertilizer can be applied without danger of fertilizer injury.
2. Labor and time are saved, particularly if the fertilizer is custom-applied.
3. Planting operations are speeded up because time-consuming fertilizer placement equipment is eliminated.
4. Cost of bags is eliminated if bulk fertilizer is used.

Fig. 4.14—Broadcasting dry fertilizer with a "Big A" flotation-type fertilizer spreader. (Courtesy, Rickel Manufacturing, Inc., Salina, Kansas)

Disadvantages of broadcast applications include:

1. The possibility of soil phosphorus fixation is increased. This is more evident on fine-textured clay soils low in phosphorus.
2. There is a dilution effect on the fertilizer applied. Crops may start off slower due to a less concentrated amount of fertilizer elements near the young plant roots. This is more evident on shallow-rooted, short-season, cool-season crops, such as Irish potatoes or onions.
3. The possibility of weed competition is increased, but the use of herbicides eliminates this problem.

4:10.3 □ Top-Dressing on Established Stands

Meadows (areas of forage cut for hay) and pastures may be top-dressed with phosphorus and/or potassium at any time during the year. However, fertilizers applied to frozen ground may be lost to the field and contaminate streams if a heavy rain should occur soon after application.

Winter grains may be top-dressed with nitrogen or with complete fertilizers where phosphorus and/or potassium are needed in addition to nitrogen. These top-dressing applications should be made during late winter or early spring.

4:10.4 □ Side-Dressing of Crops

Side-dressing is one way to apply additional nitrogen to corn. In general, plow-down or row applications of nitrogen are preferable to side-dressing. For side-dressing corn with nitrogen, the fertilizer is applied in bands 3 to 4 inches

(8–10 cm) below the soil surface in the *middle* of the corn rows. This application should be made when the corn is 6 to 12 inches (15–30 cm) tall. Deep side-dressing close to the corn plants will damage corn roots and may reduce yields.

The advantage of side-dressing is that the fertilizer can be applied close to the time of greatest need by the crop. This is a good method of applying N on sandy soils.

The disadvantage of this method is that wet soil may prevent timely application, and when the weather is dry after side-dressing, the roots may not be able to absorb the N if it is placed above the root absorption zone.

4:10.5 □ Pop-Up

Small amounts of fertilizer applied in contact with the seed at the time of seeding are called pop-up. The term *pop-up* is a misnomer because this method may hinder or help germination (Figure 4.15). If starter fertilizer is placed in di-

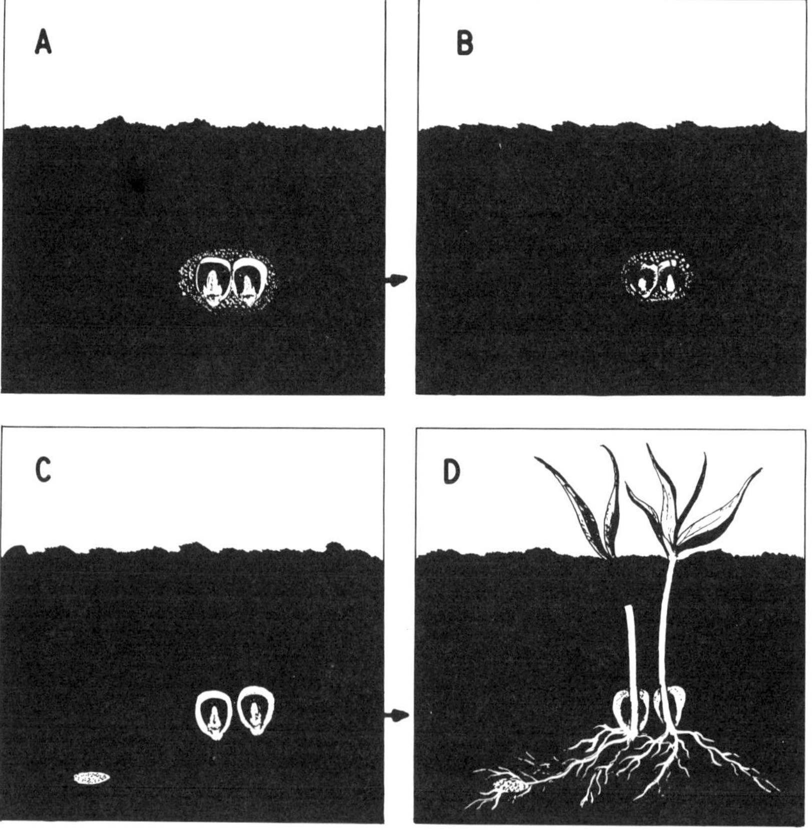

Fig. 4.15—When fertilizers are placed with the seed, as in "A," germination is hindered, as seen in "B." However, when fertilizer is placed about 2 inches (5 cm) to one or both sides and 2 inches (5 cm) below the level of the seed ("C"), germination is normal and early growth is enhanced, as in "D." (Courtesy, Potash and Phosphate Institute)

rect contact with the seed, the amount of nitrogen plus potash ($N + K_2O$) should be limited to no more than 10 pounds per acre (11 kg/ha) for row crops and 20 pounds per acre (22 kg/ha) for small grains.

4:10.6 □ Row Placement

Row placement (also called band placement) involves placing the fertilizer to one side and below the level of the seed. Research has shown that on row crops, placing moderate amounts of fertilizer in a band 2 inches (5 cm) to one side and 2 inches (5 cm) below the seed level will not generally cause injury to the germinating seed.

The main advantage of row placement of fertilizers lies in supplying readily available fertilizer to the plant, enabling it to grow faster. The method of placement also cuts down the amount of soil fixation of fertilizer elements. This is particularly true of phosphorus on fine-textured clay soils. Row placement also reduces the possibility of weed competition, since the fertilizer is concentrated in a narrow band near the crop plant.

Disadvantages of band placement include:

1. The hazard of fertilizer injury exists if large amounts are placed too close to the seed.
2. Time and labor involved are high.
3. Special equipment is needed.

4:10.7 □ Foliar Application

Foliar application of N, P, and K solutions is not usually recommended for farm crops. Foliar sprays of the primary plant nutrients are seldom practical because of the large amounts needed and their greater cost. Micronutrients and some of the secondary nutrients can be satisfactorily applied in solutions to the foliage since only small amounts are needed, and they can usually be added to an insecticide or a fungicide and applied simultaneously.

4:10.8 □ Soil Injection

Anhydrous ammonia, a liquid under pressure and a gas in the atmosphere, must be injected into the soil (Figure 4.16). Because ammonia vaporizes quickly, it must be injected directly into the soil and sealed by a press wheel. As soon as liquid ammonia is released in the soil, it changes from a liquid to a gas and then reacts immediately with soil moisture to form ammonium hydroxide ($NH_3 + H_2O \rightarrow NH_4OH$). Ammonium ions ($NH_4^+$), which have a positive charge, split off and are attached and held by negatively charged clay and organic colloids in the soil. The ammonium so absorbed is available to plants.

Anhydrous ammonia application is easy to combine with other tillage operations. The injection equipment can readily be adapted to the simultaneous appli-

Fig. 4.16—Application of anhydrous ammonia as a source of nitrogen in a fertility program. This material is in a gaseous form and is kept in pressurized tanks. It is released into the soil at a depth of 4 to 6 inches (10–15 cm) by chisel-type blades. (Courtesy, Roy Hunter Follett)

cation of ammonia during other farm operations, such as plowing, discing, and planting. With the excellent modern equipment that industry has developed, applying anhydrous ammonia is a quick labor-saving and cost-effective method of applying nitrogen.

4:10.9 □ Nitrogen Solutions

The most popular nitrogen solution contains 28 percent N and is a mixture of urea and ammonium nitrate in water. It has no ammonia vapor pressure and is generally sprayed or dribbled on the soil surface. The loss of N from surface application of 28 percent N solution is generally not considered to be too great. There are conditions however that N loss might be quite serious. If the conversion of urea in the liquid fertilizer to ammonia (NH_3) takes place on the soil surface, some NH_3 may be lost. The remainder of the ammonia may react with water on the surface to produce an alkaline condition which also promotes NH_3 loss. Conditions which are most favorable for N loss from surface-applied urea (solid or liquid) are moist and slightly acid soils, warm temperatures, and sandy soils with low cation exchange capacities. Either tillage or rain can move the surface-applied urea into the soil before it is converted to NH_3, thus reducing N losses.

4:11 ▫ CALCULATION OF FERTILIZER COSTS

Comparisons of fertilizer costs and fertility values, in general, are most reliable when based on the cost per unit of available plant nutrients such as cents per pound of N, P_2O_5, or K_2O. Where liquid fertilizers are priced by the volume, barrel, or other volume measure, the buyer must know the weight per unit volume as well as the percent of available nutrients in order to calculate the cost per unit of nutrient.

4:11.1 ▫ Determining Nutrient Cost in Straight Fertilizers

When purchasing fertilizer, the two most important things on the label to consider are the net weight and the grade (guaranteed analysis). Fertilizer prices should be compared on the basis of cost per pound of nutrients rather than cost per ton of material. The cost per pound of N, P_2O_5, or K_2O in straight fertilizer materials can be determined as follows:

$$\frac{\text{Price / ton}}{\%\ \text{nutrient} \times 20} = \text{Cost / lb. of nutrient}$$

For example, if we assume that ammonium nitrate (33.5-0-0) sold for $167.50 per ton, then the cost per pound would be:

$$\frac{\$167.50}{.335 \times 20} = 25¢\ /\ \text{lb. of N}$$

If concentrated superphosphate (0-45-0) sold for $225.00 per ton, then the cost per pound would be:

$$\frac{\$225.00}{.45 \times 20} = 25¢\ /\ \text{lb. of } P_2O_5$$

And muriate of potash (0-0-60) selling for $144.00 per ton would cost per pound:

$$\frac{\$144.00}{.60 \times 20} = 12¢\ /\ \text{lb. of } K_2O$$

4:11.2 ▫ Determining Comparative Costs of Multinutrient Fertilizers

Once the cost per pound of N, P_2O_5, and K_2O in straight fertilizers has been determined, their relative cost in mixed fertilizer can be compared as follows:

$$(\%\ \text{N} \times \text{cost / lb. N}) + (\%\ P_2O_5 \times \text{cost / lb. } P_2O_5)$$
$$+ (\%\ K_2O \times \text{cost / lb. } K_2O) = \text{Cost / 100 lb.} \times 20 = \text{Cost / ton}$$

For example, if we assumed that a 5-20-20 grade fertilizer sold for $178 per ton, then the cost could be compared as follows:

$$(5\%\ N \times 25¢\ /\ lb.\ N) + (20\%\ P_2O_5 \times 25¢\ /\ lb.)$$
$$+ (20\%\ K_2O \times 12¢\ /\ lb.) = \$8.65\ /\ 100\ lb. \times 20 = \$173\ /\ ton$$

Thus, by this technique the cost of various multinutrient fertilizers can be compared on the basis of the cost per pound of nutrient. According to the calculations, the selling price of $178 per ton is reasonable for the 5-20-20 grade fertilizer.

4:12 □ REFERENCES

Association of American Plant Food Control Officials, Inc. *1989 AAPFCO Official Publication No. 42.* Virginia Dept. of Agriculture, Richmond.

Bundy, L. G. "Soil and Applied Nutrients." University of Wisconsin, Fact Sheet No. A 2519, 1985.

Cook, Ray L., and Boyd G. Ellis. *Soil Management: A World View of Conservation and Production.* New York: John Wiley & Sons, Inc., 1987.

Follett, R. F., *et al.* "Soil Fertility and Organic Matter as Critical Components of Production Systems." SSSA Special Pub. 19, Madison, Wisconsin: Soil Science Society of America and the American Society of Agronomy, 1987.

Jackson, T. L., ed. "Chloride and Crop Production." Atlanta: Potash and Phosphate Institute, Special Bul. No. 1, 1984, 108 pp.

Mengel, D. B. "Types and Uses of Nitrogen Fertilizers for Crop Production." Purdue University, Co-op. Ext. No. AY-204, 1986.

Phosphorus for Agriculture: A Situation Analysis. Atlanta: Potash and Phosphate Institute, 1986, 90 pp.

Sanford, J. O., and J. E. Hairston. "Fertilizer Nitrogen vs. Legume Nitrogen for Corn." Mississippi State University, Res. Rpt. No. 24, 1986.

Schulte, E. E. "Soil and Applied Sulfur." University of Wisconsin, Fact Sheet No. A 2525, 1983.

Simson, C. R., and E. E. Schulte. "Sampling Soils for Testing." University of Wisconsin, Fact Sheet No. A 2100, 1982.

Soils Improvement Committee, California Fertilizer Association. *Western Fertilizer Handbook,* 7th ed. Danville, Illinois: The Interstate Printers & Publishers, Inc., 1985, 288 pp.

Spiva, C. P. "Micronutrients—A Growth Investment." *Solutions,* January 1987, pp. 30–34.

CHAPTER 5

Manures and Crop Residues

"If there be enough manure even an idiot will be a successful farmer." — Ancient Telugu Proverb (India)

OUTLINE

The original site of the Beltsville aeration-pile composting research and demonstration. (USDA's Beltsville Research Center near Washington, D.C.) (Courtesy, USDA – Agricultural Research Service)

□ □ □

5:1 □ OVERVIEW

The eight major sources of organic residues are: (1) crop residue, (2) animal manure, (3) sewage sludge, (4) logging and wood-manufacturing residue, (5) food-processing residue, (6) industrial organic residue, (7) municipal refuse, and (8) green manures. Minor sources of organic residues include septage, compost, and peat.

It is estimated that in the United States during an average year, a total of 802 million tons (728 million mt) of these seven kinds of organic residues are generated. About 54 percent of this total is crop residue and 22 percent is animal manure. However, a greater percentage of animal manure is used as a soil amendment.

Sewage sludge currently comprises only 0.5 percent of the total of all organic residues but only 23 percent of that is applied to the soil. The remaining four kinds of wastes have not been used extensively as soil amendments because of the high cost of collecting, processing, transporting, and applying; competition for other uses such as a fuel; high acidity or high alkalinity content; and the presence of nonbiodegradable material such as glass and / or toxic heavy metals.

5:2 □ USING CROP RESIDUES

There are about 383 million acres (155 million ha) of cropland in the United States, yielding an average of about 1 ton per acre (2.24 mt / ha) of crop residues. Only 68 percent of these residues are returned to the soil for soil protection and soil improvement. Sixty-eight percent of 1 ton = 0.68 × 2,000 = 1,360 pounds per acre (1,525 kg / ha) (dry-weight basis). This amount of crop residue, if actually left on each acre of cropland, would be adequate for protection against erosion on *loam* soils. Sandy soils would need 1,700 pounds per acre (1,906 kg / ha), but clay loams only 750 pounds per acre (841 kg / ha).

The probability of increasing the use of crop residues on soils in the near future is placed at the relatively low level of 5 to 20 percent (Table 5.1).

Before the "Conservation Decade" of the 1930's, crop residues were often burned or otherwise considered a waste. However, there was a demand for baled cereal straw for use as bedding in dairy barns and in feedlots.

F. L. Duley and J. C. Russell researched what they called **stubble mulching** in the 1930's and published their results in 1942. The technique consisted of developing an implement with several shallow sweeps that cut the small-grain stubble and weeds a few inches below the soil surface. This left almost all of the grain stubble on the soil surface as a mulch.

Table 5.1 — Annual Production of Organic Residues in the United States, Current Use on Soils, and Probability of Increased Usage[1]

Organic Residues	Total Production: Tons (Dry-Weight Basis)[2]	Total Production: Percent of Total of All Residues	Current Use on Soils	Probability of Increased Use on Soils
			(%)	(%)
Crop residue	430,892,000	53.7	68	5 to 20
Animal manure	174,920,000	21.8	90	55
Sewage sludge	4,367,000	0.5	23	20 to 50
Logging and wood manufacturing waste	35,698,000	4.5	5	<5
Food processing residue	3,198,000	0.4	13	5 to 20
Industrial organic residue	8,212,000	1.0	3	5 to 20
Municipal solid waste	144,934,000	18.1	1	5 to 20
Total	802,221,000	100.00		

[1] USDA. "Improving Soils with Organic Wastes. Report to the Congress in Response to Section 1461 of the Food and Agriculture Act of 1977 (PL 95–113)." Washington, D.C.: U.S. Government Printing Office, 1978, 157 pp.

[2] *Note:* Tons × 0.907 = mt.

Modern research and farm practices have been expanded to accomplish the same purpose as stubble mulching and are called by one of these 10 names: (1) conservation tillage, (2) mulch farming, (3) minimum tillage, (4) reduced tillage, (5) no-tillage, (6) zero tillage, (7) strip tillage, (8) sweep tillage, (9) plow-plant, (10) wheel-track planting.

Special benefits from using crop residues, especially as surface mulches, are as follows (see Chapter 6):

1. Less erosion by water and wind.
2. More infiltration of rain.
3. Greater water-holding capacity.
4. More snow captured (Figure 5.1).
5. More water available to plants.
6. Lower soil temperature in summer but higher in winter.
7. More soil humus.
8. Less surface crusting.
9. Fewer tillage pans.
10. Lower tractor fuel costs.
11. Better soil tilth with less tilling.
12. The same or higher crop yields.
13. More song birds and migratory ducks encouraged.
14. More beneficial animal life such as earthworms.

Note: Decomposing organic matter (humus) makes soil darker and therefore capable of absorbing more heat from the sun. However, this effect is counteracted by the characteristic of humus absorbing more water and therefore requir-

Fig. 5.1 — In cold winters in Minnesota, corn residues left on the soil surface result in as much as 65 percent *shallower* frost penetration. This means the soil under residues will warm earlier in the spring even though shaded by residues. (Courtesy, USDA - Soil Conservation Service. Reported in *Agricultural Research*, April 1987)

ing more heat units to raise the temperature. Therefore, humus has almost no effect on soil temperature.

Disadvantages of using a crop residue surface mulch include:

1. Not adapted for many north-temperate, poorly drained soils but recommended in Alaska and Minnesota because more snow is retained to reduce freezing depth and to supply more soil moisture.
2. More insects, diseases, weeds, and rodents.
3. Slower seed germination due to more wetness and coldness in the soil surface.
4. Higher costs for herbicides.

5. Slower seed germination due to water-soluble toxins in residues from sweetclover, corn, wheat, oats, and sorghums.
6. A wider carbon to nitrogen ratio and more nitrogen fertilizer is required; however, after a few years of minimum tillage, more nitrogen *is not required* because it is supplied by decomposing residues.

5:3 ▫ USING ANIMAL MANURES

Animal manure production in the United States totals nearly 175 million tons (334 million mt) on a dry-weight basis. About 90 percent of all animal manures are used for soil improvement, and the scope for increasing this practice is 5 percent (see Table 5.1).

State and federal laws specify that all animal manures and their effluents must be retained within the boundaries of each owner's property. This is in contrast to a common practice prior to the 1960's to flush or dump manures into a creek. Those days are gone forever! Manures are a valuable fertilizer and soil amendment. As chemical fertilizers become more expensive, more care will be taken to manage animal manures more efficiently. However, there are potential hazards to the environment from using animal manures.

5:3.1 ▫ Animal Manure Management Systems

There are two major types of systems of manure management: *solid* and *liquid*. The solid system is essentially the traditional one of collecting the solids, using enough bedding to absorb the liquids, and spreading the mixture on the soil. If spreading every day is not feasible, the manure should be protected from the weather to reduce leaching until it can be spread.

The newer techniques consist of liquid systems that include:

- A **debris basin** in which water and manure are held in a slurry. It is never stirred and is therefore **anaerobic** (without oxygen) (Figure 5.2).
- An **oxidation ditch** which may be any size and depth but is aerated by motors and is therefore **aerobic** (with oxygen).
- A **lagoon,** which is a large pond with more water than a debris basin or an oxidation ditch. When a deep lagoon is not stirred artificially it is anaerobic; when shallow, or deep and stirred, it is aerobic.

Periodically, in all liquid systems, the slurry of manure and water is spread on the soil. Field application of the liquid manures may consist of:

1. Injecting the slurry into the soil or spreading it over the soil surface.
2. Injecting the slurry into a sprinkler irrigation system and then spraying it on the soil with irrigation water.

Fig. 5.2 — A debris basin (*foreground*) has been built just below the paved feedlot, into which all manure is pushed by a bulldozer. The net result is no stream pollution and the conservation of all manure for use on fields. (Arkansas) (Courtesy, USDA)

Minor types of managing manures include dehydrating only and dehydrating plus composting. The products from both systems are usually bagged and sold as a fertilizer, mostly for use on home gardens.

Comparing the composition of manures managed by the solid system and the liquid anaerobic pit system, more dry matter and more major nutrients are conserved by the old-fashioned solid system of management.

5:3.2 ▫ Crop Response to Manures

In general, crop yields are higher with the use of manures as compared with the use of chemical fertilizers which contain the same amounts of nitrogen, phosphorus, and potassium. There are several reasons to explain the superiority of manures. A few are:

1. The presence of secondary nutrients and micronutrients in manure.
2. The improvement in desirable **tilth** (soil physical condition) resulting from the application of manure (Figure 5.3).
3. The release of plant nutrients by microbial decomposition of manures in harmony with the needs of growing plants; leaching losses are therefore held to a minimum.

Fig. 5.3 — Top-dressing a legume-grass rotation pasture with solid animal manure is perhaps the oldest method of efficient use; it is also one of the most popular methods today. Plowing under the manure immediately after application will prevent some loss of ammonia nitrogen. (Courtesy, New Idea)

In a continuous corn study in Nebraska, the application of 12 tons of manure per acre (27 mt / ha) per year containing 77 pounds (35 kg) of nitrogen and 115 pounds (52 kg) of P_2O_5 was compared with an annual application of 120 pounds (55 kg) of nitrogen (N) and 92 pounds (42 kg) of P_2O_5 per acre. Yields of corn were 5 percent higher from the manure alone.

Manures should not be spread on alfalfa because of the weed seeds in the manure.

5:3.3 ▫ Manures and the Environment

The U.S. Environmental Protection Agency has the responsibility of enforcing Public Law 92-500 of the Federal Water Quality Improvement Act of 1970, as amended, in cooperation with each state water quality commission or board. Because of this legal necessity, there will be an increasing percentage of animal manures spread on the land, resulting in the enhancement of water quality. More manure on the land will increase soil moisture, soil productivity, tillability, and tilth; reduce erosion and sedimentation; and increase farm income. Management practices for the disposal of animal manures to comply with state and federal water quality standards are compatible with management practices to increase soil productivity and crop yields, resulting in an increase in farm income (Figure 5.4). (See Chapter 4.)

Fig. 5.4 — Animal manure from beef cattle may increase plant growth or contaminate the environment, depending primarily on animal density and management techniques. *Top:* A stocking rate of about one animal unit per 10 acres (4 ha) is grazing a southern Utah range where annual precipitation is 16 inches (41 cm). With intelligent grazing management, the animal wastes will improve the range. *Bottom:* About 5,000 animal units on 10 acres (4 ha) in Nebraska. The potential for pollution has been magnified by more than 5,000 times. *(Top:* Courtesy, Soil Conservation Service. *Bottom:* Courtesy, USDA)

Animal wastes degrade waters in four principal ways:

1. By increasing the nitrate content of drinking water. Nitrates are a hazard, especially to young babies and to livestock.
2. By increasing the phosphorus content of water. Phosphorus encourages the excess growth of water weeds.
3. By increasing the **biochemical oxygen demand (BOD)** of water because of organic decomposition of manure which requires oxygen. This increased BOD reduces the oxygen level and causes many fish to die.
4. By increasing disease-producing organisms.

Confining the animal wastes to the farm boundaries will assure minimum pollution, except to the farm drinking water supply. Environmentally, the best method of disposal of all animal manures is to spread them on the land. Economically, this is also the best management practice.

All animal manures should be spread on fields as fast as they are produced. Fresh manures supply more nutrients and will improve soil structure more than stored or weathered manures. To protect the environment, however, manures should not be spread on steeply sloped or frozen soils.

Suggestions for reducing the contamination of the environment by animal manures include the following:

1. Locate the animal or poultry production facility on the crest of a slope, as far away from a stream as possible, and east to northeast of towns and housing developments. The summer winds in the interior of the United States are usually westerly, and livestock facilities located east or northeast of towns will result in a minimum of odors toward concentrations of people. On the Atlantic Coast and the Gulf of Mexico Coast, summer breezes are usually toward the shore.
2. Build a diversion terrace around the livestock facility to serve two purposes: (1) to *divert* water *away* from the facility and (2) to *contain* waters *within* the boundaries of the farm and *away* from streams.
3. Establish a smaller debris basin within the diversion terrace to catch manure solids. Periodically the debris basin should be cleaned and spread on cropland or pasture. (See Figure 5.2.)
4. Do not use cat and dog manure on gardens because they contain internal parasites which are transmissible to humans.

5:3.4 □ Methane Gas from Manures

With the increasing costs of all kinds of energy, recent research has demonstrated the feasibility of generating methane gas from anaerobic fermentation of animal manures. The gas so generated is about 60 percent methane (CH_4) and 40 percent carbon dioxide (CO_2). It is competitive in cost with other forms of energy. Furthermore, the manure residues after gas generation have almost the same value as a fertilizer as the fresh manure. All kinds of manure can be used, but poultry manure is the most efficient, and swine manure is the least desirable. Wheat straw, although very slow-acting, was also proven satisfactory for producing methane gas.

5:4 ▫ USING SEWAGE SLUDGE

5:4.1 ▫ Introduction

Most sewage sludge is a waste product to be disposed of at the least possible environmental and economic cost. Disposal techniques include burying in a landfill, dumping in the ocean, burning (incinerating), composting, and spreading on the land. About 23 percent is spread on soils as an organic fertilizer.

Municipal wastewaters entering the average treatment plant are 99.9 percent water and only 0.1 percent solids. Treatment must therefore consist largely of removing surplus water. When the percentage of solids reaches from 1 to 10 percent, the "liquid" sludge will flow by gravity; when 20 to 30 percent "solids," it is a semisolid and must be moved by a conveyor belt or a water-tight truck bed; and when 30 to 80 percent "solids," it can be conveyed about like moist soil or fresh manure. As a "rule of thumb," "liquid" sludge is about 5 percent solids and is transported as a fluid in a tank truck. Typically, 5,000 gallons of "liquid" sludge with 5 percent solids is equal to 1 "dry" ton of sludge (10 qts. / lb.; 20 liters / kg). Because of the variation in water content, it is customary to discuss the amounts of sewage sludge on a "dry" or "dry-weight" basis. This means "oven-dry," and is achieved by drying a sample in an oven set at 221° F (105° C) until it loses no more weight. Of course, sludges are not oven-dried before being used; they are air-dried. At air-dryness, sewage sludge may contain about 20 percent water.

The production of sludge in the United States during 1985 was about 17,650 tons *per day* (16,012 mt / day), on an oven–dry-weight basis. This sludge had a chemical-fertilizer – equivalent value of one-half million dollars per day. Who can ignore this valuable resource? The answer is "almost everyone," for at least four reasons:

1. A dislike for the odors and the association of ideas of a human effluent.
2. The cost; in most instances, equivalent nutrients in chemical fertilizers are cheaper when transportation and spreading costs are included.
3. The spread of **pathogens** (disease-producing organisms) is a very serious hazard to livestock and to humans.
4. Heavy metals content, especially cadmium, is also a threat to livestock and to humans.

5:4.2 ▫ Fertilizing Value

The chemical-fertilizer – equivalent value for 1985 of 6.4 million dry tons

(5.8 million mt) of sewage sludge was $196 million or about $30 per ton (calculated in 1980 dollars and at 1980 fertilizer prices). At 1987 fertilizer prices, for every one thousand people, human effluent was worth $4,500 ($4.50 per person per year).

On the average, a ton (or mt) of dry sludge contains 5 percent total nitrogen (N), 6 percent total phosphorus (P_2O_5), and 0.5 percent total potassium (K_2O). All of the phosphorus and potassium are available to plants during the first growing season. Of the total nitrogen, about 50 percent is available the first year. This includes all in the form of ammonium (NH_4^+) and nitrate (NO_3^-) plus about 20 percent of the organic nitrogen (N), the exact amount depending on microbial decomposition and transformation. Other plant nutrient content, as well as typical heavy metal content, is presented in Table 5.2.

Table 5.2 — Typical Composition of Fresh, Heated, Anaerobically Digested Sewage Sludge[1]

Nutrients Essential for Plant Growth	Composition (Oven-Dry Basis)	
	(%)	*(lb/ton)*
Nitrogen		
Ammonium (NH_4^+)	2	40
Organic (N)	3	60
Phosphorus (P_2O_5)	6	120
Potassium (K_2O)	0.5	10
Calcium (Ca)	3	60
Magnesium (Mg)	1	20
Sulfur (S)	0.9	18
Iron (Fe)	4	80
Zinc (Zn)	0.5	10
	Milligrams per Liter (parts per million)	
Copper (Cu)	1,000	2
Manganese (Mn)	500	1
Boron (B)	100	0.2
Nonessential Heavy Metals	**Milligrams per Liter (parts per million)**	
Chromium (Cr)	3,000	6
Lead (Pb)	1,000	2
Nickel (Ni)	400	0.8
Cadmium (Cd)	150	0.3

[1] "Utilization of Sewage Sludge on Agricultural Land." University of Illinois Soil Mgt. and Cons. Ser. SM29, 1975.

One application of 15 tons per acre (33.6 mt / ha) of sewage sludge has been very effective for at least five years in correcting iron deficiency of grain sorghum growing on calcareous (pH 7.8) soils in New Mexico.

5:4.3 ▫ Application Rate

The application rate of sewage sludge should be based on two guidelines: nitrogen and heavy metals.

- ▫ ***Nitrogen*** — The amount of nitrogen required should be calculated, and that amount should be applied. There is no way to avoid an excess application of any nutrient, for example, phosphorus, but any deficiency can be corrected by adding that nutrient as a chemical fertilizer, for example, potassium.
- ▫ ***Heavy metals*** — The environmentally safe heavy metal rate should be calculated. If it is in excess, the total application should be reduced, and any deficient nutrient should be compensated for by being added as a chemical fertilizer.

Example

Assuming that the soil does not supply any plant nutrient, a 100-bushel corn crop per acre requires 150 pounds of available nitrogen.

Table 5.2 indicates that all of the ammonium nitrogen (40 pounds per ton) is available, plus about 20 percent of the organic nitrogen (N) ($0.20 \times 60 = 12$ pounds). $40 + 12 = 52$ pounds of nitrogen available per ton of sewage sludge (dry-weight basis). This is rounded to 50 pounds per ton of sewage sludge. The corn needs 150 pounds of available nitrogen, so $150 \div 50 = 3$ tons of sludge (dry-weight basis) to be applied. To apply 3 "dry" tons would require $3 \times 5{,}000$ gallons, or 15,000 gallons, of sludge per acre with an average of 5 percent solids.

These same 3 dry tons would contain:

1. 360 pounds of P_2O_5. This is 9 times the 40 pounds of P_2O_5 required, but nothing can be done about this.
2. 30 pounds of K_2O. This is only 30 percent of the 100 pounds of K_2O required. Add 70 pounds of chemical potassium (K_2O) fertilizer to make up the deficiency. 117 pounds of 60 percent KCl will supply 70 pounds of K_2O.

Allowable heavy metal application rates must be limited by the soil pH and the soil texture. With cadmium, federal regulations, too complex to explain here, went into effect in 1979.

With other heavy metals, the law specifies the following:

> At a soil pH of 6.5 and above, the amount of allowable lead in sewage sludge that can be applied is 500 pounds per acre (560 kg / ha) on loamy sands, 1,000 pounds per acre (1,121 kg / ha) on loams, and 2,000 pounds per acre (2,242 kg / ha) on clay loams and clays. Decreasing allowable rates are specified for zinc, copper, nickel, and cadmium. At a pH below 6.5, the allowable rates are less because the solubility of the heavy metals is higher.

5:4.4 ▫ Health Hazards

In the general statement "The health hazards of a well-managed land application of sewage sludge are minimal," the key words are *well-managed.*

There have been outbreaks of human diseases resulting from the application of sewage effluents, but all of them have been the result of applying *raw* (untreated) sludge.

> Within the last century several cities . . . have utilized land application to manage municipal waste. These farms are safe, and *there have been no outbreaks of waterborne disease* either among the farm workers or in the city where land application systems are used and managed properly.

The phrase *no outbreaks* does not mean that this health hazard of pathogens in municipal sewage and its wastewaters can be ignored. Disease-producing organisms are common and must be avoided. The four groups of pathogens found in raw sludges and wastewaters are: (1) plant-like bacteria, (2) animal-like protozoa, (3) parasitic worms, and (4) viruses.

The diseases they cause are described briefly in Table 5.3.

The use of sewage sludge and excessive heavy metals, especially cadmium,

Table 5.3 — Major Organisms of Human and Animal Health Concern That May Be Present in Sewage from U.S. Communities[1]

Organisms	Disease	Reservoir(s)
Bacteria		
Salmonellae (Approx. 1,700 types)	Typhoid fever Salmonellosis	Humans, domestic and wild animals, and birds
Shigellae (4 spp.)	Shigellosis (bacillary dysentery)	Humans
Escherichia coli (enteropathogenic types)	Gastroenteritis	Humans and domestic animals
Enteric Viruses		
Enteroviruses (67 types)	Gastroenteritis, heart anomalies, meningitis, others	Humans and possibly lower animals
Rotavirus	Gastroenteritis	Humans and domestic animals
Parvovirus-like agents (at least 2 types)	Gastroenteritis	Humans
Hepatitis A virus	Infectious hepatitis	Humans and other primates
Adenoviruses (31 types)	Respiratory disease, conjunctivitis, others	Humans
Protozoa		
Balantidium coli	Balantidiasis	Humans and swine
Entamoeba histolytica	Amebiasis	Humans
Giardia lamblia	Giardiasis	Humans and domestic and wild animals?

(Continued)

Table 5.3 (Continued)

Organisms	Disease	Reservoir(s)
Helminths		
Nematodes (Roundworms)		
Ascaris lumbricoides	Ascariasis	Humans and swine?
Ancylostoma duodenale	Ancylostomiasis	Humans
Necator americanus	Necatoriasis	Humans
Ancylostoma braziliense (cat hookworm)	Cutaneous larva migrans	Cats
Ancylostoma caninum (dog hookworm)	Cutaneous larva migrans	Dogs
Enterobius vermicularis (pinworm)	Enterobiasis	Humans
Strongyloides stercoralis (threadworm)	Strongyloidiasis	Humans and dogs
Toxocara cati (cat roundworm)	Visceral larval migrans	Carnivores
Toxocara canis (dog roundworm)	Visceral larva migrans	Carnivores
Trichuris trichiura (whipworm)	Trichuriasis	Humans
Cestodes (Tapeworms)		
Taenia saginata (beef tapeworm)	Taeniasis	Humans
Taenia solium (pork tapeworm)	Taeniasis	Humans
Hymenolepis nana (dwarf tapeworm)	Taeniasis	Humans and rats
Echinococcus granulosis (dog tapeworm)	Unilocular echinococcosis	Dogs
Echinococcus multilocularis	Alveolar hydatid disease	Dogs and other carnivores

[1]Source: Vimala A. Majeti and C. Scott Clark. "Potential Health Effects from Viable Emissions and Toxins Associated with Wastewater Treatment Plants and Land Application Sites." U.S. Environmental Protection Agency, Health Effects Research Laboratory, Cincinnati, EPA-600/51-81-006, April 1981.

is also hazardous to health. Studies revealed that cadmium levels in forage crops on soils receiving sewage sludge were high; and cattle consuming the forage had high cadmium in their livers and kidneys. Swine fed a 10 to 20 percent dried sludge ration did not gain as rapidly as the swine not fed sludge.

The sludge-fed swine and their pigs had four to six times more cadmium in their kidneys. High-cadmium cattle and swine livers and kidneys were fed to mice. The result was fewer baby mice weaned and an increase in the amount of cadmium in the mouse livers and kidneys.

The salts in sewage sludge have recently been shown to be toxic to legume bacteria.

5:5 □ USING SEPTAGE

Septage is an anaerobic mixture of liquids and solids pumped from residential septic tanks. It consists of 95 to 97 percent water and 3 to 5 percent solids. Septage is very foul-smelling and contains many pathogens, including the following viral, bacterial, and parasitic diseases of humans: ***viral*** — meningitis, poliomyelitis, hepatitis, and gastroenteritis; ***bacterial*** — food poisoning, cholera, tetanus, diarrhea, tuberculosis, typhoid fever, and bacillary dysentery; and ***parasitic*** — amoebic dysentery, giardiasis, coccidiosis, hookworms, roundworms, and whipworms.

However, in perhaps two months after septage has been applied to pasture grasses or forage crops, there should be very little hazard to forage-eating livestock.

Septage is lower in all plant nutrients and lower in heavy metals than municipal sewage sludge.

The most environmentally acceptable method of disposing of septage seems to be to spread it on cropland fields or pastures where and when land owners agree. Acceptable soils for spreading septage have these characteristics:

- ***Texture*** — Sandy loam or finer. Sands should be avoided, as they will not adequately filter out pathogens. The result may be contaminated groundwaters.
- ***Soil pH*** — Between 6.5 and 7.0.
- ***Depth*** — A minimum depth of 20 inches (51 cm) to bedrock or a wet-season water table. A dry-season water table should be below 4 feet (122 cm).
- ***Slope*** — A maximum slope of 12 percent, free from sink holes and depressions to hold water.
- ***Flooding*** — Freedom from flooding.
- ***Erosion*** — Minimum water erosion.

When using septage, observe the following precautionary measures:

1. Septage should *not* be applied to home gardens or commercial gardens where vegetables, especially root / tuber crops, are grown which may be eaten raw.
2. Septage should be stored in a lagoon for at least a month before being spread on soils near humans. This will allow most of the foul odors to disappear.
3. Septage should not be applied within 300 feet (91 m) of an occupied dwelling, a well, a spring, or other water supply; within 100 feet (30 m) of a stream; within 50 feet (15 m) of a property line; or within 25 feet (8 m) of a bedrock outcrop.

   Septage can be composted in a similar manner as sewage sludge by the Beltsville aeration-pile method explained in Section 5:6.2.

5:6 □ USING COMPOST

5:6.1 □ Introduction

Evidence that making compost from leaves is an age-old practice is presented here in an ancient Telugu (India) proverb: "Leaf manure gives luxuriance."

Organic materials suitable for making a **compost** include all animal manures, sewage sludge, tree leaves, grass clippings, garden weeds, hay, straw, sawdust, peat, garbage, cotton gin trash, rice hulls, or almost anything. "Almost anything" does not mean *anything. Not* suitable are grass clippings and hay with viable grass or weed seeds or viable sprigs of grasses such as bermudagrass or quackgrass, weeds that have gone to seed, garden refuse that is diseased, and tree leaves that resist decomposition. Tree leaves that *do not* rot fast enough to be recommended for use in a compost include all oak leaves and all coniferous needles except whitecedar. Most leaves suitable for use are elm, ash, basswood, maple, hickory, alder, apple, peach, plum, and sassafras.

More recent techniques have been developed for making satisfactory compost from sewage sludge, discussed in the next section.

One recommended technique for making a compost is to build a retaining barrier with about a 2-inch (5-cm) mesh woven wire, about 4 feet (1.2 m) high and any diameter suitable to accommodate the volume of organic materials to be used. To eliminate the need for a gate, an access can be made simply by extending one end of the wire about 4 feet (1.2 m) beyond the other end, and outward about 3 feet (1 m), similar to an arc-of-a-circle. This system is superior to leaving a gap-without-a-gate, because the arc-of-a-circle access will retain more leaves against the hazard of the winds blowing them away. Another technique is to build retaining walls with cement blocks (Figure 5.5).

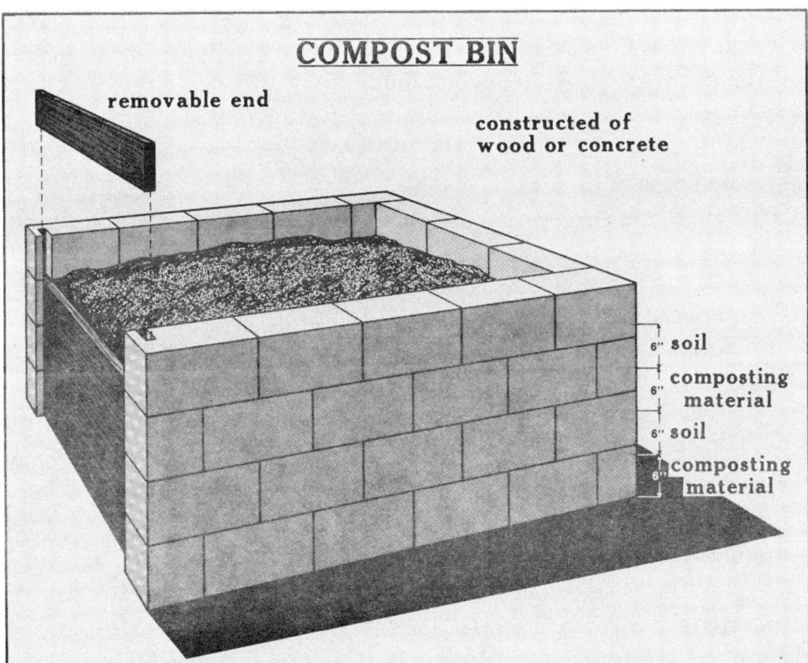

Fig. 5.5 — Making and using compost is easier when a bin is constructed with a removable end. (Courtesy, Instructional Materials Services, Texas A&M University)

Filling the compost enclosure can be haphazard, with organic material being added whenever it becomes available. About the only rule to follow is to always flatten the top or depress the middle of the pile to increase water penetration. After every 4 inches (10 cm) of organic material, chemical fertilizers sprinkled evenly over the surface will hasten decomposition. In the absence of more scientific information, a 20-10-0 fertilizer at the rate of 1 pound for every 10 pounds (1 kg per 10 kg) of organic matter (dry basis) should be satisfactory. In areas of acid soils, lime should also be applied over the organic material at the same rate as fertilizer. The correct amounts of fertilizer and lime can also be estimated by spreading them at about the same rate as you would put salt on meat or potatoes or salads. Fertilizer and lime should be applied on *separate* layers of organic material. If applied together, lime causes some of the nitrogen in the fertilizer to change to ammonia and to escape into the atmosphere as a gas. Lime also causes phosphorus to become less available. (See Chapter 3.) One man known by the authors recommends the application of yeast on the compost pile. Commercial preparations are also available for adding to compost piles, but no one knows if they serve any useful purpose. To hasten decomposition, the compost pile should be kept moist but not wet. (See Chapter 15.)

5:6.2 □ Composting and Using Sewage Sludge

The outdoor-windrow method of composting sewage sludge, tried by scientists working for the U.S. Department of Agriculture – Agricultural Research Service in an area of high-density population, was unsuccessful and had to be abandoned because of excessive foul odors. This method is satisfactory, however, in sparsely populated areas. After several failures experienced with other methods, a successful method of composting digested sewage sludge, completely eliminating the problem of odor, was developed in 1973 at Beltsville, Maryland, by the Agricultural Research Service.

The **Beltsville aeration-pile composting method** was perfected (Figure 5.6 and the frontispiece of this chapter) and is explained here.

1. A 4-inch-diameter (10 cm) perforated plastic pipe is placed in a rectangular configuration directly under what will become the pile of sludge and woodchips.
2. A 6- to 8-inch (15- to 20-cm) layer of woodchips or other similar bulking material is placed over the pipe and the area to be occupied by the pile. This layer comprises the pile base and facilitates the movement and distribution of air during composting. The base material also absorbs excess moisture that may condense and leach from the pile.
3. The mixture of sludge and woodchips is then placed loosely upon the prepared base with a front-end loader or conveyor system to form a pile with a triangular cross-section 15 feet (5 m) wide, 7.5 feet (2.5 m) high, and as long as necessary.
4. Excess woodchips are removed from around the base, and the pile is

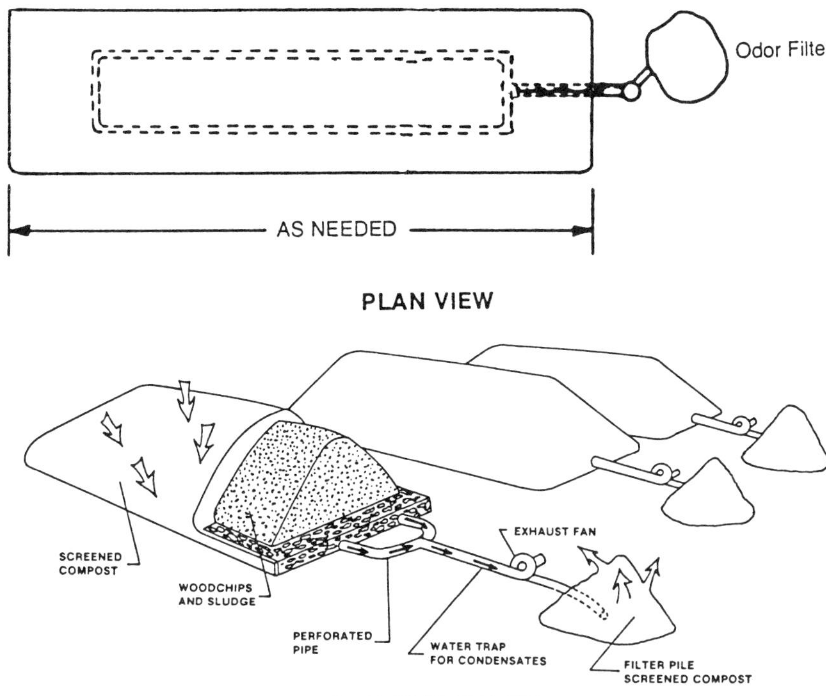

Fig. 5.6 — The Beltsville aeration-pile composting method for sewage sludge. *Top:* Plan view showing location of 4 inch (10-cm) perforated plastic pipe and odor filter. *Bottom:* Schematic diagram showing woodchips and sludge, screened compost blanket, water trap, exhaust fan, and filter pile for odor control. (Source: G. B. Willson, *et al.* "Manual for Composting Sewage Sludge by the Beltsville Aeration-Pile Method." U.S. Environmental Protection Agency, EPA-600/8-80-027, May 1980)

completely covered with a 12-inch (30-cm) layer of cured, screened compost or an 18-inch (46-cm) layer of cured, unscreened compost to provide insulation and prevent the escape of malodorous gases during composting.

During construction of the pile base, the perforated pipe is connected to a section of solid plastic pipe extending beyond the pile base. The solid pipe is connected to a blower rated at 160 cubic feet per minute (75 liters/sec.), powered by a ⅓ horsepower motor and controlled by a timer. Aerobic composting conditions are maintained by drawing air through the pile intermittently. The exact aeration schedule will depend on pile geometry and the amount of sludge to be composted. For a pile with the dimensions described, 66 feet by 16 feet by 9 feet (20 m × 5 m × 2.5 m), the timing sequence for the blower is 4 minutes on and 16 minutes off.

The effluent air stream discharged from the compost pile blower is conducted into a small, cone-shaped pile of cured, screened compost approximately 4 feet (1.2 m) high and 8 feet (2.4 m) in diameter, where malodorous gases are absorbed. These are commonly referred to as "odor filter piles." The odor filter pile should contain about 1 cubic yard (0.76 cu m) of screened compost for each 10 wet tons (9 mt) of sludge being composted.

For the sake of the environment, some municipalities are interested in making sewage sludge compost and possibly giving it away; others may want to sell it for whatever it will bring. Most municipalities would be pleased to"break-even" on the operation.

Commercial markets for composted sewage sludge would be the same as for the animal manures (see Section 5:3). For this reason, a comparison is made of the major nutrients in sewage sludge compost and manures from dairy cattle, beef cattle, poultry, and swine (Table 5.4). Digested sewage sludge compost is higher in total nitrogen content than all but poultry manure, almost equal in the percentage of phosphorus content as poultry manure, but is lower in potassium content than the least of the manures.

Table 5.4 — A Comparison of the Major Nutrients in Digested Sewage Sludge with Animal Manures Handled by the Solid System[1]

Organic Amendment	Total Composition: Nitrogen (%)	Phosphorus (% P_2O_5)	Potassium (% K_2O)
Digested sewage sludge	0.9	2.3	0.1
Dairy cattle manure	0.4	0.2	0.5
Beef cattle manure	0.6	0.4	0.5
Poultry manure	1.6	2.4	1.7
Swine manure	0.5	0.4	0.4

[1]Sources: For digested sludge: G. B. Willson, *et al.* "Manual for Composting Sewage Sludge by the Beltsville Aeration-Pile Method." U.S. Environmental Protection Agency, EPA-600 / 8-80-027, May 1980; For manures: A. L. Sutton, *et al.* "Utilization of Animal Waste as Fertilizer." Purdue University, 10–101, 1975.

Composted sewage sludge also contains a high percentage of the heavy metals (see Table 5.2) and pathogens (see also Section 5:4.4). There would, however, be a slight diluting effect because of the bulking agent used, such as woodchips. For this reason, the percentage of the heavy metals content would be slightly less (Figure 5.7). Also the heat generated by microbial decomposition of the sewage sludge should reach 130° F (55° C). If this temperature is maintained uniformly throughout the pile for a few days, most pathogens should be killed. "Cold spots," however, frequently occur, permitting some pathogens to survive. Human health hazards in making and using sewage sludge compost include those in Table 5.5.

Note: The Beltsville system of sewage sludge composting was modified by Rutgers University scientists. The modification consists of heat control from 140° F to 176° F (60° C to 80° C) to achieve uniform decomposition and to kill pathogens.

Fig. 5.7 — Scientists working for the USDA are testing leaves of soybeans for heavy metal concentration following the application of composted sewage sludge. (Courtesy, USDA - Office of Communications)

Table 5.5 — Typical Pathogens Present in Making and Using Sewage Sludge Compost[1]

Group	Example	Human Disease
Primary pathogens		
Bacteria	*Salmonella enteritidis*	Salmonellosis (food poisoning)
Protozoa	*Entamoeba histolytica*	Amoebic dysentery (bloody diarrhea)
Helminths	*Ascaris lumbricoides*	Ascariasis (worms infecting the intestines)
Viruses	Hepatitis virus	Infectious hepatitis (jaundice)
Secondary pathogens		
Fungi	*Aspergillus fumigatus*	Aspergillosis (growth in lungs and other organs)
Actinomycetes	*Micromonospora* spp	Farmer's lung (allergic response in lung tissue)

[1] G. B. Willson, *et al.* "Manual for Composting Sewage Sludge by the Beltsville Aeration-Pile Method." U.S. Environmental Protection Agency, EPA-600 / 8-80-027, May 1980.

5:7 ▫ USING LOGGING AND WOOD MANUFACTURING RESIDUES

Logging and wood manufacturing residues are estimated to total about 35.7 million dry tons (32.4 million mt) each year (see Table 5.1). Current use on soils is about 5 percent or 1.8 million tons (1.6 million mt) a year. Increase in use is estimated at less than 5 percent. However, some kinds of wood residues such as those from black walnut should not be used because of the toxin **juglone.**

Of the millions of tons of sawdust available for use as soil organic matter, only a few thousand tons are actually used as livestock bedding, to pack nursery stock for shipping, and as a mulch around shrubs and on gardens. When used directly on gardens as a mulch, its nitrogen content should be increased by adding 1 pound of nitrogen (N) fertilizer to each 10 pounds (1 kg / 10 kg) (dry weight) of sawdust. Without nitrogen fertilizer, bacteria may take nitrogen away from the crops as they decompose the highly carbonaceous and low-in-nitrogen sawdust. All common wood residues are acid-forming.

Redcedar sawdust depresses the growth of young legumes such as beans. Pine bark and incense cedar woodchips and sawdust are toxic to garden peas, and fresh walnut bark inhibits the germination and growth of many plant species. Tomatoes and beans are sensitive to sawdust regardless of its age.

Machines are now available that chip brush, limbs, and even logs as large as 10 inches (25 cm) in diameter. The chips are often made incident to clearing a highway right-of-way, electric line, or a forested urbanizing area. Woodchips are also made intentionally for use as a mulch on highway and roadside slopes to stabilize the soil until a permanent seeding of grasses becomes established. On steep slopes, woodchips are more stable against wind and water and will not decompose as rapidly as sawdust, straw, or hay (Figure 5.8). (See Chapter 20.)

Fig. 5.8 — A wood chipper in action, making chips from branches and small poles. (Courtesy, USDA - Soil Conservation Service)

5:8 □ USING MUNICIPAL, FOOD PROCESSING, AND INDUSTRIAL ORGANIC RESIDUES

Municipal solid wastes (residues) are disposed of by being dumped in the ocean, burned, buried in a landfill, or composted. However, the variable composition of the residues makes composting the least attractive of the alternatives.

Solid residues have been composted on a research basis in two ways: (1) they were screened and piled in long windrows and (2) earthworms were added to ingest (eat) the residues and excrete them in casts. The casts were then sold to gardeners.

In one recent study, composting costs varied from $25 to $30 a ton, compared with a cost of $15 a ton for burning and $6 a ton for landfilling.

Food processing and industrial organic residues have a low present use and a low potential future use as soil amendments. Some food processing wastes are more valuable for other use such as feed for livestock. Many of the industrial organic wastes are potentially toxic.

5:9 □ USING PEAT

Peat is a naturally occurring organic material that has accumulated over many centuries in wet and / or cool places. Many kinds of peats are valuable for use as mulches, soil conditioners, and acidifying agents. There are at least five types of highly organic materials that are sold on some commercial markets as *peat*.

- **Moss peat** — A product originating mainly from sphagnum moss or hypnum moss. Its characteristics are a light brown color, a high acidity, and a very high water-holding capacity. Its principal uses are for soil conditioning, for top-dressing lawns, for surface mulching, for rooting cuttings, for acidifying the soil, for mixing with a potting soil, and for packing nursery stock for shipping.

- **Reed-sedge peat** — A product originating from residues of reeds, sedges, marsh grasses, and cattails. It is brown to reddish-brown in color and has a moderate water-holding capacity. Its principal uses are as a soil conditioner, a top-dressing for lawns, and a potting soil mix.

- **Peat humus** — A product of advanced decomposition of hypnum moss peat or reed-sedge peat. It is dark brown to black in color and is used principally as a soil conditioner, a top-dressing for lawns, and a golf green soil mix.

- **Muck soil** — A product of decomposed peats. It is dark gray to black, finely divided, variable in composition, and of variable value for use as a soil conditioner, a top-dressing, or a soil mix. It is frequently sold as "topsoil," "humus," or "peat" for top-dressing lawn areas before seeding or sodding.

- **Sedimentary peat** — Originating at the bottom of certain ponds from partially decomposed algae, plankton, water lilies, and pond weeds. The color is usually dark gray to black, and the particles are so finely divided that they swell when wet and contract when dry. (When moist, they are "rubbery.") Furthermore, some sedimentary peat contains marl (lime), silt, and clay. Sedimentary peat has no positive value as a soil conditioner, a top-dressing, or a soil mix.

5:10 □ USING GREEN-MANURE CROP RESIDUES

Green-manure crops supply fresh food for beneficial soil organisms, make nutrients more available, and, if legumes, add atmospheric nitrogen. Green-manure crop residues also have unique values in reducing the severity of fungal diseases. Two such instances are:

1. In the state of Washington, the residues from a field pea – sudangrass rotation before planting Irish (white) potatoes reduced verticillium wilt on po-

tatoes by 50 percent and increased potato yield between 10 and 20 percent.

2. In the Texas Blacklands (Vertisols), where cotton root rot is very severe, green crop residues from a rotation of wheat, rye, barley, sweetclover, or crownvetch with cotton reduces the severity of the cotton root rot and increases cotton yields.

An alfalfa variety, "Nitro," was developed in Minnesota for use as an *annual* hay and green-manure crop. It has deeper roots, faster growth, and fixes 59 percent more atmospheric nitrogen than the "other" varieties of alfalfa (Figure 5.9).

Fig. 5.9 — The improved "Nitro" alfalfa cultivar (*right*) grows faster, has deeper roots, and fixes 59 percent more atmospheric nitrogen than the traditional alfalfa (*left*). (Courtesy, Dave Hansen, University of Minnesota)

In semiarid regions it is common practice to leave fields bare (no crop planted for a season) to conserve moisture. However, during this period of no crop, wind and water erosion are often severe. Instead of a rotation of wheat- or barley-bare fallow, North Dakota State University is recommending wheat- or barley-sweetclover. The sweetclover can be cut for hay and also serves as an excellent erosion control and green-manure crop.

Used successfully on Ultisols as green-manure crop residues in Georgia are these winter legumes: crimson clover, hairy vetch, and winterpea. Their residues added nitrogen from the air equivalent to 75 to 100 pounds per acre (84 – 112 kg / ha) of commercial nitrogen to the soil for use by the following corn crop.

Alfalfa, birdsfoot trefoil, and red clover in Pennsylvania each supplied adequate nitrogen to the following corn crop.

5.11 □ REFERENCES

Ahmad, F., and K. H. Tan. "Effect of Lime and Organic Matter on Soybean Seedlings Grown in Aluminum-toxic Soils." *Soil Science Society of America Journal,* Vol. 50, No. 3, May – June 1986.

Blevins, R. L. "After 15 Years of No-Tillage Corn." *Soil Science News and Views,* Vol. 6, June 1985. University of Kentucky.

Brady, Nyle C. *The Nature and Properties of Soils,* 9th ed. New York: Macmillan Publishing Co., Inc., 1984, pp. 224 – 281.

Brown, Robert, Terry Logan, Richard Dorn, Vincent Hamparian, and Abramo C. Ottolenghi. "Demonstration of Acceptable Systems for Land Disposal of Sewage Sludge." EPA 600 / 52-85-062, August 1985.

Brust, G. E., D. A. McCartney, and B. R. Stinner. "Predators Reduce Black Cutworm Damage in No-Tillage Corn." Ohio Report. The Ohio State University Agricultural Research and Development Center, May – June 1985.

Donahue, Roy L., Raymond W. Miller, and John C. Shickluna. *Soils: An Introduction to Soils and Plant Growth,* 5th ed. Englewood Cliffs, New Jersey: Prentice-Hall, Inc., 1983, pp. 113 – 162.

Duley, F. L., and J. C. Russell. *Effect of Stubble Mulching on Soil Erosion and Runoff.* Madison, Wisconsin: Soil Science Society of America, Proceedings 7-77-81, 1942.

Edds, G. T., and J. M. Davidson. "Sewage Sludge Viral and Pathogenic Agents in Soil-Plant-Animal Systems." U.S. Environmental Protection Agency, Health Effects EPA-600 / S-1-81-026, Res. Lab., Cincinnati, 1981.

Fenster, C. R., H. I. Owens, and R. H. Follett. "Conservation Tillage for Wheat in the Great Plains." U.S. Dept. of Agriculture – Extension Service, PA-1190, 1977.

Finstein, Melvin S., Frederick C. Miller, Steven T. MacGregor, and Kevin M. Psarianos. "The Rutgers Strategy for Composting: Process Design and Control." U.S. Environmental Protection Agency, EPA-600 / S-2-85-059, July 1985.

Follett, R. H. "Septage." A paper prepared for the national workshop "Utilization of Wastes on Land: Emphasis on Municipal Sewage." University of Maryland, July 15 – 17, 1980, and Anaheim, California, August 12 – 14, 1980.

Follett, R. F., *et al.* "Soil Fertility and Organic Matter As Critical Components of Production Systems." SSSA Spec. Pub. 19, Madison, Wisconsin: Soil Science Society of America and the American Society of Agronomy, 1987, 166 pp.

Follett, R. H., and B. R. Sabey. "Land Application of Municipal Sludge." *Service in Action,* Colorado State University Ext. Serv., No. 547, 1981.

Follett, R. H., Larry S. Murphy, and Roy L. Donahue. *Fertilizers and Soil Amendments.* Englewood Cliffs, New Jersey: Prentice-Hall, Inc., 1981.

Fox, R. H., and W. P. Piekielek. "Fertilizer N Equivalence of Alfalfa Birdsfoot Trefoil and Red Clover for Succeeding Corn Crops." *Journal of Production Agriculture,* Vol. 1, No. 4, 1988.

Harkin, John M., and John W. Rowe. "Bark and Its Possible Uses." U.S. Dept. of Agriculture – Forest Service. Forest Products Laboratory, Madison, Wisconsin. Res. Note FPL-091, 1971, p. 16.

Hausenbuiller, R. L. *Soil Science: Principles and Practices,* 3rd ed. Dubuque, Iowa: Wm. C. Brown Group, 1985, pp. 89 – 115.

Helsel, Zane, *et al.* "No-till Planting Systems." University of Missouri, Agricultural Guide No. 4080, 1985.

Kowal, Norman E. "Health Effects of Land Application of Municipal Sludge." U.S. Environmental Protection Agency, EPA-600 / 1-85-015, August 1986.

Malek, R. B., and J. B. Gartner. "Hardwood Bark as a Soil Amendment for Suppression of Plant Parasitic Nematodes on Container-grown Plants." *Hortscience,* Vol. 10, No. 1, February 1975, pp. 33 – 35.

McCaslin, B. D., *et al.* "Sorghum Yield and Soil Analysis from Sludge-amended Calcareous Iron-deficient Soil." *Agronomy Journal,* Vol. 79, 1987, pp. 205 – 209.

Meyer, D. W. "Sweetclover: An Alternative to Fallow for Set-aside Acreage in Eastern North Dakota." *North Dakota Farm Research,* Vol. 44, No. 5, March – April 1987, pp. 3 – 8.

Papendick, R. I., and L. F. Elliott. "Tillage and Cropping Systems for Erosion Control and Efficient Nutrient Utilization." In *Organic Farming,* ASA Special Pub. No. 46, 1984, pp. 69 – 81.

Ringenbach, Laura A. "Compendium on Solid Waste Management by Vermicomposting." U.S. Environmental Protection Agency, EPA-600 / 8-80-033, August 1980. (*Note:* "vermicomposting" means composting with the use of earthworms.)

Roder, W., S. C. Mason, M. D. Clegg, J. W. Doran, and K. R. Kniep. "Plant and Microbial Response to Soybean-Sorghum Cropping Systems and Fertility Management." *Soil Science Society of America Journal,* Vol. 52, 1988, pp. 1337–1342.

Rothenberger, R. R., and K. Hildahl. "Organic Garden Techniques." Science and Technology Guide 6220, University of Missouri – Columbia, 1981.

Runge, E. C. A., ed. "Utilization, Treatment, and Disposal of Waste on Land." Madison, Wisconsin: Soil Science Society of America, 1986, 318 pp.

Schulte, E. E. "Fertilization for Maximum Economic Yield Under Conservation Tillage." *Better Crops with Plant Food,* Vol. 69, Fall 1985.

Schulte, E. E., and K. A. Kelling. "Organic Soil Conditioners." Ser. No. A-2305, University of Wisconsin, 1981, 4 pp.

Sommerfeldt, T. G., and C. Chang. "Changes in Soil Properties Under Annual Application of Feedlot Manure and Different Tillage Practices." *Soil Science Society of America Journal,* Vol. 49, No. 4, July – August 1985, pp. 983 – 987.

Sutton, A. L., D. W. Nelson, and D. D. Jones. "Utilization of Animal Manure as Fertilizer." West Lafayette, Indiana: Purdue University, ID-101, March 1983.

Troeh, Frederick R., J. Arthur Hobbs, and Roy L. Donahue. *Soil and Water Conservation for Productivity and Environmental Protection.* Englewood Cliffs, New Jersey: Prentice-Hall, Inc., 1980.

Undersander, D. J., and Cecil Reiger. "Effect of Wheat Residue Management on Continuous Production of Irrigated Winter Wheat." *Agronomy Journal,* Vol. 77, No. 3, May – June 1985.

Unger, Paul W. "Wheat Residue Management Effects on Soil Water Storage and Corn Production." *Soil Science Society of America Journal,* Vol. 50, 1986, pp. 764 – 770.

U.S. Dept. of Agriculture. "Report and Recommendations on Organic Farming." Washington, D.C.: U.S. Government Printing Office, 1980.

U.S. Environmental Protection Agency. "Demonstration of Acceptable Systems for Land Disposal of Sewage Sludge." EPA-600 / S-2-85-062, August 1985.

U.S. Environmental Protection Agency. "Handbook: Groundwater." EPA-625/6-87-016, 1987.

U.S. Environmental Protection Agency. "Handbook: Septage Treatment and Disposal." EPA-625/6-84-009, June 1984.

U.S. Environmental Protection Agency. "Health Effects of Land Application of Municipal Sludge." EPA-600/S-1-85-015, 1985.

U.S. Environmental Protection Agency. "Protection of Public Water Supplies from Groundwater Contamination." EPA-625/4-85-016, 1985.

CHAPTER 6

Tillage Systems

"He that tilleth his land shall have plenty of bread."
Proverbs 28:19

OUTLINE

Tilling the soil is a fundamental crop production practice. Shown here are farmers in Burkino Fasso (West Africa) preparing their land for planting. Tillage pans will never develop here, but people may starve. (Courtesy, International Development Association)

□ □ □

6:1 □ OVERVIEW

Tilling the soil is a fundamental crop production practice. Soil tillage is as old as written history, and no one knows how much older. Early humans obtained food where they found it. Centuries later, wild fruits and other crops were planted in small clearings near human dwellings. To prepare a good seedbed and to keep weeds from choking out the garden and orchard crops, some form of tillage was necessary. From hand pulling to stick scratching, to brush and log dragging, to mule-drawn plows, to modern steel plows and cultivators, and to minimum tillage, the evolution of tillage has taken place (Figure 6.1).

The modern-day farm has a large choice of tillage implements, and the field tractor provides plenty of power to operate them. The investment in tillage equipment — tractor power and its operation — is high. Each individual tillage operation must, therefore, be evaluated in terms of its effects on the soils and its economic return. Every time tillage is planned, a "management decision" must be made evaluating its effect on the cost of production per unit of crop. All tillage costs must be paid for by the increase in value of the resulting crop yields.

6:2 □ PURPOSES OF TILLAGE

The primary purpose of tillage is to improve upon the natural soil conditions for more uniform emergence and greater crop growth. Stirring the soil by some kind of tillage tool has always been done for one or more of the following basic purposes:

1. To prepare a seedbed — Soils should be mellow but firm to permit uniform planting depth and an even crop emergence.
2. To control weeds — Weeds should be removed to conserve moisture and to increase plant nutrients and light.
3. To improve the physical condition of soil — Soil should be plowed and cultivated to increase aeration and permit infiltration of water.
4. To save moisture — The surface crust should be broken and the soil loosened to improve water intake, water movement, and water storage. Unnecessary tillage wastes soil moisture. In semiarid regions, early spring tillage is conducive to greater infiltration and storage of more available water for higher crop yields.
5. To break out sod crops — Sod crops should be broken in order to permit rotations of annual crops with grass and legume sod crops.
6. To manage crop residues — Crop surface residues should be incorpo-

Fig. 6.1 — *Top*: Plowing the soil as it was done 50 years ago. *Bottom*: Plowing as it is sometimes done now. The most recent technique is conservation tillage (see Section 6:5.1). (Courtesy, Bob Taylor, Cordell, Oklahoma)

rated and covered so that equipment can operate more efficiently. (This objective has been modified by the newer technology of minimum tillage.)

7. To control diseases and insects — Proper planting dates, timely plowing under of crop residues, and good crop rotations should be employed to produce early harvests and help to reduce damage from diseases and insects.
8. To control wind erosion in case an emergency arises — A cloddy and ridged surface should be created during spring and winter — especially during spring.

6:3 ▫ SOIL CONDITIONS IMPORTANT TO PLANTS

The five soil conditions that are important to the growth of young seedlings and plants are (1) soil temperature, (2) soil moisture, (3) bulk density, (4) granulation, and (5) porosity. These conditions are interrelated and all can be influenced to some extent by tillage.

6:3.1 ▫ Soil Temperature

Except when moisture is lacking, growing-season soil temperatures are seldom too high; hence, anything that can be done to raise them, especially in the spring, is desirable. Land that is wet and compact is colder than the average; whereas, open, well-aerated, well-drained soil is warmer. The effect of soil temperatures on the time required by four kinds of seed to germinate and emerge is shown in Table 6.1.

One way to combat cool spring temperatures is to delay planting until the weather is warm. This procedure is used with fast growing crops, but it may cre-

Table 6.1—Days for Emergence of Seedlings at Specified Soil Temperatures[1]

	Temperature in Degrees F and Degrees C				
	41° F (5° C)	59° F (15° C)	66° F (19° C)	77° F (25° C)	86° F (30° C)
Crop	Days to Emerge				
Beans	—	16	11	8	6
Beets	42	10	6	5	4
Carrots	51	10	7	6	6
Corn...........	—	12	7	4	4

[1]Source: Cornell Ext. Bul. 1176.

ate a problem with some crops because of a shorter growing season, uneven work distribution, midsummer water shortage, and reduced yields. In spite of its disadvantage, early planting generally produces better crops of forage, corn, spring grain, and many vegetables.

Spring tillage has a warming effect on soil. Plowing removes weeds that shade the soils, thereby exposing a larger area of soil to radiation. Numerous soil temperature readings show that within a few hours after plowing on a clear day, soil temperatures are increased by as much as 4° F (2° C).

6:3.2 ▫ Soil Moisture

There is a direct relationship between the amount of moisture in the soil and its temperature. About five times more heat units of radiant energy are needed to raise the temperature of water than of an equal weight of dry soil. Increased tillage can be employed to speed the removal of excess water from the seed zone in the early spring, and decreased tillage can delay the loss of needed water later in the summer. Soils should not be worked when they are too wet (see Section 6:4).

Increased aeration and increased infiltration as a result of tillage are, in a sense, accomplished together. A soil that permits good air circulation also encourages greater water movement both into and throughout the soil.

It is normal to have either too much or too little soil moisture at different times on the same soil in the same growing season. The excess water which follows any hard storm or irrigation is not harmful unless it is trapped on the surface or within the root zone for an extended period.

In the 20-inch (51-cm) rainfall belt in Kansas, the results of a tillage experiment over an 18-year period at three locations may be summarized as follows:

> Late spring plowing resulted in an average of 1 inch (2.5 cm) of stored available soil water at seeding time and an average yield of 8.4 bushels of wheat per acre (565 kg/ha). By contrast, early spring plowing allowed 1.8 inches (4.6 cm) of available soil water to be stored, which produced 11.5 bushels of wheat per acre (773 kg/ha). A yield of 20.3 bushels of wheat per acre (1,365 kg/ha) was obtained following a year of fallow (no crop) when there were 5.9 inches (15 cm) of available soil water at seeding time (Figure 6.2).

6.3.3 ▫ Bulk Density

The term *bulk density* is used to describe how tightly soil particles are pressed together. **Bulk density** is the density or weight of a given bulk (unit volume) of soil. The grams per cubic centimeter (gm per cc) of soil, including the pores, is the density. Bulk density has a direct bearing on the rate of water and air movement; the incidence of surface flooding, erosion, and drought; and the soil temperature.

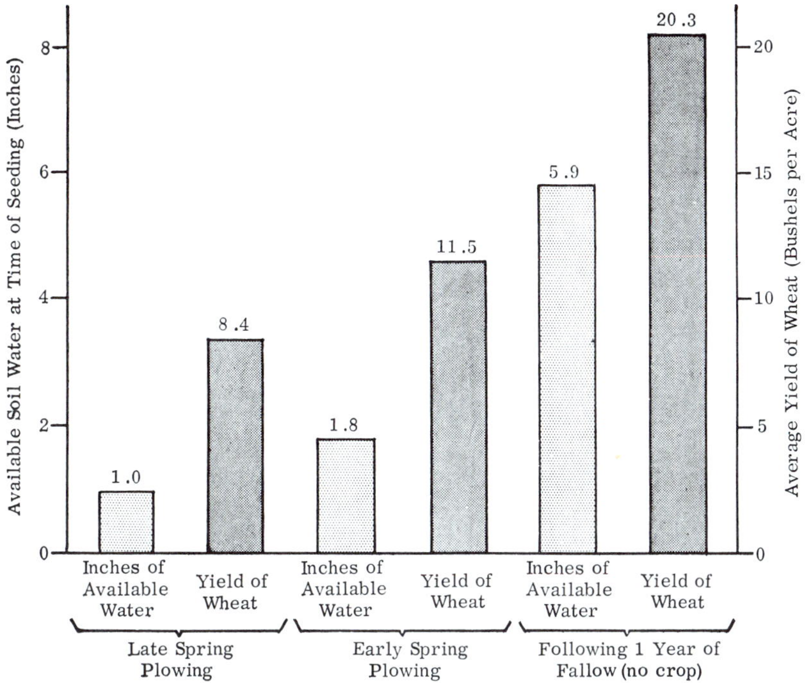

Fig. 6.2 — Tillage, available soil moisture, and yield of wheat in Kansas. (Courtesy, Kansas State University)

Note: The results are averages of three locations for a period of 18 years, in the 20-inch (51-cm) rainfall belt.

Bulk densities of most surface soils range from less than 1.0 to more than 2.0 gm per cc, depending on their condition. Freshly plowed land has a bulk density of less than 1.0, which is too loose for some small-seeded crops and for the efficient operation of some planting tools. Most plants do best at a bulk density of 1.1 to 1.4 gm per cc. At a density of about 1.6, water movement and root development are severely curtailed. Very compact subsoils may have bulk densities as high as 2.0 gm per cc or even higher and thus have no roots growing in them. (This is desirable for roadbeds but not for growing plants.)

Tillage can be employed to make rather rapid, though sometimes temporary, changes in bulk density. A plow can immediately change the density of the furrow from about 1.5 to 0.8. Four to five trips over plowed land with secondary tillage tools may repack the soil to a density of 1.4. An ordinary tractor tire will compact freshly plowed soil to the same density that it was before plowing. The soil density is also influenced by forces other than tillage equipment. The beating action of rainfall compacts the soil, while alternated wetting and drying and freezing or thawing constantly operate to decrease the density.

6:3.4 □ Granulation

The size, number, and location of soil granules are affected by management and time. Plowing increases the oxidation rate of the organic matter in the soil and speeds its decay into **humus**. Humus is a sticky substance which loosely holds individual soil particles together into aggregates like popcorn in a popcorn ball. These *naturally* aggregated particles are called **peds** and, unlike the **clods** in puddled and wet or compacted soil, readily crumble into porous granules that can be easily crushed between the fingers. This is the mark of a mellow, productive, and easily managed soil.

The best **granulation** is found on productive soil worked at the right moisture content. This desirable **crumb structure** usually lasts about 30 days after planting, then raindrop impact and natural weathering tend to **slake** (break down) the crumbs into smaller segments.

One purpose of tillage is to preserve as many of the **granules** (soil aggregates) as possible and concentrate them around the seed where a uniform environment is most important. Some larger lumps are not objectionable if they do not create large air pockets in the root zone and are not in contact with the seed. In fact, some egg-size lumps on the surface may reduce the danger of crusting, discourage weeds, and slow wind and water erosion (Figure 6.3).

The selection and adjustment of tillage tools and the timing of cultural operations play an important part in determining how well soil is granulated.

Fig. 6.3 — Moisture, management, and machine work together to develop a well-granulated seedbed. Cultivation is done on the contour with an undercutting tool (sweep) to leave stubble on the surface. (Courtesy, USDA - Soil Conservation Service)

6:3.5 ▫ Porosity

Porosity is closely related to granulation, and, like granulation, can to a large extent be controlled. The large pores are openings between granules which allow free water to move downward by gravity, thus draining the surface, wetting the lower subsoil, and recharging the underground water table. The large pores also allow air to permeate the root zone. Compaction that markedly increases the soil bulk density seriously reduces the functioning of these pores.

The small pores among the granules are the storehouse for available water and the location of plant nutrient uptake. Once this moisture is absorbed around the soil particles it does not move by gravity, and it will be available until it is used by plants and/or evaporates.

Depending on the texture and structure, the small pores in a soil to a depth of 6 feet (1.8 m) will hold from 6 to 12 inches (15 – 30 cm) of available water. In humid regions, this amount is equal to one to three months of normal rainfall and explains why plants can withstand drought. This ability of soil to hold water explains the statement "A well-managed soil is second only to a lake as a reservoir for water."

6:4 ▫ SOIL COMPACTION

Soil compaction means that soil particles are closer together. Compaction occurs by both natural and human forces. The greatest cause of compaction is tillage. When compaction occurs, there is less space for air and water among soil particles. A severely compacted layer may be formed just below the normal tillage depth. Working the soil at the same depth year after year, especially plowing it with a moldboard plow, increases the chances of a root-restricting, compacted layer developing. These compacted layers are called **hardpans, traffic pans, plowpans, wheel compactions,** and **disc pans.** Some compaction is likely to occur in most soils that are tilled.

Several factors have increased the possibilities for soil compaction in recent years, including larger and heavier farm equipment, more applications of fertilizers and pesticides which may increase trips over the field, and earlier spring seedbed preparation causing soils to be worked while they are too wet. These factors combined have the potential to greatly compact soil. Such problems may be lessened by reducing tillage and by combining operations such as planting, fertilizing, and applying pest control chemicals with one pass over the field.

A soil's value to plants is based on its ability to provide a suitable environment for plant roots. A good soil provides plants with water, air, and nutrients. A soil must be able to transmit water and heat to produce good plant growth. The ability of soils to provide a good environment for plant roots is, to a large extent, determined by the size and distribution of pore spaces among soil particles. Compaction of soil alters the distribution and reduces the size of pores.

Compaction reduces the rate of infiltration of water, causing more runoff and erosion and less water stored in the soil for plant growth.

Compaction causes surface crusting which reduces seedling emergence and root growth. Tillage and wheel traffic increases compaction. Working soils while they are wet increases compaction to depths below normal tillage where it is more difficult to break the dense layer with future tillage. Roots have difficulty growing through compacted soils. Plants with restricted roots won't reach their full yield potential.

For example, in an unrestrictive soil, cotton roots will grow 6 feet (1.8 m) deep and 4 feet (1.2 m) across. Compare that with the curtailed root development shown in Figure 6.4. These plants grew in a field with tractor wheel compaction between rows and a tight plowpan. Roots didn't grow in the soil compacted under the tractor wheels or in the plowpan on the bottom. This situation is common on much of our farmland. So, instead of growing crops in a 4- or 5-foot (1.2- or 1.5-m) deep soil, we are essentially farming in a window box — a little strip 8 or 9 inches (20 or 23 cm) deep and 13 inches (33 cm) wide. A vivid demonstration of benefits from planting a crop directly above a subsoiled break in the hardpan is shown in Figure 6.5 from the National Tillage Laboratory at Auburn, Alabama. Corn roots readily penetrated the plowpan through the subsoiled area to get water and nutrients from below the hardpan. Notice in the upper right corner how roots failed to penetrate the compacted soil where the tractor tire traveled and

Fig. 6.4 — This photograph was taken in a cotton field in Alabama. The root system is confined on the sides by soil compacted by tractor wheels and by the plowpan on the bottom. The result is the same as if the cotton were growing in a window box. (Courtesy, National Tillage Laboratory, USDA – Agricultural Research Service, Auburn, Alabama)

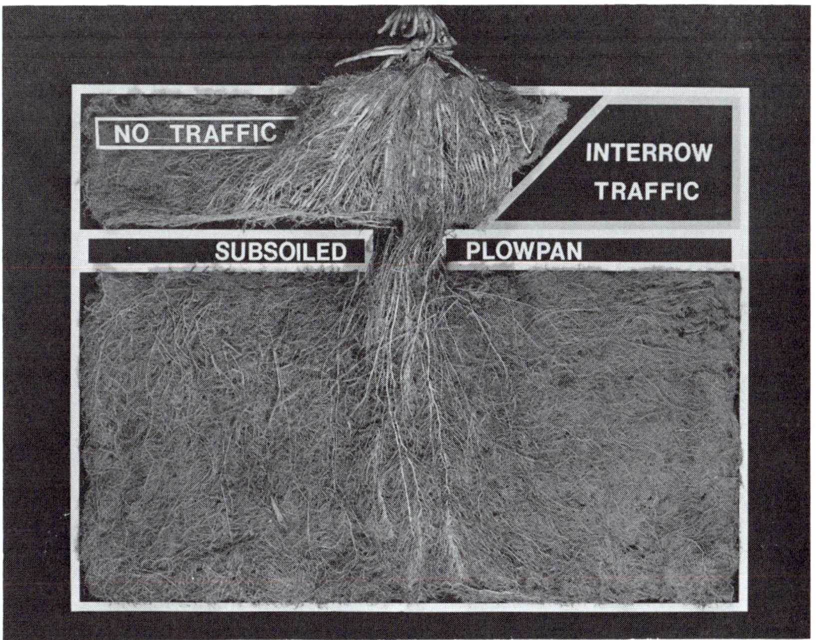

Fig. 6.5 — Row subsoiling allows roots to grow into the subsoil even though there is a hardpan present. Note that interrow traffic (tractor tire) has also interfered with root growth. (Courtesy, National Tillage Laboratory, USDA - Agricultural Research Service, Auburn, Alabama)

how roots grew profusely in the upper left corner where there was no traffic and no compaction.

In California, a study was made of tractor compaction in relation to the ease of movement of water (infiltration) into the soil. When no tractor was run over the soil, water moved into the soil at the rate of 1.6 inches (4.1 cm) per hour; whereas, when the tractor was run over the soil, the infiltration was reduced to 0.4 inch (1.0 cm) per hour, only one-fourth as much (Figure 6.6). Turning plows have long been considered a factor in compacting soil. Machine traffic is an increasing contributor to soil compaction (Figure 6.7).

Soil compaction resulting from off-road vehicles in the deserts of California was documented in the 1982 September/October issue of *California Agriculture.* Wheel-track compaction reduces the density of vegetation and causes serious erosion.

Compacted soils may be classified as follows:

- **Induced pans** — Soils in which the layer that limits water movement or root penetration is the result of the use of plows or other heavy machinery. Induced pans include plowpans that are formed about 6 inches (15 cm) deep and tillage pans that are formed below this depth by the use of heavy machinery (Figure 6.8).

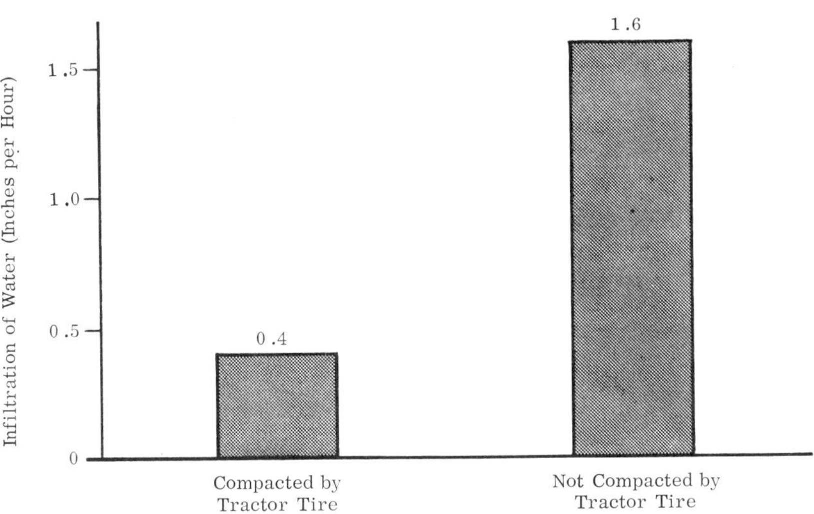

Fig. 6.6 — Tractor tire compaction reduces water infiltration in California. (Courtesy, Potash and Phosphate Institute)

Fig. 6.7 — Continuous tillage at the same depth of a fine-textured soil, especially with heavy machinery, high horsepower, and high speeds, will form a tillage pan of compacted soil just below the depth of plowing. (Courtesy, Ford Tractors)

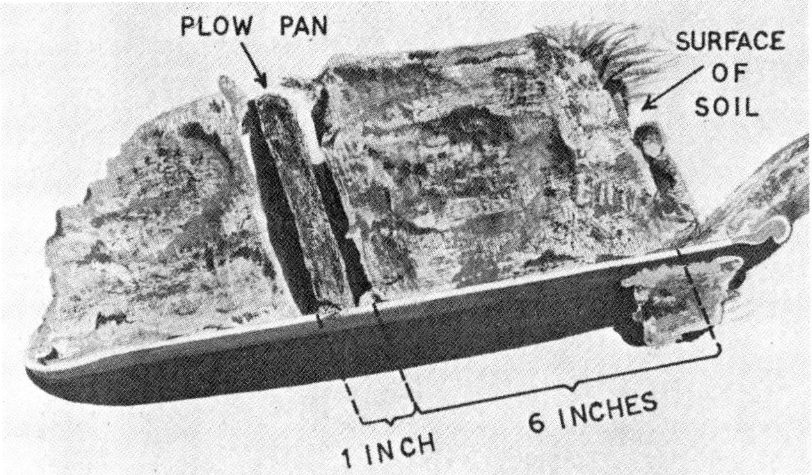

Fig. 6.8 — The 1-inch thick (2.5-cm) compacted layer in the center, occurring at a depth of approximately 6 inches (15 cm), is the tillage pan or plowpan. It is caused by the sliding and compacting action of the bottom of a turning plow, especially when the soil is fine-textured and wet when plowed. This plowpan can be shattered with a chisel that runs 7 inches (18 cm) or more in depth. (Courtesy, Oklahoma State University)

- **Induced crusts** — Soils in which the surface inch or more is crusted and compacted due to the action of beating raindrops on unprotected soil, the trampling action of livestock or humans, or off-road vehicles.

- **Natural pans** — Soils further classified into:
  - Claypan horizons due to the accumulation of clay.
  - Fragipan horizons due to the accumulation of silt.
  - Indurated (cemented) horizons due to natural cementation by iron, aluminum, silica, calcium carbonate, gypsum, or colloidal humus.

6:5 ▫ CONSERVATION TILLAGE

There is some question as to what various tillage systems are called. **Conservation tillage** as used here is very broadly defined as "any tillage which reduces erosion." Tillage implements which thoroughly work and/or invert the soil, such as a turning plow, generate the highest possible potential for erosion by water and wind (Figure 6.9). An offset disc works residues into the soil but still leaves considerable soil exposed to erosion (Figure 6.10). A chisel plow leaves some of the surface covered with mulch and less soil exposed (Figure 6.11). Sweep tillage equipment may shatter and lift the soil, thus killing fall weeds while leaving residue in place for wind and water erosion control (Figure 6.12). The major purpose

Fig. 6.9 — A moldboard plow leaves a beautiful seedbed, but since little organic residue is left on the surface, the soil has little protection against erosion. (Courtesy, Case Tractors)

Fig. 6.10 — An offset disc works residue into soil. Soil is subject to less erosion than when it is plowed with a turning plow. (Courtesy, Case Tractors)

Fig. 6.11 — A chisel plow leaves some of the residue on or near the surface to help reduce erosion. Chisel plows may also be used to help break up pans. (Courtesy, Case Tractors)

Fig. 6.12 — Sweep tillage. This subsurface tillage sweep shatters and lifts the soil, thus killing the fall weeds while leaving the residue in place for wind and water erosion control. (Courtesy, USDA - Soil Conservation Service)

of conservation tillage is to reduce erosion, and this is done largely by managing residues. Parts of conservation tillage have been identified under several different names, namely, no-tillage or zero tillage, till-plant, ridge plant, minimum tillage, mulch tillage, and stubble mulching. *Conservation tillage* is the term more frequently used to describe tillage practices that provide for control of wind and water erosion and for an increase in the effective supply of water for crops. With some systems, a surface configuration is achieved that will increase infiltration and water retention. With others, residue from the previous crop is left on the soil surface to break raindrop impact and reduce the amount and velocity of the runoff.

6:5.1 □ Types of Conservation Tillage

The conservation tillage that best fits a farm operation depends on several factors, such as crops grown, soil characteristics, and climate of the area. Significant progress has been made in developing successful conservation tillage systems for a number of climatic areas and crop sequences. Most soils, except poorly drained ones in northern latitudes, are suitable for conservation tillage. The systems listed below have all been used and have been shown to be effective in reducing soil erosion.

- □ **No-tillage** *or* **zero tillage** (Figures 6.13 and 6.14) — This system uses a fluted colter or double-disc openers to cut through or push aside residues of the previous crop, ahead of the planter shoe. No seedbed preparations precede this operation. This system leaves a maximum of residue cover.

Fig. 6.13 — No-till planter. Old cornstalks are cut or pushed aside, and the field is planted with no soil preparation. Corn is not touched until harvest time. (Courtesy, USDA - Soil Conservation Service)

Fig. 6.14 — No-till planter. As barley is being harvested, a farmer plants soybeans in the barley stubble. By using conservation tillage in a double-cropping system, the farmer is protecting his soil from erosion as well as saving time and fuel. (Courtesy, USDA - Soil Conservation Service)

- **Ridge plant** (Figures 6.15 and 6.16) — This system gives a row configuration similar to listing (described next in this list), but planting is done on the ridge year after year with no seedbed preparation preceding planting. This system has the most promise for controlling erosion when the ridges are on the contour. Rain falling on the ridge must run down the ridge into the residue which has collected in the furrow. Most of the soil sediment is deposited in the crop residue and is kept near the point of detachment.
- **Listing** — Plowing and planting are done in the same operation. Plowed soil is pushed into ridges between rows, and seeds are planted in the furrows between the ridges. When operated on the contour, this system conserves soil and water (see "Ridge plant" above).
- **Till-plant** (Figure 6.17) — With this system, wide sweep and trash bars clear a strip over the old row, and a narrow planter shoe opens a seed furrow into which a seed is dropped. A narrow wheel presses the seed into firm soil, and covering discs place loose soil over the seed. This system controls erosion most satisfactorily when done on the true contour or across the slope on the approximate contour.
- **Strip tillage** — A narrow strip is tilled with a rototiller gang or other implement. Seed is planted in the same operation. This system is applicable on soils where some tillage is desirable to aerate and warm the soil in the plant root zone.

Fig. 6.15 — Buffalo ridge cultivator being demonstrated at the 1987 Farm Progress Show in Alleman, Iowa. (Courtesy, Conservation Technology Information Center)

Fig. 6.16 — This seedbed for cotton on the Texas Blacklands is ridged on the contour to provide a less erosive, deeper, better-drained, and faster-warming soil for earlier planting. (Courtesy, Texas A&M University)

Fig. 6.17 — Buffalo till-planter, planting in old corn residue. No plowing is done prior to planting. (Courtesy, USDA - Soil Conservation Service)

- **Sweep tillage** (see Figure 6.12) — This practice is used on small-grain stubble to kill the early fall weeds. It shatters and lifts the soils, thus enhancing infiltration while leaving the residue in place for water and wind erosion control.

- **Chisel planter** (Figure 6.18) — This system breaks or loosens the soil without inversion. Most of the crop residue remains on the surface for control of water and wind erosion.

- **Plow-plant** — Planting is done directly into plowed ground with no secondary tillage. This system increases infiltration, water storage in the plow layer, and surface storage. Surface sealing is delayed because of the large clods and peds.

- **Wheel-track plant** — This system is similar to plow-plant, but is not restricted to freshly plowed ground. Planting is done in the compacted wheel tracks of the tractor or planter.

- **Paraplow or paratill** (Figure 6.19) — This system provides deep tillage with little disturbance of surface residues. Planting can be done with a no-till planter.

In determining which type of conservation tillage to use, you need to con-

Fig. 6.18 — A chisel planter planting in old corn residues. The soil surface is left rough in order to hold and absorb surface moisture and to reduce wind and water erosion. (Courtesy, USDA - Soil Conservation Service)

Fig. 6.19 — A paratill breaks up compaction and pans to a depth of 14- to 16-inch (35.6- to 40.6-cm) depths while leaving most crop residue on the soil surface. (Courtesy, The Tye Company)

sider several factors: (1) crop yields; (2) soil erosion; (3) fertilizer and lime usage; (4) water utilization; (5) disease, insect, and weed control; and (6) energy use.

6:5.2 □ Crop Yields

In many areas of the country, when adequate nitrogen was applied, corn yields have been as good or slightly better with conservation tillage than with conventional tillage on well-drained soils. On the other hand, cooler soils such as those found in the northern United States hold more water with conservation tillage than with conventional tillage and thus warm more slowly. This generally requires a later planting date, which results in lower yields because of the shorter growing season. Poorly drained soils also respond with lower yields under conservation tillage.

6:5.3 □ Soil Erosion

Soil erosion caused by both wind and water can be reduced to almost zero under no-tillage (Figure 6.20) and reduced greatly with almost all kinds of conservation tillage. This saving in erosion not only protects a farmer's most valuable resource, the soil, but also prevents ditches from filling up and helps reduce

Fig. 6.20 — This zero tillage equipment plants soybeans in wheat stubble. This leaves the surface virtually undisturbed and keeps wind and water erosion to a minimum. (Courtesy, USDA - Soil Conservation Service)

air pollution, water pollution, and the loss of nutrients. While results of erosion studies vary with soil, slope, geographic location, and climatic conditions, it certainly can be said that conservation tillage provides conditions which promote soil conservation and reduce soil sediment losses which pollute our environment. (See Chapter 7.)

6:5.4 □ Fertilizer and Lime Usage

One of the early concerns about reduced tillage or no-tillage was the lack of tillage operations to incorporate the fertilizer and lime. Phosphorus and potassium fertilizer and lime applied on the soil surface move downward very slowly. In soils of low fertility, this movement into the root zone may not be rapid enough to meet plant nutrient needs. In these situations, fertility needs can be met by plowing down fertilizer and lime before reduced tillage practices begin.

Soil testing provides the information for determining fertilizer and liming needs for all tillage systems. Soils that require lime should have the lime applied and worked into them to plow depth before being switched to reduced tillage. For reduced tillage systems, it may be advisable to take the soil sample in 3-inch (7.6-cm) increments in order to determine if the surface is becoming acid compared to the soil at the 6- to 8-inch (15- to 20-cm) depth. A surface soil pH of 6.2 or less could mean trouble, because the triazine herbicides become less effective as the soil pH decreases.

The primary inorganic form of nitrogen found in the soil is nitrate-nitrogen (NO_3^-–N). Soil bacteria convert other forms of nitrogen to the nitrate form. The nitrate form is water soluble and moves quickly within the soil. Therefore, nitrogen applied to the soil surface should be readily available for nutrient absorption by plants provided there is adequate moisture for movement of nitrate into the root zone.

Surface application of phosphorus and potassium results in the concentration of these elements at or near the surface. If the surface becomes dry, these elements become **positionally unavailable** to plants. With adequate rainfall, moist surface soil encourages active root growth near the surface and makes phosphorus and potassium available to plants. Furthermore, the minimum tillage approach with a mulch of crop residues near the surface tends to reduce evaporation and maintain high moisture levels for the active root zone near the surface. Several reports from the more humid states have indicated an excellent utilization of surface-applied phosphorus and potassium in reduced tillage systems.

Lime must come in contact with the maximum number of soil particles to achieve the quickest reaction. Continuous reduced tillage practices allow little chance for soil manipulation to incorporate lime. Therefore, there may be a need to modify the standard liming program.

The pH of the surface layer of soil may become acidic with repeated surface nitrogen applications. For this reason, plowing to incorporate the lime may be necessary after several years of reduced tillage.

6:5.5 ▫ Water Utilization

Soil water content is almost always higher in conservation tillage than in conventional tillage. The presence of a mulch on the surface of the soil reduces water evaporation and allows for better water penetration into the soil (Figure 6.21). This extra water can save soil and increase yields, especially in dry years (Figure 6.22).

6:5.6 ▫ Disease, Insect, and Weed Control

Some diseases, such as stalk rots, are less common with the use of conservation tillage than with the use of conventional tillage. Other diseases, such as early anthracnose in corn, are worse under conservation tillage. Two practices that farmers can use will help with most disease problems. First, and perhaps most important, is to rotate crops to prevent disease organism buildup and, second, to select disease-resistant varieties. This is especially important where crops are known to be exposed to disease.

Weed control is critical for many conservation tillage methods. Herbicide application must control weeds for the entire season, since no supplemental culti-

Fig. 6.21 — The mulch in the top layer of soil after it has been tilled with the chisel plow hooked in tandem with a stalk chopper leaves a desirable surface for the infiltration and retention of rainwater. (Courtesy, USDA - Soil Conservation Service)

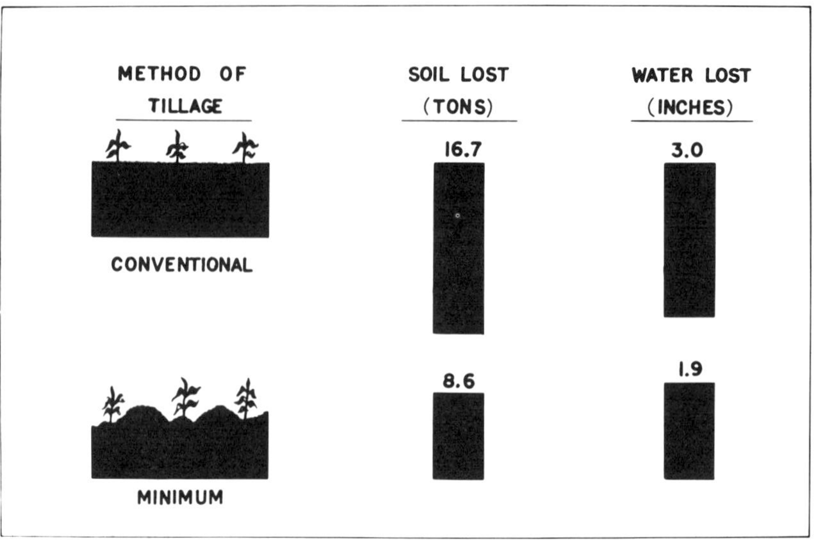

Fig. 6.22 — Minimum tillage cultivation on corn reduced water losses by more than one-third and soil losses by nearly one-half. In the Corn Belt, each additional inch (2.5 cm) of water that goes into the soil can produce 8 more bushels of corn per acre (502 kg/ha). (Courtesy, John Deere & Company)

vation is usually planned. The herbicides must be tailored to fit the weed problem and soil conditions (Figure 6.23).

6:5.7 □ Energy Use

Agriculture is a very energy-dependent industry. While agriculture uses only 3 percent of the total energy consumed in the United States, costs are an important factor in determining profit or loss to the individual farmer. Conservation tillage can reduce fuel consumption. Some of the savings may be offset by increased use of fertilizer, herbicides, insecticides, and seed. A study in Kentucky showed about 7 percent net savings on no-till corn. Since nitrogen is not added to no-till soybeans, the savings on no-till soybeans were about 18 percent (Figure 6.24).

6:6 □ SUMMER FALLOW

Summer fallowing (leaving the land fallow, or without a crop) is generally used for profitable crop production in a semiarid region with less than 20 inches (51 cm) of precipitation per year. It is practiced primarily to conserve precipitation received one year to be used for crop production the next. During the fallow year and the period of crop establishment, wind and water erosion can cause serious crop and soil damage.

Fig. 6.23 — No-tillage corn. Rapid growth is shown where corn was planted in wheat stubble and competing weeds were chemically killed at planting time. The wheat stubble is an excellent surface mulch. (Courtesy, USDA - Soil Conservation Service)

Fig. 6.24 — One of the major concerns for future agricultural production is the availability of the necessary petroleum products. Adapting minimum tillage methods reduces the per acre consumption of fuel. The corn planter shown above equipped with spring teeth is preparing a seedbed and saving fuel by planting, fertilizing, and applying insecticide and herbicide in one operation. (Courtesy, USDA - Soil Conservation Service)

Land fallowed without crop residues on the surface and by periodic surface cultivation (**black fallow**) is subject to wind and water erosion during the fall, winter, and spring, and during the early cropping season. However, land that has considerable crop residues left on the soil surface has only slight wind and water erosion, either during the fallow period or after the crop is seeded. The beneficial effect of crop residues in reducing both wind and water erosion has become widely recognized, especially throughout the Great Plains.

Stubble-mulch farming (Figure 6.25) was developed and expanded as a result of severe wind erosion which occurred in the Great Plains and Canada in the late 1930's. Surface cultivation with small sweeps or springtooth harrows tends to roughen the surface of the soil. Compared to conventional tillage, this helps prevent soil blowing between rains while increasing the intake of water during rains.

Fig. 6.25 — Stubble-mulch farming with a three-blade flexible sweep. The protective wheat stubble is undercut, and the land will lie fallow to gain soil water for the next crop, but with a minimum loss of soil from wind and water erosion. (Courtesy, USDA - Soil Conservation Service)

6:7 □ DOUBLE-CROPPING

Growing two crops on the same field in a single year has been a popular moneymaker in the South for a number of years. With the coming of minimum tillage, short-season crop varieties, and improved chemical weed control, the system of double-cropping is becoming popular throughout the United States and in southern Canada.

Soybeans following small grains seems to be one of the most common double-cropping sequences. Other combinations include small grain – sorghum; small grain – corn; small grain silage – soybeans; grass silage – soybeans; hay – soybeans; and pasture – soybeans. The advantages of time and moisture conservation can be gained from no-till planting of soybeans after a small grain harvest, according to research conducted in Missouri.

Conservation tillage, especially no-till, probably is a contributing factor in the gradual northward movement of double-cropping. It is used in conjunction with a substantial portion of double-cropped acreage.

Serious soil erosion often occurs on sloping fields where wheat and soybeans are double-cropped with conventional tillage systems. However, minimum tillage solves some of the problems because the crop residue covers the soil during these highly erodible periods between crops. Research in Illinois demonstrated that no-till planting compared to conventional tillage reduced soil losses by 93 percent on a 9 percent slope

6:8 ▫ CONTROL OF TILLAGE PANS

A tillage pan (plowpan) is caused by the repeated use of a plow or other tillage implement operating at the same depth each year and causing compaction of the soil. The result is restricted, shallow roots (Figure 6.26).

Fig. 6.26 — These cotton roots reached the plowpan and were forced to turn sideways. Cotton normally has a strong taproot that grows straight down and strong lateral roots. (Louisiana) (Courtesy, USDA - Soil Conservation Service)

The prevention and control of tillage pans and management to overcome the effect of natural pans include the following practices:

1. Tillage operations may be simplified by fewer trips being made over the field, especially when the soil is wet. This is called "minimum tillage."

   In one New York study, tractor wheels packed the soil from a density of 75 pounds to a density of 100 pounds per cubic foot (1.2 – 1.6 gm/cc). The squeezing together of one-third more solid soil particles in each cubic foot means one-third less pore space to supply air and water to plant roots. The problem is just as serious as that of a heavily loaded life raft losing one-third of its air.

   It has been estimated that a tractor travels 25 miles (40 km) in the process of producing 1 acre (0.4 ha) of potatoes. Rainwater has been observed to stand in a tractor wheel track for as long as seven days, although the soil was a loam and had pervious gravel underlying it at a depth of 30 inches (76 cm).
2. A good rotation system, including a sod crop, should be followed for as long as possible. For example, in Pennsylvania, the density of the surface soil under a bluegrass sod for many years was 68.7 pounds per cubic foot (1.1 gm/cc). Under a good farm crop rotation, the soil had a density of 78.0 pounds per cubic foot (1.25 gm/cc), or 15 percent heavier. Continuous cultivation, especially with the use of heavy machinery at high speeds, undoubtedly would have made the soil even more dense, perhaps to a density of 100 pounds per cubic foot (1.6 gm/cc), or 45 percent more dense than under bluegrass sod.
3. The organic content of the soil should be increased by applying liberal amounts of barnyard manure or sewage sludge, leaving all crop residues, and using large quantities of lime (on acid soils) and commercial fertilizers. Proper liming and fertilizing will increase root and top growth, thus making more organic matter available to cushion the soil against soil compaction caused by heavy farm implements. Moreover, proper soil management practices result in stronger root systems that are capable of penetrating compacted layers (Figure 6.27).

   Fifty years of cropping in Ohio increased the density of the plow layer from 56.2 to 68.7 pounds per cubic foot (0.9 – 1.1 gm/cc), an increase of more than 22 percent in weight due to the packing action of tillage machinery and the reduction in the percentage of organic matter by faster decomposition with continuous tillage. Since organic matter weighs less per cubic foot or per cubic centimeter than mineral soil, a reduction in the percentage of organic matter due to tillage makes any given volume of soil weigh more.
4. The soil should be chiseled when dry so as to shatter the tillage pan (Figure 6.28). To be effective for more than a year, chiseling must be followed

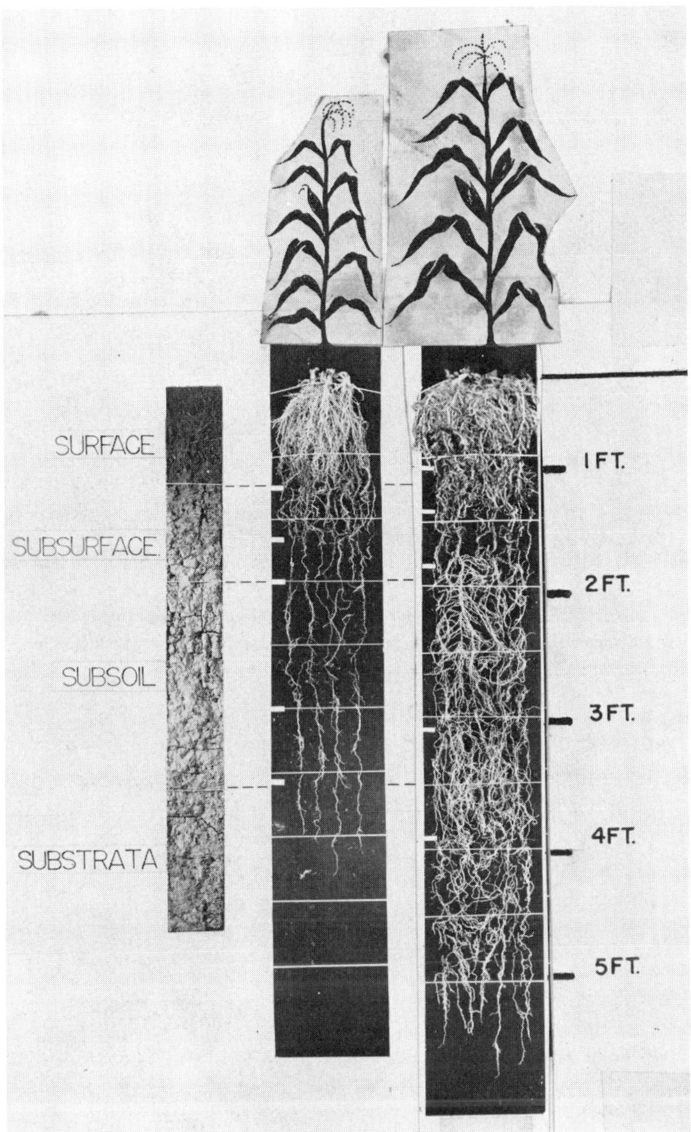

Fig. 6.27 — The yield of corn on a soil with a *naturally* compacted subsoil was increased from 20 to 75 bushels per acre (1,256 - 4,708 kg/ha), following several years of treatment with crop residues, green manures, lime, and fertilizers. *Left*: A profile of Cisne silt loam in southern Illinois, with a compacted layer starting at a depth of 18 inches (46 cm) and extending to 40 inches (102 cm). *Center*: Corn roots from a plot with no treatment. *Right*: Corn roots from a plot that had received crop residues, green manures, lime, phosphate, and potash. (Courtesy, University of Illinois)

Fig. 6.28 — Shattering subsoil to a depth of about 16 inches (41 cm) with a Vibra-tiller. Tillage pans form easily under cultivation and need to be broken up for better root and water penetration. (Courtesy, USDA - Soil Conservation Service)

by planting a deep-rooted crop, such as alfalfa or sweetclover, or an adapted perennial grass, such as orchardgrass or bromegrass.

5. The depth of tillage should be varied from one year to the next.

6:9 □ CONTROL OF SURFACE CRUSTS

Surface crusts are commonly formed by raindrops striking bare soil which has little or no stable structure and beating it into flowing mud. The beating action of the raindrops and the formation of small pools of water cause the sand to settle to the bottom. Silt comes to rest on the sand, and the clay stays in suspension in the water and is finally deposited on top. When the pools of water dry up, a surface crust is formed with clay on top. The clay crust is usually about ¼ inch (0.6 cm) or less in thickness, but is very harmful because it restricts the amount of oxygen available for plant roots. Coarse, sandy soils do not readily form surface crusts; neither do well-granulated clay soils. Upon drying, most clay soils shrink and crack enough to break up any crust. Sandy loam and loam soils low in organic matter seem to produce the thickest and most dense soil crusts.

Crusts are harmful because they prevent the emergence of seedlings, especially of small seeded crops such as grasses and legumes. Crusts are also harmful because they reduce the amount of rainwater which enters the soil. Likewise, crusts reduce the amount of necessary air exchange for proper root growth, resulting in a shallow root system. Special implements are often used to break the surface crust (Figure 6.29).

Providing a cushion of coarse organic matter, either living or dead, on the soil surface to break the impact of the falling raindrop will prevent surface crusts. Duley, in Nebraska, found that the physical condition of the soil surface was more influential in regulating water intake and reducing crust formation than soil series, slope, or soil moisture content was. Cropping systems that include sod-forming crops are desirable for reducing surface crust formation.

Fig. 6.29 — A rotary hoe is being used to shatter the surface crust on a small field. (Courtesy, Minneapolis-Moline Company)

After small seeds, such as lettuce, are planted, rain, sprinkler irrigation, or flood irrigation often causes the soil to form a crust above the seed, thus reducing germination and emergence. The University of California proved that when applied on the surface of the soil directly above the planted seed, the following materials are satisfactory in overcoming the detrimental influence of surface crusting: (1) vermiculite (mica expanded by roasting), (2) farmyard manure, (3) ground coke, (4) rock dust, (5) petroleum mulch sprayed on the surface, and (6) **krilium** (a synthetic organic chemical related to the natural slime of earthworms).

6:10 □ REFERENCES

Cook, Ken. "Conservation Tillage: Marrying for Money." *Journal of Soil and Water Conservation,* Vol. 39, No. 6, November – December 1984, pp. 368 – 370.

Ditsch, D. C. "Ridge-Tillage: Advantages and Disadvantages." *Agronomy Notes,* Vol. 7, No. 4, University of Kentucky, April 1986, 2 pp.

Donahue, Roy L., Raymond W. Miller, and John C. Shickluna. *Soils: An Introduction to Soils and Plant Growth,* 5th ed. Englewood Cliffs, New Jersey: Prentice-Hall, Inc., 1983, pp. 46 – 85.

Entz, M. H., and D. B. Fowler. "Critical Stress Periods Affecting Productivity of No-Till Wheat in Western Canada." *Agronomy Journal,* Vol. 80, 1988, pp. 987 – 992.

Frisby, James C., and Donald L. Pfost. "Soil Compaction: The Silent Thief." *Agricultural Guide.* University of Missouri – Columbia, July 1987, 3 pp.

Halvorson, A. D., and G. P. Hartman. "Nitrogen Needs of Sugarbeet Produced with Reduced-Tillage Systems." *Agronomy Journal,* Vol. 30, 1988, pp. 719 – 722.

Hargrove, W. L., F. C. Boswell, and G. W. Langdale, eds. *The Rising Hope of Our Land: Proceedings, Southern Region No-Till Conference.* Griffin, Georgia, July 16, 17, 1985, 247 pp.

Herbek, Jim, Lloyd Murdock, and Bob Blevins. "Effect of Planting Dates of No-Till and Conventional Corn on Soils with Restricted Drainage." *Agronomy Notes,* Vol. 17, No. 3, University of Kentucky, October 1984, 4 pp.

King, Kevin. "Ridge Tillage: A Cost-cutting Option." *Agriculture News,* The Ohio State University, August 1986, 2 pp.

"No-Tillage: An Old Idea with New Life." *Science of Food and Agriculture,* Vol. 3, No. 3, Iowa State University, September 1985, pp. 22 – 27.

Phillips, R. E., and S. H. Phillips, eds. *No-Tillage Agriculture: Principles and Practices.* New York: Van Nostrand Reinhold Co., 1984, 306 pp.

Pierson, B. J., C. E. Lewis, and C. A. Birklid. "Observer Differences in Determining Crop Residue Cover in the Alaskan Subarctic." *Journal of Soil and Water Conservation,* Vol. 43, No. 6, 1988, pp. 493 – 495.

Schmepf, Max, ed. "Conservation Tillage, Special Issue." *Journal of Soil and Water Conservation,* Vol. 38, No. 4, May – June 1983, pp. 134 – 319.

Siemens, J. C., D. R. Griffith, and S. D. Parsons. "Energy Requirements for Corn Tillage-Planting Systems." *National Corn Handbook,* Purdue University, NCH-24, January 1986, 4 pp.

Stein, O. R., *et al.* "Runoff and Soil Loss as Influenced by Tillage and Residue Cover." *Soil Science Society of America Journal,* Vol. 50, pp. 1527 – 1531.

Unger, Paul W. "Residue Management Effects on Soil Temperature." *Soil Science Society of America Journal,* Vol. 52, 1988, pp. 1,777 – 1,782.

Unger, Paul W. *Tillage Systems for Soil and Water Conservation.* Rome: Food and Agriculture Organization of the United Nations. FAO Soils Bul. 54, 1984, 278 pp.

Veseth, Roger, *et al.* "Effective Conservation Farming Systems." *Crop Management Series: No-Till and Minimum Tillage Farming.* Pacific Northwest Extension Pub., PNW 275, October 1986, 12 pp.

Veseth, Roger, *et al.* "Uniform Combine Residue Distribution for Successful No-Till and Minimum Tillage Systems." *Crop Management Series: No-Till and Minimum Tillage Farming.* Pacific Northwest Extension Pub., PNW 297, May 1986, 6 pp.

Wells, K. L., *et al.* "Effect of Tillage Tools on Improving Corn Yields from a Compacted Soil." *Agronomy Notes,* Vol. 19, No. 3, University of Kentucky. April 1986, 4 pp.

Wells, K. L., and J. H. Grove. "The Effects of Three Tillage Systems on Bulk Density and Porosity of a Pembroke and a Beasley Soil After Three Years." *Agronomy Notes,* Vol. 20, No. 4, University of Kentucky, May 1987, 8 pp.

Wells, K. L., L. W. Murdock, and W. W. Frye. "Intensive Cropping Effects on Physical and Chemical Conditions of Two Soils in Kentucky." *Communications in Soil Science and Plant Analysis,* Vol. 14, No. 4, 1983, pp. 297 – 307.

Note: The Conservation Technology Information Center is a clearinghouse for information on soil conservation, water conservation, and water quality practices on cropland, was established as a special project of the National Association of Conservation Districts, and is administered in cooperation with agricultural industry, governmental agencies, private foundations, organizations, and farmers.

For information, contact:

Conservation Technology Information Center
1220 Potter Drive, Room 170
Purdue Research Park
West Lafayette, Indiana 47906-1334
(317) 494-9555

CHAPTER 7

Soil and Water Conservation

In 1799, George Washington wrote plans for the eroded areas on his farm as follows: "The most broken, washed, and indifferent soils are to remain unplowed; in the Spring these fields are to be harrowed and covered with litter, weeds, or straw to cure the open 'sores.'"

OUTLINE

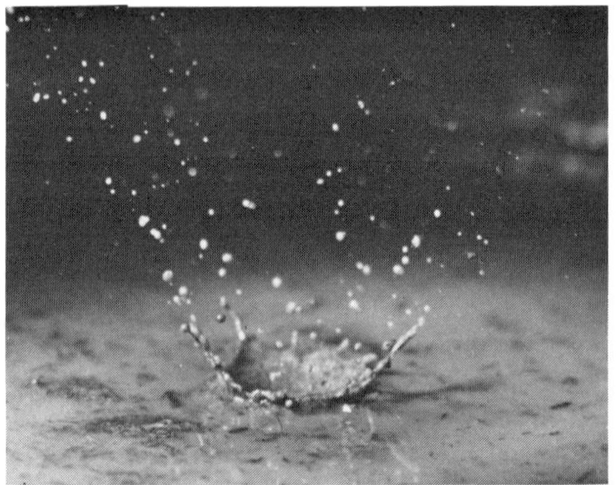

Top: A raindrop falls with a speed of 20 miles (32 km) per hour. *Center:* When the raindrop strikes wet soil, it may splash soil as much as 2 feet (61 cm) high and 5 feet (152 cm) from the spot where it hit. *Bottom:* When the raindrop hits bare soil, it causes splash erosion; continuous splash erosion results in sheet erosion. Sheet erosion is usually so slow that the farmer is seldom conscious of its existence. This picture shows that 18 inches (46 cm) of soil have been lost. (Courtesy, USDA – Soil Conservation Service)

□ □ □

7:1 □ OVERVIEW

Every hour of every day the amount of suspended sediment in the Mississippi River is equal to the topsoil from a 40-acre (16-ha) field. This much topsoil weighs about 33,000 tons (30,000 mt). At a contractor's delivered price of about \$7.50 per ton (\$8.33 / mt), the value of the 33,000 tons is \$7.50 × 33,000 = \$247,500 (rounded to \$250,000 or a quarter of a million dollars) of soil loss each hour. This is equal to a loss of \$6 million every day and \$2.2 billion every year. Please recall that this \$2.2 billion loss each year from soil erosion applies only to the Mississippi River watershed above Memphis. Annual soil erosion losses in all 50 states are 6.4 billion tons (5.8 billion mt) valued at \$7.50 per ton or \$48 billion. This amounts to a loss of \$213 per person per year for every man, woman, and child in the United States, or about \$7,000 a year for each person living on a farm or ranch. Sediment and pollution cause further problems and make for additional costs.

After more than 50 years of intensive soil conservation promotion, soil erosion continues to accelerate. In the early 1930's soil erosion in the United States was estimated at 4 billion tons (3.6 billion mt) each year. More than 50 years later, an estimated 6.4 billion tons (5.8 billion mt) annually erodes from soils in the 50 states. This is a 60 percent increase in the rate of erosion.

7:2 □ CONSERVATION PROGRESS

Although total soil erosion in the United States has increased by 60 percent since 1933 when the Soil Erosion Service was established, there has been progress. (In 1935, the "Soil Erosion Service" was changed to the "Soil Conservation Service.")

The U.S. Department of Agriculture – Soil Conservation Service provides technical assistance in the planning and application of conservation practices and systems to reduce excessive soil erosion on croplands, rangelands, pasturelands, and forestlands; to conserve water used in agriculture and improve water quality; and to reduce upstream flood damages.

The bulk of Soil Conservation Service (SCS) assistance is provided by its field offices to farmers and ranchers throughout the nation's nearly 3,000 soil conservation districts. A significant amount is provided in cooperation with other federal, state, and local government agencies, as well as private organizations.

The Food Security Act of 1985 brought several new conservation provisions. These provisions — known as the Conservation Reserve, Conservation Compliance, Sodbuster, and Swampbuster — are designed to make USDA farm and conservation programs more consistent.

7:2.1 □ Conservation Reserve Program

The **Conservation Reserve Program (CRP)** is the first of the new conservation provisions of the Food Security Act of 1985 to be implemented. It is a program designed to reduce surplus commodity production and soil loss by retiring highly erodible cropland. The Soil Conservation Service provides technical assistance to farmers who enter into CRP contracts to plant more grass, trees, and wildlife cover. Soil Conservation Service field office staffs also determine the eligibility of land submitted for the program.

It is estimated that establishment of permanent cover will reduce the average annual rate of erosion on land under CRP contracts by 27 tons per acre (60.5 mt/ha) per year. CRP signups are designed to retire 40 – 45 million acres (16.1 – 18.2 million ha) of highly erodible cropland. Participating farmers receive cost-sharing assistance from the U.S. Department of Agriculture – Agricultural Stabilization and Conservation Service (ASCS).

7:2.2 □ Sodbuster and Swampbuster Provisions

Sodbusting refers to plowing highly erodible land out of sod, unless it was plowed during the four years prior to enactment of the 1985 Farm Bill. In order to produce annual crops on these lands and remain eligible to participate in USDA programs, a farmer must certify that such land is to be used according to a conservation plan developed by the U.S. Department of Agriculture – Soil Conservation Service.

Swampbusting refers to draining and converting to cropland any wetlands on which the conversion began after Dec. 23, 1985. There are some exemptions for wetland conversion which will be determined by the U.S. Department of Agriculture – Agricultural Stabilization and Conservation Service on an individual basis.

7:2.3 □ Conservation Compliance

Highly erodible land which was used for crop production at least once during the four years prior to Dec. 23, 1985, is covered by this provision. Farmers will have until 1990 to have the Soil Conservation Service develop an approved conservation plan for crop production on such land and initiate use of the plan. By 1995, farmers must have such plans fully implemented.

7:2.4 □ Conservation Tillage

Conservation tillage is any tillage and planting system in which at least 30 percent of the soil surface is covered by plant residue after planting to reduce soil erosion by water, or, where soil erosion by wind is the primary concern, at least 1,000 pounds per acre (1,120.9 kg/ha) of flat small-grain residue is on the surface during the critical erosion period. No-till, or zero till, is a conservation tillage method in which only a narrow seedbed is disturbed for planting (Figure 7.1).

Fig. 7.1 — Zero-till planted corn in a double-cropping system quickly shoots up above wheat stubble which serves as a moisture-holding mulch throughout the growing season. (Johnson County, Illinois) (Courtesy, Tom Pouton, USDA - Soil Conservation Service)

The use of all forms of conservation tillage continues to increase for all crops. According to the National Association of Conservation Districts' Conservation Technology Information Center, farmers used some form of conservation tillage on 97.6 million acres (39.5 million ha) during 1986. This is about 33 percent of the 296 million acres (119.8 million ha) planted to crops — up from about 31.5 percent the previous year. At 43 percent, the Corn Belt was the region with the greatest percentage of its cropland in conservation tillage (Figure 7.2).

7:2.5 ▫ Agricultural Conservation Program

The Soil Conservation Service provides technical assistance to farmers and ranchers who install conservation practices under the Agricultural Conservation Program (ACP). Under long-term agreements, the Soil Conservation Service assists these farmers in installing lasting practices such as terraces and grassed waterways. Through the ACP, farmers and ranchers install water conservation

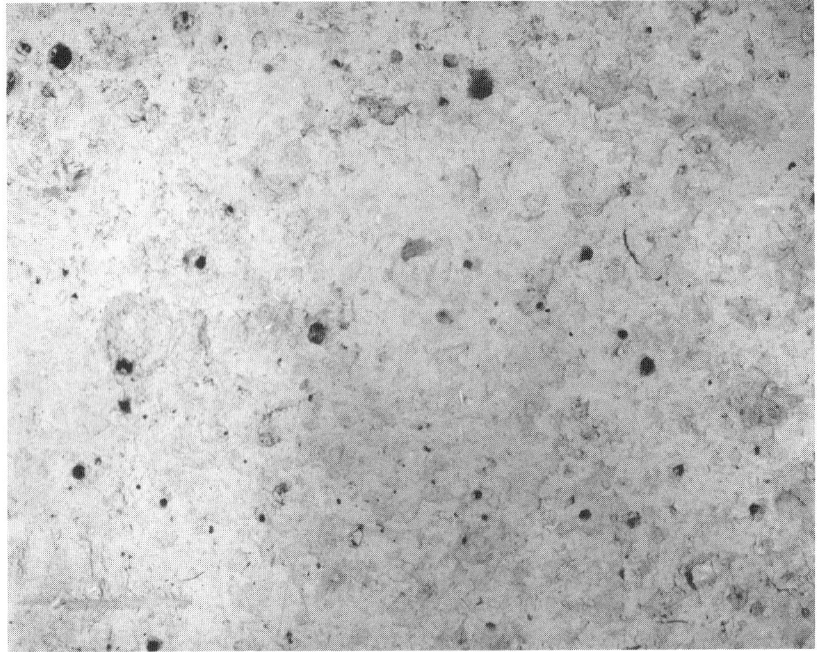

Fig. 7.2 — After 20 years of no-till corn in southern Ohio, these holes in the soil, made by earthworms and plant roots, help to make the soil more productive. Under no-till, less than 2 percent of the total precipitation ran off, whereas 17 percent ran off under conventional tillage. (Courtesy, W. M. Edwards, USDA - Agricultural Research Service)

Note: Photo was taken at a 6-inch (15-cm) depth, the usual depth of a plowpan under conventional tillage.

practices and terrace systems and apply conservation tillage. The ACP is administered by the Agricultural Stabilization and Conservation Service, which provides financial assistance to the participating land owners.

7:2.6 ▫ Other Programs

In addition to these activities, the Soil Conservation Service carries out numerous research, development, educational, and assistance programs which improve soil, water, and related resources. Each year new soil surveys are completed. For instance, in 1986 there were 82 new soil surveys published. Each survey describes the physical and chemical characteristics of the soils in the survey area — generally a county. It names and classifies the soils for various uses. Detailed maps show where each soil is located. (See Chapter 1.)

The Soil Conservation Service now has more than 15,000 career employees who provide on-the-land service in stabilizing the soil and water environment for the public good. Most of this service is provided through conservation districts.

Individual land owners or operators can obtain assistance from Soil Conservation Service personnel by signing an agreement with the local conservation district. Land use plans and designs for conservation practices recommended by the soil conservation technicians are based on soil maps interpreted according to the information in soil survey reports. Plans are made according to the choices of the cooperating land owner or operator in harmony with the capabilities and needs of the soils (Figure 7.3).

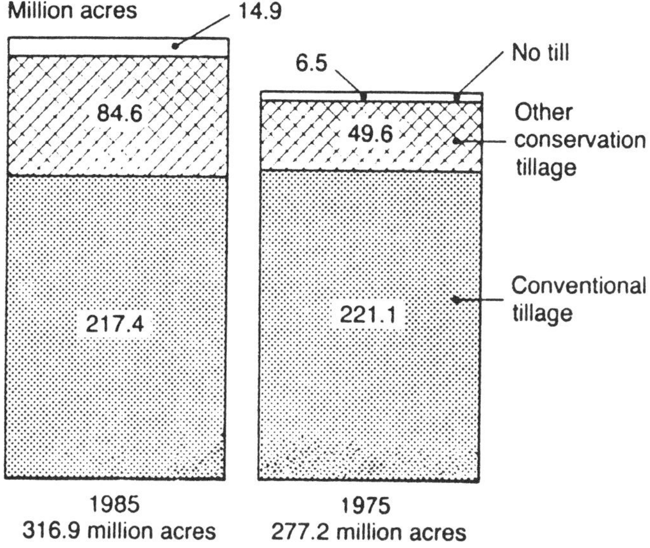

Fig. 7.3 — Conservation tillage methods, which leave additional residue on the soil surface after planting, nearly doubled in use during 1975 - 1985, due to cost and time savings. (Sources: Conservation Tillage Information Center and *No-Till Farmer*)

7:3 □ UNIVERSAL SOIL LOSS EQUATION

One useful way to estimate soil erosion by water is by using the ***Universal Soil Loss Equation (USLE).***

The Universal Soil Loss Equation is ***A*** = ***RLSKCP***, where ***A*** stands for annual average soil loss by water (expressed in tons per acre per year), and each factor is defined as follows:

- □ ***R*** — The potential of the rain that falls in a particular area to produce erosion. The R factor is based on analysis of 22 years of precipitation records in the central and eastern United States.

- □ ***LS*** — The effects of length (L) and steepness (S) of slopes on the occurrence of erosion. Erosion increases sharply as either slope length or steepness increases. Usually, a numerical value is assigned to the product of L and S in solving the equation (see p. 209).

- **K** — The erodibility of a particular soil series, some soils being more subject to erosion than others. K values were obtained for each soil by repeated analysis of erosion on plots made up of that soil series.

- **C** — The type of cropping system and agronomic practices being used. If there is no soil cover and no residual effect of prior croppings, C has a value of 1. Any crop which covers the soil during at least part of the year will reduce the erosion occurring on that soil.

- **P** — The reduction in erosion resulting from installation of conservation practices that change the flow pattern of runoff on the soil, such as contouring, strip-cropping, and minimum tillage.

Numerical values for the factors in the equation have been published in several references and are an important part of the Soil Conservation Service's *Technical Guide,* which is used in designing conservation projects for individual farms. Given these charts and tables, the equation is easy to use. Each individual looks up the value given for his or her own situation.

The equation is solved by multiplying the previously determined values for these factors. It works this way:

As stated,

$$R$$

means the erosion potential inherent in the rainfall patterns of a particular area. Next,

$$R \times LS$$

represents the water erosion potential in a specific area for a particular combination of slope length and steepness. Next,

$$RLS \times K$$

is the amount of soil that could potentially be lost from a particular field *if* it were maintained in the fallow (no crop) condition year around. Then,

$$RLSK \times C$$

indicates how much a particular cropping system and management practices will reduce the water erosion potential. Finally,

$$RLSKC \times P$$

shows how much erosion would be further reduced by conservation practices that change the flow pattern of the runoff. Thus,

$$A = RLSKCP.$$

Example[1]

A farmer lives in an area where the rainfall factor, R, is 160. He is raising continuous corn on a field with an 8 percent slope that's about 200 feet long and mapped as Oshtemo sandy loam soil with 2 to 8 percent slopes. He farms on the contour but uses conventional tillage. This particular soil has a 3-ton-per-acre **"soil loss tolerance"** (the amount of soil that can be lost each year without seriously reducing productive capability). But what is the soil loss actually likely to be on this field and what can be done to reduce it?

For the conditions described, the Soil Conservation Service's *Technical Guide* shows the slope (LS) value to be 1.40, the soil (K) value to be 0.20, the cropping-management practice (C) value to be 0.39, and the contouring (P) value to be 0.50. The product of these USLE factors reveals the following:

$$\begin{aligned} A &= R \times LS \times K \times C \times P \\ &= 160 \times 1.40 \times 0.20 \times 0.39 \times 0.50 \\ &= 8.74 \text{ tons/acre/year (19.58 mt/ha/yr)} \end{aligned}$$

That's almost three times the annual soil loss tolerance of 3 tons per acre per year (6.72 mt / ha / yr) for this soil.

To reduce this severe erosion if the farmer substituted *minimum tillage* for conventional tillage, the cropping factor (C) would be reduced from 0.39 to 0.12, with the result that average annual soil loss would drop from 8.74 to 2.69 tons per acre per year, which is within the soil loss tolerance limits of 3 tons per acre per year (6.72 mt / ha / yr) for this field.

7:4 ▫ EROSION REDUCES PRODUCTIVITY OF SOILS

Soil erosion reduces the capability of the soil to produce crops and can increase the costs of farming. Sheet and rill erosion moves more than 3.4 billion tons (3.07 billion mt) of soil on nonfederal rural land, and wind erosion moves 2 billion tons (1.8 billion mt). More than 286 million acres (115 ha) of nonfederal land are eroding at rates greater than the respective soil tolerances.

Soil erosion can never be stopped; it can only be controlled. Erosion is assumed to be controlled when the soil loss per acre per year on cropland, pastureland, and forestland averages no more than 5 tons (11.2 mt / ha). On rangeland the permissible amount of erosion is only 2 tons per acre (4.5 mt / ha) per year. Permissible erosion is less on rangeland because the formation of new

---

[1] This section is adapted from "Value of the Universal Soil Loss Equation to Water Quality Management Planning." Planning Note No. 4, Indiana Inter-Agency Information Education Committee on Nonpoint Source Pollution. *Note:* The Universal Soil Loss Equation is used to predict *water erosion.* For an equation on how to predict *wind erosion,* consult Frederick R. Troeh, J. Arthur Hobbs, and Roy L. Donahue. *Soil and Water Conservation for Productivity and Environmental Protection.* Englewood Cliffs, New Jersey: Prentice-Hall, Inc., 1980, pp. 172 – 197.

Table 7.1 — Total Water and Wind Erosion on Nonfederal Lands in the United States by Land Use and Erosion Class[1]

Land Use	3-Ton Erosion Rate per Year (million acres)	Total Erosion per Year (million tons)	10-Ton Erosion Rate per Year (million acres)	Total Erosion per Year (million tons)	25-Ton Erosion Rate per Year (million acres)	Total Erosion per Year (million tons)	Total Acres (million)	Total Erosion per Year (million tons)	Average Erosion Rate (tons/acre/year)	Percent of Average
Cropland	272	816	93	930	48	1,200	413	2,946	7.1	148
Pastureland	119	357	9	90	5	125	133	572	4.3	90
Forestland	353	1,059	12	120	5	125	370	1,304	3.5	73
	1-Ton Erosion Rate		4-Ton Erosion Rate		15-Ton Erosion Rate					
Rangeland	283	283	56	224	69	1,035	408	1,542	3.8	190*
TOTAL	1,027	2,515	170	1,364	127	2,485	1,324	6,364	4.8	100
Other Uses	**	**	**	**	**	**	175	**		
						GRAND TOTAL	1,499			

*Since permissible erosion on rangeland is only 2 tons per year, rather than 4.8, the average erosion rate of 3.8 tons is 190% of the permissible rate of 2 tons.

**"Other Uses" include surface mining, urban areas, homesteads, roads, and airports, for which no estimate has been made on erosion. All erosion sediment was assumed by the authors to come from cropland, pastureland, forestland, and rangeland.

[1] Calculated by the authors from data in "Summary of Appraisal, Parts I and II, and Program Report." Soil and Water Resources Conservation Act. USDA – Soil Conservation Service, 1980, p. 8.

soil from soil minerals in the arid and semiarid climates is much slower than in humid climates where most cropland, pastureland, and forestland exist. In one sense, permissible erosion sediment can be considered mostly geologic erosion and therefore beyond human control.

Table 7.1 shows a total of about 1.3 billion acres (0.53 billion ha) of cropland, pastureland, forestland, and rangeland from which almost 6.4 billion tons (5.8 billion mt) erode each year. This averages 4.8 tons per acre (10.8 mt / ha) of soil loss annually (Figure 7.4).

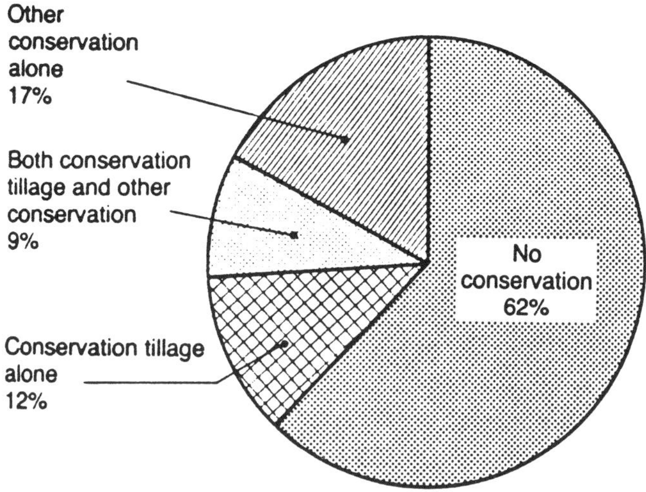

Fig. 7.4 — About 38 percent of the nation's farmers practice some kind of soil conservation, including reduced tillage methods, contouring, strip-cropping, and terracing. (1983 data. Includes only farms with cropland.) (Courtesy, Farm Production Expenditure Survey)

The authors believe the permissible soil loss of 4.8 tons per acre (10.8 mt / ha) per year on cropland, pastureland, and forestland and 2 tons (1.8 mt) for rangeland is far too high, because the rate of renewal of soil from parent materials in humid regions averages only ½ ton per acre (1.1 mt / ha) per year and is much slower in arid and semiarid regions where rangelands exist. At the rate of soil renewal of ½ ton per acre (1.1 mt / ha) per year, it takes 2,000 years to rebuild 6 inches (15 cm) of soil.

Although a permanent cover of grass or trees generally provides adequate protection against erosion, about 18 percent of our rangeland, 9 percent of our pastureland, and 6 percent of our forestland are eroding at rates above tolerance. Cropland is far more vulnerable to erosive forces — 40 percent of all cropland is eroding by sheet and rill or wind erosion at rates that exceed the tolerable rates. About 23 percent of cropland is eroding at twice the tolerable rate.

Soils differ in the ease with which they erode and, therefore, in the kind of management required to protect them from damage. Sheet, rill, and wind erosion are not hazards on some soils; about 94 million acres (38 million ha) of cropland have almost no susceptibility to damage by these forms of erosion. On 277 million acres (112 million ha) of cropland erosion is a hazard, but damage can be prevented if resource management systems are used. On about 45 percent of that land, however, current management permits erosion to exceed tolerance. About 50 million acres (20 million ha) of cropland are so highly susceptible to erosion damage that preventing damage is nearly impossible. If converted to grass or forestland, however, these soils would maintain their usefulness indefinitely.

To conclude, over the 50 states, wind and water erosion have been most severe on cropland and rangeland, followed by pastureland, and then forestland. Controlling wind and water erosion on cropland and rangeland should receive the highest priority. Conservation action programs to control erosion on a priority basis should include these practices:

1. Concentrating first on the most seriously eroding croplands, estimated to total 48 million acres (19 million ha), now eroding at an average rate of 25 tons per acre (56 mt / ha) per year. An estimated 35 percent is eroding so rapidly that it should be taken out of cultivation and planted and maintained in permanent grass or trees.
2. Concentrating second on controlling erosion on the 93 million acres (38 million ha) of cropland now eroding at the 10-ton average rate per acre (22 mt / ha) per year. Conservation practices should include establishing permanent vegetative cover on 2 percent of the acreage, minimum tillage on 82 percent, strip-cropping on 9 percent, and terracing on 7 percent of the area.
3. Improving 125 million acres (51 million ha) of the most rapidly eroding rangeland by range reseeding, rotation, deferred grazing, and fertilization.

7:5 □ CROPPING SYSTEMS TO REDUCE SOIL AND WATER LOSS

Research has provided some clues for adapting cropping systems to reduce soil and water loss.

Twenty years of research on soil and water losses in the 55-inch (140-cm) rainfall region of the southern Piedmont Plateau of Georgia resulted in these conclusions:

1. The dominant problem on 92 percent of the cropland is erosion and sedimentation.
2. The predominant soil is Cecil sandy loam, a member of the clayey, kaolinitic, thermic family of Typic Hapludults (Figure 7.5). (See Chapter 1.)
3. Sod crops such as bermudagrass, fescuegrass, vetch, rye, crimson clover, and alfalfa in cropping sequence with corn and cotton reduce soil and water loss; increase efficiency of water, lime, and fertilizer; increase

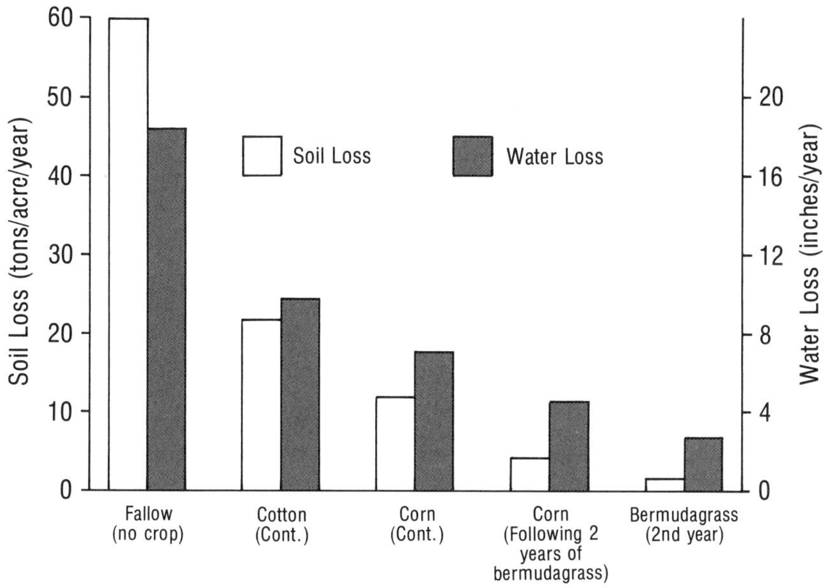

Fig. 7.5 — Average annual soil loss and water loss on a Cecil sandy loam with a 7 percent slope in Georgia where the average annual rainfall is 55 inches (140 cm).

yields of row crops; and increase net farm income. Nitrogen fixed by vetch, alfalfa, and crimson clover gives yield responses of the following crop equal to or greater than 80 to 120 pounds of nitrogen (N) fertilizer per acre (90 – 135 kg/ha) (see Section 5:10).

In the Corn Belt, land in 10 years of continuous corn soaked up 73 percent of the rain. In contrast, land in a good rotation of corn-wheat-hay permitted 84 percent of the rainfall to move into the soil. This means that 4 more inches (10 cm) of rain went into the soil where the crop rotation had been followed. The additional 4 inches (10 cm) of water can mean the difference between a good crop and a poor crop.

Research in Missouri, the pioneer state in soil erosion research, has demonstrated that under continuous bluegrass sod only a trace of soil erosion occurred. Soil under a good rotation of corn-wheat-redclover-timothy lost an average of 10 tons of soil per acre (22 mt/ha) per year. By contrast, soil under continuous corn lost 50 tons of soil per acre (112 mt/ha) per year. At this rate of soil loss under continuous corn, the surface 6 inches (15 cm) would be all gone in 18 years (Figure 7.6).

In the 30-inch (76-cm) rainfall area near Guthrie, Oklahoma, an additional 10 inches (25 cm) of water infiltrated the soil under a system of cotton-wheat-sweetclover as compared with continuous cotton. The same annual precipitation in La Crosse, Wisconsin, resulted in runoff ratios of 1:2:8 for forestland-pastureland-cropland.

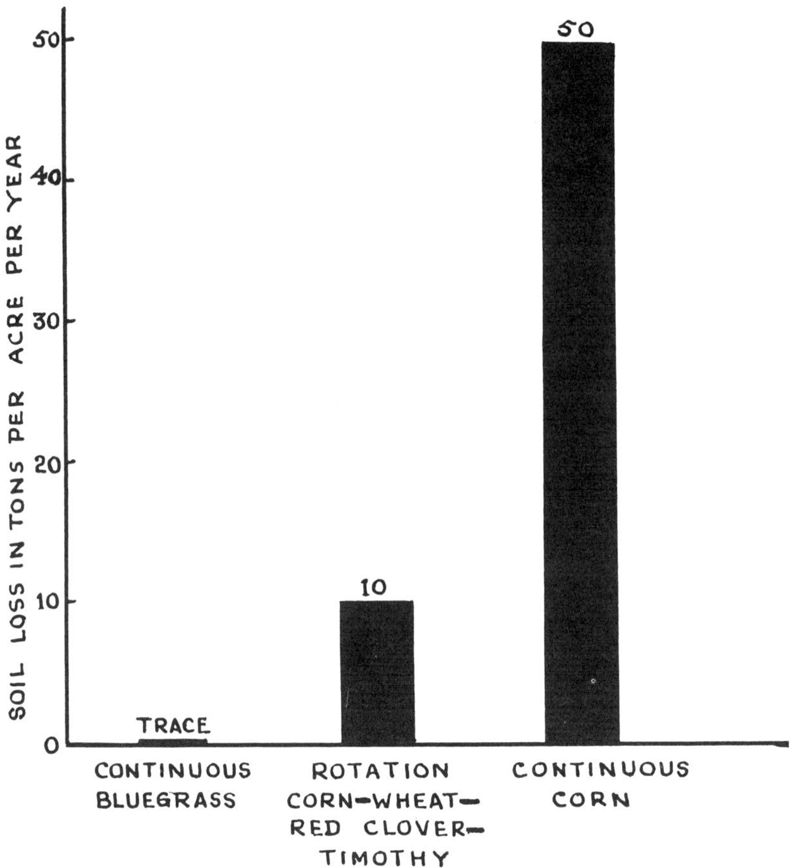

Fig. 7.6 — Soil losses by water erosion in relation to cropping systems in Missouri. *Note:* Tons per acre × 2.24 = mt/ha. (Courtesy, University of Missouri)

7:6 ▫ MINIMUM TILLAGE TO REDUCE SOIL LOSS

Minimum tillage means the least tillage necessary to grow a successful crop. With the practice of minimum tillage, crop residues serve as mulches on the soil surface.

The increasing export market for wheat, corn, and soybeans has resulted in the planting of increasing acreages of these crops. Most of the increase has been on soils too sloping for cultivated crops, and the result has been more soil and water loss and more polluting sediment in surface waters.

One hopeful response to more cultivated crops on sloping soils has been a recent increase in acreage of minimum tillage to 60 million acres (24 million ha). (See Chapter 5.) Research in southern Illinois has confirmed the fact that soil erosion is adequately controlled by minimum tillage with surface residue management on 5 and 9 percent slopes (Figures 7.7 and 7.8).

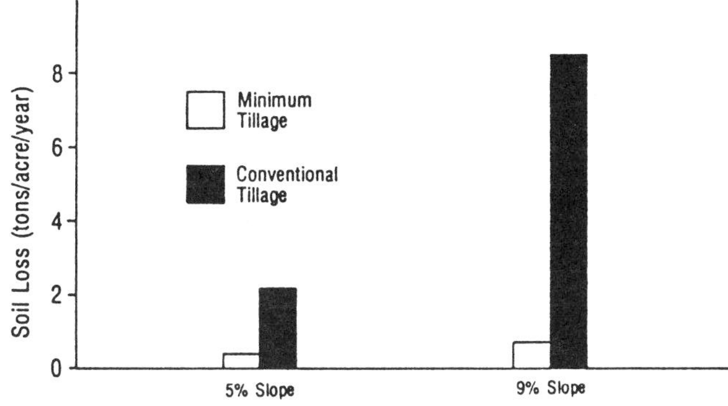

Fig. 7.7 — Soil loss: minimum tillage compared with conventional tillage on 5 and 9 percent slopes. Cropping system: wheat following corn. *Note:* Average of four years. (Courtesy, University of Illinois)

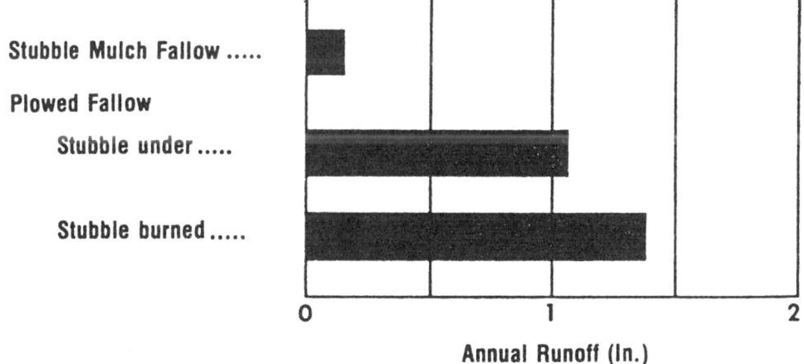

Fig. 7.8 — *Top:* Here minimum tillage includes stubble mulching. On the right the wheat stubble has been "plowed" without being buried, leaving a stubble-mulch that increases the movement of water into the soil. This results in less erosion by water and wind and an increase in crop yields. (Montana) *Bottom:* Stubble must be left *on the soil surface* to reduce annual runoff. (Pullman, Washington, 22 percent slope) Average annual precipitation is 20 inches (51 cm). (Courtesy, USDA – Soil Conservation Service)

7:7 ▫ TERRACING

A terrace is a low ridge of soil and a channel erected on the contour across sloping fields for the purpose of holding soil by slowing the movement of runoff water.

The main function of a terrace is to intercept runoff water. This water may later be absorbed, or it may be slowly and safely led off the field. More moisture in the soil means more water to support the growth of protective vegetation. In performing this function, terraces also aid in reducing wind erosion.

The rainfall over many semiarid regions is adequate for crop production, provided all of it soaks into the soil. However, spring and summer rains often fall quickly and run off rapidly, and not enough water soaks into the soil to support a crop. Special terraces catch these rains and make the difference between a successful crop and a crop failure. Evidence that terraces reduce runoff and save soil in widely scattered states such as Wisconsin, Texas, and North Carolina is given in Figure 7.9.

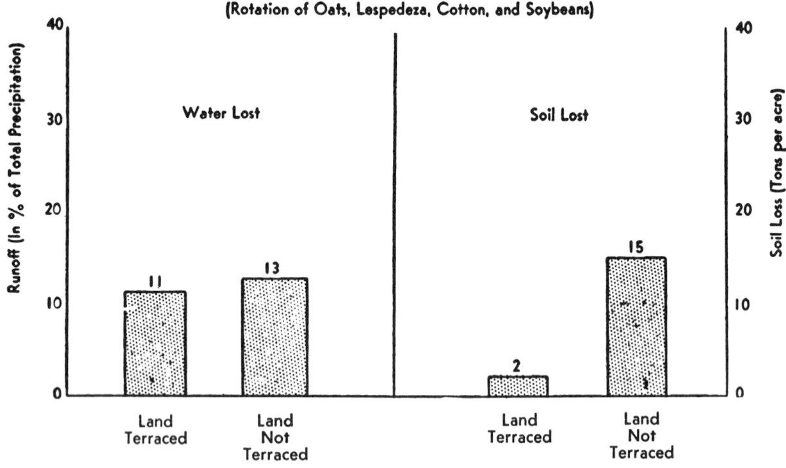

Fig. 7.9 — Terraces reduce runoff and save soil at Statesville, North Carolina. Average annual precipitation is 50 inches (127 cm), and the soil is Cecil sandy clay loam on a 10 percent slope. Almost identical percentages of soil and water were saved by terraces in Texas and Wisconsin with annual rainfalls of 32 and 40 inches (81 and 102 cm) respectively. (Courtesy, USDA)

The two main types of terraces are the ***broadbase*** terrace and the ***bench*** terrace. The broadbase terrace is constructed with wide ridges and channels so as to not interfere with tillage operations. In semiarid areas, the water channel may be level with closed ends to impound water and to increase water infiltration (Figure 7.10). In humid areas, the channel is built on a grade with open ends to permit surplus water to be discharged into a protected outlet at a nonerosive velocity.

A modification of the broadbase terrace on the contour is the parallel terrace.

Fig. 7.10 — Terraces with a broad base to permit cropping on them have been successful in temperate humid regions in controlling water erosion on sloping soils. This terrace in Missouri is seeded to wheat on 8 percent slopes. Excess water from the terrace flows through a grassed waterway at a nonerosive velocity to a stream. (Courtesy, USDA - Soil Conservation Service)

Parallel terraces eliminate partial rows (point-rows) and thus speed all cultural operations. The terrace interval is adjusted up or down hill to allow the spacing between the terraces to be multiples of the width of the field equipment used (Figure 7.11).

Bench terraces are adapted to steep slopes up to 25 percent and to very high-priced semiarid to humid lands. They are also used on slopes of about 6 percent with special modifications. Bench terraces consist of nearly level benches separated by nearly vertical back slopes (risers), similar to stairsteps.

Population density and scarcity of level lands do not as yet demand extensive cultivation of many excessively steep slopes in the United States. Some cultivation of field crops on steep, bench-terraced slopes has been practiced in sections of the Southeast for several generations, and the continuation of this practice may be necessary in hilly or mountainous sections. In the highly productive citrus and avocado districts of southern California, bench terraces have been used to a considerable extent on steep slopes. In other states, there are scattered examples of their use in connection with irrigated land, truck farming, and vineyard and orchard cultivation.

All types of terraces in humid regions need some type of outlet for the surplus water. The most common kinds of terrace outlets include artificially grassed waterways, naturally grassed or wooded areas, drop spillways made of concrete or rock, and pipe outlets. Since pipe outlets are relatively new, this system will be discussed in more detail.

A vertical intake riser pipe with perforations is located at the lowest point in each terrace channel. These risers are attached to a pipe buried about 2 feet (61 cm) that has an outlet protected by grass, trees, or a drop spillway. The design should allow for the drainage of all ponded water within 24 to 48 hours to prevent crops from drowning (Figure 7.12).

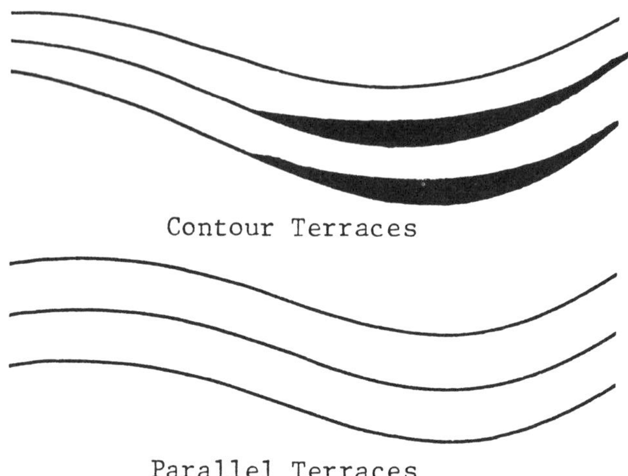

Fig. 7.11 — *Top:* Traditional broadbase terrace on the contour showing variable spacing between the terraces. This causes some rows at the widest spacings (in shaded areas) to be shorter than others and therefore cause difficulty in cultural operations. *Bottom:* Parallel terraces permit all full-length rows, and, therefore, more efficient plowing, planting, and harvesting. (Courtesy, University of Nebraska)

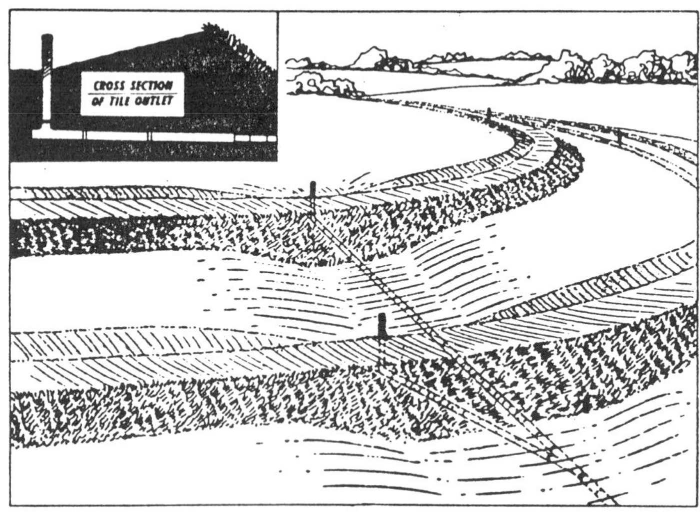

Fig. 7.12 — Terraces with pipe outlets and grassed backslopes. (Courtesy, University of Nebraska)

7:8 □ CONTOUR TILLAGE

Contour tillage should be practiced on all terraced land. This reinforces the effectiveness of the terrace because each tillage furrow acts as a miniature terrace. On less-steep land, contour tillage is also effective without terraces (Figure 7.13).

Fig. 7.13 — Contour planting of pineapple in West Oahu County, Hawaii, saves both water and soil and increases production. (Courtesy, USDA - Soil Conservation Service)

An adaptation of contour tillage is ***contour furrowing.*** This consists of pulling a series of narrow chisels on the contour through the soil at a depth of 6 to 12 inches (15 – 30 cm). The objective is to increase water infiltration. Contour furrowing is best adapted to overgrazed stolon-bearing grasses such as buffalograss and bermudagrass on ranges and pastures.

7:9 □ WATER MANAGEMENT

Agriculture is a major user of water. Agriculture also affects water quality. Land and water use and production practices have important implications not only for water supplies and quality but also for human and livestock health and for total environmental quality.

7:9.1 □ Making Optimal Use of Water

In arid regions, crop production depends almost entirely on irrigation. Irrigated agriculture accounts for more than 40 percent of all freshwater withdrawals in the United States and consumes nearly three times as much water as all other uses combined. In subhumid and semiarid regions, agricultural production depends on water storage and distribution systems. Even in humid areas, droughts during the growing season are common and can significantly reduce yields.

Under current use and management, water shortages occur in all regions. Shortages are most common and acute in the arid regions. The imbalance of supply and demand can have immediate or long-term effects — or both. Inadequate water supplies for irrigation and other offstream uses can result in the loss of a season's crops or eventual changes in land use and shifts of irrigated farming from one region to another. Diminished streamflows can interfere with hydropower generation and navigation without causing more than temporary hardship, but if aquatic life and fish habitat are even briefly affected, recovery may be lengthy and difficult. Overdrafts (mining) of groundwater soon increase the economic and energy cost of the water, with repercussions that work their way through the regional economy. If groundwater mining results in a substantial decline in the permanent water table or in the water pressure within an aquifer, the change is not likely to be reversed.

Although localized or wide-scale problems of water availability are occurring in many areas, most problems could be minimized with careful planning. Improved management of irrigation water, improved management of soil moisture, and increased storage to increase dependable supply can bring supply and demand into better balance.

7:9.2 □ Using Irrigation Water More Efficiently

Irrigation techniques will be explained fully in Chapter 9, but a brief presentation will be made here of irrigation in relation to soil and water conservation and environmental pollution.

More than 80 percent of all water consumed in the United States is used for irrigation, and more than half of the supplies originate from groundwaters. In many areas of the United States, both groundwaters and surface waters are being used for irrigation faster than they are replenished. Stated in another way: ***The waters are being mined and will soon be gone*** (Figure 7.14).

Farmers have increased their efficiency in using irrigation water. A national average of 47 percent of the water withdrawn from surface or ground sources for irrigation was consumed by crops in 1982 as opposed to 41 percent in 1975. In 1975, an average of 78 percent of the water withdrawn for irrigation reached the farm; in 1982 the average improved to 81 percent. In 1975, an average of 53 percent of the total withdrawn was applied to the crop; this on-farm efficiency increased to nearly 59 percent in 1982.

Expansion of irrigation has slowed in recent years. According to the Census

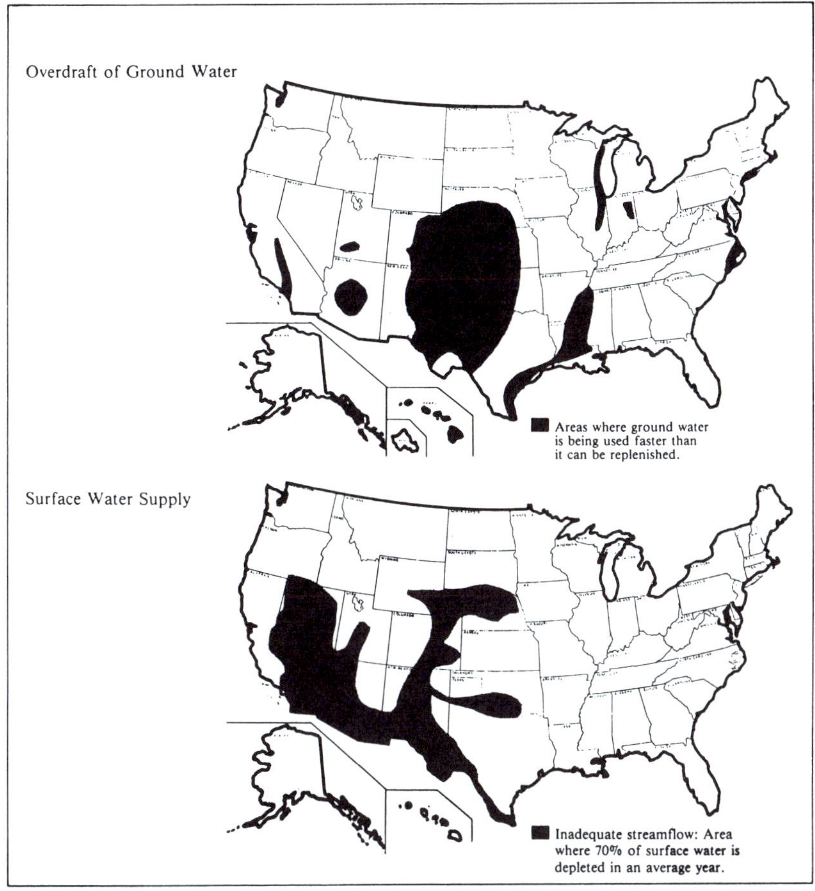

Fig. 7.14 — Groundwaters over a large part of the nation and surface waters in semiarid and arid regions are being depleted. (Source: National Association of Conservation Districts, "Soil Degradation: Effects on Agricultural Productivity," National Agricultural Lands Study, Washington, D.C., 1980)

of Agriculture, farmers reported irrigating 1.3 million less acres (0.52 million ha) in 1982 than in 1978, and 12 of 14 states that irrigated over 1 million acres (0.4 million ha) irrigated fewer acres in 1982 than in 1978. Reduction of the acres irrigated affects regional economies because irrigated farming is highly productive.

More efficient use of water and less pollution will result from more research and education on, and adoption of, these practices:

1. Breeding and selecting of plants that use less water per unit of grain/forage produced.
2. Breeding and selecting of plants that are more tolerant of brackish (salty) waters.
3. Better land leveling, more irrigation ditch linings, reuse of all **tailwaters** (end-of-field irrigation waters), and more care in land configuration to reduce erosion and to obtain more uniform depth of water infiltration.

7:9.3 ▫ Management of Soil Moisture

On nonirrigated land, especially in areas of limited rainfall, farmers can make more water available for their crops by increasing infiltration and reducing runoff. Use of effective practices for soil conservation / soil moisture management can reduce erosion and increase net returns. Such practices tend to have greater effect on costs of production than on yields, according to computer simulations.

Effective use of soil moisture conservation practices requires careful management and accurate soil information, however. Practices that increase soil moisture retention in dry years also tend to retain water in wet years. Increases in movement of soil moisture to groundwater may affect water quality, because soluble nutrients and pesticides tend to move with water through the root zone into the groundwater.

7:9.4 ▫ Improving Drainage in Cropland to Improve Water Use

Too much water, rather than too little water, is the major water management problem in some soils. In the *1982 National Resources Inventory*, the U.S. Department of Agriculture identified about 24 million acres (10 million ha) of nonirrigated cropland and 2.6 million acres (1.1 million ha) of irrigated cropland where drainage is the main improvement needed. (See Chapter 8.)

Installing drainage on wet cropland soils generally increases yields and makes farming operations easier. When drained, many wet soils are highly productive. Nearly 47 percent of all wet soils is prime farmland. (See Chapter 1.)

However, the costs of providing drainage and the effects of drainage on the environment must be considered before drainage is installed. In some cases the benefits do not outweigh adverse effects on wildlife habitat and water quality.

7:9.5 ▫ Nonpoint Source Pollutants

Sources of pollution include point sources from industrial and municipal discharges and nonpoint sources such as septic tank drain fields and runoff from lands devoted to agriculture. These nonpoint pollutants are carried into the surface waters by way of surface runoff and by attaching to eroding soil particles. All soil and water conservation practices that reduce surface flow will reduce soil loss and also reduce movement of nonpoint pollutants into streams and lakes.

The greatest nonpoint pollutant is the 6.4 billion tons (5.8 billion mt) of soil erosion sediment discharged each year from the land surface of the United States. Other nonpoint sources of pollutants contribute excess nutrients to 56 percent of the 246 drainage basins, excess bacteria to 51 percent, pesticides to 22 percent, and other toxins contribute to 32 percent of the drainage basins (Figures 7.15 and 7.16).

Fig. 7.15 — Housing developments and highway construction, the greatest causes of soil degradation, have combined here to erode soil sediments to pollute the environment. (Courtesy, Virginia Polytechnic Institute and State University)

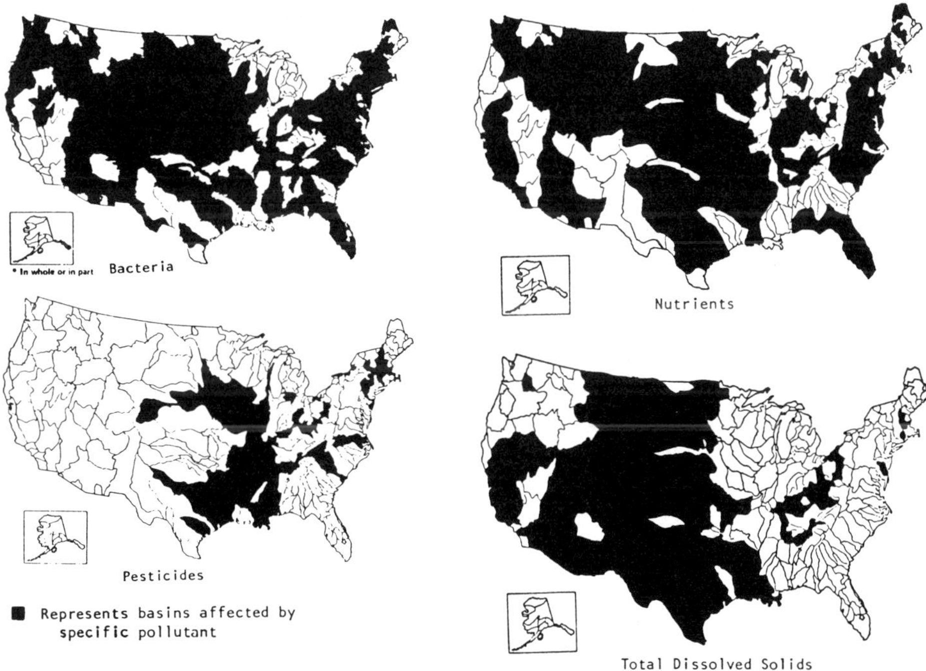

Fig. 7.16 — Extent of surface water pollution by bacteria, nutrients, pesticides, and total dissolved solids. (Courtesy, U.S. Environmental Protection Agency)

7:10 □ DESERTIFICATION

The word ***desertification*** may not be in your dictionary, but the root word *desert* is. *Desert* means "an arid region lacking moisture to support vegetation." *Desertification* means "tending to become a desert."

In the United States, about 225 million acres (91 million ha) of desertification is now very severe. This is about 10 percent of our total land area. The soils affected most severely lie between the United States and Mexico in the border states of Texas, New Mexico, Arizona, and California. Less severe are parts of Colorado, Utah, Nevada, Wyoming, and Montana.

Causes of desertification are **anthropogenic** (caused by humans), such as overgrazing, erosion, accumulating salt from irrigation, and using water faster than it is being replenished. Any pollutant that kills vegetation can also be considered a cause of desertification.

Examples of severe desertification include the Rio Puerco Basin in northwestern New Mexico, the San Joaquin Valley of California, and Gaines County in Texas. In these areas, groundwaters are being depleted rapidly. The application of recommended soil and water conservation practices can reverse the process of desertification and restore natural productivity, but only at great cost.

7:11 □ FACTORS AFFECTING ADOPTION OF CONSERVATION PRACTICES

Although we do not yet know all the answers, recent research in the area of adoption of soil and water conservation has provided useful information. We know, for example, that those who adopt conservation practices are likely to be well-educated, full-time farmers who participate in many organizations. We know that land users who possess the "conservation ethic" — that is, those who are concerned about preserving the land for future generations — are also more likely to practice conservation, although other factors, such as their willingness to take risks, may intervene. Land users who use conservation practices are generally owner-operators, their families participate in the operation, and they realize that they have erosion problems on their land. Although most of this research has involved farmers, it may be possible to expand these generalizations to include ranchers with similar characteristics.

These research findings are supported by Soil Conservation Service evaluations of agency programs and activities. A recent evaluation of the Soil Conservation Service's Conservation Technical Assistance (CTA) Program, for example, shows that, in most areas, half of CTA planning assistance in 1983 was provided to those who had been conservation district cooperators for at least three years. Owner-operators received almost three-fourths of CTA direct assistance. Where the Soil Conservation Service had redirected its activities to focus on the most critical problem areas, a higher proportion of assistance was directed to landlords, nonfarmers, and those who were not conservation district cooperators.

Research confirms that the U.S. farming community (and therefore, the conservation program clientele) is less and less the homogeneous group of full-time family farmers that comprise the popular U.S. image of "farmer," and more and more a heterogeneous group that includes corporate farmers, part-time farmers, professional farm managers, and absentee land owners. Each of these groups has unique values, needs, and attitudes that influence its conservation decisions.

7:12 □ REFERENCES

Arakeri, H. R., and Roy L. Donahue. *Principles of Soil Conservation and Water Management.* New Delhi, Bombay, Calcutta, India: Oxford and IBH Publishing Co., 1984, 254 pp.

Cook, Ken. "Conservation Tillage: Marrying for Money." *Journal of Soil and Water Conservation,* Vol. 39, No. 6, November – December 1984, pp. 368 – 370.

Crowder, Bradley M., Marc O. Ribando, and C. Edwin Young. "Agriculture and Water Quality." Agr. Inform. Bul. No. 548, USDA – Economic Research Service, Washington, D.C., 1988, 6 pp.

Donahue, Roy L., Raymond W. Miller, and John C. Shickluna. *Soils: An Introduction to Soils and Plant Growth,* 5th ed. Englewood Cliffs, New Jersey: Prentice-Hall, Inc., 1983.

Environmental Progress and Challenges: EPA Update. U.S. Environmental Protection Agency, Washington, D.C., August 1988, 140 pp.

Hargrove, W. L., F. C. Boswell, and G. W. Langdale, eds. *The Rising Hope of Our Land: Proceedings, Southern Region No-Till Conference.* Griffin, Georgia, July 16, 17, 1985, 247 pp.

Herbek, Jim, Lloyd Murdock, and Bob Blevins. "Effect of Planting Dates of No-Till and Conventional Corn on Soils with Restricted Drainage." *Agronomy Notes,* Vol. 17, No. 3, University of Kentucky, October 1984, 4 pp.

Losing Ground: Iowa's Soil Erosion Menace and Efforts to Combat It. Des Moines, Iowa: U.S. Dept. of Agriculture – Soil Conservation Service, 1986, 24 pp.

Phillips, R. E., S. H. Phillips, eds. *No-Tillage Agriculture: Principles and Practices.* New York: Van Nostrand Reinhold Co., 1984, 306 pp.

Russnogle, John. "The Conservation Crackdown: If You Don't Comply, Kiss Uncle Sam's Money Goodbye." *Farm Journal,* January 1987, p. 22-D.

Schmepf, Max, ed. "Nonpoint Water Pollution, Special Issue." *Journal of Soil and Water Conservation,* Vol. 40, No. 1, January – February 1985, pp. 1 – 178.

Soil Conservation: Assessing the National Resources Inventory. Committee on Conservation Needs and Opportunities, Board on Agriculture, National Research Council, National Academy Press, Washington, D.C., 1986. Vol. 1, 114 pp.; Vol. 2, 314 pp.

Stein, O. R., *et al.* "Runoff and Soil Loss as Influenced by Tillage and Residue Cover." *Soil Science Society of America Journal,* Vol. 50, 1986, pp. 1527 – 1531.

Troeh, Frederick R., J. Arthur Hobbs, and Roy L. Donahue. *Soil and Water Conservation for Productivity and Environmental Protection.* Englewood Cliffs, New Jersey: Prentice-Hall, Inc., 1980.

Unger, Paul W. *Tillage Systems for Soil and Water Conservation.* Rome: Food and Agriculture Organization of the United Nations. FAO Soils Bul. 54, 1984, 278 pp.

Veseth, Roger, *et al.* "Uniform Combine Residue Distribution for Successful No-Till and Minimum Tillage Systems." *Crop Management Series: No-Till and Minimum Tillage Farming.* Pacific Northwest Extension Pub., PNW 297, May 1986, 6 pp.

Wheaton, Rolland Z., and Edwin J. Monke. "Terracing as a 'Best Management Practice' for Controlling Erosion and Protecting Water Quality." *Agricultural Engineering.* West Lafayette, Indiana: Purdue University, AE-114, June 1981, 3 pp.

CHAPTER 8

Drainage Systems

"When Lamont E. Teuller recently retired, after being a county agent in Utah for 34 years, he noted that his early work was trying to get irrigation water on dry land while his later work was attempting to reclaim the same land by drainage." — *Farm Technology*

OUTLINE

This laser-guided machine is digging a trench at a predetermined depth and laying a continuous 8-inch (20-cm), perforated, corrugated plastic tube to drain the field. On the right is a tractor with a bucket that is dumping fine gravel into a hopper that leads the gravel around the drainage tube to facilitate water movement into the tube. (North Dakota) (Courtesy, Bureau of Reclamation, U.S. Department of the Interior)

□ □ □

8:1 □ OVERVIEW

The public concept of the word ***drainage*** is ***surface water runoff.*** However, in soil science, *drainage* means artificial drainage — the removal of surplus water from the root zone **(rhizosphere)** of the soil by means of open ditches or subsurface conduits.

Surplus water in the rhizosphere may result from a high, true water table, a perched water table, or an inflow of excessive surface or subsurface (spring-fed) waters.

Most benefits from artificial drainage result from greater aeration **(oxygenation)** and faster warming of a drier soil in the spring. In irrigated soils, better drainage permits excess salts and excess sodium to be rendered less toxic.

Soil surveys can be used to predict the need for artificial drainage. Such soils are shown with a "w" in the map symbol. In the United States 309.3 million acres (125 million ha) are so mapped, 75 percent of which are used for agriculture. Drainage classes are determined by the depth of mottling. Soils that are classified as "somewhat poorly drained" or wetter should be drained for common farm crops. Most crops can tolerate a higher permanent water table on sandy soils than on clay soils because of faster exchange of air in the former. When drained, wet soils make better agricultural soils because they are level and more fertile.

The most common drainage systems are classified as sub-surface (tile, tube) and surface drainage (drainage-by-beds, open-ditch, and sump-and-pump). *Surface* drainage accounts for about 65 percent of all farm areas artificially drained.

8:2 □ BENEFITS FROM ARTIFICIAL DRAINAGE

Wet soils that have been satisfactorily drained artificially are better aerated, better granulated, and more productive. Crops can be planted earlier, and heavy machinery can be used to harvest later in the fall. A well-drained soil is also less liable to be compacted by heavy machinery or to develop a plowpan. Plant roots grow deeper after artificial drainage. Furthermore, a well-drained soil is less droughty, warms faster in the spring, and **heaves** (plants pushed out of the ground by ice crystals) less.

The benefits to be expected from draining land that is too wet for the maximum production of crop plants are itemized as follows (Figure 8.1):

1. Improved aeration and more oxygen available to plant roots.
2. Less loss of nitrate nitrogen in the form of nitrogen gas **(denitrification).**
3. Increased efficiency of phosphorus fertilizer.

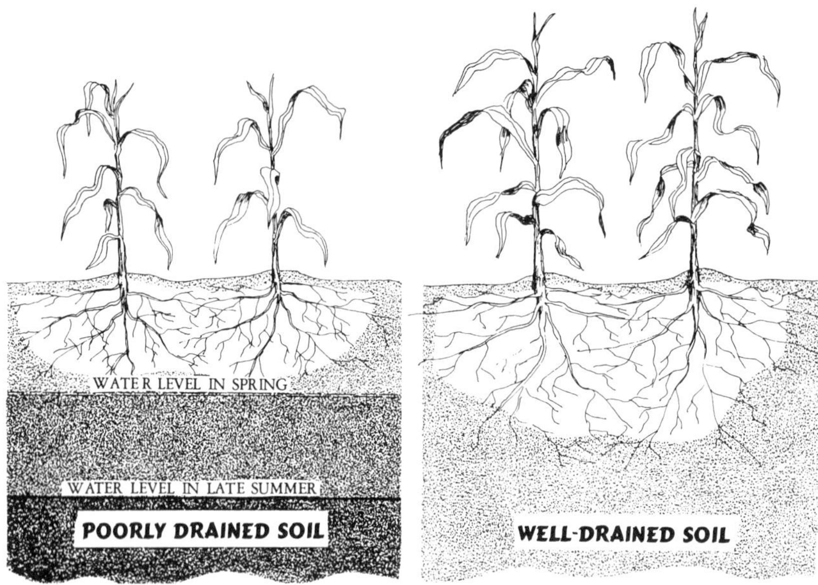

Fig. 8.1 — Poorly drained soil in humid regions (*left*) should be artificially drained to increase soil aeration and root extension. (Courtesy, Instructional Materials Services, Texas A&M University)

4. Increased activity of earthworms and consequently more holes in the soil to let more oxygen diffuse to plant roots and through which roots grow.
5. Reduced weed infestation.
6. Improved workability of the soil for a greater number of days.
7. Improved soil structure.
8. Increased depth of rooting.
9. A warmer soil earlier in the spring, resulting in earlier planting, a longer growing season, and higher yields.
10. Less damage to roots by frost heaving.
11. Less drought damage because of deeper rooting.
12. Reduction in the accumulation of harmful salts and toxic sodium.
13. Decrease in plant diseases such as root rots.
14. Wider adaptation of the drained soil to a greater number of crops.
15. Improved health of humans and their animals, because mosquitoes and many parasites are killed by drying. For this reason, sewage sludge should be applied only on naturally or artificially well-drained soils.
16. Less hazard of formation of surface crusts and tillage hardpans.
17. Increased crop yields.
18. More farm income.

Thirteen years of research in Ohio on Toledo silty clay soil compared corn and soybean yields on undrained soil, with surface, tile, and a combination of surface-and-tile drainage. (See Section 8:9 for definitions.) The combination of drainage systems resulted in 38 percent higher yields of both corn and soybeans. On the average, for each $1.00 invested in drainage systems, the value of the *increase* in crop yields varied from $2.50 to $3.00.

In a 16-year experiment with Nappanee silty clay soil in Ohio, it was shown that subsurface drains spaced 30 feet (9 m) apart removed 43 percent more water annually than drains spaced 60 feet (18 m) apart.

8:3 ▫ DRAINAGE ENTERPRISES

Drainage enterprises in the United States are almost as old as the country. Some of the original colonists drained, by open ditches, some of the wet spots and wet fields in the original 13 states. Today the states ranking high in acreage artificially drained are Indiana, Minnesota, Michigan, Ohio, Iowa, Illinois, Arkansas, Texas, and Florida, in the order named. Also, all irrigated soils throughout the United States must have drainage systems established to handle surplus water and to dilute excess soluble salts (Figures 8.2 and 8.3).

About one-third of the total cropland acres in the United States (110 million acres [44.6 million ha]) are artificially drained. However, an estimated 25 percent (20 million acres [8.1 million ha]) of the soils on which corn is grown are not adequately drained.

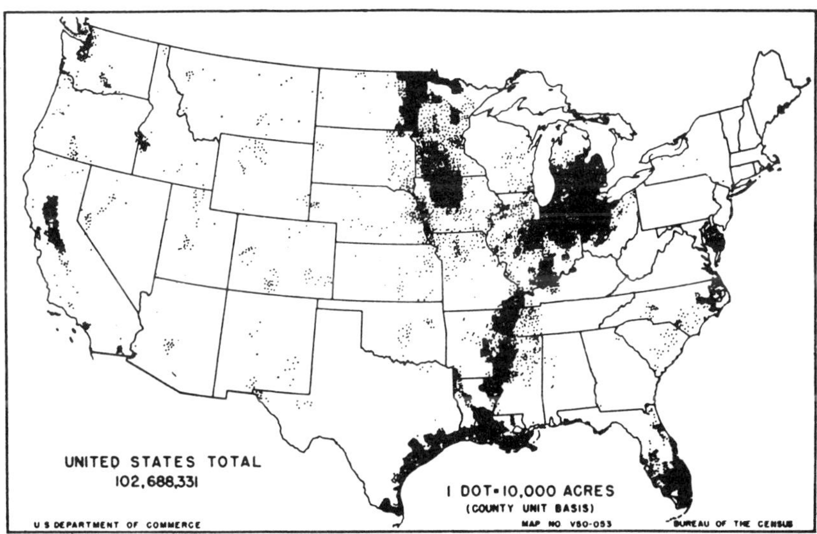

Fig. 8.2 — Map of the United States showing the areas of soils that have been artificially drained. Although new drainage areas have been established, some old areas have been abandoned; the net result of which is about 100 million acres (40.5 million ha) now effectively drained. Drainage systems are now established on about one-third of U.S. cropland. However, still in need of artificial drainage are 26.6 million acres (10.77 million ha), 11 percent of which are irrigated soils. (Courtesy, U.S. Bureau of the Census and USDA – Soil Conservation Service)

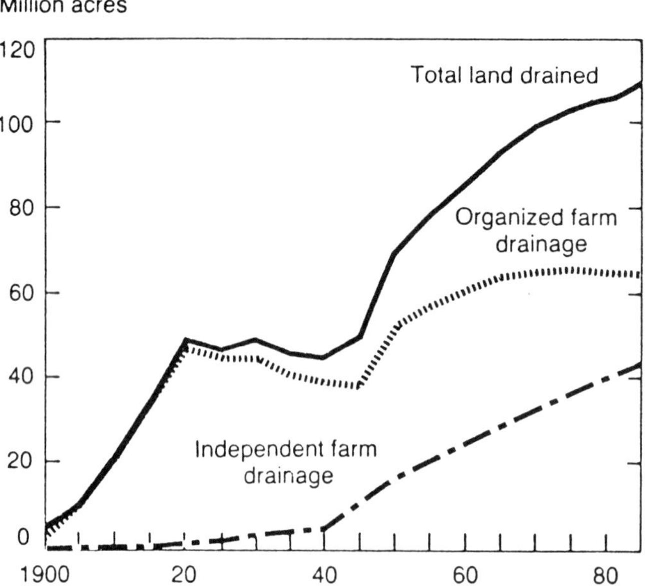

Fig. 8.3 — Of the 100 million acres (40.5 million ha) of U.S. farmland drained, 60 percent rely on drains installed by townships, counties, and drainage districts. These organized farm drainage enterprises are not increasing as rapidly as those established by independent land owners. (Source: USDA Handbook No. 663, 1986)

8.4 □ WATER TABLES

In the humid regions of the world, there is almost always a water table; that is, you could dig, bore, or drive a hole in the earth until you reach a depth where water would stand in the hole. This is the method by which wells to supply water for human and livestock use are established. If you were to select a low place to dig the hole in wet weather, the water level would be at or above the soil surface. The top of the body of water is called the **water table.** In arid and semiarid regions, a water table may not exist or it may be several hundred feet deep; the source of the water may be from rain and snow from the surrounding mountains or it may be **fossil water**. Fossil water is water that was stored during geologic times from melting glaciers or increased precipitation. The best example of fossil water is in the Ogallala Formation, which extends from the High Plains of Texas to Nebraska. Like coal, fossil water, once used, will not be replaced. When the water table is near the surface, the soil may be ideal for establishing a recreation pond or a farm pond. Wet soils are also ideal soils on which to grow rice. Rice culture is most successful when 2 to 4 inches (5 – 10 cm) of water are kept over the surface of the soil. To harvest the rice, however, the rice field must be drained to permit access for a combine. A rice field must also be drained between crops of rice to reduce the accumulation of toxic substances such as hydrogen sulfide, iron sulfide, or an excess of soluble manganese.

For the satisfactory growth of most crops, the water table should be at a depth below 2 to 5 feet (61 – 152 cm), depending on soil texture and the crop. In irrigated soils, drainage facilities are usually deeper in order to leach salt below the rhizosphere. A water table at a shallower depth will usually not permit enough oxygen from the air to enter the roots. (See Figure 8.1 and Table 8.1.)

Water in the soil that is in excess of the needs of plants should be removed by drainage to increase plant growth. This surplus water is called **gravitational water** because it moves in response to the force of gravity when it is not restricted by barriers such as fine clay soils, natural hardpans, plowpans, or tight bedrock. To permit surplus gravitational water to flow by the pull of gravity, you must construct an open ditch, open tile, or pipe system of drainage.

Soils may have a surplus of water for maximum plant growth for one of three reasons:

1. A **true water table** may exist at or near the surface of the soil. A true water table means that from the top of the water table downward for many feet, free water exists. The top of the water level in an open well is an example of a true water table. In depressed areas during wet periods, the true water table may be at or above the soil surface.
2. A **false, suspended, or perched water table** may exist during a period of heavy precipitation in soils that have some "waterproof" (slowly permeable or impermeable) layers in their profile. A waterproof layer is one through which water moves very slowly and may consist of natural hardpans, plowpans, or other tillage pans caused by heavy machinery operating on moist clay soils. A shallow, impervious layer of bedrock can also cause a suspended water table.
3. **Wetland,** which is a soil that lies in a land-locked, "blind" valley or depression that may have no natural outlet drainage and may become too wet at times for plant growth because of surface inflow from the adjoining slopes or from springs.

Whatever the reason for the presence of a surplus of gravitational water, this water must be removed from fields for the satisfactory production of most farm crops except rice and cranberries.

8:5 ▫ SOIL SURVEYS: AN AID TO DRAINAGE PREDICTIONS

The local extension agent or district conservationist can offer guidance in the use of a soil survey for obtaining a scientific appraisal of the need for artificial drainage of a house lot, proposed shopping center, or farm.

From the time of the first soil survey in the United States in 1899, internal soil drainage has been one dominant **criterion** (standard) for differentiating the soil-mapping units. Each year soil classification and mapping become more sophisticated, but the soil characteristics resulting from differences in internal soil drainage have remained as a differentiating criterion in separating one mapping unit from another.

The modern name of a mapping unit on a county soil map is a **phase of a series.** A soil series is a natural body of soils having similar texture, structure, acidity, thickness, and color of the various layers **(horizons).** It is the soil color of the subsurface horizons that indicates the relative wetness of a soil-mapping unit and, therefore, its probability of requiring artificial drainage for achieving maximum yields of all upland crops.

In the U.S. system of soil classification and mapping, there are seven natural (internal) **soil-drainage classes.** This term refers to the condition of frequency and duration of periods of saturation or partial saturation that existed throughout the time of the formation of a particular soil.

The seven natural internal drainage classes that are officially recognized and mapped are characterized by the color and thickness of the subsurface horizons as follows:

- **Excessively drained soils** are very porous, freely permeable, and permeable to great depths. No **mottling**[1] occurs in the profile. Such soils would never need artificial drainage.
- **Somewhat excessively drained soils** are ones through which water and air move freely but slower than in excessively drained soils. No mottling occurs in the profile.
- **Well-drained soils** are usually sandy or intermediate texture such as loam and are almost free of mottling except near the deep water table. These soils do not need artificial drainage. (See "Plainfield Sand" in Table 8.1.)
- **Moderately well-drained soils** have slower internal drainage and mottling at the bottom of the B- (subsurface) horizon. These soils almost never need artificial drainage except possibly if alfalfa is to be grown. (See "Brems Sand" in Table 8.1.)
- **Somewhat poorly drained soils** have mottling in the B-horizon and are normally wet on the surface for many weeks in a year. These soils almost always need artificial drainage, especially for sensitive crops such as alfalfa. (See "Morocco Loamy Sand" in Table 8.1.)
- **Poorly drained soils** may be mineral or organic and are wet for many months in the year and are mottled in the A- (surface) horizon. Artificial drainage is always necessary for crops such as corn and cotton (Figure 8.4).

[1]**Mottling** is a spot-by-spot mixture of soil colors, usually of gray, yellow, red, and brown. The more gray color present, the wetter the soil has been during its formation and the more likely it is that artificial drainage is necessary.

Fig. 8.4 — These two fields are examples of soils classified as *poorly drained* and both need artificial drainage for enhancing crop production. *Top* (Louisiana). *Bottom* (Texas). (*Top:* Courtesy, USDA - Soil Conservation Service. *Bottom:* Source: *The Paris News,* Paris, Texas)

- **Very poorly drained soils** may be mineral or organic and are wet nearly every month in the year. Mineral soils are usually uniformly gray in color but are sometimes mottled near the surface. Such soils always require artificial drainage for upland crops, but they lie so low that a suitable gravity outlet ditch may be difficult to build. Some of these soils may need to be drained by the sump-and-pump system. When adequately drained by artificial means, they are usually very productive.

8:6 ▫ WATER TABLES, SOIL MOTTLING, AND ARTIFICIAL DRAINAGE GUIDELINES

Water table depths were measured throughout the year on three sandy soils mapped in three different drainage classes in northeastern Indiana. The results were as follows:

1. Plainfield sand — Well-drained (average depth to water table, 68 in.; range, 64 – 70 inches).
2. Brems sand — Moderately well-drained (average depth to water table, 31 in.; range, 21 – 40 inches).
3. Morocco loamy sand — somewhat poorly drained (average depth to water table, 14 in.; range, 8 – 25 inches) (Table 8.1).

Depths to mottling were then determined and the conclusions were:

1. Soil mottling generally corresponded to the ranges in depths of the water table throughout the year.
2. The more poorly drained the soil, the greater the range in water table depths during the year.
3. Soils in drainage classes "Somewhat Poorly Drained," "Poorly Drained," and "Very Poorly Drained" should be artificially drained for growing crops such as alfalfa and other upland crops.

Specific drainage guidelines are recommended by Purdue University for the somewhat poorly drained Morocco loamy sand as follows:

> For water table control, use **field laterals** (open ditches) spaced 600 feet (183 m) apart and 2.5 to 4.0 feet (0.76 to 1.2 m) deep; bottom width, 4 feet (1.2 m); and side slopes, 2:1 (2 feet horizontal to 1 foot vertical) or flatter. Since **over-drainage** may make these soils droughty, consider controlling water table depth at 24 to 36 inches (60 – 90 cm) for crop production. Encase subsurface drains with **filter materials** (hay, straw). Depths of subsurface drains 3 to 4 feet (0.9 – 1.2 m), and spacing (space between subsurface drain lines), 100 to 150 feet (30.5 – 45.75 m).

In summary:

1. *Both* surface and subsurface drainage techniques are recommended.
2. Recommended depths of surface and subsurface drainage lines are *below* the deepest water table measured [25 inches (0.6 m) shown in Table 8.1].

Table 8.1 — Water Table Depths in Inches Throughout the Year in Sandy Soils in Three Drainage Classes as Mapped in Northeastern Indiana[1]

Month	Plainfield Sand (well-drained)	Brems Sand (moderately well-drained)	Morocco Loamy Sand (somewhat poorly drained)
January	70	35	15
February	65	28	8
March	66	28	8
April	68	28	8
May	65	21	10
June	64	27	14
July	69	39	20
August	70	40	25
September	70	30	15
October	70	30	18
November	70	32	20
December	69	35	16
Average	68	31	14
Range	64 - 70	21 - 40	8 - 25

[1] D. P. Franzmeier, *et al.* "Water Table Levels and Water Contents of Some Indiana Soils." Dept. of Agronomy, Purdue University, and USDA - Soil Conservation Service, RB 976, January 1984, 49 pp.

Note: Inches × 2.54 = cm.

Table 8.2 — The Relationship Between Soil Texture and Depth of Water Table for Maximum Yields of Selected Crops[1]

Soil Texture	Depth to Water Table				
	0.5 ft (15 cm)	1.0 ft (30 cm)	2.0 ft (61 cm)	3.0 ft (91 cm)	5.0 ft (152 cm)
Clay	Timothy Reed canary-grass		Grain sorghum Potato	Pea	Bean Alfalfa (to 8 ft.) Corn Sugar beet Wheat Barley Oat Tomato
Loam	Bean Fescuegrass Ladino clover	Orchard-grass	Carrot Pea	Potato Cabbage Tomato Squash Most pasture grasses	Grape orchards
Sandy loam			Cabbage Corn Squash		Cantaloupe Watermelon

[1] Sources: Jans Wesseling. "Crop Growth and Wet Soils." In Jan Van Schilfgaarde, ed., *Drainage for Agriculture*, No. 17, *American Society of Agronomy, 1974,* pp. 7 - 37, and "Irrigation on Western Farms." USDA - Agr. Inf. Bul. 199, 1959.

8:7 □ WATER TABLES AND SOIL PRODUCTIVITY

The common field and garden crops have various soil wetness adaptations at which they produce their highest yield. Between rains or irrigations, the relative wetness is determined by the water-holding capacity of the soil and the amount of water that rises by capillarity from a water table or a wetter soil below. Because water moves higher by capillarity in clay soils than in sands, it can be concluded that the water table should be deeper in clay soils than in sands to have the same degree of soil wetness in the plant root zone. Research data portrayed in Table 8.2 support this fact.

As shown in Table 8.2, maximum yields of corn varied with depth of water table and soil texture. When the water table was 2 feet (0.6 m) in depth, corn yields were highest on a sandy loam soil, but with a 5-foot (1.5-m) water table depth, corn yielded most on a clay soil. Also as shown in Table 8.2, reed canarygrass and timothy seem to grow best when the water table is at the surface during May and June (Figure 8.5). By contrast, sweetclover, alfalfa, and crested wheatgrass must have a well-drained soil (Figure 8.6). For most varieties of rice, the water table during the growing season should be maintained at a depth of 2 to 4 inches (5 – 10 cm) *above* the surface of the soil.

Fig. 8.5 — Timothy *(left)* and reed canarygrass *(right)* grow best in poorly drained soils. (Courtesy, USDA)

Fig. 8.6 — Sweetclover *(top)* and crested wheatgrass *(bottom)* need large amounts of oxygen in the soil; therefore, they do well only on soils that are naturally or artificially well-drained. (*Top:* Courtesy, University of Illinois. *Bottom:* Courtesy, USDA)

8:8 □ DRAINAGE SYSTEM SELECTION

Drainage systems may be classified as subsurface (tile or tube) and surface drainage (drainage-by-beds, open-ditch drainage, and sump-and-pump drainage) (Figure 8.7).

The selection of a system of artificial drainage depends upon the following factors: (1) lay-of-the-land **(topography)**, (2) drainability of the soil, (3) crops to be grown, (4) water quantity and quality (in irrigated areas), and (5) economics such as cost-benefit ratio.

The topography must be assessed in great detail before a suitable system of drainage can be selected. A **topographic map** should be available that shows contour lines at intervals of 1 foot (30 cm) or less. From this **contour map**, the principal drainage outlets can be located as well as the lateral lines feeding into the principal outlets. Based on the topographic survey, the entire drainage design as well as the depth of all ditches or tile lines can be determined. If the soil is flat, clayey, and sloping gently toward a creek, open ditches or drainage-by-beds may be the most appropriate system of drainage. An interior basin with no natural outlet may have to be drained by a sump-and-pump system. With a drainable loam soil that is fairly level, a tile- or tube-drainage system may be the most suitable.

The amount of water to be carried by the drainage system is important in the design criteria of any of the drainage systems. The size and spacing of the drain-

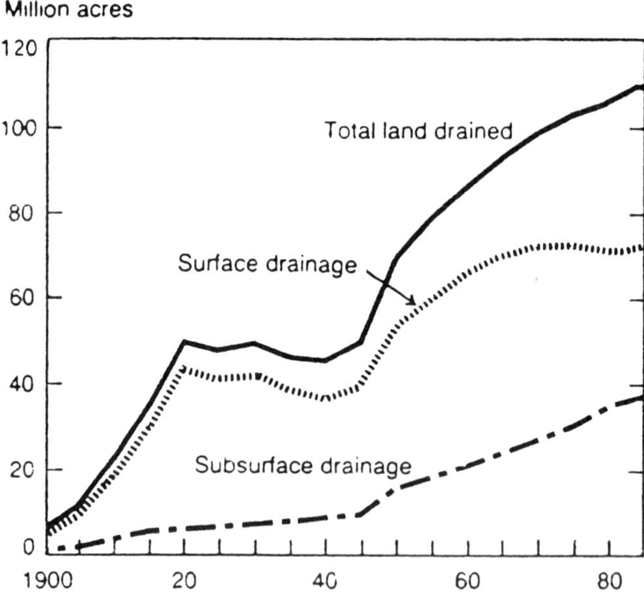

Fig. 8.7 — As prime farmland becomes scarcer and higher priced, subsurface tile- or tube-drainage systems are better options than surface drainage because crops can be grown over the tile or tube lines. (Source: USDA Handbook No. 663, 1986)

age lines or capacity of the sump pump must be determined by the volume of water to be discharged over a given time. Information to help in designing such capacity may be obtained from the director of the county agricultural extension service or from the district conservationist of the Soil Conservation Service.

Drainability of the soil also is a factor in determining the kind of drainage system. Water will not move through some fine-textured clay soils fast enough to make feasible the use of a tile- or tube-drainage system. Such clay soils must be drained by open ditches, drainage-by-beds, or a sump-and-pump. For example, the Sharkey-clay phase of soil series in the Mississippi Delta contains so much fine clay with such small capillaries that water will not move through it fast enough to discharge water by a tile- or tube-drainage system. **Swelling clays** such as montmorillonite cannot be drained adequately by buried tiles or tubes. These must be drained by open ditches or a sump-and-pump.

8:9 ◻ TILE DRAINAGE

A **tile drainage system** consists of burying sections of tile at suitable depths and slopes to carry surplus water from the soil. This practice of using tile for draining soils in the United States started at Geneva, New York, in 1835 and is still popular today. The early Romans used a similar system which included the burying of tree limbs or rocks in a trench that sloped toward a larger outlet such as a creek.

Today the standard tile consists of a 1-foot (30-cm) section of straight-sided pipe made usually of clay and then fired. Concrete is also used in making sections of tile, but its use is restricted to neutral or alkaline soils because soil acids dissolve the concrete. Other materials used in making tile are perforated plastic and bituminous fiber. For use in peat and muck soils, at least 2-foot (61-cm) sections of tile are highly recommended because of the uneven settling and the greater hazard of the short tiles becoming out of line. Tile with a bell on one end (sewer tile) is satisfactory for use in a tile-drainage line but the cost is greater.

The diameter of the tile should be at least 4 inches (10 cm) for lateral lines and 6 or 8 inches (15 or 20 cm) for main drainage lines.

The recommended depth and spacing of the lines of tile will vary, depending on the level of the main outlet creek or ditch, the soil texture, the height of the water table, the amount and intensity of the rainfall or irrigation, and the kinds of crops to be grown. A common practice is to place the tile lines 3 to 4 feet (91 – 122 cm) deep and 50 to 100 feet (15 – 20 m) apart. On a field where the soil is fairly level and uniformly sandy to a depth below the tile-drainage line, the tile lines may be shallower and spaced wider than on fine-textured soils such as clay. In high rainfall areas where the intensity is also high, narrower spacing is necessary. Lastly, if reed canarygrass or timothy is to be grown, the tile lines can be smaller in diameter, the depth shallower, and the spacing of the lines wider than on fields where crops such as wheat, sugar beets, or snap beans are to be grown. (See Table 8.2 and Figure 8.5.)

The bottom of the tile-line ditch must have a gentle slope (grade) toward the outlet. Perhaps $^{1}/_{10}$ of a foot (1.2-cm) fall for each 100 feet (30 m) is a typical tile-line grade (0.1 percent grade). When the tile lines are on a 1-foot (30-cm) fall for each 100 feet (30-m) (1.0 percent grade), the water will usually flow so fast in them that the tiles will wash out (Figure 8.8). When the tile-line gradient changes abruptly, a small "breather" pipe should be extended from the tiles to the soil surface to prevent a vacuum from forming, thus reducing the water flow in the tile lines.

The 1- or 2-foot (30- or 61-cm) sections of tile are laid in the bottom of a trench, with the ends just touching. In mineral soils, there is an attempt to leave a

Fig. 8.8 — An 8-inch (20-cm) diameter bell-tile system is being laid. Each tile is 2 feet (61 cm) long. A 4-inch (10-cm) gravel envelope will surround the line of tiles before soil is backfilled. (Wyoming) (Courtesy, Bureau of Reclamation, U.S. Department of the Interior)

⅛-inch (0.3-cm) space between all tile sections, and a ¼-inch (0.6-cm) space in peat and muck soils to allow water to enter.

When the tile lines are laid in fine-textured soils such as clays, it is often a standard practice to place an envelope of 4 inches (10 cm) of gravel around the entire line. This gravel hastens the movement of drainage water from the subsoil into the drain tile. In silty and fine sandy soils, filtering materials should be added over the gravel to keep fine sand, silt, and clay from seeping into the drainage tile. The usual filtering materials are straw, hay, and woodchips.

The pattern in which the tile lines are laid will mostly depend on the topography and the area needing draining. The usual types of tile-line or tube-drainage systems are (Figure 8.9): (1) parallel, (2) gridiron, (3) herringbone, (4) random, (5) grouping, and (6) double main.

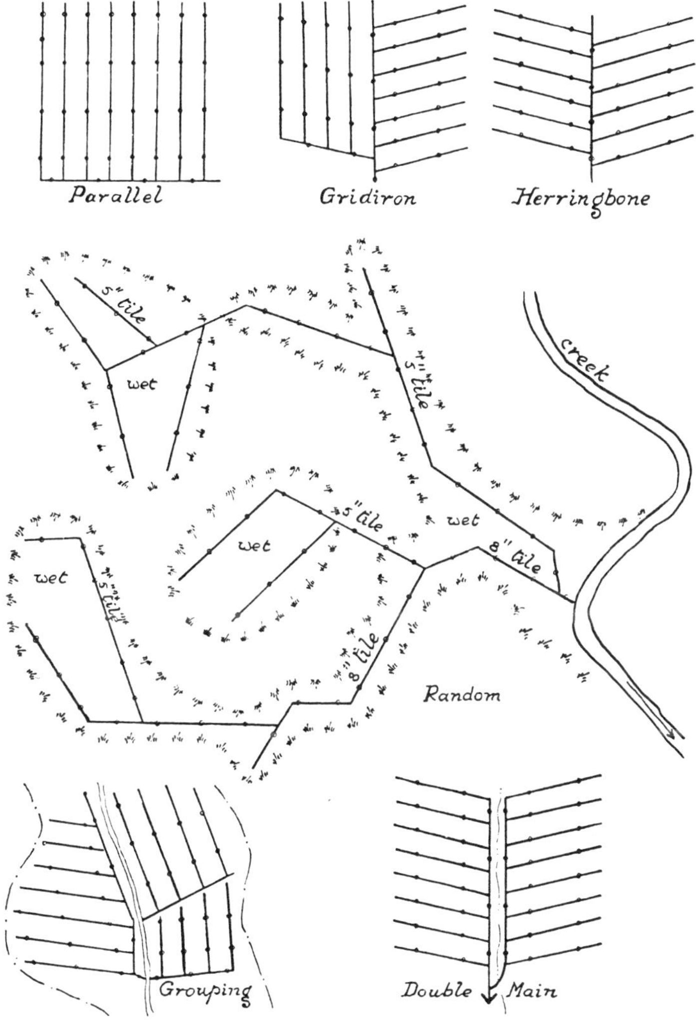

Fig. 8.9 — Types of tile- or tube-drainage systems. (Courtesy, USDA)

All tile-line outlets should be screened to keep out insects, snakes, and rodents. The last 10 feet (3 m) of the tile line before the outlet should have all joints cemented to prevent washout and collapse of the line.

Maintenance of all tile lines should include inspection at least twice a year — in the spring and in the fall. Every tile line should be walked over to observe:

1. A vertical hole, meaning a broken or washed-out tile.
2. Wetness along a tile line that is plugged.
3. Trees or shrubs whose roots may be filling the tile lines. Roots of willow and elm trees are especially such a hazard.
4. The outlet, which should have the screen cover removed and the end of the tile cleaned of trash.
5. Signs of erosion below the outlet where the outflow runs. If erosion is present, rock should be laid in to break the energy of the flowing water.
6. The rate of effluent after the soil has been irrigated, if the tile line serves as drainage in an irrigation system. Water should flow for several days after irrigation water has been applied.

Preventive maintenance is even cheaper than maintenance. This includes these practices:

1. Heavy vehicles should be kept off the tile-drained field when the soil is wet. The machinery compacts the soil and reduces its drainage capacity.
2. As many deep-rooted and close-growing crops as possible, such as alfalfa and sweetclover, should be grown. Roots, both living and dead, provide channels for water to move rapidly downward and into the tile lines.
3. Establish minimum tillage on the artificially drained field. This will increase the rate of percolation and therefore the speed of water removal from the soil. (See Chapters 5 and 6.)

Properly installed and annually inspected and repaired tile-drainage systems should last up to 100 years or more. However, a study of 95 tile-drainage systems in Ohio showed that one-fourth of the drains 1 to 10 years old needed repair, two-thirds of the 11- to 20-year-old drains needed repair, and all drains over 40 years old needed repair.

8:10 □ TUBE DRAINAGE

Tube drainage refers to the use of a continuous, flexible tube with drainage holes that is laid in a trench. Also used are rigid 10-foot (3-m) sections of plastic pipe with holes along the sides near the bottom. This system differs from tile drainage primarily by the substitution of a flexible (or sometimes rigid) plastic pipe for the 1-foot (30-cm) to 2-foot (61-cm) sections of tile. Machines have been made that dig a trench, lay flexible pipe, and cover the trench, all in one continuous operation. The depth, spacing, size of pipe, and efficiency of operation are similar to those for tile drainage. The claim for superiority of the plastic pipe system is that the installation costs are less and that, being one long piece, uneven settling of the soil does not get the pipe out of alignment. Because of greater uneven settling in peat and muck soils, the tube-drainage system may be better suited to them than tile (Figure 8.10).

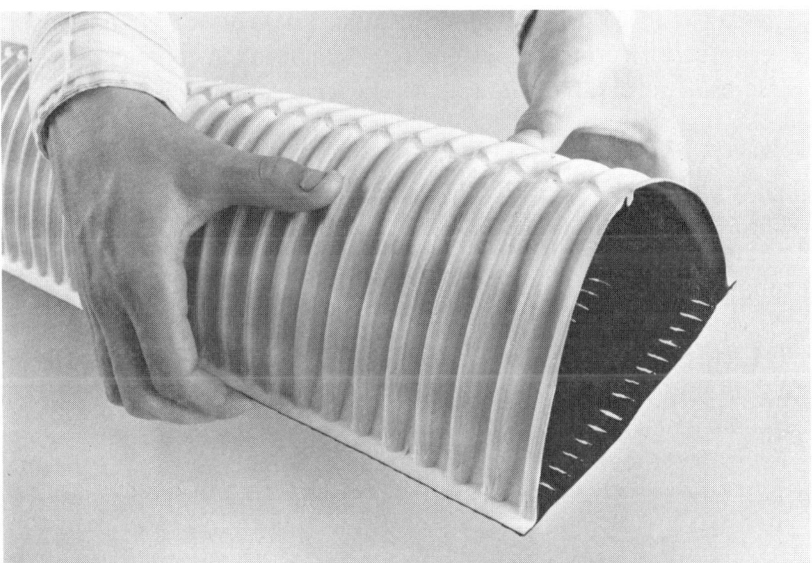

Fig. 8.10 — *Top:* A tiling machine laying a plastic, circular cross-section tube-drainage system. If this were a tiling operation, all of these people would be laying the tile instead of watching. *Bottom:* A newer type of plastic conduit that folds nearly flat in transport but opens to an approximate half circle with perforations in the bottom when it is laid. (*Top:* Courtesy, Michigan State University. *Bottom:* Courtesy, Hancor, Inc., Findlay, Ohio)

8:11 ▫ DRAINAGE-BY-BEDS

The system of **drainage-by-beds** is adapted best to clay soils on nearly level topography, with slopes up to 1.5 percent. To operate successfully, the tile- or tube-drainage systems must be established on soils that permit water to move through them at a rate fast enough to drain the fields within a few days. Some fine-textured soils are naturally so impervious (or have been made so impervious by heavy vehicles) that either the open-ditch system or the drainage-by-beds system must be used.

In cross-section, the drainage-by-beds system resembles the back of a turtle, viewed from side to side. Rainwater and melted snow move sideways to open shallow ditches, which then drain to a larger ditch or a stream. Practical widths of the beds are from 50 to 200 feet (15 – 61 m) depending on the volume of water to be removed and the height of the crown. The crown is normally about 1 foot (30 cm) higher than the sideslope drains on a 100-foot (30-m) wide bed. In general, the wider the bed, the higher the crown.

The width of beds and the height of the crown are determined by these factors:

1. Slope of the soil. The flatter the soils, the narrower the beds and the higher the crown.
2. Permeability of the soil. The less permeable the soil, the narrower the beds and the higher the crown.
3. Kind of cropping system or land use. A cropping system that is primarily one of continuously cultivated crops requires narrower beds and a higher crown than a cropping system that includes deep-rooted grasses and legumes, a permanent pasture, hay meadow, or minimum tillage.

Rows are laid out parallel to the side ditches. The crown of approximately 1-foot (30-cm) height must be maintained by starting the plowing there and turning a **back furrow** at the ridge of the crown. All turning plow furrows should then be turned toward the crown. The side ditches are the **dead furrows.**

At the ends of the rows a lateral collection drain must be established. The sides of the drain should permit a tractor to cross them readily, with grades between 6:1 and 8:1 (horizontal to vertical ratio). Beyond the collection drain a tractor turn strip about 18 feet (5.5 m) wide must be left between the lateral collection drain and the fence (Figure 8.11).

A lower-cost modification of the system of drainage-by-beds is **drainage-by-ridges-and-furrows.** This consists of establishing by a turning plow or a lister, beds 2 to 3 feet (61 – 91 cm) wide from furrow-to-furrow and 1 to 2 feet (30 – 61 cm) high above the bottom of the furrows. The bottom of the furrows should slope toward a prepared outlet to carry away surplus water. Crops are planted on top of the ridges (Figure 8.12).

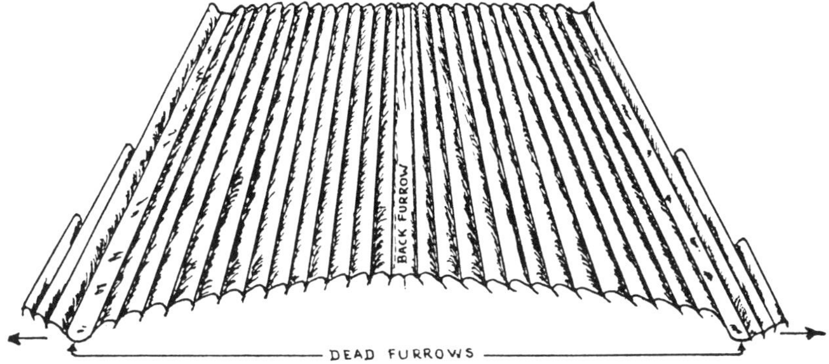

Fig. 8.11 — A system of drainage-by-beds, showing how it is maintained by the direction of plowing. The center, or crown, should be the back furrow, and the side drains should be the dead furrows. The width of the beds varies from 50 to 200 feet (15 - 61 m), and the crown should be about 1 foot (30 cm) higher than the dead-furrow side drains. (Courtesy, Cornell University)

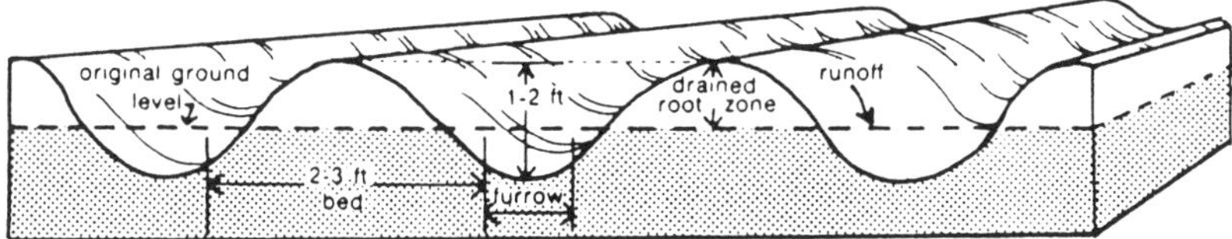

Fig. 8.12 — Drainage-by-ridges-and-furrows consists of establishing a system of furrows and ridges similar to those used in furrow irrigation. Crops are then planted on top of the ridges where the soil is drier. (Source: William H. Doucette, Jr., "Coastal Land Drainage for Agriculture and Forestry," North Carolina State University Pub. AG-223, 1980, 11 pp.)

8:12 ▫ OPEN-DITCH DRAINAGE

Open-ditch drainage may be laid out at random, from wet spot to wet spot, or be parallel on level wetland.

The cross-section of random open ditches should have side slopes of about 8:1 to facilitate crossing with tillage machinery. By contrast, on more level and uniformly wet soils, parallel open ditches may have noncrossable side slopes as steep as 2:1 for loam soils, 1½:1 for clays, and 1:1 for sands, peat, and muck soils (Figure 8.13).

A suitable fall for large farm ditches may vary from almost level to 3 inches (8 cm) per 100 feet (30 m), while for small ditches it may be advisable to use grades up to 4 or 6 inches (10 or 15 cm) per 100 feet (30 m).

Open ditches must be maintained on an annual basis. When weeds choke a ditch enough to reduce the efficiency, the ditch can be cleaned with the use of herbicides or by being mowed, grazed, burned, or backhoed.

Fig. 8.13 — Open-ditch drainage is often the cheapest system for draining large, nearly level areas. (South Carolina) (Courtesy, USDA - Soil Conservation Service)

8:13 ▫ SUMP-AND-PUMP DRAINAGE

The system of **sump-and-pump drainage** is ideal for locations such as wet basements and low wet areas in fields that cannot be drained by gravity flow.

The world's most successful, large-scale use of the sump-and-pump system of drainage is the drainage and reclamation of soils below the level of the Zuider Zee in the North Sea in the Netherlands.

Here dikes have been built around a part of the North Sea, the sea water pumped out, and the salty soils reclaimed for use in growing farm crops. To maintain the proper level of the water table, pumps must operate almost continuously from a low sump area. This system of drainage permits half of the people in the Netherlands to live *below* sea level.

In the United States, the sump-and-pump drainage system is adapted to large flat areas with high water tables and permeable subsoils and where energy sources for power are relatively low in cost. The permeability of subsoils over extensive areas can be determined best by boring holes in the soil to the depth of the proposed water table, then pumping from the hole where the water table is nearest the surface. If the water levels in all test holes recede satisfactorily, the sump-and-pump system can be established. In irrigated areas with low salt content in the water, surplus irrigation water can be pumped from the drainage ditch back to a level where it can be used again for gravity flow irrigation.

One possible source of failure of the sump-and-pump system of drainage is the lack of suitable support to maintain the sump pit. To avoid collapse of the sides of the sump and to reduce the transport of sediment in the water, the sump hole should be lined with sand, gravel, and then rock. The sump pump should then be mounted rigidly at the proper level above the rock. The motor should be activated by an automatic float switch set at the predetermined level of the water in the sump hole.

8:14 ▫ REFERENCES

"Benefits from Drainage." *Ohio Report on Research and Development,* November – December 1980, p. 91.

Donahue, Roy L., Raymond W. Miller, and John C. Shickluna. *Soils: An Introduction to Soils and Plant Growth,* 5th ed. Englewood Cliffs, New Jersey: Prentice-Hall, Inc., 1983, Chapter 17.

Drablos, C. J. W., *et al.* "Illinois Drainage Guide." Urbana – Champaign: University of Illinois, Co-op. Ext. Serv., Cir. 1226, 1984.

Edwardson, Steven, David Watt, and Lowell Disru. "Laser-controlled Land Grading for Farmland Drainage in the Red River Valley: An Economic Evaluation." *Journal of Soil and Water Conservation,* Vol. 43, No. 6, 1988.

Fausey, N. R., and R. Lab. "Drainage Effects on Soil Physical Properties." *1988 Agronomy Abstracts.* American Society of Agronomy, Crop Science Society of America, and Soil Science Society of America, Madison, Wisconsin.

Franzmeier, D. P. "Water Table Levels and Water Contents of Some Indiana Soils." Purdue University, RB 976, January 1984, 48 pp.

Nolte, B. H. "Life Expectancy of Ohio Tile Drainage Systems." *Ohio Report on Research and Development,* May – June 1980, pp. 38 – 40.

Palmer, M. L., and N. R. Fausey. "Factors Affecting Drainage on Corn Land." Purdue University, NCH-38, July 1986.

Schwab, G. O., *et al.* "Tile and Surface Drainage of Clay Soils, Parts IV – VII." The Ohio State University, Ohio Agricultural Research and Development Center, Res. Bul. 1166, January 1985.

Sinclair, H. R., *et al.* "Indiana Drainage Guide: Part 1. Soil Drainage Recommendations." Purdue University, ID-160, December 1985, 11 pp.

Troeh, Frederick R., J. Arthur Hobbs, and Roy L. Donahue. *Soil and Water Conservation for Productivity and Environmental Protection.* Englewood Cliffs, New Jersey: Prentice-Hall, Inc., 1980, Chapter 15.

U.S. Dept. of Agriculture – Soil Conservation Service. *Engineering Field Manual for Conservation Practices.* Unnumbered publication, Chapter 14, "Drainage" (updated periodically).

CHAPTER 9

Irrigation Systems

" . . . the desert shall rejoice and blossom as the rose." — Isaiah 35:1

OUTLINE

"Snow water keeps the arid West alive." Water from mountain snowmelt is used to irrigate this potato farm. (Idaho) (Courtesy, USDA – Soil Conservation Service)

□ □ □

9:1 □ OVERVIEW

Irrigating crops dates back thousands of years and has been practiced in some form by farmers throughout the world. Irrigation is basically an agricultural operation, supplying the need of a plant for water. Irrigation is necessary in a dry climate, and to the agriculturalist, it is a component of successful crop husbandry. Agriculture is the nation's largest consumer of water. Four times more water is consumed to produce food and fiber than for all other purposes combined. Although grown on only about 12 percent of the nation's cropland, irrigated crops account for about 25 percent of the total value of crop production in the United States. Figure 9.1 shows where irrigation is concentrated in the United States, and Figure 9.2 compares the acreages of irrigated and nonirrigated cropland by crop production region.

Conservation of irrigation water through proper handling is becoming more and more important as demands on our water supplies increase. Water management is equally important whether it is in a large irrigation project, in farm ditches, or on the land.

During the last 25 years, better and easier methods of water application have been developed. Control of irrigation water, from a dam on the river or from a private well, is not complete until the water reaches the end of the last field or row and any water left over is taken care of through reuse or adequate surface drainage. Efficiency in transporting and spreading water on the land is only the first

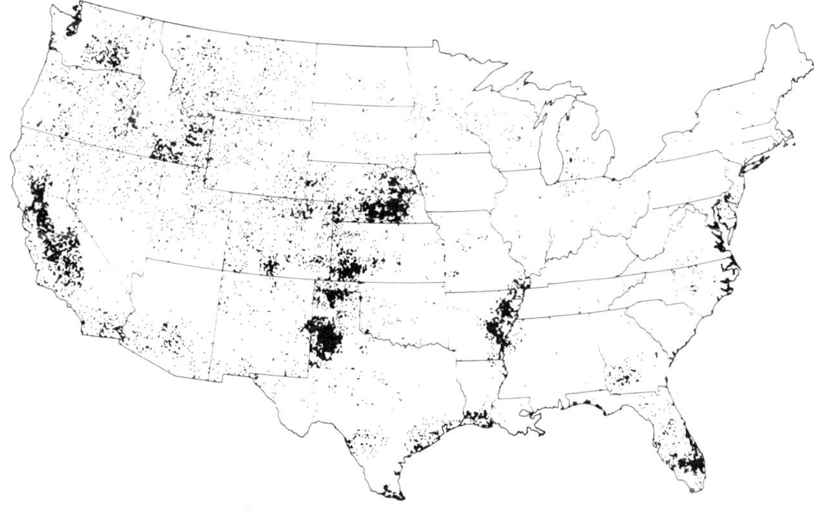

Fig. 9.1 — Irrigated acreage in the United States. One dot equals 8,000 acres (3,200 ha) where irrigation facilities are in place. (Source: *National Resources Inventory*, USDA - Soil Conservation Service)

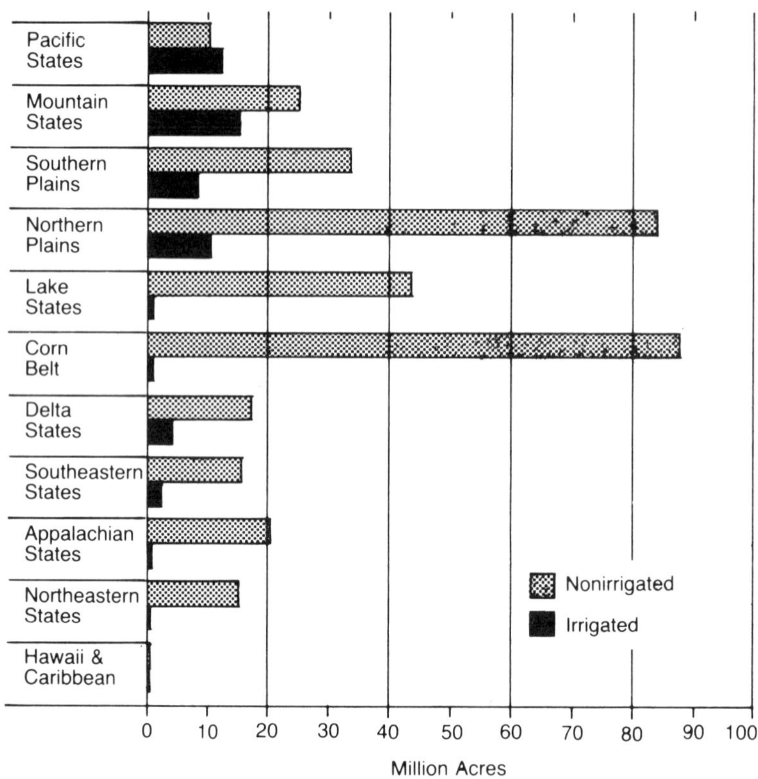

Fig. 9.2 — Comparison of irrigated and nonirrigated cropland by crop production region. (Source: USDA - Soil Conservation Service)

Note: Acres × 0.405 = hectares.

step in supplying the moisture needs of growing crops. Wasted water often results in wasted soil through erosion, leached fertilizers, drainage troubles, and possible salt accumulation.

9:2 □ THE IMPORTANCE OF WATER

The demand for water by the plant varies considerably from day to day, depending upon conditions. On a hot, dry, windy day the loss of water will be great. On a cool, humid day the amount of water loss is much less. If the roots can provide as much water as the leaves lose, the plant will remain turgid and full, like a fire hose with water flowing through it under pressure. But let the amount of water escaping through transpiration exceed the quantity entering through the roots, and the plant will lose its rigidity, start to wilt, and may die for lack of water. Water constitutes up to 90 percent or more of the fresh weight of actively growing plants.

The longer the rate of water loss exceeds the rate of water absorption, the greater the reduction in water content of the plant. The reduced water content interferes with the normal growth processes as well as with the food-making power of the plant, and if this loss of moisture proceeds far enough, part of the plant or the entire plant will wilt and may eventually die. Plants are most susceptible to damage from moisture deficiency at the blossoming, pollinating, and heading stages.

A plant absorbs through its roots and transpires from its leaves from 300 to 500 pounds (136 – 227 kg) of water for every pound (0.45 kg) of dry weight added. An acre (0.4 ha) of corn requires roughly a half million gallons (1.9 million liters) of water to grow — as much as a day's requirement for a city of 5,000 people.

It is important to maintain water balance within a plant's system. High transpiration rates are not damaging so long as a corresponding high rate of root absorption is maintained. In general, transpiration exceeds absorption from the early part of the day until noon.

A plant that enjoys feast then famine (lots of moisture, then not enough) will grow, but the harmful effects will nevertheless show in the final reduction in yield. The growth denied it during the periods of starvation or unquenched thirst is never made up. The plant will simply be stunted and its productivity reduced accordingly.

Plants absorb water mainly from their root tips. The small rootlets that extend through the soil pore spaces contact and absorb water. Water dissolves plant nutrients and carries them into the plant, chiefly through the cell walls in the roots.

9:3 □ PLANT AND SOIL MOISTURE RELATIONSHIPS

A few years ago, **water requirement** meant the number of pounds of water transpired by a plant to produce a pound of plant dry matter; these figures were determined under artificial conditions. During more recent years, however, research under field conditions has furnished us with a better "yardstick."

Instead of measuring water requirement in pounds, the most useful method of reporting it is in terms of acre-inches or hectare-centimeters. When 1 inch of rain falls on 1 acre (0.4 ha) of land, it will cover the acre to a depth of 1 inch, assuming no evaporation or infiltration. When irrigation water is added to the land, a volume of 1 acre-inch would cover a hectare to a depth of about 1 centimeter.

The water requirement of the most common plants has been determined by scientists. Although the water requirement varies somewhat because of differences in temperature, relative humidity, wind movements, and soil fertility, the values shown in Figure 9.3 were determined under average field conditions.

A glance at Figure 9.3 shows that truck crops require 16 acre-inches (16 ha-cm) of water, the least of any of the plants shown. Next in water requirement are cotton and the small grains such as wheat and oats. These plants need 18 acre-inches (19 ha-cm) of water to produce good yields. Corn is next, with a 20 – acre-inch (21 – ha-cm) requirement, followed by a 24 – acre-inch (25 – ha-cm) need for sugar beets, lawn grasses, shrubs, and flowers. Deciduous fruit trees require about 30 acre-inches (31 ha-cm) of water. Highest in water requirement are alfalfa and irrigated pasture grasses and clovers, with an annual need of 36 acre-inches (37 ha-cm) of water.

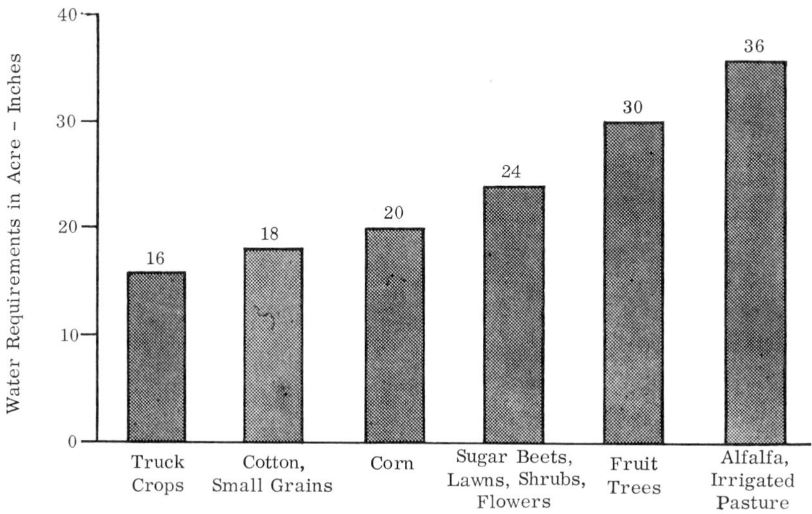

Fig. 9.3 — Annual water requirement per acre for good yields of the common crops. (Courtesy, USDA)

Note: Acre-inches × 1.028 = ha-cm.

The amount and frequency of rainfall or irrigation water received is only a part of the full story of efficient water use in growing abundant crops. The other part consists of the significant variation in available water capacity of different soils.

Shortly after an irrigation a soil is saturated with water. This water drains downward and plants take up very little of it because of the short period they have access to it. About 24 to 48 hours after irrigation, the root bed (rhizosphere) of the soil is usually at the upper limit of its available moisture range. The moisture content of a well-drained soil at the upper limit of the available moisture range is termed **field capacity.** The lower limit of available moisture range is known as the **wilting coefficient** or **wilting point** (Figure 9.4). **Available water** for plant growth is all of the water in the range between field capacity and wilting point.

Moisture storage characteristics of a soil are very important in irrigation. The amount of moisture a soil can store for plant use (its available water capacity) is

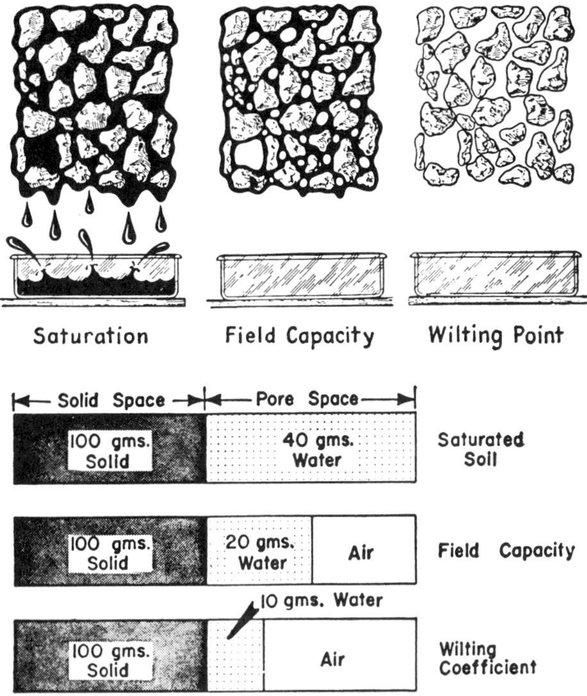

Fig. 9.4 — Diagram showing the solids-water-air relationships in a well-granulated silt loam soil when the water volumes are at *saturation, field capacity,* and *wilting point* (wilting coefficient), respectively. Assuming 100 grams of soil solids (mineral matter plus organic matter): at saturation there may be 40 grams of water and no air (oxygen); at field capacity there may be 20 grams of water and more air; and at the wilting point there may be 10 grams of water and the most air. Water in the soil that plant roots can absorb is the 10 grams between the field capacity (20 grams) and the wilting point (10 grams). Most upland plants cannot absorb water from the field capacity to saturation because there is not enough air. (Courtesy, USDA and the U.S. Department of the Interior)

determined by the size and arrangement of the soil particles, the amount of organic matter, and the depth of the soil to a hardpan or to bedrock.

Field capacity of a soil is the term used to describe the maximum amount of water left in the soil after losses to the forces of gravity have ceased and no surface evaporation has occurred. The ***wilting point*** is the point at which the plant can no longer obtain enough moisture to supply its transpiration needs, and the plant permanently wilts and dies.

It is necessary to maintain a level of moisture content between the field capacity and the wilting point. The amount of available water that a soil will hold is very important when irrigating because it governs the amount of water that can be effectively applied at each irrigation and influences the time interval between irrigations. Table 9.1 can be used as a guide. For example, a silt loam soil has a field capacity of approximately 3.5 inches (8.9 cm) per foot (30 cm) depth of soil, a wilting point of 1.4 inches (3.6 cm) per foot (30 cm) of soil, and an **available water capacity** of 2.1 inches (5.3 cm).

Table 9.1 — Water-holding Characteristics of Various Soil Textural Classes[1]

Soil Textural Class	Approximate Inches of Water per Foot of Soil Depth		
	Field Capacity	Wilting Point	Available Water Capacity
Sand	1.2	0.3	0.9
Fine sand	1.5	0.4	1.1
Sandy loam	1.9	0.6	1.3
Fine sandy loam	2.5	0.8	1.7
Loam	3.2	1.2	2.0
Silt loam	3.5	1.4	2.1
Sandy clay loam	3.7	1.6	2.1
Clay loam	3.8	1.8	2.0
Silty clay loam	3.8	2.1	1.7
Clay	3.9	2.4	1.5

[1]Courtesy, USDA.

Note: To convert from inches per foot to centimeters per meter, multiply by 8.33.

9:4 ◻ WHEN TO IRRIGATE

A simple method for determining when to start irrigating consists of obtaining a handful of soil at a depth of approximately 1 foot (30 cm), squeezing it in the hand, and then noticing if it forms a stable ball. If the soil is a loam and it forms a stable ball, it is *not* time to apply irrigation water; if the ball crumbles, it *is* time to irrigate (Figure 9.5).

Another very simplified method used to determine when to start irrigating consists of observing when the crop begins to wilt at midday. Between 12:00 noon and 3:00 P.M., plants usually exhibit maximum wilting. When wilting is first observed on the driest part of the field, it is time to apply more irrigation water.

Moisture meters (tensiometers), long the subject of experimentation, now take the guesswork out of irrigation, and one can tell at a glance the exact moisture content of the soil. Tensiometers should be installed at depths and locations where the roots are most actively absorbing water. For row crops, the instruments are usually installed in the row at two depths within the active root zone. In orchards, tensiometers should be installed at two or more depths and should be placed about one-third the distance from the drip line of the tree to the trunk.

Soil moisture measurements with tensiometers often disclose amazing variations in available soil moisture in different sections of even relatively small acreages, due to unsuspected variations in kinds of soil and subsoil. Irrigation treatments are then adjusted accordingly.

The beginning of the silking period is the most critical time for irrigating corn. Research has demonstrated that regardless of the total amount of irrigation water applied during the season, the earlier the irrigation water was applied during the silking period, the greater was the yield of corn. Soybeans use the most water per

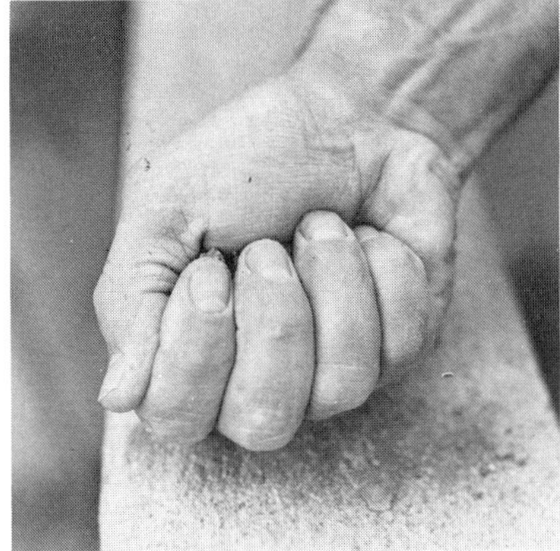

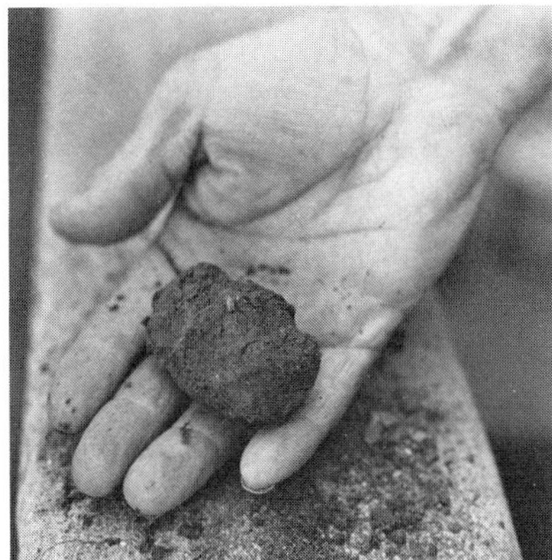

Fig. 9.5 — To determine if it is time to start irrigating, obtain a sample of soil from the depth of 1 foot (30 cm) and squeeze it firmly in the hand *(top left)*. If the soil is a loam and it forms a stable ball *(top right)*, no water is needed; however, if the soil crumbles *(bottom left)*, it is time to start irrigating.

day during the rapid vegetative growth period and during the reproductive stage (Figure 9.6).

9:5 ▫ CROP ROOTS AND MOISTURE EXTRACTION PATTERNS

The root system of plants is fixed by heredity. Some plants have deep tap roots while others have shallow primary and lateral roots. Table 9.2 shows the normal maximum root depths of a few mature irrigated crops. The table also shows the depth in which about 70 percent of the plant's roots are concentrated.

Root development and penetration can be altered by soil factors such as compacted layers, shallow soils over gravel or hardpan, high water table, rising water table, dry layers, and salt concentrations.

Along with Table 9.2, consider the fact that the soil moisture is actually extracted from the soil according to Figure 9.7. Observe that 70 percent of the mois-

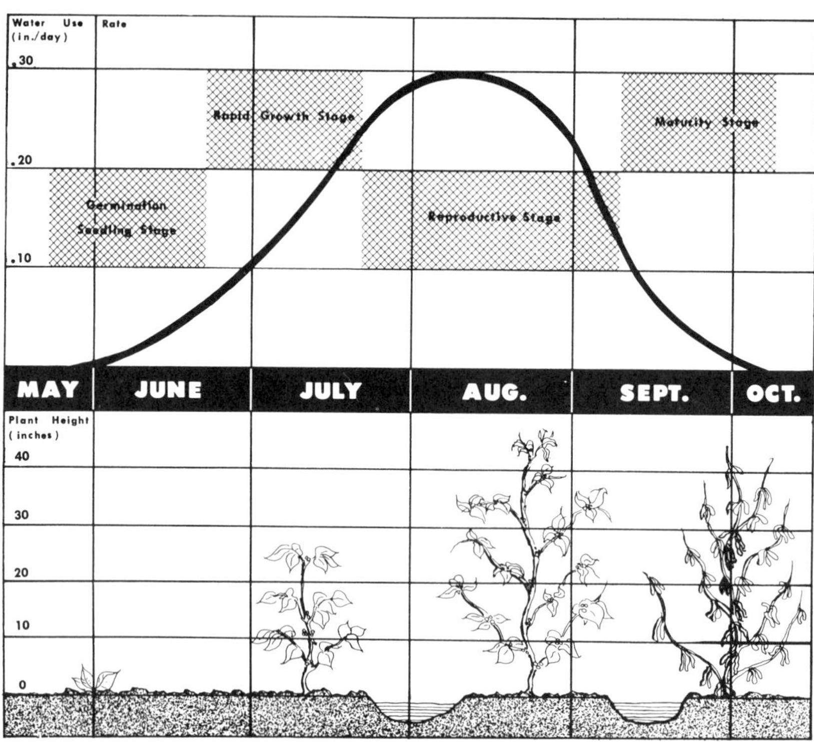

Fig. 9.6 — Characteristic growth and water use pattern of soybeans. (Full season varieties) (Source: *Irrigating Soybeans,* Irrigation 5, Kansas State University)

Table 9.2 — Crop Rooting Systems in a Normal Deep Soil

Crop	Maximum Root Depth (feet)	Concentration of 70% of Roots (feet)
Alfalfa	9.0	5.0
Corn	8.0	4.0
Grain sorghum	7.0	4.0
Red clover	6.0	3.0
Wheat	5.0	3.0
Soybeans	5.0	3.0
Tomatoes	5.0	3.0
Grass	5.0	2.5
Potatoes	3.5	1.5
Cucumbers	3.0	1.5

Note: Feet × 30.5 = cm.

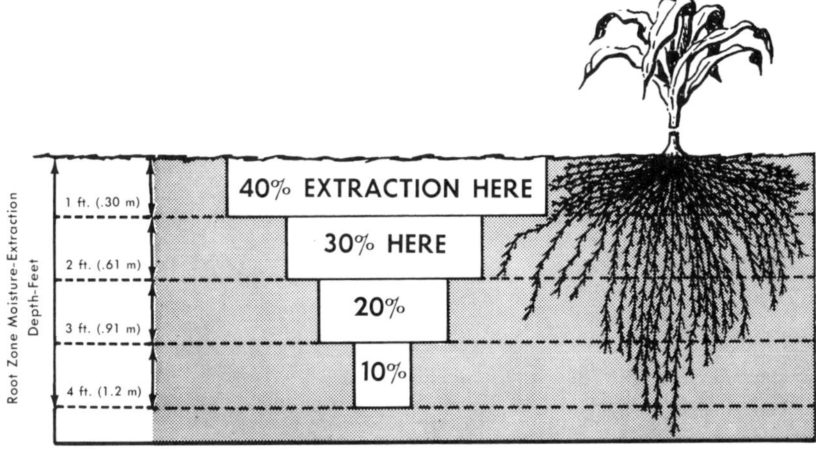

Fig. 9.7 — Average moisture extraction patterns of plants. (Adapted from *SCS National Engineering Handbook,* USDA)

ture used is removed in the top half of the root zone. As an example, if normal root depth is 4 feet (122 cm), 70 percent of the consumed moisture is taken from the top 2 feet (61 cm) of soil. This figure is quite accurate when moisture extraction patterns of plants growing in a soil without restrictive layers and with adequate moisture throughout the root zone are described.

9:6 ▫ MOISTURE USED BY IRRIGATED CROPS

Total moisture used on an irrigated field is called **consumptive use** and includes: (1) water used in plant transpiration (movement of water through roots, stems, and stomata of leaves), (2) water evaporated from the surface of plant leaves, (3) water evaporated from the soil surface in the field, and (4) water in the plant.

Consumptive use is affected mainly by plant density, plant height, sunshine, air temperature, relative humidity, wind speed, available soil moisture, and soil texture. Water for consumptive use can be supplied to the root zone by precipitation and / or irrigation.

Peak period consumptive use rates given in Table 9.3 are normally used for irrigation design. The peak period consumptive use is defined here as "the peak moisture use by the crop per day through the **irrigation frequency,** where the frequency is the time that it takes to cover the field with the irrigation system." For example, if corn uses 0.25 of an inch (0.64 cm) per day for 14 days, the irrigator must provide 0.25 × 14 or 3.50 inches of water to restore soil moisture to the desired percentage in the root zone. This 3.50 inches (8.9 cm) is the net water needed by the plant and does not include the irrigation water lost by evaporation, seepage, or wastage during the irrigation. Under poor management, these losses may sometimes amount to 25 to 60 percent.

Table 9.3 — Estimated Consumptive Water Use by Crops

Crop	Seasonal Use (inches)	Peak Period Use (inches per day)
Alfalfa	26	0.25
Tamegrass pasture	23	0.24
Corn	21	0.25
Grain sorghum	18	0.25
Soybeans	18	0.22
Potatoes	17	0.20
Small grains	16	0.20
Melons	15	0.17

Note: To convert from inches to centimeters, multiply inches by 2.54.

9:7 ▫ SELECTION OF AN IRRIGATION SYSTEM

The proper selection and design of the irrigation system is important in determining the success of the irrigation enterprise. Regardless of the system selected, it should fit the conditions of the field where it will be used.

Factors that affect the irrigation system include topography; soil characteristics; source, amount, and quality of water supply; crop to be grown; and labor and capital available. System selection and design assistance is available from many sources, including equipment dealers and manufacturers, private consultants, the Soil Conservation Service, and the Cooperative Extension Service.

These guidelines are suggested for evaluating and selecting an irrigation system:

1. Uniform distribution of water.
2. Minimum erosion or other damage to the soil.
3. Maximum application and water use efficiency.
4. Practical and economical performance from the standpoint of the crop, labor requirements, energy requirements, cost of land preparation, and capital and maintenance costs.

Other considerations include:

1. ***System capacity*** — There should be enough water and equipment to meet the peak water demand of the crop unless soil moisture storage is high and can be depleted without depressing crop yields.
2. ***Economical pipe size*** — There should be an economic balance between pipe cost, energy cost, and labor cost.
3. ***Application rate*** — Sprinklers should be designed so that application rate matches as nearly as possible the basic intake rate of the soil. When application rate exceeds the soil intake rate, runoff and erosion will occur.

9:8 ▫ IRRIGATION METHODS

The most common methods of irrigation are flood, furrow, corrugation, border, sprinkler, trickle, surge, and cablegation. The reason for so many different types of irrigation systems is the wide variety of field conditions, soils, and cropping systems to which the irrigation system must be adapted. It is important to select the method that best fits the needs of the field and crop(s) that require irrigation. The final selection of one irrigation system may very likely be based on the total annual cost and the expected return.

9:8.1 ▫ Flood Irrigation

Flood irrigation consists of covering an area of land with several inches or centimeters of water at one time. To do this requires some type of retaining wall on the borders of the field. This wall is usually a low terrace-like border which is just high enough to hold the anticipated amount of irrigation water. In some places, gently sloping land is first leveled in a series of flat benches and a border is built around each bench. On flat land, the area to be flooded is divided into basins. The flooding method of irrigation is the cheapest per acre-inch or hectare-centimeter of water applied but it takes more water to produce the same results. **Wild flooding** consists of allowing water to flow over a field without previously leveling the land and without establishing any water-retaining borders. This method of irrigation is very wasteful of water (Figure 9.8).

Fig. 9.8 — An example of the bad irrigation practice of wild flooding. (Courtesy, USAID - India)

Some form of surface leveling or land forming is essential to maintain surface drainage and optimum flood depths. Precision land grading to a plane surface is the preferred method (Figure 9.9). Land forming means reshaping the surface of the land with a land plane to ensure orderly movement of water over a field or section of a field and to provide a more uniform distribution of water in the soil. Land planing involves smoothing, ditching, filling low spots, and establishing recommended grades (slopes) if none exist.

Fig. 9.9 — Before land is irrigated, it should be made smooth to permit a uniform application of irrigation water. Land is being leveled with a self-loading type scraper. (Courtesy, USDA - Soil Conservation Service)

9:8.2 ▫ Furrow Irrigation

Furrow irrigation is the oldest kind of irrigation system. In this method, water flows by gravity from a main ditch and down each furrow, usually between each two rows of crops (Figure 9.10).

Crops that are planted in double rows on beds are irrigated by directing the water between the beds. Crops planted in a wide spacing, such as berries or grapes, as well as crops planted in an orchard, usually have two furrows for irrigation between each two rows of plants.

In general, soil erosion is excessive when the furrow method of irrigation is used on rows that have a slope of more than 2 percent. An aim should be to keep the slope of the furrows less than 0.25 percent. On slopes greater than 2 percent, running the rows on the contour helps to reduce erosion.

The aim of the irrigator should be to obtain the maximum flow of water down each furrow without causing excessive erosion. In this way, water will soak into

Fig. 9.10 — A cotton field near Phoenix is irrigated by the furrow method with water from the watersheds of the national forests. (Courtesy, USDA - Forest Service)

the soil at a fairly uniform rate all along each furrow. As a rule of thumb, if it takes four hours to add sufficient water to a furrow, the water should be turned into the furrow in such volume that it will flow to the end of the furrow in one hour (one-fourth of the elapsed time).

The common methods of controlling the distribution of water in furrow irrigation make use of the following:

1. Field lateral ditch with small equalizing ditches leading directly to each furrow.
2. Field lateral ditch with siphon tubes leading to each furrow (Figures 9.10 and 9.11).
3. Field lateral ditch with spiles (small, straight pipes) leading directly to each furrow.
4. Irrigation pipe with large openings (gates) emptying into each furrow (Figure 9.12).
5. Buried pipe to carry the water to the field, with risers emptying into each furrow or series of furrows.

9:8.3 ▫ Corrugation Irrigation

Corrugation irrigation is a modified furrow system that is adapted to sod-like crops. Narrow and closely spaced furrows are made on appropriate contours across a field, and water is allowed to soak into the soil along the furrows.

Fig. 9.11 — Siphon tubes are placed in position to irrigate corn by the furrow method. (Kansas) (Photo by Roy Hunter Follett)

Fig. 9.12 — Furrow irrigation with gated pipe on a corn field. Water is released from the large pipe through openings (gates) seen at the right. (Kansas) (Photo by Roy Hunter Follett)

9:8.4 □ Border Irrigation

The **border irrigation** method is used on gentle slopes. Narrow strips of land, 20 to 50 feet (6 – 15 m) wide, are leveled (across the slope), a low ridge is built up and down the slope as a border to hold water, and each strip is irrigated by flooding (Figure 9.13).

Border irrigation is most satisfactory under these conditions:

1. When the surface soil is deep enough to permit the land to be leveled without leaving areas of unproductive subsoil.
2. When the infiltration rate of the soil is intermediate. Sandy soils would not permit a uniform depth of water penetration, while an impervious clay soil would permit an excessive amount of water to be lost by evaporation before it soaked into the soil.
3. When the slope is satisfactory to permit the construction of borders with a slope of 2 percent or less. Land to remain in sod crops can be steeper without causing excessive erosion.
4. When the land is planted to crops which are not injured by temporary flooding.

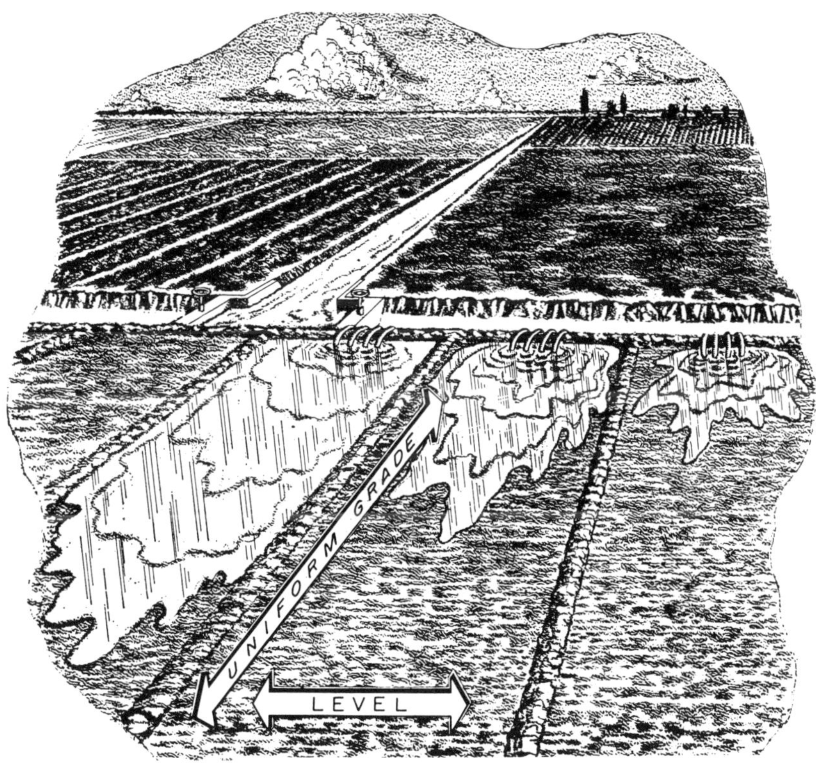

Fig. 9.13 — The field has been prepared for border irrigation by building small levees around each leveled area. Then the areas are flooded in rotation to irrigate them. (Courtesy, USDA)

9:8.5 □ Sprinkler Irrigation

Sprinkler irrigation can be used where surface irrigation is not practical because of excessive slopes, irregular topography, erosive soil, unfavorable intake rates, coarse-textured, deep sandy soils, or combinations of these factors.

The advantages of the sprinkler system are:

1. Land-leveling is not necessary.
2. Drainage problems are decreased.
3. Erosion is kept to a minimum.
4. Fewer special skills are required.
5. It is adapted to fields with uneven topography.
6. It is adapted to soils too sandy for other methods.
7. It can be used also as frost protection.
8. Fertilizers can be spread with the irrigation water. (See Section 9:9.)

Some disadvantages of the sprinkler system are:

1. The initial investment is high.
2. Power costs are tremendous.
3. More labor is required to move the pipe.
4. Wind prevents a uniform distribution, making it often necessary to irrigate at night.
5. Evaporation losses of water are greater than with other methods of irrigation.
6. Because of wet foliage, more plant diseases, such as halo blight, anthracnose, black rot, downy mildow, leaf spot, early blight, and bacterial spot, are encouraged.

The **center-pivot sprinkler irrigation** system is a self-propelled, continuously moving lateral that rotates around a center-pivot well to irrigate a circular area as large as 130 acres (53 ha). Center-pivot systems are propelled by electric-, oil-, or water-driven systems. Oil- and electric-driven systems have higher initial investments than water-driven systems but have the capability of instant reversing and can move forward or backward without applying water. Center-pivot can operate on rolling topography and have variable rotation speeds, allowing a variation in amounts of water applied (Figures 9.14 and 9.15).

9:8.6 □ Trickle (or Drip) Irrigation

Trickle irrigation (sometimes called **drip irrigation**) is delivered by plastic tubing and near zero water pressure. Water is applied on a frequent, often daily, basis to prevent moisture stress in the plant by maintaining favorable soil moisture conditions. Water "drips" or "trickles" from **emitters (orifices)** at selected spacings along the tube. The tube itself may be porous, allowing water to "seep"

Fig. 9.14 — A self-propelled center-pivot sprinkler system irrigating alfalfa. (Source: *Irrigation Age Magazine*)

Fig. 9.15 — Aerial view of a 1,300-foot (390-m) center-pivot sprinkler system. In one revolution, the center-pivot sprinkler irrigates 130 acres (53 ha) in 33 hours and is set to apply 0.6 inch (1.5 cm) of water. (North Dakota) (Courtesy, U.S. Department of the Interior - Bureau of Reclamation)

into the soil. Trickle irrigation uses less water than the more conventional irrigation systems do. Because less water is lost by evaporation, trickle irrigation requires less water than is normally required in sprinkler irrigation. The trickle irrigation system is best adapted to crops that are widely spaced, such as those in orchards (Figure 9.16).

The main advantage of trickle over other types of irrigation is the excellent control of water application which it provides. Water is applied daily at a rate as close as possible to the rate of consumption by the plant. Evaporation from the soil surface is minimal, and deep percolation loss can be almost entirely avoided.

Fig. 9.16 — A new windbreak of juniper, blue spruce, and honey locust trees is watered with a trickle irrigation system. The system includes 150 feet (46 m) of a ½-inch (1.27-cm) supply line and 375 feet (114 m) of "spaghetti" hose, which runs to each tree. (Source: *Soil and Water Conservation News,* Vol. 1, No. 5, August 1980)

9:8.7 □ Surge Irrigation

Surge irrigation is a relatively new practice that can be described as intermittent application of water to furrows in a series of pulses or surges, rather than a continuous furrow stream. Usually the water is alternated (varying from one to two hours) between two sets of furrows until the irrigation is completed. Switching is accomplished with a surge valve and automatic controller. Surge valves are typically located between two sets of furrows. Gated pipe is used to distribute the irrigation water to each of the two furrows from opposite sides of the surge valve. The concept is shown in the schematic drawing of surge irrigation (Figure 9.17).

With continuous flow furrow irrigation, a large furrow stream is needed to rapidly advance the stream from the head to the tail for a uniform application of

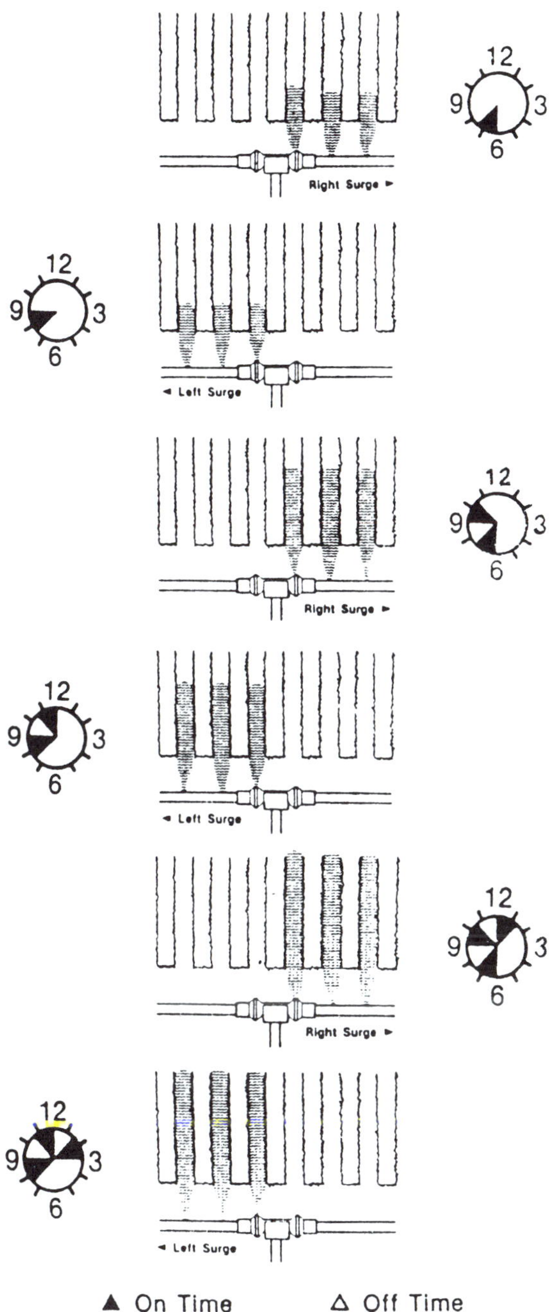

Fig. 9.17 — This represents a schematic drawing of surge irrigation. The water is alternated between two sets of furrows until the irrigation is completed. (Source: USDA - Soil Conservation Service)

water to minimize deep percolation. The large stream will usually result in excess runoff because it is larger than what is needed to satisfy infiltration.

Cutback furrow irrigation has never been widely accepted because of the increased labor and management involved. Researchers, in their attempt to automate cutback irrigation, found it was easier to alternately turn gated pipe gates on and off rather than attempting to turn them half way off. During the testing, they observed an increased advance rate with the on-and-off flow, the beginning of surge irrigation. Surge irrigation has the water conservation advantages of cutback furrow irrigation without the labor and management problems.

The reason the furrow stream advances faster under surge is not fully understood. Experience indicates that the alternating wetting and "resting time" for each surge slows down the intake rate of the wet furrow and produces a smoother and hydraulically better surface. This allows the next surge to travel more rapidly down the wet furrow until it reaches the dry furrow.

The water conservation advantage of faster advance with surge on many soils and field conditions, coupled with the capability of cutback, makes surge an effective management practice. It can improve the uniformity of application and reduce the amount of tailwater on many field conditions.

9:8.8 □ Cablegation — An Automated Furrow Irrigation System

Cablegation was developed during the late 1970's at the U.S. Department of Agriculture – Agricultural Research Service's Snake River Conservation and Research Center in Kimberly, Idaho. Cablegation can be used on either new or presently irrigated fields with some modification.

The irrigation pipe is installed at the head of the field. The pipe is rotated upwards so the gates are at about a 30-degree angle from the top of the pipe. All pipe gates are open, but water is discharged only through a set number of gates behind a moving "plug" inside the pipe.

Inside the pipe is a 2-foot (61-cm) long plug composed of two bowls fastened together. The plug, which is pushed through the pipe by water pressure, is restrained by a cable hooked to a *braking wheel,* thus, the name *cablegation.*

The braking wheel, which is 5 feet (1.5 m) in diameter, is located at the high point in the lateral. The wheel controls the speed of the plug by regulating how fast the cable unrolls from a reel attached to the wheel.

Proper slope of the pipe is critical in cablegation. There has to be at least a 6-inch (15-cm) drop per 100 feet (30.4 m) because the system works from the gravity flow of water rather than water being pumped under pressure. The slope allows the head to move the water through the pipe and push the plug along with it (Figure 9.18).

When the system is operating, water backs up behind the plug and spills out of the pipe through the gates. Flow rates are highest closest to the plug. But because the water level remains constant in the sloping pipe, pressure and flow gradually taper off farther back in the line. Smaller and smaller amounts of water

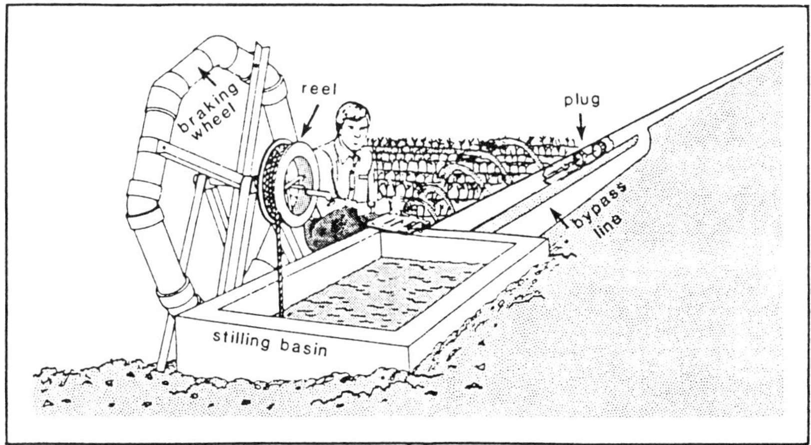

Fig. 9.18 — Cablegation irrigation system. The braking wheel controls the speed of the plug by regulating how fast the cable unrolls from a reel attached to the wheel. (Source: USDA - Soil Conservation Service)

flow from each successive gate back to a point where the water level is below the holes, and no water is released.

Proper plug speed is the key to the system operation and control. Determining the plug speed through the pipe determines the irrigation time. By controlling the plug speed, a farmer can match gate flow rates to soil intake rates more closely, making cablegation more efficient.

Potential runoff is reduced because of the constant reduction of inflow rate into the furrow. With cablegation, the water flow into the furrow is heavy at the start to force water down the row and then tapers off to provide only enough water for a good soaking without excessive runoff.

9:9 □ FERTILIZER APPLICATION THROUGH AN IRRIGATION SYSTEM

The application of fertilizer in the form of dry materials, dissolved materials, liquids, and gases through irrigation systems has been a common and recommended procedure in various areas of the country, principally in California. The word ***fertigation*** has been coined to designate this practice.

Various types of metering devices have been developed for introducing the desired quantity of fertilizer solution into the irrigation system at **weir boxes** (dams to raise the level of water) or in open canals that supply water to the particular area being fertilized. Liquid fertilizers, usually nitrogen, also are supplied to overhead irrigation systems through commercial metering devices connected to the water supply line, and in some cases by homemade equipment.

Many fertilizers *can* be applied through a sprinkler irrigation system but this does not necessarily mean that they *should* be applied in this manner. To be a safe practice, the fertilizer should dissolve and stay in solution and should not be a corrosive hazard to the irrigation system.

Fertilizers That Can Be Applied Through Sprinkler Systems

15-60-0 (ammonium polyphosphate)	Ammonium nitrate
13-39-0 (ammonium phosphate)	Ammonium sulfate
11-48-0 (ammonium phosphate)	Urea
21-53-0 (diammonium phosphate)	Urea-ammonium nitrate
8-24-0 (liquid ammonium phosphate)	Muriate of potash
25-35-0 (urea ammonium phosphate)	Potassium sulfate
	Water-soluble Borax or Borate (boron fertilizers)

Fertilizers That Should Not Be Applied Through Sprinkler Systems

Phosphoric acid	This is water-soluble and can be applied through a sprinkler system but it will cause considerable corrosion of metal.
Anhydrous ammonia (NH_3)	This material is water-soluble but a large percentage of the nitrogen would be lost in the air.
Ammonia solutions (aqua ammonia)	This material is water-soluble but a large percentage of the nitrogen would be lost in the air.

Fertilizers That Cannot Be Applied Through Sprinkler Systems

Superphosphates Elemental sulfur Some mixed fertilizers Gypsum Lime	These materials will not dissolve thoroughly in water.

The nitrogen in fertilizer solutions escapes rapidly into the atmosphere as ammonia when either the irrigation water or the soil is strongly alkaline. This is true whether the source of nitrogen is ammonium sulfate, ammonium nitrate, urea, ammonium phosphate, or liquid ammonia.

For example, there is almost no loss of nitrogen when the irrigation water or the soil is neutral (pH 7.0). At a pH of 8.0, however, approximately 10 percent of the nitrogen in all sources of nitrogen fertilizers may be lost by escaping into the atmosphere as ammonia. At a pH of 9.0, approximately one-third of the nitrogen may be lost, and at a pH of 10.0, about 60 percent is likely to be lost.

Anhydrous ammonia can be applied along with the irrigation water in some systems, saving energy, time, and expense of conventional application. Cylinders of anhydrous ammonia are commonly seen along many irrigation ditches.

From these cylinders the anhydrous ammonia under pressure is metered into irrigation water through pipes that discharge it below the surface. Injecting and applying anhydrous ammonia through irrigation systems takes careful management however, and it will not work for everyone. Presently it is not recommended for sprinkler systems, but it is recommended for surface irrigation systems with a system for reuse of wastewater (tailwater). (See Section 9:10.)

Anhydrous ammonia is not recommended for sprinkler systems because of the volatility of the ammonia when sprayed into the air and because it precipitates the salts in the irrigation water which can cause a buildup in the pipe and plug up the sprinkler nozzles.

9:10 ▫ IRRIGATION TAILWATER MANAGEMENT

Responsibility for the proper management of the waters diverted for irrigation rests with the irrigator. This responsibility includes the management of the waters applied to the field as well as the runoff waters which result from the applied waters. The term ***tailwater*** is commonly used in the gravity irrigation industry to describe waters accumulating at the lower (tail) end of the irrigation run. Irrigation tailwater is considered to be the easiest component of surface irrigation return flow to manage and control. It is said that if tailwater is controlled and / or reused, irrigation application efficiency would be improved, water and energy would be conserved, and at the same time discharge of pollutants would be substantially reduced.

Two important factors justify the installation and use of a tailwater management system. First, it enables the irrigator to obtain the full economic benefits from the water. Second, it assists in the conservation of a limited natural resource — water.

There are several methods of using the tailwater that is recovered. It can be used in any of the following ways: (1) to irrigate more acres of land, (2) to use on the same field again, (3) and to establish a pond for recreation.

A tailwater recovery system should be designed for each individual irrigation system (Figure 9.19). The tailwater must be collected from the ends of the irrigated fields and be transported to the recovery pit. The size or volume of the pit is determined by the amount of tailwater produced and by the operation plan of the system. A small pit is satisfactory if an automatic pump is used, but a larger pit is necessary for a system using a long holding time between pumpings.

The reuse of tailwater can have the effect of increasing the irrigation water supply by the amount of water recovered. Most irrigation systems require about 10 to 30 percent of the water applied to the field to be lost as runoff from the field. This water is lost to any beneficial use unless it can be recovered and reused. With an ever-increasing emphasis on the environment, it is important that tailwater be captured and thus prevent agriculture from being the cause of pollution of downstream rivers, lakes, and reservoirs.

Fig. 9.19 — A tailwater recovery pit for 680 acres (275 ha) that are under irrigation. (Kansas) (Courtesy, USDA - Soil Conservation Service)

9:11 ▫ REFERENCES

Black, R. D., and D. R. Hay. "Soil-Water-Plant Relationships." Kansas State University, MF–466, 1980.

"Cablegation Automated Furrow Irrigation System." U.S. Dept. of Agriculture – Soil Conservation Service, Leaflet No. 12, undated.

Eck, Harold V. "Winter Wheat Response to Nitrogen and Irrigation." *Agronomy Journal,* Vol. 80, 1988, pp. 902 – 908.

Israeli, Israel. "Irrigation Scheduling." Colorado State University, Serv. in Action No. 4.708, 1987.

Jensen, M. E., ed. *Design and Operation of Farm Irrigation Systems.* ASAE Monograph No. 3, American Society of Agricultural Engineers, 1983, 829 pp.

Kemper, W. D., *et al. Cablegation Systems for Irrigation: Description, Design, Installation and Performance.* U.S. Dept. of Agriculture – Agricultural Research Service, ARS-21, 1985, 208 pp.

Shawcroft, R. W., and R. Croissant. "Irrigation of Winter Wheat in Colorado." Colorado State University, Serv. in Action No. 556, 1986.

Tanji, K. K., *et al.* "Irrigation Tailwater Management." U.S. Environmental Protection Agency, EPA-600 / S-2-81, July 1981.

Thomas, James G. "Scheduling Irrigations by Electrical Resistance Blocks." Kansas State University, MF-465, 1978.

Turner, J. H., ed. "Planning for an Irrigation System." Athens, Georgia: American Association for Vocational Instructional Materials, 1980, 120 pp.

Walker, W. R., and G. V. Skogerboe. *The Theory and Practice of Surface Irrigation.* Utah State University, 1986, 514 pp.

Withers, B., and S. Vipond. *Irrigation: Design and Practice.* Ithaca, New York: Cornell University Press, 1980, 306 pp.

CHAPTER 10

Reclaiming Salt-affected Soils

Great Nations may spring from the salt of the earth, but crops won't." — *Agricultural Research,* Vol. 22, No. 10 (1974)

OUTLINE

A farmer walks across a natural saline seep in an irrigated field, an area rendered useless for agriculture by pools of salty water that have seeped up from beneath the earth's surface. (North Dakota) (Courtesy, USDA – Agricultural Research Service)

□ □ □

10:1 □ OVERVIEW

Saline and sodic soil conditions reduce the value and productivity of a considerable area of land throughout the world. More than one-fourth of the irrigated farmland in the United States is affected to some extent by soil salinity. In Canada, salt-affected soils occur on about $^1/_{10}$ of the irrigated land. In general, soil salinity and its related problems occur in the arid and semiarid climates and in coastal regions in all climates where ocean tides salinize soils and groundwaters. Rainfall leaches salts out of soils in humid regions, making salt problems important but rare and often transitory in some areas.

The ions that contribute to soil salinity include Ca^{2+}, Mg^{2+}, Na^+, Cl^-, SO_4^{2-}, HCO_3^-, and (rarely) NO_3^- and K^+. The salts of these ions occur in highly variable concentrations and proportions. They may be indigenous from geologic formations, but more commonly they are brought into an area in the irrigation water or in waters draining from adjacent areas. Natural surface and subsurface drainage are often so poorly developed in arid regions that salts collect in inland basins rather than being discharged into the sea. The Great Salt Lake in Utah is an example.

The term **alkali**[1] is often used to refer to soils that are light in color with snow-white surface crusts. The term *alkali* also is used to imply that the affected soil is high in exchangeable sodium. Soils affected by salts have been given nonscientific descriptive names such as **white alkali, black alkali, gumbo,** or **slick spots** (Figure 10.1). These names come from land surface appearances as soils be-

Fig. 10.1 — Gumbo spots, or slick spots, though small in area, restrict the efficient management of entire fields. (Courtesy, USDA - Agricultural Research Service)

[1]Although technically obsolete, this term is still used by some people.

come salt contaminated. If the soil pH is high enough, sodium-dispersed organic colloids color the surface water black so that they appear like puddles of oil. Upon drying, the soil has black crusts over its surface. These terms, as commonly used, usually include several kinds of salt-affected soils that differ not only in composition and soil properties but also in use suitability, productivity, ease of reclamation, and management needs.

10:2 ▫ CLASSIFICATION OF SALINE AND SODIC SOILS

Salt-affected soils may contain (1) an excess of water-soluble salts **(saline soils)**, (2) an excess of exchangeable sodium **(sodic soils)**, or (3) an excess of both salts and exchangeable sodium **(saline-sodic soils)**. The classification depends on the total soluble salts (as measured by conductivity), soil pH, and the exchangeable sodium percentage, as summarized in Table 10.1. It is important to understand these differences because they determine, to a great extent, how these soils should be reclaimed and managed.

Table 10.1 — Summary of Salt-affected Soil Classification

Classification	Conductivity (mmhos/cm)	Soil pH	Exchangeable Sodium Percentage
Saline	Greater than 4.0	Less than 8.5	Less than 15
Sodic	Less than 4.0	Greater than 8.5	Greater than 15
Saline-sodic	Greater than 4.0	Less than 8.5	Greater than 15

10:2.1 ▫ Saline Soils

All soils contain water-soluble salts. Soils containing water-soluble salts in amounts sufficient to be harmful to seed germination and plant growth are called saline or salty soils (Figure 10.2). Saline soils usually have a surface crust of white salts, especially during the summer, when the net movement of soil moisture is upward. The salts dissolved in the soil moisture move to the surface where they are left as a crust when the water evaporates. These white salts are mostly sulfates and chlorides of calcium, magnesium, and/or sodium.

Soils are classified as saline if the solution extracted from a saturated soil paste (soil-saturation extract) has an electrical conductivity value of 4 or more **millimhos per centimeter (mmhos/cm)** (milli-reciprocal ohms per centimeter, the usual method of reporting total soluble salts in soils). This information is obtained on a special conductivity bridge, patterned after a common Wheatstone bridge. The amount of exchangeable sodium in saline soils is lower than 15 percent; as a result, the soil pH is below 8.5.

Fig. 10.2 — A "salt spot" in a field of barley in the Imperial Valley of California. (Courtesy, USDA)

Saline soils occur commonly on nearly level or low-lying areas with poor drainage and with seasonally high water tables or in areas with groundwater seepage. A water table, especially one that occurs at depths of 5 feet (152 cm) or less, is a major factor in the development of many saline soils. Water rises in the soil above the water table by capillary action (a wick-like process) and carries dissolved salts upward in the soil. The salts become concentrated at the soil surface or in the root zone as the water evaporates or is removed by plants.

Saline soils also may result from drying of lakes and ponds that contain salty waters, or they may occur in areas where salts are blown onto adjacent soils from dry saline lake bottoms. In a few cases, saline soils have resulted from overflow or uncontrolled flow of saline artesian wells, waste saline waters discharged when drilling oil or gas wells, or from flooding by other saline waters. Some soils along ocean beaches have become saline from salty ocean sprays; and soils near highways may become saline from road salt ($NaCl$, $CaCl_2$) applied to melt road ice.

10:2.2 □ Sodic Soils

Sodic soils are relatively low in soluble salts but are high in exchangeable sodium. These soils tend to remain in a dispersed condition, almost impermeable to both rain and irrigation water. The degree of saturation of the soil exchange complex with sodium is called exchangeable sodium percentage (ESP). The exchangeable (adsorbed) sodium in sodic soils often exceeds 15 percent of the total cation exchange capacity. The sodium causes these soils to be dispersed and strongly alkaline (pH is greater than 8.5). A "dispersed" soil is one that is extremely sticky when wet and that tends to crust and become very hard and cloddy when dry. Water intake is usually severely restricted, especially in soils with a high percentage of silt and clay. Locally, many of the areas are known as "slick spots," because when slightly wet soil is plowed with a turning plow it turns over in slick, rubbery furrow slices.

Sodic soil conditions are unfavorable for plant growth because of both chemical conditions and physical properties of the soil. Chemically, the high content of alkaline (sodium) ions may be toxic or harmful to plant roots, and the availability of some plant nutrients may be low. Physically, dispersion caused by the sodium may result in poor physical soil conditions, and, as a consequence, water and air will not readily move through the soil. Some plants, such as beans, clovers, rice, and citrus, are more sensitive to low concentrations of sodium than are other crops, such as barley, sugar beets, alfalfa, and cotton.(See Table 10.6.)

10:2.3 □ Saline-Sodic Soils

Saline-sodic soils contain high amounts of total soluble salts as well as more than 15 percent of exchangeable sodium. These soils may range from near neutral to strongly alkaline in reaction. As long as an excess of all salts is present, the physical properties of these soils are good and are similar to those of saline and nonsaline soils.

When rains occur or after irrigation with low-salt (good-quality) water, most of the soluble calcium and magnesium are leached out of the surface soil and the sodium remains attached to the clay and humus. The pH rises above 8.5, the soil becomes dispersed, and permeability to water virtually ceases. Rainwater often stands on these soils for many days until it evaporates. The danger of excess sodium often goes unrecognized until these events occur. Such soils require amendments and leaching to remove the excess sodium. (See Section 10:9.)

10:3 □ EFFECTS OF SALTS ON PLANT GROWTH

Plants absorb water and dissolved nutrients from the soil through the root hairs partly by the physical process of **osmosis.** Water can move from the soil into the root hairs only so long as the osmotic pressure within cells of the root hairs is greater than that of the soil water. Salts, sugars, nutrients, and other soluble materials contribute to the osmotic pressure or solute content of the cell sap, not only in the roots but also in all plant tissue. Any increase in the soluble salt content of the soil raises the osmotic pressure of the soil solution. As a result, less water flows from the soil into the plant, thus reducing the amount of soil water available to the plant. In the nonsaline soil, about half of the total soil water could be used before the plant wilts. At moderate and high salinity, smaller and smaller amounts of soil water are available to plants; therefore, plants wilt sooner (Figure 10.3).

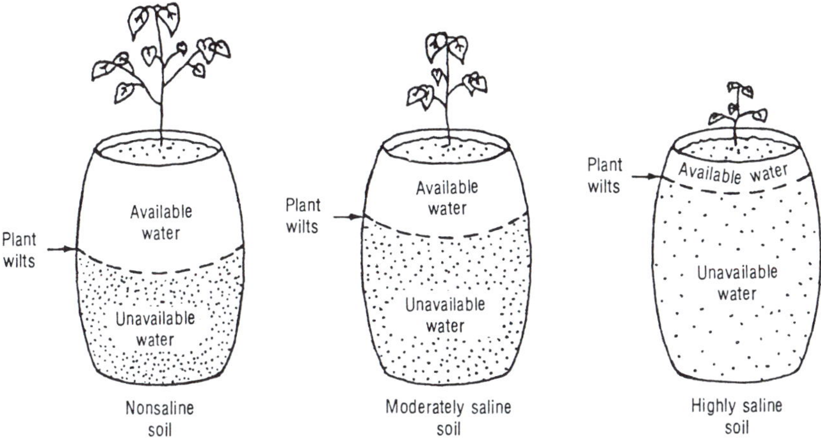

Fig. 10.3 — Diagram showing that as the salt content of the soil increases, the availability of water to plants decreases. In a nonsaline soil, about ½ of the water is available, whereas in a highly saline soil, perhaps only 1/10 is available. (Courtesy, USDA)

10:4 □ SELECTING PLANTS TOLERANT OF SALINITY

Proper selection of plants is an essential part of successful management of salt-affected areas. Many investigators have evaluated the **salt tolerance** of a wide variety of plants in both field and greenhouse research. **Salt tolerance ratings** are based on the yield reduction on salt-affected soils when compared with yields on similar soils not affected by salt. The significance of the mmhos / cm as a salinity unit is best understood in relation to its effect on plant growth (Table 10.2).

Table 10.2 — Effect of Various Degrees of Soil Salinity on Crop Plants

Soil Salinity Class	Conductivity (mmhos / cm)	Influence on Crop Plants
Nonsaline	0 to 2	Salinity effects are negligible
Very slightly saline	2 to 4	Very sensitive plants are restricted
Moderately saline	4 to 8	Many crops are restricted
Strongly saline	8 to 16	Only tolerant plants yield satisfactorily
Very strongly saline	16+	Only a few very tolerant plants grow

The salt tolerance of certain crops varies with the stage of development (Table 10.3). For example, the salt tolerance of alfalfa is poor during the germination stage but is good after the crop is established. With rye and corn the reverse is true.

Salinity affects some crops more than other crops (Figure 10.4). Of the field crops, barley is known to have the greatest tolerance. Sugar beets and beets are

Table 10.3 — Tolerance of Crops to Salts at Two Stages of Growth

Crops	Germination Stage	Established Stage
Barley	Very good	Good
Rye	Good	Poor
Corn	Good	Poor
Wheat	Fairly good	Fair
Alfalfa	Poor	Good
Sugar beets	Very poor	Good
Beans	Very poor	Very poor

fairly tolerant once the plants are well-established. Rye, wheat, oats, flax, and lettuce are moderately tolerant provided fertility levels are suitable and adequate moisture is available during germination. Beans and peas are very sensitive to soil salinity.

Forage crops tend to tolerate salt better than most cereal crops do. Because most forage crops are perennials, they shade the soil longer than cereal crops do. The shade reduces evaporation from the soil surface; therefore, less salt rises with capillary water to the surface. Furthermore, the fibrous roots of forage crops improve the physical condition of the soil because they add organic matter to it; thus, increased permeability aids water infiltration to keep salts leached below the rhizosphere.

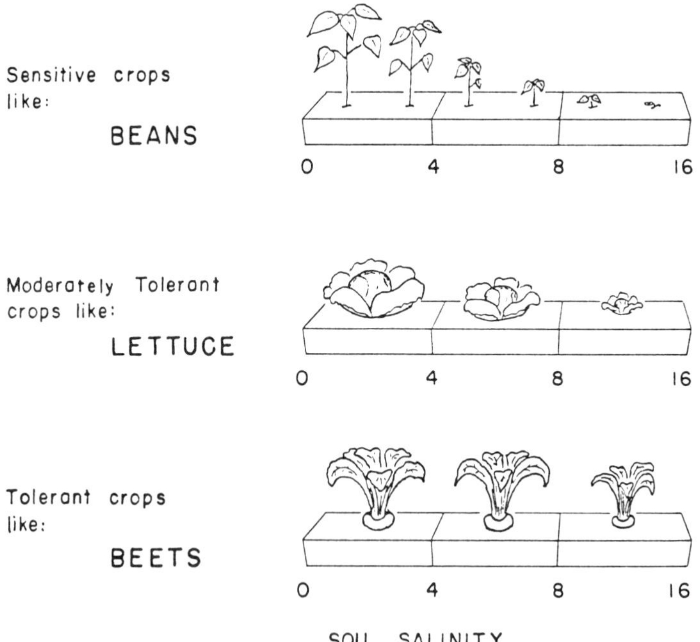

Fig. 10.4 — How does salinity (in millimhos / cm) affect different crops? Some are affected much more than others. (Courtesy, USDA)

Fruit trees are more sensitive to salinity than most field crops are. Many trees are sensitive to salts, especially seedlings and transplants of evergreen and **deciduous** trees (trees that lose their leaves in the fall). Because most trees are extremely sensitive, they should be planted only in areas that are comparatively salt-free.

The relative tolerances of various plant species are rated into what are called **salt tolerance groups:** that is, sensitive, moderately tolerant, tolerant, and highly tolerant. These tolerance ratings are rather general because other factors such as management practices, quality of irrigation water, environmental conditions, crop variety, and type of rootstocks used in propagation of fruit trees can affect tolerance. The tolerance ratings (Table 10.4) should be used only as a general guide. In this respect they can be quite useful.

Table 10.4 — General Salt Tolerance Ratings of Various Crops[1]

Sensitive (0 - 4 mmhos/cm)	Moderately Tolerant (4 - 6 mmhos/cm)	Tolerant (6 - 8 mmhos/cm)	Highly Tolerant (8 - 12 mmhos/cm)
	Field Crops		
Field bean	Soybean	Wheat (grain)	Barley (grain)
	Castorbean	Oats (grain)	Rye (grain)
	Sesbania	Safflower	Sugar beet
	Rice	Cotton	
	Flax	Sunflower	
	Guar	Triticale	
	Sorghum (grain)		
	Corn (field)		
	Forage Crops		
White clover	Reed canarygrass	Hardinggrass	Bermudagrass
Dutch clover	Oats (hay)	Kleingrass	Crested wheatgrass
Alsike clover	Orchardgrass	Alfalfa	Barley (hay)
Red clover	Bromegrass	Birdsfoot trefoil	Rye (hay)
Ladino clover	Big trefoil	Hubam sweetclover	Panicgrass
Crimson clover	Grama grasses	Dallisgrass	Alkali sacaton
Meadow foxtail	Sour clover	Tall fescuegrass	Rhodesgrass
Kentucky bluegrass	Milkvetch	White sweetclover	Saltgrass
	Timothy	Yellow sweetclover	Western wheatgrass
	Sudan-sorghum hybrids	Perennial ryegrass	
	Sorghum (forage)	Wheat (hay)	
	Corn (forage)	Johnsongrass	
	Vegetable Crops		
Carrot	Lettuce	Tomato	Asparagus
English pea	Corn (sweet)	Beet	
Radish	Potato	Kale	
Celery	Squash	Spinach	
Green bean	Onion	Broccoli	
Lima bean	Sweet potato	Cabbage	
Kidney bean	Bell pepper	Cauliflower	
Cucumber	Blackeyed pea	Watermelon	
Rhubarb	Muskmelon		

(Continued)

Table 10.4 (Continued)

Sensitive (0 - 4 mmhos / cm)	Moderately Tolerant (4 - 6 mmhos / cm)	Tolerant (6 - 8 mmhos / cm)	Highly Tolerant (8 - 12 mmhos / cm)
	Fruit, Nut, and Vine Crops[2]		
Grapefruit	Pecan	Pomegranate	Date palm
Orange	Peach	Fig	
Lemon	Apricot	Olive	
Avocado	Grape		
Pear	Quince		
Apple			
Cherry			
Plum			
Walnut			
Blackberry			
Raspberry			
Strawberry			
Boysenberry			
	Ornamental Shrubs		
Viburnum	Spreading juniper	Oleander	Purple sage
	Arborvitae	Bottlebrush	Saltcedar
	Lantana		
	Pyracantha		
	Privet		
	Japonica		

[1] Primary Source: Publications of the U.S. Salinity Laboratory, Riverside, California.
[2] Ratings may vary somewhat depending on the particular rootstock used for propagation.

It is important to select grasses which are salt tolerant for roadside use, particularly where salt is used for ice removal. For example, more than 24,000 pounds (12 tons) (6.8 mt / km) of sodium chloride per mile of four-lane highway have been used per year for ice removal on some sections of Interstate 80 in Iowa. Although salt tolerance of mature grass is important, seedling tolerance and rate of seedling development are critical in areas where new highways are being built. Highways may be open for traffic and pavements salted before roadsides are seeded. Thus, salt contamination of soils may be more important in establishing new turf than in maintaining older stands. Grasses which have been used effectively on roadsides are Kentucky 31 (tall) fescuegrass, western wheatgrass, slender wheatgrass, intermediate wheatgrass, Russian wild-rye, and reed canarygrass. Of these grasses, Kentucky 31 (tall) fescuegrass and western wheatgrass are the most salt tolerant.

It is sometimes necessary to consider degree of salinity when choosing a suitable turfgrass species. The salinity tolerance of various turfgrass species is given in Table 10.5. Precise judgments of plant tolerance to salinity are not simple to obtain because some plants are more affected by salinity at one stage of development than another. Even highly tolerant plants may be acutely affected at some particular stage. In general, well-established plants are more tolerant of

Table 10.5 — Relative Salinity Tolerance of Selected Turfgrasses[1]

Soil Salinity Class	Conductivity (mmhos / cm)	Adapted Species
Nonsaline	0 to 2	All turfgrass species
Very slightly saline	2 to 4	Kentucky bluegrass, red fescuegrass, colonial bentgrass
Moderately saline	4 to 8	Alta fescuegrass, perennial ryegrass
Strongly saline	8 to 16	Zoysiagrass, St. Augustinegrass, bermudagrass, seaside bentgrass
Very strongly saline	16 +	Saltgrass, switchgrass, alkaligrass

[1]R. H. Follett, "Salinity Problems." *Proceedings of the 42nd Annual International Turfgrass Conference and Show, 1971.*

salt than new transplants. This fact is of considerable importance for plants propagated by transplanting.

10:5 ▫ SELECTING PLANTS TOLERANT OF SODIUM

The nutritional problems in sodic soils are usually related to the relative amounts of calcium and sodium accumulated by the plants. The tolerance of crops is not so closely related to the absolute amount of **exchangeable sodium** (ES) in the soil as to the **exchangeable sodium percentage (ESP)** (the percentage of the total exchangeable cations that are sodium ions). Some crops are much more tolerant of a soil low in calcium or high in sodium than other crops.

At a given ESP, the physical condition of a soil depends on the soil texture. For example, the structure of a high-sodium, fine-textured (clay) soil is usually worse than that of a high-sodium, coarse-textured (sandy) soil. This difference in structure is related to the greater number of **colloids** (fine clay and humus particles) in a clay soil that are dispersed by the sodium.

The retarded growth of sensitive and even moderately tolerant crops is due in part to the nutritional problems associated with sodic soils and in part to the adverse structural characteristics of such soils. The most tolerant crops are apparently not affected nutritionally at moderately high ESP values, but growth is retarded as a result of poor physical condition of the soil. Therefore, the reduced growth of crops grown on sodic soils may be due to adverse nutritional factors, adverse physical conditions, or a combination of both. The crops listed in Table 10.6 are arranged approximately in order of increasing tolerance to exchangeable sodium.

Toxicity of sodium soils can occur with almost any crop if concentrations are high enough. Sodium toxicity often accompanies and is complicated by soil salinity and / or reduced permeability. Plant symptoms of sodium toxicity occur first on the oldest leaves because a period of time (days or weeks) is normally required before plant accumulations reach toxic concentrations. Symptoms usually appear as a burn or drying of tissue at the outer edges of the leaf, progressing inward between veins toward the leaf center as severity increases.

Table 10.6 — Tolerance of Selected Crops to Exchangeable Sodium Percentage (ESP)[1]

Crop	Tolerance to ESP and Range at Which Affected	Growth Responses Under Field Conditions
Deciduous fruits Nuts Citrus Avocado	Extremely sensitive (ESP = 2 - 10)	Sodium toxicity symptoms even at low ESP value
Beans	Sensitive (ESP = 10 - 20)	Stunted growth at low ESP values even though the physical condition of soil is good
Clovers Oats Tall fescuegrass Rice Dallisgrass	Moderately tolerant (ESP = 20 - 40)	Stunted growth due to both nutritional factors and adverse soil physical conditions
Wheat Cotton Alfalfa Barley Tomato Beets	Tolerant (ESP = 40 - 60)	Stunted growth usually due to adverse soil physical conditions
Crested wheatgrass Tall wheatgrass Rhodesgrass	Most tolerant (ESP = more than 60)	Stunted growth usually due to adverse soil physical conditions

[1] Primary Source: Publications of the U.S. Salinity Laboratory, Riverside, California.

10:6 □ IDENTIFICATION OF SALINE AND SODIC SOILS

The appearance of growing crops is a useful clue to the presence of saline or sodic soil conditions (Figure 10.5). Generally, plants which are affected by soil salinity have a bluish-green appearance. Of the common field weeds, Russian thistle, kochia, wild barley, and goosefoot species are commonly found in areas of high salt concentration. Crop stands often present a very "spotty" appearance on these soils because the intensity of the conditions often varies considerably in short distances. The spotty effect may become apparent as early as the seedlings' emergence stage, or the crop may make good early growth with differences in stand becoming more apparent later in the growing season. In extreme cases, the crop has a very uneven appearance with stunted growth; growth may be entirely eliminated where the saline or alkali condition is the most severe. The effects are most noticeable in dry years.

Certain tolerant plants are useful indicators of saline and sodic soil conditions (Figure 10.6). Some plants that are typical of salt-affected soils are saltgrass, alkaligrass, shadscale, greasewood, and alkali sacaton. On highly saline and sodic areas, a sparse growth of these plants may be the only vegetation present. On soils of slight to moderate salt or sodium content, these plants may occur intermixed with less tolerant species.

Fig. 10.5 — This farm is suffering serious loss of production because of saline spots in the field. (New Mexico) (Courtesy, USDA - Soil Conservation Service)

Fig. 10.6 — The type of vegetation can sometimes be used as an indicator of saline soil conditions. The vegetation on this saline subirrigated range site consists of inland saltgrass, annual grasses, sedges, weeds, and saltcedar (tamarisk). (Kansas) (Courtesy, USDA - Soil Conservation Service)

Strongly to very strongly saline soils are often devoid of any type of vegetation. The surface may have a powder-like or crusted layer of light gray or white salt crystals, especially when the soil is dry. Less strongly saline soils may not have surface salts, but small gray or white clusters and flecks of salts may be exposed by digging into the soil. Salts may be in solution and not visible when saline soils are moist or wet. Therefore, the absence of visible salts is not a positive means of identifying a soil as nonsaline.

A sodic soil condition may be suspected in areas in which gumbo spots occur. Other clues are extreme stickiness and a tendency of the soil to flow together when wet and when dry to become crusted, very hard, and almost impermeable to water. On drying, the surface of these soils often has a crusted grayish appearance. Sodic soils support very little vegetation if the exchangeable sodium percentage (ESP) is more than 15 percent.

The only positive way of identifying the various salt-affected soils is by laboratory analysis. Laboratory analysis of soil samples can be used to determine if a soil is saline, sodic, or saline-sodic and if a soil is slightly, moderately or strongly affected by salts or sodium. Many state-supported soil laboratories as well as commercial soil testing laboratories can provide the soil tests necessary to identify and evaluate soil salinity problems. (See Section 4:9.)

Salinity varies with depth and from place to place in a field. At each location, samples should be taken from at least two depths in the root zones, such as 0 to 8 inches (0 – 20 cm) and 8 to 24 inches (20 – 61 cm).

10:7 □ WATER QUALITY AND SALINE SOILS

In many instances, salt-affected soils are the result of the application of low-quality irrigation water. All waters used for irrigation carry varying amounts of dissolved salts and other constituents. Some of the dissolved constituents can improve crop growth if present in small to moderate amounts, but can harm soils and restrict plant growth if they are present in excessive amounts. Water applied for irrigation is usually lost by evaporation and plant use, but almost all of the salts in the water remain to increase the salt concentration in the soil. To maintain a safe salt balance in the soil, the quality and quantity of water used for irrigation must be considered. Many soil testing laboratories also provide a water analysis service to evaluate the quality and relative usefulness of water for irrigation.

The quality of water is determined mainly by four characteristics: (1) the total concentration of soluble salts, (2) the amount of sodium in relation to calcium-plus-magnesium, (3) the amount of bicarbonate, (4) and the presence of boron.

The soluble salts in irrigation waters are mainly the cations: calcium, magnesium, and sodium. Anions consist of sulfates, chlorides, and bicarbonates. The higher the salt concentration in irrigation water, the greater the hazard of toxic accumulations in the soil.

10:8 ▫ RECLAIMING SALINE SOILS

Saline soils can be reclaimed for crop production if adequate amounts of low-salt irrigation water are available. The main solution to the problem is to leach most of the salts downward and out of contact with plant roots and subsequent irrigation water (Figure 10.7).

Fig. 10.7 — Continuous leaching by ponding has been developed through research to benefit farmers with soil salinity problems in the Imperial Valley, California. After leaching is completed, the land must be releveled before irrigation can continue. (Courtesy, USDA - Agricultural Research Service)

The amount of water required to leach salt-affected soils depends on the salt content of the soil and the final salinity level desired. The rule of thumb suggests that to reduce the salinity level by 80 percent, it is necessary to leach with 1 foot (30 cm) of low-salt water for each foot (30 cm) of saline soil. For example, if the average salinity in a 3-foot (91-cm) depth of a field is 20 mmhos / cm and it is necessary to reduce this to 4, then 3 feet (91 cm) of water must be leached ($.80 \times 20 = 16$; $20 - 16 = 4$). Fifty percent of the salt can be removed by leaching with 6 inches (15 cm) of water for each foot (30 cm) of soil, but to remove 90 percent of the salt, 2 feet (61 cm) of water must be used for each foot (30 cm) of soil to be leached.

Good land-leveling and recommended irrigation methods help reduce salt accumulation. High spots in a field generally do not receive enough water. This favors the buildup of salts and may result in poor yields. Therefore, high spots should be leveled down and the salt removed by irrigating heavily. Studies by the U.S. Department of Agriculture in California show that intermittent flooding leaches salty soils more efficiently in some cases than does either continuous ponding or continuous sprinkling.

Adequate drainage is necessary for the removal of salt. In most cases, salinity problems are associated with a ground water table close to the soil surface. Where a high water table exists, water penetration through the soil profile is severely restricted or may be completely stopped. Under such conditions, soluble salts leached from the soil surface by irrigation water soon return to the surface with water through upward **capillary water movement.** (Capillary water movement is similar to the upward movement of kerosene in a lamp wick.) This produces a salt crust when the salty water evaporates at or near the soil surface in the seed germination and root zone.

Drainage must be provided to ensure permanent salt removal. Drainage results in lowering the high water table, improving water percolation through the soil, and removing dissolved salts in the drainage water. Open drain ditches or tile can be used. In some situations, drainage may be accomplished by pumping out accumulating water. If seasonal or periodic water table fluctuations occur, salt removal from the root zone can be accomplished by applying leaching water when the water table is low. However, this will not accomplish permanent reclamation, but it will help to prevent the accumulation of excessive salts near the surface and will keep the salts distributed within a greater depth. Providing drainage is often a very costly operation. The economic feasibility of drainage should be investigated thoroughly before any reclamation treatments are applied. (See Chapter 8.)

Reclamation of saline soils may be hastened by the application of a surface organic mulch. Organic matter added to a soil seems to reduce the harmful effects of salts on crops, possibly by adsorbing some of the salt and by decreasing surface evaporation and therefore reducing capillary rise of salty water. High rates of manure application, 20 to 30 tons per acre (45 – 67 mt / ha), have improved crop establishment and crop yields on saline soils. The manure applications increase the organic matter content and water-holding capacity of the soil. In this way more water is available to plants even though salts are present. Manure also adds nutrients and improves the structure and tilth of the soil. (See Chapter 5.)

The reclamation of strongly saline soils is usually not practical under dryland farming conditions. These soils probably are best suited to use as native grass pasture and range or for seedings of mixtures of salt-tolerant grasses. These forage plants begin growth early in the spring before the soil becomes dry and before salt injury becomes severe. They also recover from salt injury if moisture conditions become favorable for growth later in the season.

10:9 □ RECLAIMING SODIC AND SALINE-SODIC SOILS

The reclamation of sodic soils is another story. In sodic soils, the exchangeable sodium may be so great as to make the soil almost impervious to water. The sodium must first be replaced by another cation and then leached downward and out of reach of plant roots. This is accomplished by these three steps:

1. The installation of an effective drainage system. (See Chapter 8.)
2. The application of a soil amendment such as calcium sulfate (gypsum).
3. The flushing of the soil with water to remove the sodium displaced from the soil by calcium (Figure 10.8).

In many cases, the common practice is to apply sufficient gypsum to remove most of the adsorbed sodium from the top 6 to 12 inches (15 – 30 cm) of soil. This improves the physical condition of the surface soil within a period of time and permits the growing of crops (Figures 10.9, 10.10, and 10.11). By the continued use of good-quality irrigation water, good irrigation methods, and cropping practices, further removal of adsorbed sodium, especially in the subsoil, usually takes

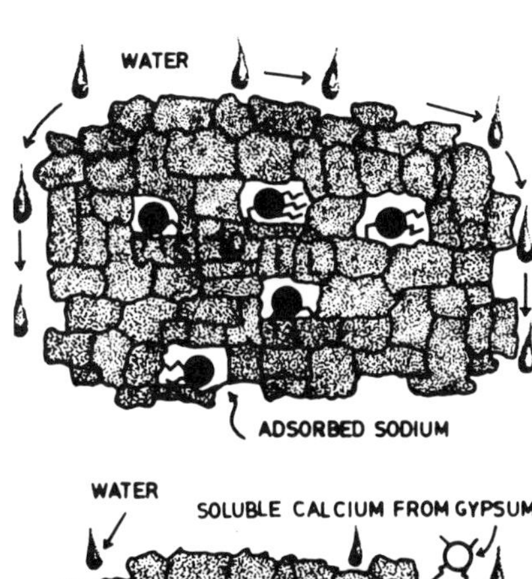

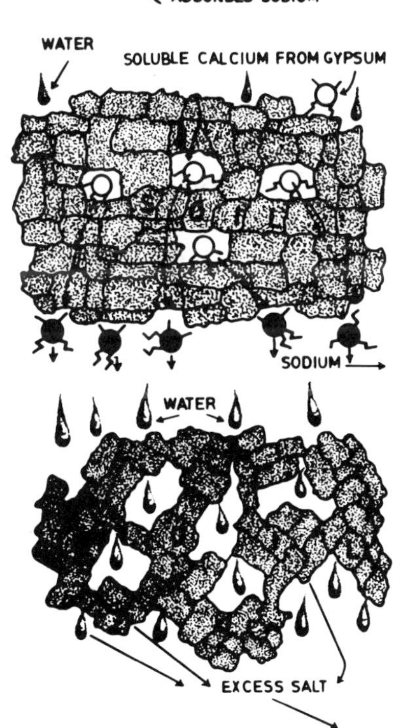

Fig. 10.8 — How gypsum ($CaSO_4$) reclaims sodic soils. *Top:* When there is a large amount of adsorbed sodium in the soil, the soil becomes hard and compact like a brick. The compacted soil will not permit water to pass downward readily, and, as a result, plant growth is retarded. *Center:* Gypsum supplies soluble calcium to the soil and the excess calcium replaces the sodium. At this time, if a large amount of irrigation water is added to the soil, the water will leach the sodium downward and out of reach of plant roots. *Bottom:* The surface soil is now full of calcium, which causes the clay particles to rearrange themselves into loose, open clusters. The result is a more open soil which contains sufficient water and air for normal plant growth. (Courtesy, University of California)

Fig. 10.9 — This is an example of the effect of gypsum on a saline-sodic soil in Reno County, Kansas. The gypsum was applied in 1965 and the picture was taken in 1971. Dryland wheat yields were increased an average of 10 bushels per acre (673 kg / ha) per year during that period of time. (Courtesy, D. A. Whitney, Kansas State University)

Fig. 10.10 — Total soluble salts and sodium on this soil (Toy soil series) were so high that nothing would grow until 20 tons per acre (44.8 mt / ha) of gypsum had been applied. (See Figure 10.11.) (Courtesy, University of Nevada)

Fig. 10.11 — This saline-sodic soil (Toy soil series) produced a fair crop of barley only after 20 tons per acre (44.8 mt / ha) of gypsum had been applied and after most of the sodium had been leached below the rhizosphere. (Courtesy, University of Nevada)

place. In some high-sodium, high-clay, and high – water table situations, it may be necessary to reclaim to greater depths in order to obtain adequate drainage and satisfactory root penetration.

Sulfur, which oxidizes to sulfuric acid, is also used in the reclamation of sodic soils. The sulfuric acid will dissolve soil lime (calcium carbonate). This will provide soluble calcium to replace adsorbed sodium. However, calcium carbonate must be present in the soil if sulfur is to be effective as an amendment.

10:10 ▫ PREVENTION AND CONTROL

Level land is essential for good irrigation water management because uniform distribution and penetration of water is necessary for prevention of salt and sodium accumulations. For this reason land should be "touched up" (leveled) frequently with a land plane.

Water in excess of that used by the current crop must be applied to prevent salts from accumulating in the root zone. All irrigation water contains some salt which becomes more concentrated in the soil solution as plants selectively absorb more water than salt. Flushing accumulating salts to below the root zone in drainage water is necessary to maintain or reduce the present soil salt level. However, overirrigation to the extent that it raises the water table to within the plant rooting zone is undesirable.

In situations in which water for irrigation or leaching is in short supply, mulching with straw or other crop residues reduces evaporation from the soil surface; therefore, fewer surface salts will accumulate. Rain or leaching water removes salts more effectively when used in conjunction with organic mulching.

10:10.1 ▫ Planting

Generally, crop seeds should be planted in such a way as to avoid a salt buildup in the immediate zone of seed placement. When a dry soil is wetted, the soluble salts already in the soil move with the water because they dissolve and are carried along with it. Germinating seed and seedlings are usually sensitive to salt. Farmers in saline areas avoid salt accumulations by planting seeds on the *shoulders* of the beds (Figures 10.12 and 10.13).

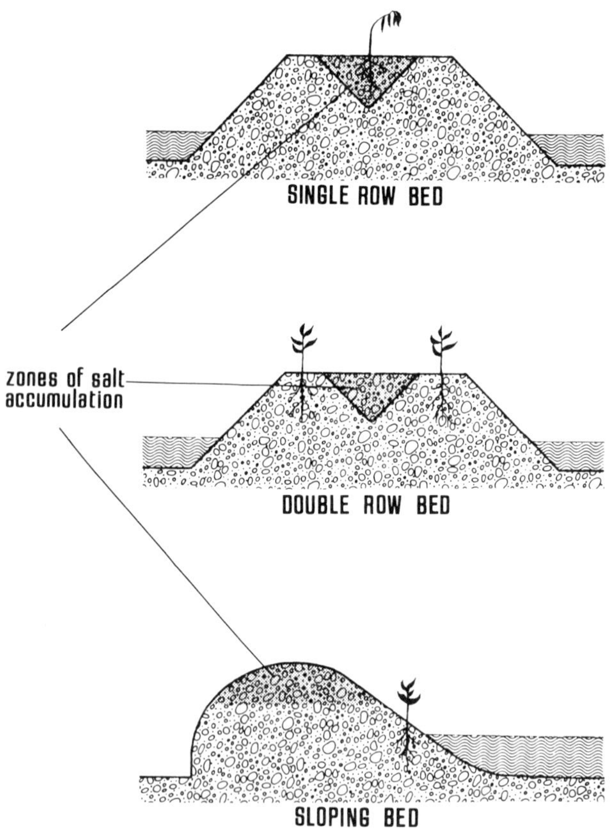

Fig. 10.12 — Seeds will sprout and grow only when they are placed so as to avoid excessive salt buildup around them. (Courtesy, Instructional Materials Services, Texas A&M University)

Fig. 10.13 — The small lettuce plants planted on the *shoulder* of the beds (at arrows) avoided the high zone of salt concentration at the top of the bed. The lettuce continued to grow and produce a good crop. (Colorado) (Courtesy, Roy Hunter Follett)

10:10.2 ▫ Tillage

Salts tend to accumulate closer to the soil surface in nonirrigated drylands than in irrigated lands. Tillage speeds up the desalination by mixing the easily soluble salts deeper in the soil and loosening the dense subsoil. Subsoil tillage operations, such as chiseling or moldboard- or disc-plowing land with compact layers, hardpans, or cemented layers, frequently will improve infiltration and uniformity of water and root penetration. This will aid in removing salt and sodium as well as in preventing their accumulation. When no such restrictive layers are present, subsoil operations are not recommended.

In some cases, it may not be economically feasible to improve drainage or undertake other practices to reduce salt and sodium levels in the soil. The alternative will be to plant crops that are less sensitive to such conditions.(See Sections 10:4 and 10:5.)

10:10.3 ▫ Organic Matter and Soil Structure

Incorporation of crop residues or plowing under a green-manure crop in soils adversely affected by sodium will improve the tilth, tend to rebuild structure, and result in better water infiltration. The regular addition of organic matter, whether from crops, manure, sludge, or compost, is the best possible insurance of good soil physical conditions and a safeguard against the adverse effects of sodium.

When soils contain plenty of rapidly decomposing organic matter, their desirable structure is less readily broken down by high sodium or compacted by heavy farm equipment. They remain open and permeable to water and air because the organic matter physically prevents the dispersed soil from compacting.

Crop residues left on the soil or worked into a rough, cloddy soil surface often improve water penetration. The more fibrous crop residues such as from cereals and sudangrass and deep-rooted crops such as alfalfa increase water penetration. Presumably, the cereal and sudangrass straw along with crop roots physically keep the soil porous by maintaining channels and voids that improve water penetration. To be very effective, however, relatively large quantities of crop or other organic residues are usually needed. Rice hulls, cotton gin wastes, sawdust, shredded bark, and many other waste products in large volumes (10 to 20 percent by volume in the upper 6-inch [15-cm] depth) have been used with varying degrees of success. From a long-term standpoint the regular return of crop and animal residues to the soil is considered to be beneficial in that this helps to maintain granular soil structure as well as to return needed nutrients to the soil. (See Chapter 5.)

10:10.4 ▫ Managing Fertility While Controlling Salinity

Proper plant nutrition will aid in minimizing yield depressions due to excessive salt and sodium. Under conditions of high salinity there may be little or no response to applied fertilizers because excess salts may be the first factor limiting growth. When excess salts are removed by leaching, much of the soluble fertilizer is leached out with them. The problem then becomes how to remove soluble salts and still maintain necessary fertility levels. This is not easy to do, particularly on sandy soils.

The best way is to leach out most of the accumulated salts before fertilizer is applied. In most cases, this is done with a heavy pre-plant water application in the spring. After planting, nitrogen (and potassium, if needed) is top-dressed or side-dressed before the first summer irrigation. Phosphate can be applied any time because it is not leached from soils.

However, when yield depressions are large because of severely salt-affected soils, fertilization will be of small consequence and likely will be uneconomical. Excess fertilization is even more of a hazard on saline soils because fertilizers themselves are salts. (See Chapter 4.)

10:11 ▫ SALINE SEEPS

Saline seeps are formerly productive nonirrigated soil areas that became too wet and salty for crop production (Figure 10.14). Saline seeps may develop near hilltops, on sidehills, at the base of hills, or on lowlands. Usually seeps occur

Fig. 10.14 — A saline seep area which grew a normal crop three years ago is now nearly sterile because of excess soluble salts. (Montana) (Courtesy, USDA - Agricultural Research Service) (See also the photo facing the first text page of this chapter.)

where there is a sudden change in the slope of the land, but not necessarily at the low point in the topography. In Montana alone, there are more than 80,000 acres (32,376 ha) of saline seep areas, the acreage increasing in size each year by almost 10 percent. It is estimated that nearly $5 million in Montana farm crop income is lost annually — not including the value of the land itself.

Saline seeps also salinize groundwater, making it undrinkable for humans and livestock. Seeps frequently dissect fields, creating a patchwork of soggy areas that make mechanized seeding and harvesting difficult.

Present crop-fallow farming practices in the Great Plains are a principal cause of today's saline seep problems, particularly in the years of above-average precipitation. When precipitation exceeds the needs of a current crop, water percolates or trickles below the roots of crops and accumulates above a dense, nearly impermeable layer of clay. Unable to penetrate farther, the water starts moving laterally through permeable layers of degraded sandstone, siltstone, and / or lignite.

The lateral movement ends when the water-conducting soil layers meet material with less permeability, causing the water table to rise. Capillary forces move the water upward from the water table to the soil surface, much as a towel absorbs water against the pull of gravity. Once on the surface, the water eventually evaporates, leaving precipitated salts collected along the water's journey.

Research and farmer experience indicate that if susceptible areas are cropped more frequently and less land is allowed to be idle during the summer, the development of new saline seeps will be significantly reduced. The more use made of rainfall by plants, the less the water remaining for developing saline seeps. Perennial grasses and legumes grown on a permanent or semipermanent basis in long-term crop rotation will also help to reduce the problem. Plants growing throughout the season absorb large amounts of water and permit very little precipitation to percolate beneath the root zones to cause saline seeps.

10:12 ▫ REFERENCES

Bernstein, Leon. "Effects of Salinity and Sodicity on Plant Growth." *Annual Review of Phytopathology,* Vol. 13, No. 295, 1975, p. 312.

Branson, R. L. "Gypsum for Soil Improvement." University of California, Leaflet No. 2149, 1976.

Cardon, G. E., and J. Letey. "Comparison of Seasonal Crop Production Simulations for Saline Conditions Using Several Crop-Water-Production Models." *1988 Agronomy Abstracts.* American Society of Agronomy, Crop Science Society of America, and Soil Science Society of America, Madison, Wisconsin.

Follett, R. H., Larry S. Murphy, and Roy L. Donahue. *Fertilizers and Soil Amendments.* Englewood Cliffs, New Jersey: Prentice-Hall, Inc., 1981.

Follett, R. H., and P. N. Soltanpour. "Irrigation Water Quality Criteria." Colorado State University, Serv. in Action No. 506, 1985.

Follett, R. H., and W. T. Franklin. "Management of Salt and Sodium Affected Soils." Colorado State University, Serv. in Action No. 504, 1985.

Follett, R. H., W. T. Franklin, and R. D. Heil. "Salt-affected Soils." Colorado State University, Serv. in Action No. 503, 1985.

Franklin, W. T., and R. H. Follett. "Crop Tolerance to Soil Salinity." Colorado State University, Serv. in Action No. 505, 1985.

Hagood, M. A., A. I. Dow, and J. H. Griffin. "Managing Saline Soils in Central Washington." Washington State University, EB0909, 1981.

Halvorson, A. D., and A. L. Black. "Saline-Seep Development in Drylands of Northeastern Montana." *Journal of Soil and Water Conservation,* Vol. 29, No. 2, 1974, pp. 77 – 81.

Hausenbuiller, R. L. *Soil Science: Principles and Practices,* 3rd ed. Dubuque, Iowa: Wm. C. Brown Group, 1985, pp. 466 – 489.

Hoffman, G. H., R. M. Mead, P. B. Catlin, V. C. Davis, and L. H. Francois. "Yield and Foliar Injury Responses of Mature Plum Trees to Salinity." *1988 Agronomy Abstracts.* American Society of Agronomy, Crop Science Society of America, and Soil Science Society of America, Madison, Wisconsin, p. 277.

Holm, H. M. "Soil Salinity — A Study in Crop Tolerances and Cropping Practices." *Saskatchewan Agriculture,* 1983, 17 pp.

Holm, H. M., and J. L. Henry. "Understanding Salt-affected Soils." *Saskatchewan Agriculture,* 1982, 14 pp.

Rush, D. W., *et al.* "Salt-tolerant Crops." *Crops and Soils Magazine,* October 1981, pp. 12 – 16. American Society of Agronomy, Madison, Wisconsin.

Soil Bul. 31. "Prognosis of Salinity and Alkalinity." Rome: FAO, 1976.

U.S. Salinity Laboratory Staff. "Diagnosis and Improvement of Saline and Alkali Soils." U.S. Dept. of Agriculture Handbook 60, 1069.

CHAPTER 11

Judging Soils

"Take care that you choose a soil that is naturally strong." — Cato (234 - 149 B.C.)

OUTLINE

Land judging is one of the most effective means of teaching soil characteristics as they relate to proper land use. (Courtesy, Purdue University)

□ □ □

11:1 □ OVERVIEW

Soils are one of the most valuable resources of a nation. In agricultural production they are an integral part of the ecological system which produces our food, feed, lumber, and fiber. Although it is a common practice to associate the importance of soils with agricultural production, because of greater intensity of use, soils may be equally as important in urban development, engineering, and recreation.

In urban areas, soils act as the foundation materials for houses and buildings. They also work as purification systems for septic tank effluent as well as the media for the growth of lawns, shrubs, and gardens. Soils in many urban areas in the United States contribute to the severe problems of supporting building foundations and driveways, adequately draining septic fields, and growing grass on barren lawns. A thorough understanding of the soil and proper planning prior to building are essential for establishing a new urban area. A knowledge of past errors in judgment can help to avoid future mistakes.

Land judging has been used for many years to evaluate soil and its use for agricultural purposes. Land judging for homesites has now been added to many land judging schools and contests to emphasize the importance of soils and their limitations for high-value essential nonagricultural uses.

11:2 □ WHY JUDGE SOILS?

Land judging is a recent judging event in the field of agriculture. Land can be judged much like animals and farm or horticultural crops. In judging crops, you look at the size, shape, and quality and determine which is the best. Similarly, when judging land, you look for clues that tell how well the land can produce crops or be used for other purposes. Soil characteristics, climate, and topography are good clues to the soil's capabilities, but close examination of the soil texture, structure, depth, permeability, reaction, degree of erosion, slope, drainage, and flooding are necessary to classify land into capability classes. In land judging, the major factors affecting how the land can be used must be determined. These factors are used to correctly recommend conservation practices and fertilizers for conserving the soil.

Land judging can help you to:

1. Understand basic soil differences.
2. Know how soil properties affect crop growth.
3. Know why soils respond differently to management practices.

4. Realize the influence of land features on production and land protection.
5. Select suitable soil and water conservation practices.
6. Determine land capability class.
7. Determine proper use and treatment.
8. Reduce erosion by water and / or wind.

11:3 □ CONDUCTING LAND JUDGING AND HOMESITE EVALUATION CONTESTS

Land judging contests or schools have been used for many years as a tool to teach youth and adults about soil and its use for agricultural purposes. The first land judging school in the United States was conducted in 1941 at Red Plains, Oklahoma, by Ed Roberts and Harley Daniel. **Homesite evaluation** has been added to emphasize the importance of soils and their limitations for nonagricultural uses.

Adding homesite evaluation will benefit a majority of people who will never engage in farming but will be selecting homesites. They will be vitally interested in proper land use for purposes such as sanitary landfills, housing developments, parks and playgrounds, roads and streets, recreation and wildlife areas, public and private utilities, and flood control facilities.

These schools and contests provide an introduction to the basic science of soil classification and soil survey. They provide a guided program of soil study to aid in the understanding and importance of:

1. Basic soil differences as they affect crop production and erosion on agricultural lands.
2. Different soil and water conservation practices for the effective management of land for both agricultural and nonagricultural uses.
3. The effect of different land features on limitations in the use of land for nonagricultural purposes.

Land is judged in much the same way that livestock are judged. A scorecard is used in both cases. Decisions are reached by sight and by touch. The land judging scorecard that was developed in Oklahoma for use in the latest international contests is divided into two parts. The land class factors are identified in Part 1 and the recommended land treatments are made in Part 2. The land judging procedures and scorecards that are used for the International Land, Pasture, and Range Judging Contests in Oklahoma will be given here as an example to use in land judging schools and contests (see Figure 11.7).

Some modification of the land judging and homesite evaluation procedure shown in this chapter can, without doubt, be used in any state in the nation and in any country in the world. The scorecard and land judging procedures should be modified to emphasize the important physical features of the land and soil of a particular region, state, or country. Also, many foreign countries are becoming interested in land judging. There is every reason to believe that these procedures can be adapted easily to use anywhere in the world to dramatize better land use and to increase food production.

11:4 □ SELECTING THE SITES

The fields used in the contest should be selected by the soil scientists of the land-grant university in cooperation with those of the Soil Conservation Service and other qualified personnel.

Each site should have an excavation to show the soil profile, samples of subsoil and topsoil, flags or stakes to indicate field boundaries, stakes or flags to indicate a 100-foot (30-m) distance for estimating slopes, and bottles of water for moistening soil to assist in texture determination, if needed and allowed. The sites should be within walking distance of each other. They may be located in crop fields, meadows, pastures, or woodlands (Figure 11.1).

Students should be given a scorecard for each field being judged. Each site should have certain information posted, giving relevant soil data (Figures 11.2 and 11.3).

Fig. 11.1 — Each site for land judging should have an excavation to show the soil profile. The soil pit being excavated here is for the district's annual land judging contest. (Washington) (Courtesy, USDA - Soil Conservation Service)

CONDITION OF FIELD

Land Judging Field No.____

1. Soil tests show:
   a. Lime needs (pH) ____
   b. Phosphorus (lbs./A. P) ____
   c. Potassium (lbs./ A. K) ____
   d. Nitrogen (p.p.m. N) ____
   e. Other ____________
2. Pay no attention to present mechanical practices or field use.
3. Original topsoil was___inches.
4. Size of field is______acres.
5. Treat for most extensive use.
6. Other factors: ____________

____

Fig. 11.2 — Example of placard to be posted on each field to be used for land judging.

CONDITION OF FIELD

Homesite Evaluation Field No.__

1. Pay no attention to present land use or mechanical practices.
2. Judge on basis of most severe limitation for each feature of the site.
3. Size of Lot:______acres.
4. Depth of water table: ______inches.
5. Original topsoil:______inches.
6. Flooding:____________
7. Other:____________

____

Fig. 11.3 — Example of placard to be posted on each field to be used for homesite evaluation.

11:4.1 ▫ Size of Field

The portions of fields to be judged should be a minimum of 100 feet by 100 feet (30 m × 30 m) in size. Flags or stakes must be set to indicate the boundary of the area to be considered (Figure 11.4). A pair of different colored flagged stakes should be used by the contestants in determining the percentage of slope.

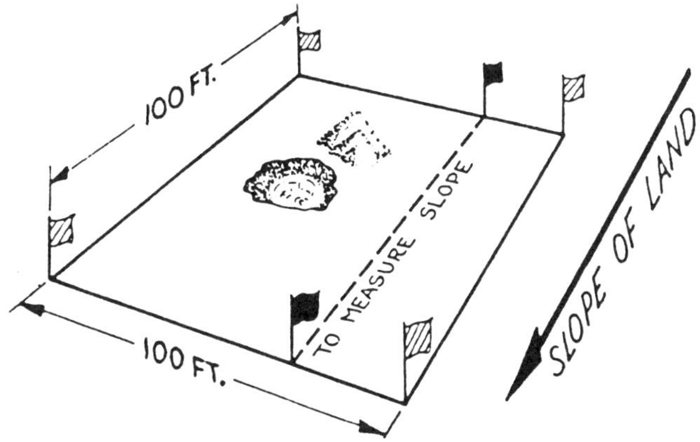

Fig. 11.4 — A group of fields should be selected for judging. Within each field, a uniform sample area of not less than 100 feet by 100 feet (30 m × 30 m) should be selected to represent the field. Sample areas do not have to be square. Slope stakes should be set 100 feet (30 m) apart.

11:4.2 ▫ Observing the Soil Profile

Soils result from the interaction of *climate* and *organisms* on *geologic materials* as conditioned by *topography* over a period of *time*. These soil-forming factors vary widely and produce many kinds of soils, differing as to profile features and other properties. The term *soil profile* is an important concept to learn. It is a sideview or vertical cross-section of the soil as seen in a ditch bank or dug pit that allows the topsoil and subsoil to be examined. When we look beneath the surface of the soil, we see that the soil is divided into layers or "horizons." These layers differ in color, physical properties, chemical composition, and biological characteristics. This is the soil profile. It has four major parts or horizons: (1) the topsoil, or A-horizon, (2) the subsoil, or B-horizon, (3) the parent material, or C-horizon, and (4) sometimes the R-horizon (underlying bedrock). A hypothetical soil profile is shown in Figure 11.5.

The trench or cut should be deep enough to show the topsoil, the subsoil, and the soil depth. Roadside cuts and ditches may be used if they expose the soil profile. The students (contestants) must be able to see the soil in order to determine important characteristics (Figure 11.6) such as the thickness of the topsoil, its texture, and the permeability of the subsoil.

In 1967, a group of soil specialists met to revise the Oklahoma land judging scorecard; the result was an international land judging scorecard. The scorecard that is shown here is used at the International Land, Pasture, and Range Judging Contests which are held in Oklahoma (Figure 11.7).

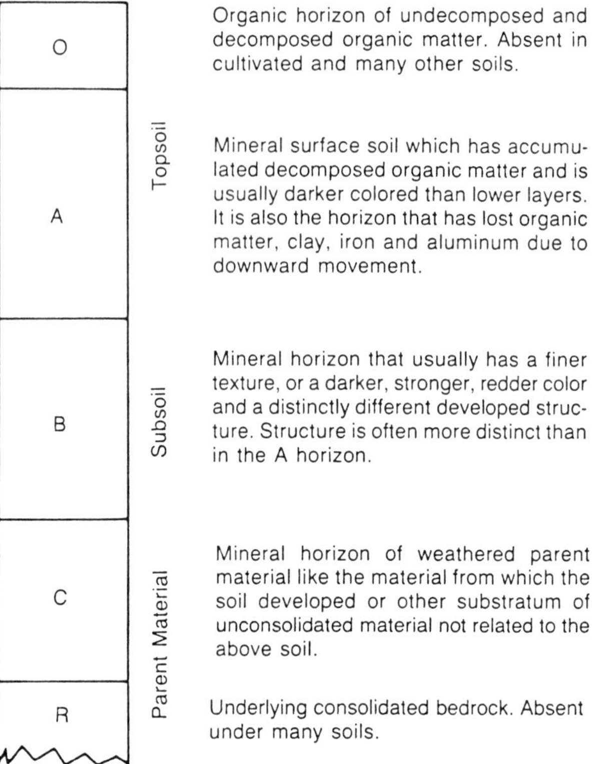

Fig. 11.5 — Hypothetical soil profile showing the letter designation used in describing the major kinds of horizons usually present.

Fig. 11.6 — Trenches should be dug to expose the soil in advance of the contest. The contestants can then estimate the thickness of the topsoil, determine its texture, and determine the permeability of the subsoil. The trench should be deep enough to show the topsoil, the subsoil, and the soil depth. Shown here are contestants at the International Land, Pasture, and Range Judging Contest near Oklahoma City. (Photo by Roy Hunter Follett)

11:5 □ LAND JUDGING SCORECARD

LAND JUDGING SCORECARD

Cooperative Extension Service
Oklahoma State University

Contestant No.__________________ Field No.____________
Name______________________________
Address______________________________
County/School__________________ Organization____________

SCORE	SOIL FACTORS · PART I Check Appropriate Square
	A. Texture Sur. Sub. □ □ 1. Coarse □ □ 2. Moderately coarse □ □ 3. Medium □ □ 4. Moderately fine □ □ 5. Fine
	B. Depth of Soil □ 1. Deep □ 2. Moderately deep □ 3. Shallow □ 4. Very shallow
	C. Slope □ 1. Nearly level □ 2. Gently sloping □ 3. Moderately sloping □ 4. Strongly sloping □ 5. Steep □ 6. Very Steep
	D. Erosion-Wind & Water □ 1. None to slight □ 2. Moderate □ 3. Severe □ 4. Very Severe
	Interpretations of Soil Factors
	E. Permeability □ 1. Rapid □ 2. Moderate □ 3. Slow □ 4. Very Slow
	F. Surface Runoff □ 1. Rapid □ 2. Moderate □ 3. Slow □ 4. Very Slow
	G. Major Factors that keep area out of Class I □ 1. Texture □ 5. Permeablity □ 2. Depth □ 6. Runoff □ 3. Slope □ 7. Wetness □ 4. Erosion □ 8. Flooding □ 9. None
	H. Land Capability Class □ 1. Class I □ 5. Class V □ 2. Class II □ 6. Class VI □ 3. Class III □ 7. Class VII □ 4. Class IV □ 8. Class VIII

SCORE	RECOMMENDED LAND TREATMENTS PART 2
	A. Vegetative □ 1. Row crop/occasional soil conserving crop □ 2. Row crop/frequent soil conserving crop □ 3. Row crops not more than 2 out of 4 years □ 4. Row crops not more than 1 out of 5 years □ 5. Return crop residue to the soil □ 6. Practice conservation tillage □ 7. Establish recommended grass or grasses and legumes □ 8. Proper pasture and range management □ 9. Protect from burning □ 10. Control grazing □ 11. Plant recommended trees □ 12. Harvest trees selectively □ 13. Use only for wildlife or recreation area
	B. Mechanical □ 14. Control brush or trees □ 15. Terrace and farm on contour □ 16. Maintain terraces □ 17. Construction diversion terraces □ 18. Install drainage system □ 19. Control gullies □ 20. No mechanical treatment needed
	C. Fertilizer & Soil Amendments □ 21. Soil Amendments □ 22. Phosphorus (P_2O_5) □ 23. Potassium (K20) □ 24. Nitrogen (N) □ 25. Fertilizer or soil amendments not needed

Score Part I (Possible 45)________ Score Part II (Possible 30)________ Total Score (Possible 75)________

Fig. 11.7 — The International Land Judging Scorecard. (Courtesy, Oklahoma State University)

The authors hope that with the land judging scorecard shown here for the international contest, anyone can adapt a scorecard for any area in the world and conduct land judging schools. The result should be a greater application of successful soil and water conservation practices and higher crop yields.

11:6 ▫ SOIL FACTORS

Soil judging is learning how to evaluate a soil for its best use through an examination of its properties. Although many sophisticated techniques and instruments are used to determine and measure soil properties, several important soil characteristics can be determined readily by anyone.

In judging soils, it is necessary first to determine the various soil properties and then to assess the potential of the soil for a particular use or uses.

11:6.1 ▫ Soil Texture

Soil texture refers to the relative proportion of sand, silt, and clay particles in a specific soil mass. It is easiest to determine when the soil is moist. Sand feels gritty when rubbed by the thumb and finger. Silt feels floury or velvety. Clay is usually sticky and plastic when wet and when pinched between the thumb and finger forms a flexible "ribbon."

Three major textural groups are recognized. Texture of a soil is an expression of the relative amounts of sand, silt, and clay in a soil body. There are 12 recognized soil textural classes, but for land judging purposes they can be grouped into three very broad or five broad textural groups. Schools may wish to use the 3, 5, or 12 textural groups as follows:

Sandy soils	Coarse-textured very sandy soils	Sands Loamy sands
	Moderately coarse-textured soils	Sandy loam
Loamy soils	Medium-textured soils	Loam Silt loam Silt
	Moderately fine-textured soils	Clay loam Sandy clay loam Silty clay loam
Clayey soils	Fine-textured soils	Sandy clay Silty clay Clay

Coarse-textured soils are loose and very friable, and the individual sand grains can be readily seen or felt. When the soil is squeezed between the thumb and forefinger, it is very gritty. Squeezed when dry, it will fall apart as pressure is released. When it is moist, a mold which is unstable may be formed and will crumble as the soil is handled.

Moderately coarse-textured soils are gritty but contain enough silt and clay to make moist soil hold together. The individual sand grains can readily be seen and felt. Squeezed when dry, the soil will form a mold which breaks readily upon handling. If it is squeezed when moist, a mold which can be carefully handled without breaking can be formed.

Medium-textured soils have a slightly gritty, smooth, or velvety feel when moist. Squeezed when dry, the soil will form a mold that will bear careful handling. The mold formed by squeezing when moist can be handled freely, without breaking. When the moistened soil is squeezed out between the thumb and forefinger, it will not "ribbon."

Moderately fine-textured soils usually break into clods or lumps when dry. When the moist soil is squeezed out between the thumb and forefinger, it will form a short "ribbon" which will tend to break or bend downward. The soil may also feel slightly gritty or velvety when moist.

Fine-textured soils form very hard lumps or clods when dry and are quite plastic and sticky when wet. When the moist soil is squeezed out between the thumb and forefinger, it will form a long ribbon which will support itself. The soil may also feel slightly gritty or velvety when moist.

The ***surface texture*** is normally determined from a sample at plow depth or 7 inches (18 cm); however, erosion may have removed the surface to such an extent that only 1 or 2 inches (2.5 or 5.0 cm) may remain. For contests, a boxed sample of surface soil and subsoil will be provided to judge texture (Figure 11.8).

Fig. 11.8 — Soil samples should be placed in boxes for judging soil texture. (Courtesy, Roy Hunter Follett)

11:6.2 □ Depth of Soil

The depth of soil is determined by the total thickness of soil layers that are significant to soil use and management. This is the depth to which plant roots are normally expected to extend. Underlying this depth are rock, clay, or shale beds or unaltered alluvial materials.

- □ ***Deep soils*** — Deep soils are those more than 36 inches (91 cm) deep.
- □ ***Moderately deep soils*** — Moderately deep soils are those 20 to 36 inches (61 – 91 cm) deep.
- □ ***Shallow soils*** — Shallow soils are those 10 to 20 inches (25 – 61 cm) deep.
- □ ***Very shallow soils*** — Very shallow soils are those less than 10 inches (25 cm) deep.

11:6.3 □ Slope

Land slope is important in land use and soil management decisions because of its influence on erosion and on the use of farm machinery. **Slope** has to do with the lay of the land.

Slope is the number of feet of rise or fall in each 100 feet[1] of horizontal distance. The following are definitions of slope terms to be used in the contest:

- □ ***Nearly level*** — Less than 1 foot of rise or fall in each 100 feet (< 1 percent slope).
- □ ***Gently sloping*** — 1 to 3 feet of rise or fall in each 100 feet (1 – 3 percent slope).
- □ ***Moderately sloping*** — 3 to 5 feet of rise or fall in each 100 feet (3 – 5 percent slope).
- □ ***Strongly sloping*** — 5 to 8 feet of rise or fall in each 100 feet (5 – 8 percent slope).
- □ ***Steep*** — 8 to 12 feet of rise or fall in each 100 feet (8 – 12 percent slope).
- □ ***Very steep*** — More than 12 feet of rise or fall in each 100 feet (> 12 percent slope).

11:6.4 □ Erosion — Wind and Water

Erosion is the loss of soil by water or wind. The following are definitions in this contest:

---

[1]*Note:* Feet × 0.305 = meters.

- **None to slight** — Less than 25 percent of the surface soil has been removed by erosion and no gullies are in evidence.
- **Moderate** — 25 to 75 percent of the surface soil has been removed by erosion, with or without gullies in evidence; if gullies are present they are crossable with farm machinery. Moderate erosion may or may not change a capability class from Class II to Class III, but it is always a factor to prevent an area from being in Class I.
- **Severe** — 75 percent or more of the surface soil has been removed by erosion with or without occasional uncrossable gullies (gullies more than 100 feet apart) and / or severe blowouts and accumulations by wind.
- **Very severe** — 75 percent or more of the surface soil has been removed by erosion with frequent uncrossable gullies (gullies less than 100 feet apart) and / or very severe blowouts and accumulations by wind.

11:7 ▫ INTERPRETATION OF SOIL FACTORS[2]

Soil factor interpretations include soil permeability, surface runoff, restrictive factors that prevent a field from being judged Class I land, and land capability class.

11:7.1 ▫ Permeability

Soil permeability is the rate of movement of both water and air through the soil. This characteristic is important in determining land use and soil management practices. It has a great influence on root penetration, rate of water penetration, rate of water absorption, internal drainage, and nutrient leaching. Permeability is related to both texture and structure of the soil.

- **Rapid permeability** — These soils have loose or open sandy subsoils.
- **Moderate permeability** — These soils have highly granular, blocky, or columnar clay loam subsoils.
- **Slow permeability** — These soils have clayey subsoils and are medium, blocky, or plate-like to somewhat crumbly under ideal moisture conditions. Slowly permeable soils frequently have a deep surface and a thick transition zone from the A-horizon to the most clayey zone in the B-horizon.
- **Very slow permeability** — These soils have dense, fine clay, or claypan subsoils with a plate-like, coarse, blocky, or massive structure. When wet, they are plastic and sticky. Sodic soils are extreme examples. Root growth is restricted to cracks between the blocks or plates.

[2]See Chapters 1 and 7.

11:7.2 □ Surface Runoff

Surface runoff is the result of a combination of natural factors including land slope and water infiltration rate of the soils in relation to rainfall intensity. It refers to the relative rate water disappears from an area by infiltration and by flowing over the surface of the soil.

- □ ***Rapid*** — Surface water flows rapidly. A considerable amount of rainfall is lost from the surface which increases the hazards of erosion and droughty conditions. Fields with slopes of 3 percent and above (except rapidly permeable soils) would be placed in this category.
- □ ***Moderate*** — This is considered as "normal" runoff from soils with slopes of 1 to 3 percent, excluding rapidly permeable soils.
- □ ***Slow*** — Surface water flows away slowly. Surplus water on clayey soils is an occasional problem. This category includes nearly level areas with moderately permeable subsoils and nearly level, slowly permeable sandy surface soils.
- □ ***Very slow*** — Surface water flows away very slowly. Slopes may be nearly level to concave; they can be deep sandy soils with rapidly permeable subsoil; or they can be nearly level sandy soils with moderate permeability. Some soils may remain wet for long periods. (See Chapter 8.)

11:7.3 □ Major Factors That Keep a Field from Being Class I Land

This section is generally self-explanatory. Any one or more factors that would keep the area out of Class I land will be checked. The contestant's decision will determine the ones to check.

- □ ***Surface texture*** — Surface soil texture is not a major factor except for sandy soils. Sandy soils can be no better than Class III because of erosion hazards of both wind and water that are very difficult to control.
- □ ***Depth of soil*** — Only shallow or very shallow soils will be a major factor.
- □ ***Slope of land*** — Any slope greater than 1 percent will be considered a major factor.
- □ ***Erosion*** — All conditions except none to slight will be considered as major factors.
- □ ***Permeability*** — Only two conditions, rapid and very slow, will be major factors.
- □ ***Surface runoff*** — Only very slow and rapid runoff conditions will be considered as major factors.

11:7.4 ▫ General Guide for Selecting Land Capability Classes

Soil Factor	*Best Land Class*
Texture	
Coarse-textured	III
Moderately coarse, medium	I
Moderately fine and fine	I
Depth	
Deep or moderately deep	I
Shallow	III
Very shallow	VII
Slope	
Nearly level (0 – 1 percent)	I
Gently sloping (1 – 3 percent)	II
Moderately sloping (3 – 5 percent)	III
Strongly sloping (5 – 8 percent)	IV
Steep and very steep (8 – 12 percent) and over	VI
Erosion	
None to slight erosion	I
Moderate	II
Severe or very severe	VI
Permeability	
Rapid	III
Moderate and slow	I
Very slow	II
Surface runoff	
Rapid	III
Moderate and slow	I
Very slow	II

11:7.5 ▫ Land Capability Classes

Land is classified by the U.S. Department of Agriculture – Soil Conservation Service on the basis of permanent limitations or hazards in its use from the standpoint of keeping the soil permanently productive. The soil features of a particular area are all considered when determining the land capability class. There are eight recognized classes of land. They are divided into cultivatable and noncultivatable.

Cultivatable

- ***Class I*** — Soils in Class I have few limitations that restrict their use. They can be safely cultivated year after year with no special treatment to control runoff or conserve the soil. They are nearly level, and the risk of erosion by either wind or water is low. They are deep, well-drained, and easy to work (Figure 11.9).

Fig. 11.9 — On Class I land, risk of damage by either wind or water erosion is low. (Courtesy, USDA)

- ***Class II*** — Soils in Class II have some limitations that reduce the choice of adapted plants or soils that require moderate conservation practices. The limitations are few and the practices are easy to apply, but the land user has fewer alternatives than with Class I land (Figure 11.10).
- ***Class III*** — Soils in Class III have severe limitations that reduce the choice of plants, or that require special conservation practices, or both. The lower number of practical alternatives and the extra effort and cost to conserve soil and water distinguish Class III from Class II lands (Figure 11.11).
- ***Class IV*** — Soils in Class IV have very severe limitations that restrict the choice of plants, require very careful management or special conservation practices, or both (Figure 11.12).

Noncultivatable

- ***Class V*** — Soils in Class V have few or no erosion hazards, but they have other limitations that are difficult to remove by practical means. The limitations prevent tillage of crops with standard farm equipment. Class V lands are restricted largely to pasture, range, woodland, wildlife food and cover, recreation, or watershed protection. Though the land is nearly level or gently sloping, it is wet and stony (Figure 11.13) and is often flooded by streams. Usually, the growing season is too short for most field crops.

Fig. 11.10 — On Class II land, conservation practices, such as contour tillage, shown here, are needed. They are easy to apply. (Courtesy, USDA)

Fig. 11.11 — Class III land showing strip-cropping and a diversion terrace planted to barley. (Arrow points to diversion terrace.) (Courtesy, USDA - Soil Conservation Service)

Fig. 11.12 — This Class IV land shows severe irrigation erosion on tomato plots. Erosion could have been reduced by terracing and irrigating on the contour. (Courtesy, USDA - Soil Conservation Service)

Fig. 11.13 — This relatively level Class V land cannot be cleared of rocks by practical means. This land is not suited for cultivation but is suitable for pasture. (Courtesy, USDA)

- ***Class VI*** — Soils placed in Class VI have severe limitations that, for the most part, make them unsuitable for tillage and restrict their use largely to trees, pasture, range, recreation, watershed protection, or wildlife food and cover (Figure 11.14).
- ***Class VII*** — Soils in Class VII have very severe limitations that make them unsuited for cultivation and that largely restrict their use to grazing, woodland, or wildlife (Figure 11.15).

Fig. 11.14 — This Class VI land is an untreated pasture with an 11 percent slope and shows some moderately severe erosion. It is too steep for safe cultivation but is suitable for pasture if never overgrazed. (Courtesy, USDA)

Fig. 11.15 — Class VII land with steep, rocky, eroded hillsides. (Courtesy, USDA)

- ***Class VIII*** — Soils and land forms in Class VIII have limitations that preclude their use for commercial plant production and restrict their use to wildlife, recreation, water supply, or esthetic purposes.

11:8 ▫ RECOMMENDED LAND TREATMENTS (PART 2 ON THE LAND JUDGING SCORECARD)

11:8.1 ▫ Vegetative

For cropland use on Classes I through IV:

1. Row crop with occasional soil conserving crop — applicable to Class I land.
2. Row crop with frequent soil conserving crop — applicable to Class II land.
3. Row crops not more than two of four years — applicable to Class III land.
4. Row crops not more than one of five years — applicable to Class IV land.
5. Return crop residue to the soil.
6. Practice conservation tillage — provides for a protective cover by leaving crop residue of any previous crop as a mulch on or mixed in the surface (first few inches) of the soil. At least 30 percent residue should remain on the soil surface after planting.

For pasture, range, wildlife, or commercial woodland on Classes V, VI, VII, and VIII:

1. Establish recommended grasses and legumes. This practice is used when permanent vegetation is needed. Because of differences in interpretation, this practice will be used on all Class V, VI, VII, and VIII lands, except where tree plantings are made.
2. Practice proper pasture or range management. The application of practices to keep plants actively growing; to encourage the growth of desirable grasses and legumes while crowding out weeds and brush; and to minimize soil erosion.
3. Protect from burning.
4. Control grazing. Carry out a system of deferred or rotational grazing and proper stocking that will maintain or improve desirable vegetation on pasture or range. The practice should not be used where tree plantings are made. (See Chapters 12 and 13.)
5. Plant recommended trees for farmstead and field windbreaks and commercial woodland plantings. (See Chapter 19.)

6. Harvest trees selectively. A system of cutting in which single trees, usually the largest, or small groups of such trees are removed and reproduction secured under the remaining stand. (See Chapter 19.)
7. Use only for wildlife or recreation area. This means protection or the development of areas that cannot be used for grazing, forestry, cultivation, or urban development.

11:8.2 □ Mechanical

1. Control brush or trees. This may be accomplished by spraying with chemicals and / or use of machinery. The purpose is to improve the desirable vegetative cover by removing or killing undesirable brush and trees or removing timber so that land in Classes I through IV could be cropped. This practice should not be used when brushy material can be controlled by normal farming operations.
2. **Terrace** and **farm on the contour.** A terrace is an embankment or ridge of earth constructed across the slope to control runoff and minimize erosion. Conduct farming operations approximately on the contour, at right angles to slope direction and parallel to the terraces.
3. Maintain terraces. Practices that keep terraces working effectively.
4. Construct **diversion terraces**. A diversion terrace is a channel with a supporting ridge on the lower side; it usually has a greater horizontal and vertical spacing and is constructed to handle a larger flow of water than normal field terraces. (See Chapter 7.)
5. Install drainage system. To remove excess surface or groundwater from land by means of surface or subsurface drains. Used where wetness is a limiting factor. (See Chapter 8.)
6. Control gullies. Use one or more conservation practices that will adequately control runoff and erosion. Used any time *active* gullies are within the field area. They should be at least 6 inches (15 cm) deep and 12 inches (30.5 cm) wide. Irregularities in a field that are grassed over with no signs of erosion are not considered as needing control.
7. No mechanical treatment needed — use when brush and trees, erosion, gullies, drainage, or off-site water is not a problem.

11:8.3 □ Fertilizer and Lime

Information relative to the lime and fertility requirements will be given at each site. The decision of the contestants as to whether to add lime or fertilizer will be based on the soil test data and their knowledge of lime and fertilizer recommendations. No set of limits with regard to lime or fertility requirements will begin to fit all soils of the country and all crops. Each has its own requirements. Each state has its own set of guidelines relative to lime and fertilizer requirements, and these must be followed. (See Chapters 3 and 4.)

11:9 □ SELECTING A HOMESITE

The selection of a homesite requires considerable care and consideration of many soil and landscape features. Many homes have been or are being built on sites where serious problems will occur if difficult soil or drainage conditions are not considered. Such critical problems include wet basements, poor drainage from the site (Figure 11.16), excessive cracking of the foundation, and several types of septic tank drain field system failures.

Fig. 11.16 — A soil survey report would have indicated that this was not a desirable site on which to build a church. (Courtesy, Virginia Polytechnic Institute and State University)

In land judging for agriculture, soils are evaluated to determine their suitability as cropland, pastureland, or forest sites. To do this, a number of soil features are identified, a land capability class is determined, and recommended land treatments are suggested.

The same soil features judged for agricultural use, plus two or three more, can provide the information necessary to evaluate a soil judging site as a homesite. Instead of determining a land capability class as in agricultural judging, the site is given a degree of limitation of *slight, moderate, severe,* or *very severe* homesite rating. These ratings reflect the number and severity of soil problems found on the site. They are similar to ratings used by the Soil Conservation Service (Figure 11.17).

Potential home builders usually look for lots that are level and clean and that

Fig. 11.17 — Soil properties are perhaps more important to the selection of good homesites than they are to recommendations for agricultural uses of the land. A power auger is being used by the Soil Conservation Service to examine the soil profile in survey operations near Farmington, New Mexico. Soil surveys can help in the selection of good building sites. (Courtesy, USDA - Soil Conservation Service)

command a scenic view, but such choice lots are not always available. Hillside lots can be quite attractive, but they are usually more expensive to improve and maintain. Steep banks are subject to erosion, and in some cases, retaining walls will be required. Grass is more difficult to establish and to cut and trim. Steep roads and walks covered with ice and snow are quite hazardous during the winter. Other factors that should be considered are soil, rocks, and trees, and the location of a driveway that will permit a convenient and safe entrance and exit from the street or highway.

11:9.1 □ Size of Lot

Many counties and municipalities exercise some control over lot size and position of the house on the lot by means of codes and ordinances. The **zoning ordinance** usually establishes the "set back" distance between the house and street as well as the minimum distances permitted between the house and the side or rear boundaries.

A lot for a small house in an area that has a public water supply and public sewage disposal system may not require more than 8,000 square feet (743 sq m). However, a three- or four-bedroom ranch-type house in a suburban area may require 15,000 to 20,000 square feet (1,394 – 1,858 sq m). A large house in a rural area may require its own on-site water supply and sewage disposal facilities. This type of house may need more than 1 acre (0.4 ha) with perhaps 200 feet (61 m) of width or highway frontage.

Soil problems, correctable when the lot size is 1 or 2 acres (0.4 or 0.8 ha), may be too great if the lot size is ½ acre (0.2 ha) or less. On small lots some practices would be extremely difficult to apply (e.g., tile drainage or extra large filter fields for septic tanks).

The system of land judging developed for this chapter assumes that the lot size is considered to be from 1 to 5 acres (0.4 – 2 ha). The judging system also assumes that the home will be built over a basement rather than on a slab or over a crawl space. Basements are included since this type of construction makes the relationship between soil factors and homesite location more apparent.

11:9.2 ▫ Flood Hazards and Wet Basements

A lot in a low area or one that straddles a natural drainage channel and collects storm water from a highway, culvert, shopping center, parking lot, or undeveloped land should be avoided (Figure 11.18). Special diversion ditches or storm sewers may be required to control the runoff that follows a heavy rain or the fast melting of deep snow.

Fig. 11.18 — Cleanup work in a flooded basement. The flooding was caused by excess storm water from a nearby construction site. The house straddled a natural drainage channel. (Virginia) (Courtesy, USDA)

It is important to divert water away from the house. A sufficient slope (2 to 3 percent) away from the house should be provided to carry away roof water and lot runoff. Location of the home on the lot and construction at a sufficient elevation to allow surface water and water from rain gutters to run away from the house are basic to controlling homesite water problems.

The old cliché "We had to drain 5 acres to get a dry basement" is literally true on some soils. If for some reason a decision is made to build a house on such soils, the house should not have a basement at all (Figure 11.19). This is especially true if it is necessary to depend upon a sump pump to remove the water from the drains. Sump pumps that run for 24 hours per day quite naturally have a tendency to break down, and if they do, the basement will very likely be a wet catastrophe. On wet and poorly drained sites, it will be necessary to take special precautions to keep water from seeping through the foundation. Water table and drainage problems may not be apparent except during wet times of the year.

Fig. 11.19 — Excavation for a basement in soil with high seasonal water table. Providing drainage for dry basements on this kind of soil is difficult or almost impossible; it is also expensive. (Michigan) (Courtesy, USDA)

11:9.3 ▫ Good Topsoil Is Necessary

Good topsoil is necessary for the homeowner who plans to have a garden, shrubs, and a beautiful lawn. The topsoil should be at least 6 to 8 inches (15 – 20 cm) deep and a uniform dark gray to brown color throughout. Some contractors follow the excellent practice of stockpiling the surface soil removed from the construction area before excavation begins. After construction, when the final grading of the lot is done, the stockpiled surface soil is returned. This gives the homeowner a good material in which to establish a lawn and plantings. The natural topsoil is a very valuable commodity as are the trees already growing on the lot. All too often the building contractor destroys all of the trees and leaves only subsoil for the lawn and garden. A lawyer may be needed to draw up a contract to assure that topsoil is stockpiled and later spread.

11:9.4 ▫ Maintain Soil Cover During Construction

The period of construction is the critical time for erosion in suburban areas (Figure 11.20). Soil losses from areas under construction can make a large contribution to sediment loads in streams near construction sites. Construction on sloping sites should be done so as not to disturb the original vegetation. However, this may not be possible in some cases. For example, land scraping for cuts and fills may be necessary for establishing subdivision roads or drainage ways.

Fig. 11.20 — Sediment and erosion damage which occurred after 6 inches (15 cm) of rainfall in one week. Temporary erosion control measures were not used by the builder while houses were being constructed. (Maryland) (Courtesy, USDA)

If the removal of vegetation is necessary, the construction area should not be permitted to remain bare, especially over winter and spring when rains can cause severe erosion. Annual ryegrass or other suitable temporary vegetation should be planted in the fall to reduce soil loss. Perennial vegetation should be established during the proper season. (See Chapter 20.) This practice would reduce considerably the costs of removal of eroded soil from streets and would improve the appearance of the construction area.

11:9.5 ▫ Building Foundations — Know Your Soil

Buildings require foundations that can withstand the rains, the winds, and the waters. It is not essential to put house foundations on rock, but the wise person looks for good soils and the foolish person continues to build on soils poorly suited for foundations.

It is especially important in selecting a homesite to avoid soils with low-bearing strength, those with high clay content, those that slip or slide, and those that are high in organic matter. Avoid Psamments, Histosols, and Vertisols if possible. (See Chapter 1.)

Soils with Low-bearing Strength

The average home does not put heavy loads on the underlying soil. Most soils can support home foundations with no special problems. However, certain kinds of soils will compress or settle, even under very light loads. This settlement will crack interior and exterior walls, warp and twist windows and doors, and may cause other serious damage to homes. Many home subdivisions are developed on sloping soils. In these areas, part of the home foundation rests on cut areas with a high-bearing strength; the other part of the home foundation may rest on the fill areas with a low-bearing strength. Settlement of these fill areas may severely damage the superstructure. Selection of the fill material should be under the direct supervision of a licensed civil engineer or a certified professional soil scientist or soil surveyor (Figure 11.21).

Soils with High Content of Clay

Certain kinds of soils, such as Vertisols, have layers or horizons high in clays that shrink when dry and swell when wet. When home foundations, carport slabs, or driveways are put on these clays, they are often subject to shrink-swell pressures which can cause serious cracking. The clay literally shrinks away from the foundation when it dries and builds up tremendous pressures as it expands when wet. This movement will crack interior and exterior walls on houses and break up concrete slabs touching the clay. The shrinking and swelling of clay occurs only when water is removed from the clay or added to the clay. Damage is worse during seasons of prolonged drought followed by wet seasons, or vice versa.

Fig. 11.21 — Settling of the foundation due to unstable soil is the cause of severe damage to this home. (New York) (Courtesy, USDA)

Soils That Slip or Slide

Hillside lots may be covered with unstable soil that has a tendency to slip or slide down the slope when it gets wet. Some steep hillsides are converted to streets and homes, with little or no thought given as to how the soil will react. In some cases whole yards have been known to slide downhill, leaving the house hanging on the bank. The damage to homes resulting from flooding and mud slides near Santa Cruz, California, in January 1982 is a catastrophic and spectacular example of soil misuse.

In these areas, the assurance of permanent homes requires more than the operation of a bulldozer and the building of the homes. If the soil characteristics are known and evaluated by technical experts, these areas may be safely used for homesites. Usually the local agricultural extension agent or the soil scientist with the U.S. Department of Agriculture – Soil Conservation Service can provide information relative to the soil, trees, or topographic conditions in any specific area. The application of this free advice can eliminate the necessity of employing costly corrective measures.

Soils with High Organic Matter

In developing subdivisions and lots for individual homes, it is often necessary to clear the area of vegetative growth. This vegetation sometimes includes beautiful shade trees growing on the potential homesite. Foundations for structures or carport slabs and driveways that are constructed on these lots frequently settle for a number of years as buried vegetation rots. If the soil settles uniformly, the damage may be negligible; otherwise, serious damage will result. Organic material, whether placed there recently as a buried log or accumulated over many years as peat or muck (Histosols), is not a satisfactory support for foundations of homes. (See Chapter 1.)

11:9.6 □ Sewage Disposal

As a rule, those homes constructed outside the municipal sewage service area depend on septic tanks and drain fields to treat the sewage. A much better job of designing and installing a septic system can be done if soil conditions such as surface and internal drainage, depth to bedrock, and permeability are known. Many septic system failures occur each year because soil conditions were not properly evaluated. Often unfiltered sewage makes its way into road ditches and streams; as a result, the whole community is polluted and lawsuits may result.

In any method of on-site sewage disposal, the most important concern should be that of effluent disposal. It is important to know how much of the lot must be used to get rid of sewage effluent and where the trenches for the tile lines' drainage fields will be located. If the soil is mellow and well-drained, a relatively small part of the lot may be satisfactory; but if poorly drained or subsoils are tight, a much larger area is needed for the drainage field.

Some soils are too permeable and may allow the effluent to travel downward long distances without adequate filtration, thus polluting the groundwater. If a home has both a well and a septic tank, it might be a good idea to discuss this matter of filtration with a certified professional soil surveyor and a licensed sanitary engineer.

The best place for the disposal field is usually an open, sunny part of the lawn. The disposal field should not be located in a wooded area. The disposal trenches should be kept away from trees and large bushes because the roots will grow into the joints or holes between or along the tiles or pipes and plug the drainage lines.

11:10 □ LAND JUDGING FOR HOMESITES

Homesite evaluation has been added to the International Land Judging Contest held each year in Oklahoma. This part of the contest is designed to emphasize the importance of soils and their limitations for nonagricultural purposes.

While limited to homesite evaluation, it is also meant to emphasize the importance of soil suitability for other nonagricultural uses of land such as roads, streets, parks, and playgrounds (Figure 11.22). Many of the features used in judging soils for agricultural use are also used in evaluating an area for a homesite.

Fig. 11.22 — Use soil information in planning private and public facilities. (Tuttle Creek Reservoir, Kansas) (Photo by Roy Hunter Follett)

In order to ensure that the contests are not lengthened too much by the addition of homesite evaluation and that grading the score sheets does not become too burdensome, several alternatives are possible in the selection of areas for judging agricultural land and soil for homesites.

Three such alternatives are: (1) judge three agricultural land sites and two homesites; (2) judge four sites with agricultural land judging and make homesite evaluation interpretations of at least two of the sites; and (3) use other combinations.

The only concern is to make sure that there are enough interpretative uses required to test the contestants' skills in homesite evaluation.

11:11 ▫ HOMESITE EVALUATION SCORECARD

The land judging procedures and scorecard (Figure 11.23) that are used for the International Homesite Evaluation Contest in Oklahoma are used here.

1. The total perfect score on each site is 97 points.
2. The total perfect score for Part 1 is 27 points; for Part 2, 70 points.
3. Part 1 has to do with those factors that the contestant must determine about the site. With the exception of shrink-swell, water table, and flooding, the factors are the same as those for land judging; however, while the factors are the same, different interpretations must be made. The contestant should exercise caution and not be too hasty in making a straight check-off from the land judging scorecard to the homesite scorecard.
4. After Part 1 is completed, the contestant should determine the severity of limitations that the existing soil conditions impose on the planned use for homesites as listed on Part 2 of the scorecard.
5. The final evaluation of the homesite is determined by the *worst* degree of limitation found for the particular planned use.

11:12 □ FEATURES OF THE HOMESITE TO BE JUDGED

The features of the homesite being considered are related to the evaluation of limitations for a specific use. The limitations are defined as follows:

- □ ***Slight limitations*** — Those soils or locations that have properties favorable for homesites and present little or no problem.
- □ ***Moderate limitations*** — Those soils or locations that have properties only moderately favorable for homesites. Limitations can be overcome or modified with special planning, design, or maintenance. Special treatment of the site for the desired use may be necessary.
- □ ***Severe soil limitations*** — Those soils or locations that have one or more properties unfavorable for homesites. Limitations are difficult and costly to modify or overcome for the use desired.
- □ ***Very severe limitations*** — Those soils or locations that have one or more features so unfavorable for a particular use that overcoming the limitation is very difficult and expensive. For the most part, these kinds of soils should not be used for homesites.

11:12.1 □ Surface Texture

Surface texture refers to the texture of the surface soil. It is not a factor for sewage lagoons and septic systems because lagoons are dug below the surface and most of the surface soil is used in making the berm around the lagoon.

- □ ***Coarse: moderate limitations*** — May require stabilization with organic material and / or loamy topsoil to improve moisture and nutrient holding

HOMESITE EVALUATION CARD
Oklahoma State University Extension

Contestant No. ______________ Field No. ______________

Name ______________________________

Address ______________________________

County/School ______________ Organization ______________

PART 1 Land Factors			PART 2 Planned Use—Family Dwelling Site. Without Basement. Interpretations of Limitations in Terms of: (2 pts. each)			
SCORE	Features of the Site Being Considered (3 pts. each)	Degree of Limitation	Foundations for Buildings	Lawns and Landscaping	Septic System	Sewage Lagoon
A.	TEXTURE - SURFACE					
	☐ Coarse	Slight	☐	☐		
	☐ Mod. Coarse, Medium, Mod. Fine	Moderate	☐	☐		
	☐ Fine	Severe	☐	☐		
B.	PERMEABILITY					
	☐ Very Slow < 0.06"/hr.	Slight		☐	☐	☐
	☐ Slow 0.06—0.6"/hr.	Moderate		☐	☐	☐
	☐ Moderate 0.6—2.0"/hr.	Severe		☐	☐	☐
	☐ Rapid 2.0/hr. +	V. Severe		☐	☐	☐
C.	DEPTH OF SOIL					
	☐ V. Shallow < 10"	Slight	☐	☐	☐	☐
	☐ Shallow 10 — 20"	Moderate	☐	☐	☐	☐
	☐ Mod. Deep 20 — 40"	Severe	☐	☐	☐	☐
	☐ Deep 40 — 72"	V. Severe	☐	☐	☐	☐
	☐ V. Deep over 72"					
D.	SLOPE					
	☐ N.L. to Gentle 0 — 3%	Slight	☐	☐	☐	☐
	☐ Moderate 3 — 5%	Moderate	☐	☐	☐	☐
	☐ Strong 5 — 8%	Severe	☐	☐	☐	☐
	☐ Steep 8 — 15%					
	☐ V. Steep 15% +					
E.	EROSION					
	☐ None — Slight — Moderate	Slight	☐	☐	☐	☐
	☐ Severe	Moderate	☐	☐	☐	☐
	☐ Very Severe	Severe	☐	☐	☐	☐
F.	SURFACE RUNOFF					
	☐ Slow	Slight	☐	☐	☐	
	☐ Moderate	Moderate	☐	☐	☐	
	☐ Rapid	Severe	☐	☐	☐	
G.	SHRINK-SWELL (heaviest layer)					
	☐ Low	Slight	☐		☐	☐
	☐ Moderate	Moderate	☐		☐	☐
	☐ High	Severe	☐		☐	☐
H.	WATER TABLE (permanent or temporary)					
	☐ Deep > 72"	Slight	☐	☐	☐	☐
	☐ Mod. Deep 40"-72"	Moderate	☐	☐	☐	☐
	☐ Shallow < 40"	Severe	☐	☐	☐	☐
I.	FLOODING					
	☐ None	Slight	☐	☐	☐	☐
	☐ Occasional < 1 in 2 yrs.	Moderate	☐	☐	☐	☐
	☐ Frequent > 1 in 2 yrs.	Severe	☐	☐	☐	☐
	FINAL EVALUATION					
	All factors none to slight	Slight	☐	☐	☐	☐
	One or more factors mod. None severe	Moderate	☐	☐	☐	☐
	One or more factors severe None V. Severe	Severe	☐	☐	☐	☐
	One or more factors Very severe	V. Severe	☐	☐	☐	☐

SCORE PART 1 __________ (Possible 27) SCORE PART 2 __________ (Possible 18 18 18 16

PART 1 AND 2 — TOTAL SCORE ______________ (Possible 97)

Fig. 11.23 — Homesite Evaluation Scorecard. (Courtesy, Oklahoma State University)

and supplying capacity for desired plant growth. Washing and blowing may be a problem during construction.

- ***Moderately coarse, medium, moderately fine: none to slight limitations*** — Care should be exercised during construction to be sure the surface soil is not covered by less desirable material.
- ***Fine: severe limitations*** — For uses other than sewage lagoons. Soil is sticky when wet, hard when dry, and difficult to work when used for flower beds and gardens. The soils crack when dry, swell when wet, requiring frequent and low rate of watering for plant growth.

11:12.2 ▫ Permeability

Permeability normally refers to the rate of water and/or air movement through the most restricted (least permeable) layer in the soil profile (Figure 11.24). This may be considered internal drainage. Laterals for septic systems may be located below such layers in some soils. For that reason this should serve as a warning, and final design should be based on a **field percolation test** to determine water infiltration rate.

Permeability is an important factor in deciding between a septic tank system and a sewage lagoon. Soil percolation tests would be required before making

Fig. 11.24 — Land judgers examine homesite for soil depth and permeability at the Kansas State Land Judging and Homesite Evaluation Contest. (Photo by Roy Hunter Follett)

further plans. ***Special note:*** For septic systems, consider the permeability below 30 inches (76 cm) soil depth. The four categories of soil permeability are:

- ***Very slow permeability*** — Very severe limitations for septic tank systems. Water movement generally is less than 0.06 inch (0.15 cm) per hour. This would require a prohibitively large field of laterals or necessary costly modifications. Septic systems generally are not recommended. None to slight limitations for lagoons. Subsoils generally are of a clayey texture and break into sharp, angular chunks or clods that are plastic and sticky when wet and very hard when dry. The clods are usually coated with clay, which restricts water movement. Shrink-swell potential is generally high, presenting severe limitations for foundations and basement construction and moderate limitations for lawns, shrubs, and gardens. The clayey subsoils may be underlain by partially weathered, very slowly permeable shales or impermeable bedrock.
- ***Slow permeability*** — Severe limitations for septic tank systems. Soils generally would be on the fine side of the loamy groups such as clay loams and silty clay loams with a columnar to subangular blocky structure. Problems are generally similar to the very slowly permeable soils, but the modifications required for use are not as great. Water movement can range from 0.06 to 0.60 inch (0.15 – 1.50 cm) per hour. Percolation tests would be required for the design of a septic tank disposal field. Should a soil be in the slowest part of this range or 0.06 inch (0.15 cm) per hour, the cost of modification or size of filter field necessary would generally be prohibitive. Limitations would be none to slight for sewage lagoons and moderate for basements, foundations, walks, flowers, and shrubs.
- ***Moderate permeability*** — Slight to moderate limitations. Soils are generally loams and sandy clay loams. No severe restricted layers with granular to fine blocky structure. None to slight limitations for septic systems and homesites, lawns, shrubs, and gardens. Permeability ranges of 0.6 to 2.0 inches (1.5 – 5.1 cm) per hour would indicate a need for a percolation test if thought to be on the low side of the range. Moderate limitations for sewage lagoon. None to slight limitations for foundations, basements, and walks.
- ***Rapid permeability*** — Soils generally are loams to sandy loams throughout. Permeability is greater than 2.0 inches (5.1 cm) per hour. Slight limitations for septic system disposal field or foundations and basement construction. Moderate limitations for lawns and shrubs. Severe limitations for sewage lagoons.

Note: If subsoil texture is sandy, permeability is probably greater than 6.0 inches (15.2 cm) per hour, and the soil would have a very severe limitation for sewage lagoons. Seepage from lagoons would cause pollution of groundwater. In subsoils with very rapid permeability, a lagoon can be made satisfactory if about 6 inches (15 cm) of fine clay is used to line the lagoon.

11:12.3 □ Soil Depth

Soil depth refers to the vertical depth of a soil to bedrock such as sandstone, limestone, shale, or consolidated clays that restrict root and water movement (Figure 11.25). Severity of limitations, because of depth, varies greatly for different uses.

Fig. 11.25 — Deep permeable soils (*top profile*) have fewer problems and less expense in design, construction, and maintenance of homesites than do shallow soils (*bottom profile*). (Scales are in feet; feet × 0.305 = m.) (Courtesy, USDA - Soil Conservation Service)

11:12.4 □ Slope

Slope refers to the steepness of the surface or the vertical rise or fall in feet for each 100 feet (30 cm) of horizontal distance, expressed in percent. Broader and different slope ranges apply to homesite use consideration than normally apply to consideration for agricultural lands. In a study of a detailed soil survey report, each soil unit mapped will have the slope ranges given that are common for the particular soil unit. (***Note:*** N.L. = nearly level.)

11:12.5 □ Erosion

Erosion of surface soil can increase the expense of landscaping. Severe gullies will impose additional limitations on septic tank disposal fields. The three categories of erosion recognized in this activity are:

- ***None to slight and moderate erosion*** — None to slight limitations for any use.
- ***Severe erosion*** — Moderate limitations may require leveling, modification of surface, or bringing in topsoil for flower beds.
- ***Very severe erosion*** — Severe limitations — severely gullied soils require much filling or leveling, extra cost for septic tank disposal systems, and extensive modification for flower beds and lawns. Time of development should be selected for the least erosive time of year. An organic mulch such as straw or woodchips can be used to stabilize eroding soil until the season is proper for sodding or seeding perennial grasses.

11:12.6 □ Surface Runoff

Surface runoff is generally a factor of importance in connection with drainage, permeability, and erosion. Special attention needs to be given to surrounding areas. Runoff from adjacent areas onto the contemplated building site and the possibility of ponding and water accumulations around the house site need consideration. Surface runoff is not a factor for sewage lagoons because they should be protected from outside water by a levee or berm around them. The three categories of surface runoff are:

- ***Rapid runoff*** — Usually occurs on slopes greater than 5 percent, except on deep sands. Severe limitations exist, requiring care to maintain and control erosion on lawns. None to slight limitations exist for basement or septic system.
- ***Moderate runoff*** — None to slight limitations for any use.
- ***Slow and very slow runoff*** — Occurs on nearly level to very gently sloping areas and deep sands. Severe limitations may require costly modification

for basements and foundations and special design of septic tank disposal fields. On deep sands slow runoff would present none to slight limitations.

11:12.7 ▫ Shrink-Swell

This factor is implied in the permeability and texture of a soil. Because it is important in foundations for buildings, the design should have special consideration. The finest (most clayey) layer in the profile is generally considered in shrink-swell limitations. Shrink-swell is generally not a factor for lawns and landscaping.

- ***Low*** — Coarse- and moderately coarse-textured soils have *none to slight limitations* for all uses.
- ***Moderate*** — Medium- and moderately fine-textured soils have *moderate limitations* for all uses.
- ***High*** — Fine-textured soils have *severe limitations* for all uses.

11:12.8 ▫ Water Table

The **internal wetness** of an area is influenced by most of the factors previously discussed. Generally, **internal drainage** is a reflection of permeability and inflow from surrounding areas. For example, a very slowly permeable soil exhibits poor to very poor internal drainage. The presence and depth of a water table are not necessarily a reflection of soil permeability because the water may have flowed in from upslope. It must be evaluated on the basis of depth and permanency, frequently requiring study during different seasons of the year. Unless obvious, **water table depth** is given to contestants at the time of judging. Categories of depth to water table specified in this presentation are:

- ***Deep*** — None evident above 72 inches (183 cm). None to slight limitations for all uses.
- ***Moderate*** — Temporary or permanent water table present at depths of 40 to 72 inches (102 – 183 cm). Moderate limitations for all uses.
- ***Shallow*** — Temporary or permanent water table at depths less than 40 inches (102 cm). Severe limitations for all uses except for lagoons.

11:12.9 ▫ Flooding

The occurrence of floods is a factor frequently overlooked. Flooding may not happen in an area for many years, then a serious flood may occur. Because of waterproof roads and roofs, housing and other urban developments on the watershed of a stream can increase runoff up to 75 percent, thus greatly increasing flood hazards. Soils can give an indication of flood hazard, but records must be studied to determine the longtime conditions. Position in the landscape and proximity to nearby streams are good indications of the frequency of flooding. Unless obvious, flooding information is given for contest purposes. The three categories of flooding hazard are:

- ***None*** — Limitations *none to slight* for all uses.
- ***Occasional*** — Flooding less frequent than *one year in two. Severe limitations* for foundations for buildings. Moderate limitations for septic system absorption field. *None to slight limitations* for sewage lagoon and lawns and landscaping.
- ***Frequent*** — Flooding more frequent than one year in two. *Severe limitations* for all uses.

11:13 ▫ REFERENCES

Birdwell, Bobby T. "Some of the World's Best Farmland Is Here." In *Our American Land, The 1987 Yearbook of Agriculture.* U.S. Department of Agriculture, Washington, D.C., pp. 91 – 95.

Bohannon, R. A., R. H. Follett, and W. M. Eberle. "Land Judging and Homesite Evaluation." Kansas State University, Co-op. Ext. No. MF-224, 1978, 28 pp.

Eberle, W. M. "Choosing a Homesite." Kansas State University, C-574, 1977, 11 pp.

Knuti, Leo L., David L. Williams, and J. C. Hide. *Profitable Soil Management,* 4th ed. Englewood Cliffs, New Jersey: Prentice-Hall, Inc.,1984, pp. 249 – 267.

McCormack, A. D. "Picking a Homesite: Soils Play a Big Role." U.S. Dept. of Agriculture, *Handbook for the Home. The 1973 Yearbook of Agriculture,* pp. 106 – 112.

Olson, Gerald W. *Soils and the Environment: A Guide to Soil Surveys and Their Applications.* New York: Chapman and Hall, 1981, 178 pp.

Rommann, L., *et al.* "Pasture and Range Judging in Oklahoma." Oklahoma State University, AGR-C1003, 1979, 32 pp.

Steinhardt, G. C., D. P. Franzmeier, and J. D. Yahner. "How to Make Miniature Soil Monoliths." Purdue University, Co-op. Ext. Agronomy Guide, AY234, 1981.

Stiegler, J. H. "Land Judging in Oklahoma." Oklahoma State University, AGR-C1002, 1986, 15 pp.

CHAPTER 12

Rangeland Management

"Grass is the forgiveness of nature — her constant benediction. Fields trampled with battle, saturated with blood, torn with the ruts of cannons, grow green again with grass, and carnage is forgotten. Forests decay, harvests perish, flowers vanish, but grass is immortal." — John J. Ingals, U.S. Senator, Kansas, 1873 - 1891

OUTLINE

Top: Heavy grazing has almost eliminated the desirable grasses. *Bottom:* Moderate grazing has permitted the desirable grasses to maintain themselves. (Courtesy, U.S. Forest Service, Manitou Experimental Forest, Colorado, Rocky Mountain Region)

□ □ □

12:1 □ OVERVIEW

Rangeland is the largest single class of land in the world, occupying nearly 44 percent of the earth's land surface. Vast expanses of prairie and savanna grassland ranges are found in North and South America, Africa, Australia, and Eurasia. Grass, shrub, and woodland-grass rangelands provide major habitats and forage resources for much of the world's wildlife and domestic livestock. ***Rangeland*** is most often defined as "land on which the natural plant cover is composed principally of grasses, forbs (nongrasslike herbs), or shrubs suitable for grazing or browsing by livestock and big game animals." Rangelands are generally unsuitable for crop production or "improved" or irrigated pasture because of topographic, climatic, soil, or sometimes economic limitations. Soil is widely regarded as the basic resource which, with climatic influences, determines the kind, yield, and quality of renewable natural resources — forage, water, wildlife, and recreation — produced by our rangelands. Minerals, fossil fuels, and wood products are also important resource values in many range areas of the world (Figure 12.1).

Fig. 12.1 — Rangeland supplies the world with forage for livestock and wildlife, plus water and recreation. In addition, minerals, fossil fuels, and wood products may be important products in some areas. (Utah) (Courtesy, USDA - Soil Conservation Service)

The settlement, development, and prosperity of Middle and Western America during the nineteenth and early twentieth centuries depended in large part on rangeland resources. Originally, about 1 billion acres (405 million ha) of native range supported a multitude of bison and other game animals, which were replaced under hunting pressure and westward expansion with an enlarging cattle, sheep, and horse population. Abundant "free" forage and demand for meat, wool, and leather created by the Civil War and the Gold Rush attracted large investments in a thriving range livestock industry. Ranges were grazed primarily by the trailing of cattle and sheep to railheads for marketing, their growth and fattening accomplished during the months required in the process. However, indiscriminate cultivation of prairie grasslands following the Homestead Act of 1862, shortages of forage due to heavy grazing by excessive livestock numbers, and a severe winter and drought of the late 1800's led to range depletion by the turn of the century. Range wars, an oversupply of beef and lamb, and the consequent low livestock prices further led to a depression of the early livestock industry.

The need for range revegetation and grazing control was widely recognized during the early 1900's. Livestock use of federal lands was brought under control with the creation of the forest service and national forest system in 1905, and passage of the Taylor Grazing Act in 1934 and concurrent formation of the grazing service (now Bureau of Land Management) for administration of grazing on the public domain lands. Slowly, livestock use of millions of acres of publicly owned rangeland has come under management on an allotment basis in which season and intensity of permitted use are regulated, and local cattle and sheep owners are assessed nominal fees for the privilege of grazing. Management of private lands was undertaken on a "home-ranch" basis in which the needed supplemental feeds and forages and livestock and range management practices were recognized as essential.

The Cooperative Extension Service, established in the land-grant colleges by the Smith-Lever Act of 1914, and the Soil Conservation Service, established in the U.S. Department of Agriculture in 1935, provide educational and technical assistance to farmers and ranchers in these efforts.

Depending on definition, from one-third to one-half of the land area of the United States is rangeland. Certainly, more than half is grazed by livestock and big game animals. There are roughly 1 billion acres (405 million ha) grazed by livestock in the United States (excluding Alaska and Hawaii) of which 58 percent is grass and shrub rangeland, 25 percent is forestland, and the balance is improved pasture and crop residue. The total area of rangeland in the United States is projected to decrease by about 7 percent by the year 2030 due to cropland conversion and development for other uses.

Over 90 percent of our rangelands are in 22 states which can be divided for discussion into three groups (Table 12.1). The 11 far western states — Montana through New Mexico and westward to the Pacific — are often referred to as the "Public Rangeland States." They include nearly half the total rangeland of the United States and half the federally owned rangelands. For more than half a cen-

Table 12.1 — The Major Rangeland States

States	Total Land Area[1] (million acres)	Federal Rangeland[2] (million acres)	Nonfederal Pasture & Rangeland[3] (million acres)	Total Pasture & Rangeland (million acres)	Land in Pasture & Rangeland (%)	Rangeland in Federal Ownership (%)
Public Rangeland States						
Arizona	73	20.9	35.1	56.0	77	37
California	100	25.3	18.7	44.0	44	58
Colorado	66	8.7	25.4	34.1	52	26
Idaho	53	15.6	7.7	23.3	44	67
Montana	93	11.0	41.5	52.5	56	21
Nevada	70	52.6	7.6	60.2	86	87
New Mexico	78	15.1	42.5	57.6	74	26
Oregon	62	13.1	11.9	25.0	40	52
Utah	53	23.1	10.0	33.1	62	70
Wyoming	62	20.4	26.9	47.3	76	43
Washington	35	1.7	7.3	9.0	26	19
Subtotal	745	207.5	234.6	442.1	59 (av.)	47 (av.)

(Continued)

Table 12.1 (Continued)

States	Total Land Area[1] (million acres)	Federal Rangeland[2] (million acres)	Nonfederal Pasture & Rangeland[3] (million acres)	Total Pasture & Rangeland (million acres)	Land in Pasture & Rangeland (%)	Rangeland in Federal Ownership (%)
Private Rangeland States						
Arkansas	33	0.0	5.9	5.9	18	0
Florida	35	0.2	8.5	8.7	25	2
Kansas	52	0.2	19.0	19.2	37	1
Kentucky	25	0.0	5.7	5.7	23	0
Missouri	44	0.2	12.9	13.1	30	2
Nebraska	49	0.5	24.9	25.4	52	2
North Dakota	44	1.5	12.1	13.6	31	11
Oklahoma	44	0.2	23.3	23.5	53	1
South Dakota	49	1.6	24.6	26.2	53	6
Texas	168	1.3	114.2	115.5	69	1
Subtotal	543	5.7	251.1	256.8	47 (av.)	2 (av.)
Alaska	363	225.2	6.3[2]	+231.5	64 (av.)	97 (av.)
All Other States (38)	612	0.4	62.6	63.0	10 (av.)	1 (av.)
Grand Total	2,263	438.8	554.6	993.4	44 (av.)	44 (av.)

[1] *National Resources Inventory.* USDA, 1987.
[2] RPA, Resources Planning Act, 1980. A report to Congress on the Nation's Renewable Resources. USDA, 209 pp.
[3] RCA, Resources Conservation Act. Appraisal 1980. USDA.

Source: Compiled by John L. Artz, Range Scientist, University of Nevada, Reno.

Note: Acres × 0.405 = hectares.

tury, the economics of these states as well as their way of life have developed to a great extent through coordinated use of private and federal grazing lands. The Bureau of Land Management and the U.S. Forest Service, the principal federal land managers, play an important role.

The second group of major range states includes the tier of states — the Plains states — that adjoin the public land states but also includes some other important range states to the east, such as Florida. These 10 states make up about one-fourth of the total U.S. rangeland but little more than 1 percent of the federal lands. Thus, they can be called the "Private Rangeland States." The area includes some of the most productive and intensively managed livestock ranches in the world.

The third "group" is a single state — Alaska. Alaska includes over half of the federal rangelands and nearly a quarter of the total U.S. rangelands. In spite of the vastness of its rangelands, Alaska has very few cattle or sheep.[1] It does have a substantial reindeer industry, and other wildlife resources are rich and abundant.

The renewable resource of plant life on our grazing lands transforms solar energy to billions of calories of digestible organic energy through photosynthesis. Cattle, sheep, deer, and other big game animals are ruminants, which, through the action of microbes inhabiting a compartment **(rumen)** in their stomach, convert fibrous forages into high-protein red meat, wool, leather, and a variety of other essential animal products. Range forage represents a significant agricultural crop as each 10 to 26 pounds (4.5 – 11.8 kg) grazed is converted to a pound or kg of live animal. Harvest and useful conversion of native vegetation on our rangelands and forestlands by any practical means other than livestock grazing is unknown.

12:2 ▫ RANGE MANAGEMENT AS A SCIENCE AND PROFESSION

Range management as a science originated in the United States in the early 1900's and has evolved with the necessity for management and improvement of our rangelands. The job of the **range manager** on either public or private lands is to promote ecological stability among the components of the **ecosystem** — soil, climate, animals, vegetation, fire, topography, and humans — and to conserve and optimize the productivity of the various rangeland resources. On public lands, the objectives of management center on satisfying the needs or returns desired by society. The private land owner, rancher, or resource manager for a landholding company is concerned primarily with economic benefits.

Range management requires a sense of stewardship and propriety for the natural resources, a thorough knowledge of the natural sciences, and an under-

---

[1]*Note:* In 1985 there were 9,500 cattle and 200 sheep in Alaska.

standing of range-livestock husbandry. The professional range person — conservationist, scientist, or manager — is employed primarily by agencies of the federal government. In the Department of Agriculture, these include the Forest Service, the Soil Conservation Service, and the Cooperative Extension Service; and in the Department of the Interior, they include the Bureau of Land Management, the Bureau of Indian Affairs, and the Fish and Wildlife Service.

12:3 ▫ RANGELAND AREAS OF THE UNITED STATES

Rangeland vegetation cover is classified under many schemes, all based on the appearance and distribution of the principal forms of plants — grasses, shrubs, and trees — and one, two, or three predominant species. Most often, plant cover is named thereby, as "types" or "communities." Recently, plant groupings have been identified as "ecosystems" or "range sites" in which consideration is given to the environment with which they are associated. Rangeland, according to the U.S. Department of Agriculture, includes natural grasslands, **savannas** (grasslands with scattered trees), shrublands, most deserts, tundra, alpine communities, coastal marshes, and wet meadows that have less than 25 percent canopy of forest trees at maturity. These grazed ecosystems can be grouped simply as grasslands and shrublands. A third group, forested lands, while not rangeland *per se*, includes wooded types that provide significant grazable forage. Together, these three areas occupy about 44 percent of the 48 conterminous United States. (See Chapter 19.)

12:3.1 ▫ Grasslands

Grassland ranges extend over 293 million acres (119 million ha), producing two-thirds of the total natural forage resource in grazing areas of the United States. The eight **grassland types** that have been identified are plains, prairie, mountain, mountain meadow, alpine, annual, wet, and desert.

Plains

The plains type, known as the short grass or Great Plains, is the largest single plant ecosystem in the United States (Figure 12.2). Called the "Great American Desert" by the early explorers, the plains occupy a broad belt of high land from 5,500-feet (1,676-m) elevation on the eastern slope of the Rocky Mountains to 1,500 feet (457 m) in the Central states. The plains grasslands are almost entirely in private ownership. Annual precipitation is from 10 inches (25 cm) at the Canadian border to over 25 inches (64 cm) in central Texas. Short, warm-season grasses, primarily blue gramagrass and buffalograss predominate, with lesser amounts of western wheatgrass and needlegrass present, mainly in the Dakotas. Plains grasses are palatable, nutritious, and resistant to heavy grazing, but may be in short supply in the early spring or during droughts, which occur frequently.

Fig. 12.2 — The plains grasslands are a productive resource in the central part of the United States. Commonly called the "short grass" country, this grassland is a relatively dry area with extremes in temperature, which led the original settlers to call it the "Great American Desert." (Courtesy, Colorado State University)

Cattle are grazed from four to eight months in the north to year-long in the southern plains. Much of the area is in cropland producing wheat, grain sorghum, forages, and other crops under both dryland and irrigated farming. The plains grasslands are utilized by cow-calf herds and are managed to promote good cover to reduce wind and water erosion.

Prairie

The prairie type is second to the plains in extent and productivity and is located between the eastern forests and the short-grass plains on the west from Canada to parts of Texas (Figure 12.3). Originally spreading into what is now the Corn Belt, the prairie grasslands were treeless due to grazing, drought, and recurring natural fires. With settlement, wildfires were limited, and the eastern forest type and introduced species have invaded, thus reducing the extent of the prairie. The prairie type is almost entirely in private ownership. Precipitation varies from 20 to 40 inches (51 – 102 cm), and elevations vary from 500 to 1,500 feet (152 – 457 m). Bluestems (big, sand, little) contribute most of the forage, reaching as much as 6 feet (2 m) in height and making excellent warm-season grazing for cattle for about four to nine months. But the value of the bluestem grasses is low during the late fall and winter when supplemental feeds need to be provided. Heavy grazing and drought decrease the productivity of the grass.

Mountain, Mountain Meadow, and Alpine

In the western states, associated with the northern Rocky, Sierra, and Cas-

Fig. 12.3 — Once a treeless prairie, the prairie type still remains treeless under wise and thoughtful management. As a result of a lack of understanding of the management requirements, a significant part of the prairie is heavily invaded by woody plants and poor-quality grasses. (Courtesy, P. D. Ohlenbusch, Kansas State University)

cade mountain ranges, are three grassland types important as summer feed for cattle, sheep, and big game, and for watershed protection and recreation. These are the mountain, mountain meadow, and alpine grasslands, occurring in this order, from intermediate to very high elevations of about 11,000 feet (3,353 m). All commonly border on coniferous forests of central Colorado, eastern California, western Montana, northern Idaho, western Wyoming, eastern Utah, and northeastern Oregon and Washington. These types receive from 20 inches (51 cm) of precipitation at low areas to 50 inches (127 cm) at high elevations where most precipitation occurs as snow. The mountain type supports perennial fescuegrasses and wheatgrasses, and a variety of palatable forbs utilized for spring, summer, and fall grazing. This bunchgrass range may be invaded along its lower borders by big sagebrush after disturbances such as overgrazing. The mountain meadow and alpine grasslands are dominated by low-growing, perennial grasses (bentgrass, bluegrass, hairgrass), sedges, and rushes. Dwarf willows and forbs are also common. Snow melt is a primary source of soil moisture for growth of grasses which is restricted to a brief summer period. Concentrated use by livestock and recreationists around water courses and on fragile upland soils has led to critical management problems for the U.S. Forest Service, the principal administrator of mountain meadow and alpine grasslands.

Annual

Annual-type grasslands are confined to uncultivated valleys and foothills below 2,000 feet (610 m) in elevation between the Sierra and Cascade ranges and Coast ranges of California. Rainfall is typically 10 to 30 inches (25 – 76 cm), confined primarily to the mild winter months. This promotes growth of cool-season annual grasses and forbs, grazed mainly from October to July. Dominant plants are wild oats, barley, and bromegrasses, and a diversity of forbs including bur clover and filaree. This grassland is very productive of beef and lamb, but forage yield is highly variable due to management in relation to corresponding fluctuations of seasonal and annual rainfall and variations of temperatures.

Wet

The wet grasslands, although smallest in extent, are most productive of forage per acre (hectare). They form narrow belts of wet prairies and marshes along most of the southern and eastern coastal areas and adjacent to major inland rivers of the southeastern United States. This type includes mainly cordgrass and saw grasses, tules, bullrushes, switchcane, and a scattering of palms and other trees. Management has involved control of surface water and construction of walkways for access by grazing cattle, range improvement, and recreational developments.

Desert

Of lowest forage productivity are the desert grasslands found on plateaus of the southwest, mainly in New Mexico, northeastern Arizona, and parts of Texas and Utah. Elevations range from 2,000 to 7,000 feet (610 – 2,135 m), and precipitation is low, about 5 to 12 inches (13 – 30 cm) a year. Highly nutritious galleta, black gramagrass, and tobosagrass dominate, with some low-value desert shrubs occurring in drainages or overgrazed areas. Exclusion of fire has promoted invasion by mesquite and other shrubs. Desert ranges are grazed year-round by sheep and cattle at very low stocking rates.

12:3.2 □ Shrublands

Nearly 295 million acres (119 million ha) of the western United States are covered by shrub rangelands, collectively producing about 26 percent of the forage grazed by livestock. Seven types are (1) sagebrush, (2) Texas savanna, (3) southwestern shrub-steppe, (4) desert shrub, (5) pinyon-juniper, (6) chaparral – mountain shrub, and (7) shinnery.

Sagebrush

By far the most extensive and important for range livestock production is the northern desert shrub, also termed *sagebrush*, occupying vast areas in the northwestern and intermountain states at elevations from 2,000 to 10,000 feet (610 – 3,048 m) (Figure 12.4). This type is second in area only to the Great Plains grass-

Fig. 12.4 — Northern desert shrubland is the largest type of shrubland (also known as the sagebrush type). Sagebrush and other woody plants make up the overstory beneath what can be a highly productive grass cover. (Courtesy, K. L. Johnson, Utah State University)

land (see Figure 12.2) in extent and represents nearly 45 percent of all shrublands. Most of this shrubland is found on plains and plateaus having only 5 to 12 inches (13 – 30 cm) of annual precipitation. A growing season of 3 to 4 months follows the early spring precipitation. Sagebrush and associated shrubs form the predominating aspect of the plant cover, but an understory of wheatgrasses, fescuegrasses, needlegrasses, and bluegrasses often provides strong feed for cattle and sheep in the spring and fall. Sagebrush and other relatively unpalatable shrubs have increased at the expense of palatable grasses on many of these ranges due to reduction in fire, improper grazing, and other factors. Cheatgrass (downy chess), a useful annual spring forage, along with a number of undesirable annual grasses and forbs, has invaded millions of acres of overgrazed range and abandoned croplands. At lower elevations, sagebrush plant communities serve as winter range for livestock and wildlife. At higher elevations, sagebrush ranges are used as summer ranges. In general, this type responds well to management practices aimed at reducing sagebrush and reestablishing more palatable forage species.

Texas Savanna

Texas savanna, the most productive per acre of the shrublands, is characterized by dense to open stands of mixed trees and shrubs, primarily mesquite, oaks, and acacia. It is associated with bluestem, the gramagrasses, buffalograss, and three-awn grass. It is a continuous type in extreme south-central Texas, receiving 20 to 30 inches (51 – 76 cm) of annual rainfall over a long growing season. Absence of fire and heavy grazing by cattle have caused large increases in woody species. Sheep, goats, and deer are managed over much of this shrubland type.

Southwestern Shrub-Steppe

Adjacent to the Mexican border in southwestern Texas, much of southern New Mexico and portions of Arizona are the southwestern shrub-steppe rangeland. This semidesert shrub-grass type is characterized by 10 to 18 inches (25 – 46 cm) of annual precipitation and a variable mix of yucca, mesquite, and creosote bush with black gramagrass, three-awn grass, and tobosagrass. This type is closely associated with the desert grassland, desert shrub, and the southern extent of the pinyon-juniper type but is more productive of forage year-round.

Desert Shrub

The southern desert shrub type is encountered primarily in southeastern California, southwestern Arizona, and southern Nevada. It also occurs in much of Utah and in portions of four neighboring states wherever annual precipitation is under 10 inches (25 cm) and high temperatures cause extremely high evaporation rates. Greasewood, creosote bush, saltbushes, and winterfat provide nearly all of the scant cover. The latter shrubs, along with sparse grasses such as Indian ricegrass and galleta and forbs, provide forage all year long. A flush of growth occurs with rainfall in winter and late summer, but the amounts of forage are undependable, causing uncertainty in livestock and range management practices for desert ranchers.

Pinyon-Juniper

The most complex in distribution of the shrub types is the pinyon-juniper (known also as **pigmy forest**), widely interspersed with or bordering all other shrublands and grasslands of the 11 Western Range States. Most of the "pigmy forests" occur in conjunction with sagebrush ranges, but on poorer soils. Forage yields are low, consisting mainly of scattered wheatgrasses, bluegrasses, gramagrasses, cheatgrass, and Indian ricegrass. Pinyon-juniper ranges have a history of heavy spring grazing by sheep and cattle and are considered the most depleted of all range types. Many species of juniper occur mostly on poor soils. Some species such as Western juniper will invade better sites when management does not include periodic fire.

Chaparral – Mountain Shrub

Chaparral – mountain shrub rangelands are similar to pinyon-juniper in distribution, occurring commonly between grasslands and forestlands. Largest communities are found in California, Arizona, and Colorado and exhibit a wide tolerance of climatic conditions. Common shrubs and small trees are oaks, manzanita, mahogany, and some juniper. Forage grasses present are common to the associated forestlands or grasslands. Management centers around prescribed burning to reduce old forage growth and to increase green grass and the sprouting of woody species for browsing by livestock and deer.

Shinnery

Shinnery shrublands are primarily confined to sandy areas in eastern New Mexico and the Texas Panhandle at 15 to 25 inches (38 – 64 cm) of annual precipitation. Vegetation is mostly open stands of Harvard oak and shinnery oak, sand sage, and mesquite with highly productive grasses of the prairie and plains. Where this type has not been cultivated, grazing with cattle is the main use.

12:3.3 □ Forested Grazing Lands

Nearly 600 million acres (243 million ha) of the 48 contiguous states are classified as forest, comprising 19 different ecosystems or types. Of this resource, 247 million acres (100 million ha) are grazed, providing about 10 percent of the total native forage consumed by livestock. Only four of the forested types justify description here due to their magnitude or productivity for range grazing: (1) oak-hickory, (2) longleaf-slash pine, (3) ponderosa pine, and (4) western hardwoods. Generally, deciduous, broadleaved forests, or pine types of lower elevations and latitudes have greater value than other forests for cattle and sheep, particularly in areas of low tree density. (See Chapter 19.)

Oak-Hickory

The oak-hickory type is the largest forest community and one of the three most extensive vegetation types in the United States. It extends from Canada to southern Texas and east to the North Atlantic Coast. Many of the watershed areas of the Mississippi, Ohio, lower Missouri, and central Arkansas river systems support the oak-hickory type at 30 to 50 inches (76 – 127 cm) of annual precipitation. Less than 40 percent of this forest is grazed. The bluestems, switchgrass, panicums, and a variety of palatable forbs and woody plants supply understory forage during the snow-free seasons. Beef and dairy cattle, recreation, timber, and coal industries are the primary users.

Longleaf-Slash Pine

Longleaf-slash pine forests characterize the low-elevation and nearly level southern Gulf Coastal Plains (Figure 12.5). Wiregrasses, bluestems, panicums,

Fig. 12.5 — Longleaf-slash pine forest is an excellent example of forested rangeland in the southeastern United States. By grazing livestock under the growing pine trees, multiple use is obtained together with greater economic benefits. (Louisiana) (Courtesy, USDA – Soil Conservation Service)

paspalums, and dropseed grasses provide an abundance of forage in open stands of pine. This type is a major cattle-producing area, used in conjunction with improved pastures and supplemental feeding. Controlled winter burning, tree thinning, fertilizing with nitrogen and phosphorus fertilizer, and seeding of grasses are practices used to increase forage production. (See Chapter 19.)

Ponderosa Pine

The ponderosa pine type is the largest western forest in the United States. Ponderosa and associated Jeffrey and sugar pines are widely adapted in areas of moderate elevations (3,000 to 6,000 feet) (915 – 1,830 m) and 15 to 30 inches (38 – 76 cm) of annual precipitation in parts of 14 western states. In grassy "parks" and open stands of pine, an excellent cover of perennial wheatgrasses, fescuegrasses, muhlys, bluegrasses, oatgrasses, and sedges provide prime summer range for livestock, while forages at lower elevations mature and become unpalatable. Palatable shrubs, especially bitterbrush, ceanothus, and snowberry, enhance the value of this plant type to big game and domestic livestock. Timber thinning and control of grazing intensity and distribution are key range management practices.

Western Hardwoods

Western hardwoods include two principal subtypes: one in California in which liveoak and blueoak occur and the other in the Rocky Mountains where quaking aspen is adjacent to conifer and chaparral types. The climate, topography, and associated species of each are similar to neighboring plant communities. California hardwood stands are high producers of winter – spring annual grassland forage species. Rocky Mountain aspen forests and the understory grass-forb cover are important summer forage for livestock and wild game.

12:4 ▫ MANAGEMENT OF GRAZING

Grazing by livestock is a primary use of grassland and shrubland ranges of the United States. Vegetation is of utmost concern to the range person as it provides forage and habitat for animals, soil stabilization, watershed protection, and recreational attributes. Different kinds of rangeland must be recognized and their present productivity assessed relative to the maximum which can be expected or achieved by improved management. The determination of proper stocking rate and other grazing practices is possible only with an understanding of plants and their growth requirements. Thus, the management essentials include the identification of plant species and knowledge of their **ecology** (the relationship between plants and their environment) and relative grazing values.

12:4.1 ▫ Range Plant Characteristics

Range forage plants are grouped as: (1) grasses, (2) grass-like plants, (3) forbs, and (4) browse. **Grasses** are a principal source of palatable and nutritious forage for cattle, sheep, elk, and horses; bluestem grasses, blue gramagrass, and wheatgrass are examples. **Grass-like plants,** represented by the sedges and rushes, resemble grasses. Sedges have solid triangular stems without joints, and rushes have solid round stems; both are somewhat less valuable for grazing than grasses. **Forbs** are broadleaved, nonwoody flowering plants, many of which are prime forage for sheep, goats, and big game. Included are palatable species such as filaree, wild dandelion and the clovers, and some poisonous and injurious plants. **Browse plants** are woody shrubs and trees which provide palatable leaves, flowers, fruit, and twigs, especially for sheep, goats, and a variety of wildlife. Oaks, willows, and wild currants represent this group.

Plants have different life spans, primary seasons of growth, and adaptability to different environments. Species that complete their life cycle from seed germination through flowering, maturation, seed bearing, and death in a year or less are called **annuals.** Those that live two years, producing seed and dying the second year, are known as **biennials.** When the life of a plant spans three or more years, made possible by dormancy during unfavorable seasons and reproduction by either or both seed and vegetative means, it is known as a **perennial.** Plants are also often classed as **cool-season** — those making primary growth in cooler seasons (usually winter and / or spring), or **warm-season** — those that make their growth in warm summer periods. Plants do not occur by chance. They live only where the environment satisfies their requirements for growth and reproduction. Therefore, certain species are adapted only to specific habitats, or ecosystems, called **range sites.**

12:4.2 □ Range Sites and Conditions

Different kinds of rangelands can be identified as communities or types because of differences in plant cover. However, since the actual and potential presence of certain plants and their productivity are determined by the total environment, consideration must be given to soil, climate, and topography. Thus, range sites, each a particular combination of rather uniform soil, topography, and climate, are recognized. Each range site has a potential productive capacity for forage, water, livestock, and wildlife. Some sites are naturally more productive than others, for example, a fertile, moist bottomland versus a barren, dry upland. Sites are named (mainly by the U.S. Department of Agriculture – Soil Conservation Service) according to easily recognized soil and topographic features, such as "loamy upland," "sandy plains," "wetland," and "alkali bottomland."

The potential site productivity, and the plant and soil conditions which characterize it, is called **range condition.** Condition is classed as excellent, good, fair, or poor. The evaluation of condition for productivity of livestock forage requires use of site standards. These standards list all plant species known to be adapted to the site and group them as **decreasers** (desirable plants on the decrease), **increasers** (less desirable plants on the increase), and **invaders** (alien, mostly unpalatable species). These groupings are usually based on plant palatability to cattle, sheep, and horses and on reaction to grazing pressure. **Plant composition** — the relative proportions of the different species in the cover — is determined by percentage on key representative areas within each site. These data are assessed, along with evaluation of plant vigor, litter (plant residue), and soil conditions — degree of exposed soil, erosion, compaction, and other factors — to derive **range condition class.** Note that all of the criteria evaluated reflect past grazing use. Ranges with greater proportions of desirable plants and better soil conditions are rated higher in condition; that is, they are more productive. Apparent change in condition known as **range trend** can be decided over a period of years as condition evaluations are repeated. Therefore, apparent trend is "up" if soil and vegetation conditions are improving, which gives an indication of increasing productivity. Resource management decisions, in part, can be made only after identification, description, mapping, and evaluation of all sites within a range unit have been compiled (Figure 12.6).

12:4.3 □ Proper Stocking Rate

The essence of range management for livestock and wildlife production is that **stocking** — the number of grazing and browsing animals per unit area —

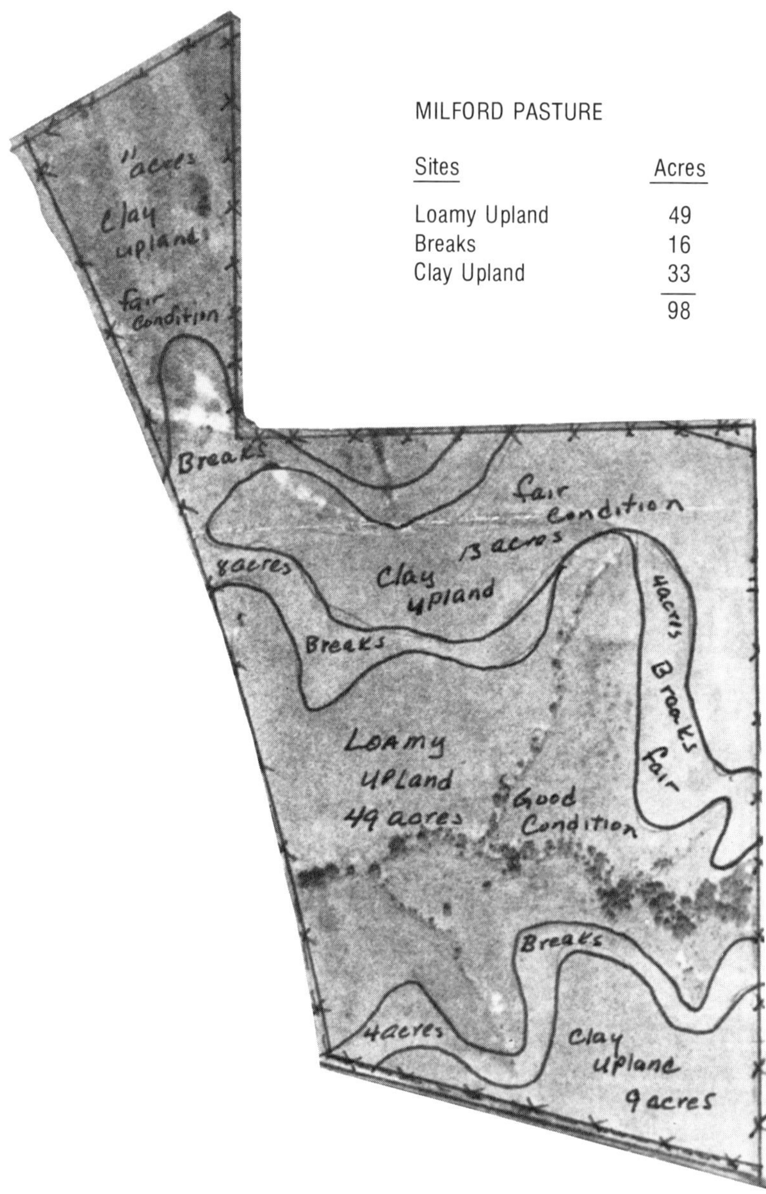

Fig. 12.6 — A range site and condition survey includes an aerial map such as this for use in making managerial decisions. The survey will also include a report that describes each site to enable the manager to understand the survey. (Kansas) (Courtesy, Kansas Cooperative Extension Service and USDA - Soil Conservation Service)

must be kept in balance with the available forage supply. This involves not only the proper number but also the kind of animals, season of use, and distribution of grazing together with flexibility to adjust to seasonal and annual changes. Improper stocking leads to vegetation and soil damage and consequent loss of productivity as reflected in **downtrend of range condition.** Underuse, while often justified for improvement purposes, results in wasted forage and often an uneconomic livestock operation. Prolonged nonuse may also result in degradation of the plant community. Numerous studies have demonstrated that the greatest yield of livestock products is normally obtained under a moderate stocking rate.

Forage supply is a function of three main factors: (1) basic site productivity (available soil moisture and fertility), (2) variation in seasonal precipitation and temperature, and (3) grazing history (past grazing management). Proper stocking varies, therefore, among sites and years and grazing systems on the same site. There is no absolute, singular **grazing capacity** figure for any specific range. Records of history of grazing use, range condition class, and trend serve as guides to estimating or adjusting stocking and grazing systems. Ranges on the downtrend or in poor condition must be grazed for a shorter period or with fewer head per unit area or in a different season to enhance livestock performance and to allow for range improvement (Figure 12.7).

Fig. 12.7 — Proper stocking must be varied, depending on climate and available forage. (Wyoming) (Courtesy, USDA - Soil Conservation Service)

Usable (grazable, available) forage is determined and expressed in pounds per acre (kg / ha), or in **animal unit months (AUMs).** An animal unit month is the amount of forage needed to support one animal unit (AU) for 30 days. One AUM is usually considered equivalent to about 800 pounds (363 kg) of air-dry forage. An AU is generally accepted as a mature cow of approximately 1,000 pounds (454 kg) liveweight or her equivalent. Equivalents are based upon animal species, stage of production, and dry matter requirements relative to body weight. For example, a calf being weaned equals 0.5 AU, a ewe 0.2 AU, and a deer, 0.15 AU.

Range inventories include the mapping of all areas suitable for grazing so that the grazable acreage in each site within a fenced unit can be determined. Stocking rates for each site are commonly expressed in acres per AUM (A / AUM) (ha / AUM) once an estimate and record of forage production has been made. Total AUMs for a site equals grazable acres (hectares) divided by A / AUM. Thus, by an accounting of the condition, estimated stocking rate, and grazable area within a fenced range unit, proper stocking can be established when the kind and class of livestock and season of use is known. This procedure is summarized in Table 12.2. It must be remembered that adjustments in animal numbers, grazing season, and length of grazing period are as necessary as the evaluations of site condition, trend, and grazing utilization.

12:4.4 □ Grazing Utilization

Good range management requires an understanding of the needs of plants, particularly under grazing or browsing stress. Knowing when and how much

Table 12.2 — Example Worksheet for Determination of Estimated Proper Stocking Rate

RESOURCE AREA Southeastern Plains YEAR 19--
RANGE UNIT / ALLOTMENT Indian Wells
CLASS OF STOCK Stocker Steers (400 – 600 lbs.) PERIOD OF USE 4 / 1 – 9 / 30

RANGE SITE	CONDITION CLASS	EQUIV. STOCKING RATE A / AUM	TOTAL ACRES	GRAZABLE ACRES	TOTAL AUMs
Loamy upland	Poor	2.5	520	520	208
Clayey bottom	Good	0.5	180	180	360
Shallow upland	Good	3.0	350	220	73
Gravelly hills	Fair	5.0	350	280	56
Nonrange	—	—	150	—	0
Totals	—	—	1,550	1,200	697

COMPUTATION OF ESTIMATED PROPER STOCKING RATE:

(1) $\frac{\text{Total AUMs of forage}}{\text{Total months of use}} = \frac{697}{6} = 116 \text{ AUs}$

(2) $\frac{\text{Total AUs of livestock}}{\text{Animal Unit equivalent}} = \frac{116}{0.5} = 232 \text{ HEAD}$

foliage can be grazed during a plant's life cycle is critical to productivity and persistence of the species. Plants are dependent upon the relative amount of stored foods — **photosynthetic rate** (food manufacture) — and metabolism of food for root and leaf growth and for reproduction. Plants of warm-season species should not be used heavily during reproduction (seed production). Food reserves for spring growth and stress periods are stored during this period. Plants of cool-season species build food reserves in the fall with initiation of vegetative growth and need careful management during hot summer weather to prevent overutilization.

Enough plant material must be allowed to remain on key forage species during and after the grazing season to maintain the plant, protect the soil against erosion, and provide a food resource for wildlife. Heavily grazed plants suffer reduced food reserves, growth, leaf area, and reproductive ability. As a result, key forage species cannot compete successfully for light, moisture, and nutrients with associated less palatable plants, and "weedy" species may "take over" the range in a few years if grazing pressure is not reduced (Figure 12.8).

Degree of grazing is expressed as a percentage or by the terms *ungrazed, light, moderate* (proper), *close,* and *severe* (overgrazed). Definitions of these may vary with differences among ecosystems and sites.

Fig. 12.8 — Overgrazing depletes better forage species, allowing less desirable species to dominate. This fenceline contrasts overgrazing on the left as compared to proper use on the right. (Texas) (Courtesy, USDA - Soil Conservation Service)

"Take half and leave half" is a good guide for determining the correct degree of grazing. However, the percentage of proper use by weight or height varies, depending upon the area, species, and time and frequency of use. Proper use of erect-growing perennials in cold or dry climates, or on steep sites, is much less than that of annuals and prostrate (sod-forming) perennials occupying favorable sites. Generally, proper or **moderate grazing** is achieved when:

1. Allowable use, usually between 10 and 55 percent by weight of the current year's growth, is reached on "decreaser" forage species.
2. Remaining plant stubble / litter / residue is "patchy" in appearance but sufficient to maintain soil stability and water retention of rainfall and snowfall.
3. Enough forage seed is left to meet reproduction needs of the site.
4. "Hard-to-get-at" areas around rock outcrops and under dense shrubs are not closely grazed.

Proper season of utilization is an important aspect of range management. The particular time during the year and the duration of the grazing period both relate to proper stocking rate and affect livestock performance. Grazing too early on wet, early spring range or on new seedings promotes injury to young plant growth and causes soil compaction. On seasonal ranges at higher elevations in the west, proper "turn-out" and "gather" dates are paramount to maintaining or improving range condition. Improvement of forage plant vigor, reproduction, food storage, and yield, as well as soil conditions, can be accomplished by deferment of grazing for a critical period of up to a year. A variety of specialized systems exist involving manipulation of time and intensity of use. Periodic "rest," "deferred," "rotation," and "alternate" grazing systems may be employed in varying combinations. However, any system must justify the additional investment of money for fencing, water development, and labor required for its implementation when compared to "continuous," or uninterrupted, year-long grazing.

Grazing utilization is also concerned with proper distribution of livestock. The diversity of plant cover types, slope, forage quality, and natural water locations, and the tendency of stock to be **gregarious** (bunch together), create problems in securing uniform use of all available forage on a range unit. Flats, gentle slopes, and open areas are often overgrazed, while steeper or "broken" country further away may not be grazed. Improved grazing distribution can be accomplished where feasible, by:

1. Salting and supplemental feeding in undergrazed but accessible areas away from water (Figure 12.9).
2. Developing water sources — springs, wells, catchments, or piped supplies — in areas of abundant forage.
3. Closing water sources when proper grazing has been obtained.
4. Cross-fencing to protect overused areas and poor condition sites and to hold stock where only light grazing has occurred.

Fig. 12.9 — Grazing distribution is necessary to ensure uniform use of the grazing unit to meet both plant and livestock management needs. Salt and mineral supplements can be moved to underutilized areas of the unit to attract livestock to those areas. (Nebraska) (Courtesy, USDA - Soil Conservation Service)

5. Herding and riding to control the grazing patterns of sheep and cattle.
6. Developing stock trails through natural barriers, giving access to under-used locales.
7. Changing the kind of grazing animal, for example, from cattle to sheep or both cattle and sheep.

12:5 ▫ RANGE IMPROVEMENT PRACTICES

Most rangelands are not producing forage, high-quality water, wildlife, and other products to the maximum of their capability. Ranges can be improved with the use of management techniques, such as prescribed burning, mechanical and herbicide treatments, reseeding, fertilizing, and other practices.

12:5.1 ▫ Prescribed Burning

Fire was and is a natural factor in the development of many grasslands. The use of fire by humans dates back to the early American Indian who used it to attract game animals and improve hunting. Scientists virtually have eliminated fire and allowed changes in vegetation types to occur. In many cases, eliminating fire has allowed woody species to increase and sometimes to dominate large areas that were formerly dominated by grasses.

Today **prescribed (controlled) burning** is again being utilized as a technique in range management. The principal objectives of prescribed fire are to:

1. Recycle nutrients in excessive litter.
2. Stimulate production of highly palatable and nutritious *new* forage growth.
3. Control unwanted woody and herbaceous plants (Figure 12.10).
4. Improve wildlife habitat and increase yield of water from the watershed.
5. Provide an ash seedbed to fertilize range seedings.
6. Reduce soil and plant diseases and insects.
7. Reduce the hazard of uncontrollable fires.

Fig. 12.10 — Prescribed burning of rangeland is a valuable practice to improve rangeland. Here, fire is used to control juniper trees which have invaded an otherwise productive grassland site. (Courtesy, T. E. Bidell, Oregon State University)

Hundreds of studies have thoroughly demonstrated the advantages of prescribed burning programs and resulting benefits to the range economy and the range resource in many areas. However, knowledge and care must be exercised to minimize the hazards of dealing with fire and to ensure that burning is the proper treatment for the particular site. The conduct of a controlled burn, behavior of the fire, and degree of success depends upon the season, the kind and amount of fuel (plant material), and weather (temperature, wind, humidity) conditions. To manage a successful burn, management needs to select a site carefully, obtain a burning permit and a designated burn day from local authorities, construct adequate firebreaks, employ correct ignition procedures, and conduct a thorough follow-up.

Prescribed fire is used extensively in chaparral (mountain brushland) and sagebrush ranges of the Great Basin, Rocky Mountains, and southwestern United States and grasslands of the central and southern Plains states. Increases in forage quality for cattle and sheep result, especially during the early spring. Reduction of dense brush and small trees "opens up the country"; increases opportunities for recreation; produces browse for big game and feed for game birds; and makes livestock management easier. Properly controlled burning is perhaps the least expensive means of improving rangeland (Figure 12.10).

12:5.2 □ Mechanical and Herbicide Treatments

Agricultural equipment has been widely used with modification on ranges. For example, bulldozers with blades are employed to construct catchments for livestock water and for contour diversions to spread and thereby detain rainwater. In addition, a variety of heavy implements have been developed or modified to mow, chop, plow, crush, uproot, and pile brush and small trees in preparation for burning and grass seeding programs (Figures 12.11 and 12.12). Range "pitting" which makes shallow depressions in compact, arid soils to retard runoff is accomplished with an eccentric disc. Each method has been developed for a particular soil, plant, environmental condition, range site, and economic situation. This practice is used on ranges with summer precipitation and short grass to retain precipitation, thus increasing available moisture for range forage growth.

Herbicides are used extensively to reduce species which are wasteful of water and are injurious, unpalatable, or poisonous. Undesirable plants compete with desirable forage plants for light, moisture, and nutrients. Therefore, the reduction of undesirable plants increases forage yield and permits a greater stocking rate. Many safe chemicals are now available for application in the form of sprays or granules which selectively kill broadleaved plants but do not harm grasses in the plant community. Sprays at varying dilutions are applied directly to the foliage during the active growth period; dry granules are spread on the soil during the dormant season where they are dissolved and absorbed by roots of the "target" species. Both enter the plant through the roots during the active growth period and kill the unwanted plants by entering the metabolic system. Thousands of acres (hectares) of sagebrush-, chaparral-, and mesquite-infested rangelands are treated by aircraft or ground equipment each year under prescription with a permit from appropriate federal or state authorities. Herbicides may also be applied to localized patches of unwanted plants with hand- or truck-mounted equipment. Rather precise time and rate of application is needed in any program in order to achieve the desired results. Proper selection and use of herbicides always requires care in following directions given on the label, as required by the U.S. Environmental Protection Agency.

Fig. 12.11 — Chaining juniper with an anchor chain between two tractors pulls the unwanted trees out of the ground. With seeding and proper management, the desirable grasses will replace the trees. (Texas) (Courtesy, USDA - Soil Conservation Service)

Fig. 12.12 — Discing with heavy discs is a common method of reducing sagebrush and preparing a seedbed for desirable grasses. (Utah) (Courtesy, USDA - Soil Conservation Service)

12:5.3 ▫ Reseeding

Unfortunately, significant portions of our ranges have been improperly grazed, unwisely cultivated, damaged by wildfire, or otherwise disturbed. Such occurrences often result in the need for artificial revegetation. On sagebrush, chaparral, and mesquite ranges, "brush-to-grass" restoration programs often require extensive seeding. Spot-seeding is also used for erosion control on disturbed sites and to enhance wildlife habitat and feed in critical areas. Most seeding is limited to rangelands receiving more than 8 inches (20 cm) of precipitation annually. In the Great Plains, marginal or abandoned cropland can be reclaimed through seeding adapted grasses.

As a result of extensive research, testing, and experience, numerous species and varieties of range grasses, legumes, other forbs, and browse have been identified and developed for use in seeding to improve forage quality and production. Five basic considerations are important to a range reseeding plan. Each of these involves decisions critical to successful establishment of the new stand, as follows:

- ***Site characteristics*** — A thorough familiarity with precipitation, temperature, soils, topography, and potential and resident plants of the site is necessary in order to make correct decisions concerning the reseeding procedures which follow.
- ***Adapted species and time of seeding*** — The selection of one species, or a mixture of species most suitable to the site, and the decision concerning when to seed are made with the advice of specialists and through experience. The local extension service, the agricultural experiment station of the state land-grant university, the Soil Conservation Service, and neighboring ranchers should be consulted to obtain local information. The correct seeding time is determined by soil and weather conditions specific to each site.
- ***Seeding method and rate*** — Generally, seed is applied by either "drilling" or broadcasting. Drilling involves tractor-drawn implements of various designs that place the seed about ¼ to 1½ inches (0.6 – 4 cm) deep and then cover it. On suitable sites, this method is preferred to broadcasting in which seed is scattered over the soil surface. Unless intentionally covered by some means, broadcast seed depends on the nature of the ground surface as a habitat for germination and protection from predators such as birds and mice. Rates of seeding vary from 1 to 20 pounds per acre (1.1 – 22.4 kg / ha), depending upon site quality, species, seed size, and seeding method.
- ***Seedbed preparation*** — Good seedbed preparation is perhaps most important to success, since preparation of a favorable environment for germination and seedling growth is paramount to plant establishment. Good

seedbed preparations control resident plant competition and provide a uniform, firm soil surface, preferably with a standing plant cover to maintain moisture. Stubble or fallowed soil of old cropland, new ash following a burn, and disturbed surfaces resulting from mechanical treatments of brush stands are all suitable seedbeds. In some situations, preparations may consist only of herbicide application. Seeding, in this case, will usually be done with a rangeland drill.

- ***Follow-up management*** — Once established, newly seeded stands must be protected from grazing for from one to three growing seasons and then followed by planned grazing to ensure root, seed, and food reserve development. Consideration should be given to periodic herbicide use, mowing, burning, and / or fertilizer treatments to suppress weed growth and stimulate increased forage productivity (Figure 12.13).

Fig. 12.13 — Careful planning of a seeding program will result in a higher productivity as seen in this contrast. The area on the right was rootplowed and seeded to grass with proper follow-up grazing management. (Texas) (Courtesy, USDA - Soil Conservation Service)

12:5.4 □ Fertilizing

The practice of range fertilization offers livestock operators in many grassland types an opportunity to increase profits through the beneficial effects fertilizer can have on forage. A number of advantages have been identified since intensive studies of range fertilization began in the late 1940's. With a properly planned fertilizer program, benefits include:

1. Increased total forage yield and stocking rate.

2. Increased palatability of forage leading to more complete and uniform grazing utilization, especially when applied on underutilized sites.
3. Earlier growth in the late winter and spring, resulting in reduced supplemental feeding and earlier gains by livestock.
4. Increased levels of plant protein, in most cases, when nitrogen is applied.
5. More efficient use of available soil moisture.
6. Improved performance of cattle and sheep resulting in greater seasonal gains per head and production per acre (hectare).
7. Increased percentage of phosphorus in forage when phosphorus fertilizer is applied.

As a general practice, range fertilization is limited to grassland sites of desirable plant composition and soils in areas of from 10 to 30 inches (25 – 76 cm) of annual precipitation. Fertilizers containing nitrogen, phosphorus, and sulfur have proven most profitable following a single application at the start of the growing season. Decisions on which elements or a combination of elements are needed, the rate of their application, and in what form they should be applied are critical. These can be made based on preliminary analyses of field plot trials, soil and plant tissue tests, and cost-benefit projections.

12:5.5 □ Other Improvement Practices

Fencing, water development, and the use of livestock handling facilities and erosion control structures are other practices used throughout the range areas. In the application of these practices, economics, land ownership, and state laws may govern when and how they may be applied. Each practice is applied only after a determination is made as to how it will increase the efficiency of the existing management of the range unit.

12:6 □ RANGE MANAGEMENT PLANNING

Successful management of rangeland requires a knowledge of all physical, biological, and cultural features of the ranch. A survey-inventory of all rangeland resources is therefore essential for sound management decisions and planning on private or public lands.

12:6.1 □ Resources Inventory

The **range resources survey-inventory** may be carried out on private lands by personnel of the Soil Conservation Service, land-grant university specialists, or private consultants. On public lands, trained range technicians work full-time in survey work and preparation of multiple-use management plans (Figure 12.14).

Fig. 12.14 — Planning the management of rangeland requires a thorough understanding of the resources gained by survey and inventory. Private land managers receive help from the Soil Conservation Service and similar personnel in formulating the management plans. (South Dakota) (Courtesy, USDA – Soil Conservation Service)

Basic techniques of the survey-inventory are:

1. Familiarization with the ecology and history of use on the area.
2. The ability to use an aerial base soil map which shows soil map units, roads, buildings, streams, lakes, and other physical features. On a clear, plastic overlay, land ownership, fences, constructed water facilities, range sites, salting areas, and other information necessary for making up a range management plan should be drawn in.
3. Evaluation of range vegetation, soil conditions, and animal utilization of all grazable sites.
4. Collection and recording of field data and observations concerning existing wildlife, watershed, recreational, timber and mineral resources, and potential uses.
5. Identification of existing and potential problems such as erosion, poisonous plants, overuse of critical feeding areas, and structural repairs needed.
6. Determination of acreages for all sites and fenced units, the estimation of forage yield and proper stocking rates for each kind of livestock, and consideration of wildlife needs.
7. Development of recommendations for problem solving and measures for resource improvement and management indicated on a map overlay and in a written plan.

After completion of the resources-inventory and study of compiled information, a management program can be drawn up.

12:6.2 ▫ Planning for Resource Conservation and Multiple-Use Management

Consideration must be given to all resources and priorities established by the land owner or manager. Production records, estimates of potential yield, and utilization of the various resources should be analyzed. Limitations on use and alternative management procedures should be recognized, including any legal restrictions and environmental protection standards which may apply. The economics of each of the proposed alternatives and improvements should be evaluated as to initial cost and expected returns over time. Availability of agricultural loans and programs for cost-sharing of proposed improvements should be investigated. On government-owned lands, grazing **permittees** (those having a permit) often share investments with the particular agency in charge. On private lands, assistance may be provided through the U.S. Department of Agriculture – Agricultural Stabilization and Conservation Service (for cost-sharing) and Soil Conservation Service (for technical assistance).

In a typical **range management plan,** proposed practices may include:

1. Establishment of stocking rate and periods of grazing to maintain about half of the forage at the end of the growing season (graze half and leave half).
2. Construction of new fences where needed.
3. Development of new salting locations and watering facilities to improve grazing distribution.
4. Implementation of a grazing system to provide periods of deferment to maintain or improve range condition class and grazing distribution.
5. Application of fertilizer (based on a soil test) to selected sites to provide additional early forage.
6. Utilization of hay stubble, crop residues, and surplus fall range forage in good years to "carry over" additional replacement heifers or ewes.
7. Seeding of a deteriorated range site on productive soils (based on a soil survey and a soil test) with improved forage species to increase production and reduce erosion.
8. Spraying of undesirable trees, shrubs, and forbs to enhance growth of desirable forage species.
9. Development of additional water sources and protected areas to improve livestock use and for improvement of wildlife habitat.
10. Installation of **diversion dams** (Figure 12.15), gully control, and spot-seeding to reduce soil erosion on disturbed, unstable areas.
11. Development of a combined fishing and overnight camping area with deferred grazing during the peak recreation use period.

Fig. 12.15 — A diversion dam has been constructed in the semiarid region of Montana to intercept runoff water and spread it over 60 acres (24 ha) of native grassland. (Courtesy, USDA - Soil Conservation Service)

12. Prescribed burning of brush-infested grassland and slash removal following thinning and pruning for timber stand improvement to increase forage growth for livestock and big game.

All practices must be integrated into a well-planned program to be successful in getting protection and efficient use of all resources (Figure 12.16). A multiple-use plan requires flexibility so that resources are managed in a coordinated, harmonious manner. Any good plan establishes short- and long-term goals, shows how to reach them, and should be frequently reviewed and updated as management needs and ecological conditions change.

12:6.3 ▫ Implementing the Plan

Once a plan has been developed, the land owner or manager must then incorporate the practices into the present management program. Each practice is applied as the labor, time, and economic resources allow. As each new practice or management change is made, the results are carefully evaluated. If problems are encountered, adjustments are made. Implementing the range management plan is the most difficult part of the process.

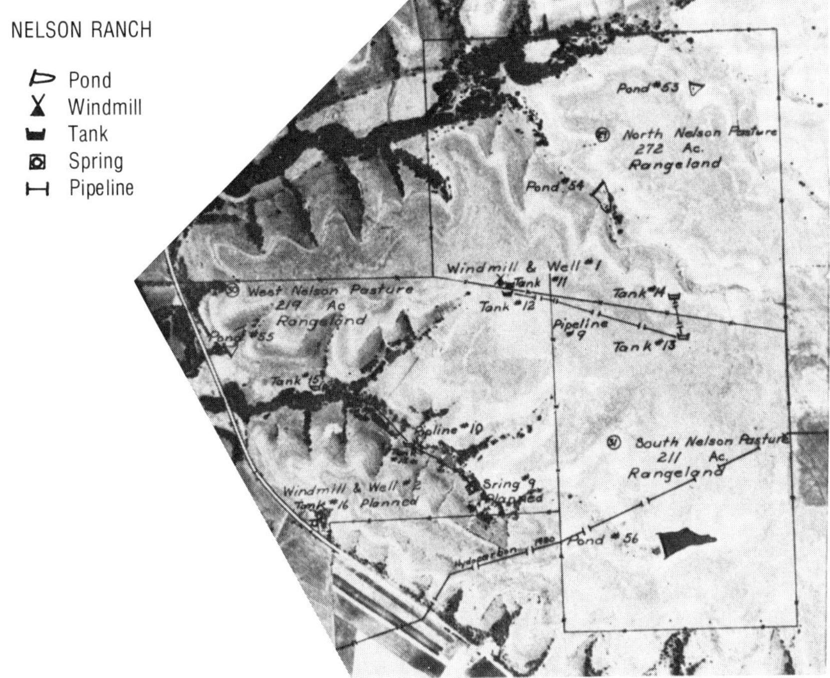

Fig. 12.16 — A good base map showing the physical features of the ranch, plus a range site and condition survey, is essential to planning and implementing the operations of a ranch. (Kansas) (Courtesy, Kansas State University and USDA - Soil Conservation Service)

12:7 ◻ REFERENCES

"An Assessment of the Forest and Rangeland Situation in the United States." U.S. Dept. of Agriculture – Forest Service, R.P.A. FS-345, 1980.

Bedunah, D. J., and E. Sosebee. "Influence of Mesquite Control on Soil Erosion on a Depleted Range Site." *Journal of Soil and Water Conservation,* Vol. 41, 1986, pp. 131 – 134.

Berg, W. A., S. J. Smith, and G. A. Coleman. "Management Effects on Runoff, Soil, and Nutrient Losses from Highly Erodible Soils in the Southern Plains." *Journal of Soil and Water Conservation,* Vol. 43, No. 5, 1988.

Grasslands of the United States. Ames: The Iowa State University Press, 1974, 201 pp.

Journal of Range Management, Vol. 1. Denver: Society for Range Management, 1948 to present (bimonthly).

"National Range Conference Proceedings." Oklahoma City, November 6 – 8, 1985, 159 pp.

Platou, K. A., *et al.* "Soil Properties Associated with Six Common Grasses of the Great Basin." *Journal of Soil and Water Conservation,* Vol. 41, 1986, pp. 417 – 421.

Rangelands, Vol. 1. Denver: Society for Range Management, 1979 to present (bimonthly).

Robberecht, Ronald, *et al.* "An Information Guide to Range Science." Moscow, Idaho: University of Idaho, 1986, 101 pp.

Rommann, Loren, *et al.* "Pasture and Range Judging in Oklahoma." Oklahoma Co-op. Ext. Ser., 1979, 32 pp.

Stechman, John V. *Common Western Range Plants,* 2nd ed. San Luis Obispo, California: Vocational Education Productions, 1977, 164 pp.

Vallentine, John F. *Range Development and Improvement.* Provo, Utah: Brigham Young University Press, 1980, 516 pp.

Workman, J. P. *Range Economics.* New York: Macmillan Publishing Co., Inc., 1986, 217 pp.

In addition to the preceding references, career opportunities, additional information, and films can be obtained through the following sources:

Extension Range Management Specialist
School of Renewable Natural Resources
University of Arizona
Tucson, AZ 85721

Range Specialist
Extension Agronomy and Range Science
University of California
Davis, CA 95616

Society for Range Management
1839 York Street
Denver, CO 80221

Extension Range Specialist
Range Science Department
240C Natural Resources
Colorado State University
Fort Collins, CO 80523

Range Specialist
School of Forestry Resources and Conservation
University of Florida
Gainesville, FL 32611

Extension Specialist
Livestock and Range Management
Kamuela Extension Office
P.O. Box 237
Kamuela, HI 96743

Range Utilization Specialist
1330 Filer Avenue East
Twin Falls, ID 83301

Extension Specialist
Range and Pasture Management
Kansas State University
Manhattan, KS 66506

Extension Range and Pasture Specialist
North Platte Station
University of Nebraska
North Platte, NE 69101

Extension Range Specialist
Renewable Resources Center
University of Nevada
1000 Valley Road
Reno, NV 89512

Extension Range Specialist
Box 34E
New Mexico State University
Las Cruces, NM 88003

Grassland Management Specialist
State University Station
c/o Hultz Hall
Fargo, ND 58105

Extension Range Specialist
Department of Agronomy
Oklahoma State University
367 Agricultural Hall
Stillwater, OK 74078

Extension Range Specialist
Rangeland Resources Department
Oregon State University
Corvallis, OR 97331

Range Management Specialist
SDSU Research and Extension Center
801 San Francisco Street
Rapid City, SD 57701

Extension Range Specialist
Department of Range Science
Texas A&M University
Room 225, AI Bldg.
College Station, TX 77843

Extension Range Specialist
Utah State University
UMC 52
Logan, UT 84322

Extension Range Specialist
Department of Forestry and Range Management
Washington State University
Pullman, WA 99163

Extension Range Specialist
Division of Range Science
University of Wyoming
University Station, Box 3354
Laramie, WY 82071

Program Leader
Range Management
Extension Natural Resources
5925 South Agricultural Bldg.
Washington, DC 20250

CHAPTER 13

Pastures: Soil, Water, and Fertility Management

"The history of man is in the story of a hungry animal in search of food. History is also the story of man's constant search for grass, the basic food source in the universe." — Hendrick Van Loon (1882 - 1945)

OUTLINE

Pastures provide some of the best and cheapest feed for livestock. *Top:* Holstein cattle on a Kentucky bluegrass - white clover pasture in Wyoming. *Center:* Jersey cows on bermudagrass in Florida. *Bottom:* Hereford cattle grazing tall fescuegrass in Kentucky. (*Top* and *bottom:* Courtesy, USDA - Soil Conservation Service. *Center:* Courtesy, Florida Department of Agriculture)

□ □ □

13:1 □ OVERVIEW

About half of the farmland in the United States is used for pasture. This pastured area includes open woodlands as well as croplands used for pasture. Such grazing provides about one-third of all nutrients consumed by livestock.

With proper soil, water, and fertility management, livestock nutrients from pastures can be *the most economical*, partly because the domestic animals do the harvesting.

The primary purpose of producing forage crops is to provide an economical feed which can be converted by cattle, sheep, and goats into protein-rich products for humans. With proper management, high yields of very nutritious forage are easy to produce. The best grass, legume, or grass-legume mix to grow depends on the region, the soil, moisture, crop adaptation, and intended use. No single species is best in all environments. The species chosen should have proven value for a particular local environment.

Grasses are classified as cool-season and warm-season as well as annual (life cycle completed in one year) and perennial. Legumes provide a higher quality of forage for livestock than grasses provide.

Grass tetany, nitrate toxicity, and hydrocyanic acid poisoning are all negative forage quality hazards to avoid.

Some grasses and legumes are low-growing and are adapted to continuous grazing with a low-stocking rate. Other grasses and legumes are tall and are best managed by rotational and deferred grazing. Regardless of height, it is always wise management to graze half and leave half of the forage for reserve and residue.

Pasture renovation by sod-seeding is a new, recommended technique of pasture management.

There have been many dollars wasted on buying good seeds for establishing pastures, only to have them fail because of infertile soils. For years, some farmers have been looking for superior "poor land" pasture plants, and later discovered that such plants make only inferior forage.

Assuming that everyone is interested in producing the greatest amount of the highest quality of forage, here are six steps necessary to attain this goal:

1. Select the best soil available.
2. Prepare the land well. (See Chapter 6.)
3. Use lime and fertilizers called for by a soil test and then apply them with timeliness and precision. (See Chapters 3 and 4.)
4. Select the recommended seeding mixture and then plant at the proper time on a firm seedbed.

5. Use herbicides and insecticides to control troublesome weeds, brush, and insects.
6. Manage wisely by mowing and by supplying adequate lime and fertilizers, regulating the grazing, and renovating when yields decline (Figure 13.1).

13:2 ◻ IMPORTANT FORAGE GRASSES

It is difficult to attempt to place a monetary value on forage grasses because they are marketed through livestock; therefore, they appear indirectly in cash re-

Fig. 13.1 — Even barren hillsides can be made into good pastures by proper soil, water, and fertility management practices. The same field before (*top*) and after (*bottom*) establishing a grass-legume pasture. (Courtesy, Tennessee Valley Authority)

Fig. 13.2 — Overgrazing on the left has resulted in severe erosion; right — properly grazed. (Pendleton County, Kentucky) (Courtesy, USDA - Soil Conservation Service)

ceipts as a return from the animal products marketed. In addition to serving as an excellent feed for livestock, the forage grasses stabilize the soil and maintain soil fertility. Grass, when properly used, counters the devastating influence of erosion (Figure 13.2).

There are over 5,000 grass species in the world, and approximately 1,500 of these species are found growing in the United States. Of this number, there are less than 100 which are of economic importance.

Most of the grasses grown under humid conditions in the United States have been introduced into the country, whereas those adapted to the arid regions are predominately native species. There are two major groups of perennial grasses: the cool-season grasses and the warm-season grasses. The cool-season grasses, such as Kentucky bluegrass, bromegrass, orchardgrass, reed canarygrass, tall fescuegrass, and wheatgrasses, make most of their growth during the cool days of spring and fall. The warm-season grasses, such as bermudagrass, bluestems, johnsongrass, and dallisgrass, start their growth in the late spring or early summer and make their main vegetative growth in warm mid-summer. The most important warm-season *annual* grasses used for forage include sudangrass, pearl-millet, grain sorghums, sorghum-sudangrass crosses, and sweet sorghums (sorgos). The cool-season annual grasses commonly used for grazing include rye, wheat, barley, and oats. A list of the more important perennial forage grasses for the United States is given in Table 13.1. Figure 13.3 indicates the regions of the United States in which various grasses are well-adapted and of primary importance.

Table 13.1 — Important Forage Grasses, Their Season Classification, Primary Use, Soil Adaptation, and Areas Where They Are Best Adapted[1]

Name	Season Classification	Primary Use	Soil Adaptation	Where Best Adapted
Bahiagrass	Warm	Used as a pasture grass on coastal plains from North Carolina to Texas.	Does best on sandy soils of the coastal area.	Florida and lower coastal plains of southeastern U.S.
Bermudagrass	Warm	Used in southern states for hay, pasture, lawns, silage, general purpose turf, and erosion control.	Grows best on well-drained fertile soils; responds exceptionally well to nitrogen.	Southern United States — Maryland through Texas and New Mexico where rain or irrigation is adequate.
Blue and sideoats gramagrasses	Warm	Grazing and erosion control in Great Plains; form dense sod and are constituent of many range pastures.	Well-adapted to clayey, rolling, upland soils; very drought resistant.	Through Great Plains (blue grama more drought resistant).
Bluestems	Warm	Leafy forage palatable to all kinds of livestock in Great Plains; used for both hay and pasture.	Well-drained loam soils.	Eastern and central U.S. extending into drier parts of the Great Plains.
Dallisgrass	Warm	Valuable pasture grass of the South.	Grows best on bottom lands with high fertility and ample moisture.	Carolinas to Texas; irrigated regions of southwestern U.S.
Italian (annual) ryegrass	Cool	Winter and early-spring pastures in the South.	Sandy loams to clays of medium to good fertility.	Grown extensively in southeastern U.S. and in the humid Pacific Coast area.
Johnsongrass	Warm	Used for hay and/or pasture in some areas of the lower South.	Best adapted to fine clay soils of high fertility and water-holding capacity of the lower South.	The entire Cotton Belt and southern areas of the Corn Belt.
Kentucky bluegrass	Cool	Important pasture grass both alone and in mixtures; an important lawn grass in northern and central U.S.	Sandy loams to clays of high productivity.	Northern two-thirds of U.S. where moisture is plentiful.

(Continued)

Table 13.1 (Continued)

Name	Season Classification	Primary Use	Soil Adaptation	Where Best Adapted
Orchardgrass	Cool	Used alone and in combination with a legume for pasture and hay.	Most soils except sands; not adapted to poorly drained soils.	North central and northeastern U.S., Northern Pacific Coast, and Great Plains where moisture is adequate.
Perennial ryegrass	Cool	Primarily for lawns; sometimes used for early grazing.	Sandy loams to clays of medium to high fertility.	Grown extensively in southeastern U.S. and in the upper Pacific Coast area.
Redtop	Cool	Hay crop on wetlands; used in pasture mixtures.	Grows best on all soils including those that are wet; does well on acid soils.	All of United States except dry areas.
Reed canarygrass	Cool	Used for grassed waterways; pasture and hay on upland soils. Contains tryptamine, an alkaloid toxic to cattle.	Thrives on fertile, moist, even swampy soils; but makes good growth on fertile, well-drained soils.	North central and northeastern U.S., humid Pacific Coast, and Great Plains where moisture is adequate.
Smooth bromegrass	Cool	Used alone and in mixture with legumes for pasture and hay.	Adapted to a wide variety of soils; resistant to drought and extremes in temperatures.	Adapted to the Corn Belt and somewhat further north.
Tall fescuegrass	Cool	Provides winter pastures in the South; useful as a high-yielding pasture alone or in combination with a legume. Sometimes toxic due to fungal endophyte.	Adapted to a wide range of soils although it does poorly on coarse, sandy soils.	Central and southeastern U.S. and humid Pacific Coast.
Timothy	Cool	Provides very palatable early pasture, hay, and silage; will not withstand close grazing.	Does poorly on sandy soils; thrives on clay and clay loams.	Cool, humid climates of the Northeast; does not thrive under hot, dry conditions.
Wheatgrasses	Cool	Early-season forage in the West; used for wind and water erosion control.	Most soils except sands.	Central and northern Great Plains, intermountain regions, and higher altitudes of the mountains.

[1] Sources: *Forage Handbook*, Amoco Oil Company, and A. A. Hanson, "Grass Varieties in the United States," *Agr. Handbook No. 170*, 1972, 124 pp.

Fig. 13.3 — Regions of the United States in which various perennial grasses are well-adapted and are of primary importance. (Courtesy, Amoco Oil Company)

13:3 □ IMPORTANT FORAGE LEGUMES

Legumes are important in providing forage for livestock since they increase the supply of high-protein feed. Legumes indirectly increase the production of other farm crops because of their nitrogen-fixing ability and resulting enrichment of the soil. In general, legumes are superior in feeding value to nonlegumes because of their high level of good-quality protein. In addition, they are relatively high in phosphorus and calcium and are excellent sources of vitamins A and D.

Legumes have the ability to probe deeply into the lower soil horizons for moisture and nutrients. They are generally vigorous-growing tap-rooted plants. The roots of legumes bear enlargements called "nodules" which are caused specifically by the activity of *Rhizobium* species of bacteria. Each nodule contains hundreds of bacteria that fix atmospheric nitrogen for use by the legume and companion plants.

A two-year cattle grazing experiment in Minnesota with four legumes concluded that:

- □ ***Alfalfa*** (bloating, generation of excessive gas in the digestive tract of cattle) — Average daily gain was 0.60 pound per acre (0.67 kg/ha).
- □ ***Birdsfoot trefoil*** (nonbloating) — Average daily gain was 0.72 pound per acre (0.81 kg/ha).
- □ ***Sainfoin*** (nonbloating) — Average daily gain was 0.71 pound per acre (0.80 kg/ha).
- □ ***Cicer milkvetch*** (nonbloating) — Average daily gain was 0.37 pound per acre (0.42 kg/ha). Furthermore, this species made the heifers sensitive to sunburn under white hairs on the body.

A list of the more important forage legumes is given in Table 13.2. Figure

Table 13.2 — Important Forage Legumes, Their Primary Use, Soil Adaptation, and Areas Where They Are Best Adapted[1]

Name	Primary Use	Soil Adaptation	Where Best Adapted
Alfalfa	Hay, silage, and pasture; performs well in pasture mixtures with orchardgrass, bromegrass, or tall fescuegrass.	Fertile, nonacid soil with good surface and internal drainage.	All regions where moisture is sufficient.
Alsike clover	Especially suited for acid wetland.	Most soils except sands; will tolerate poor drainage and acid soils.	North of Ohio and Potomac rivers, westward to Dakotas; northwest U.S.
Birdsfoot trefoil	Valuable permanent pasture legume; productive, palatable, and continues in good balance with Kentucky bluegrass; has no bloat hazard.	Does well on soils of low fertility.	Eastern Kansas and Nebraska to New York and the Atlantic Coast; south to Ohio River; Oregon and California.
Crimson clover	Main use as a winter annual legume for pasture; often seeded into established stands of bermudagrass, dallisgrass, bahiagrass, or johnsongrass.	Thrives on soils of medium acidity and on both sandy and clay soils.	Southeastern U.S. and humid or irrigated West Coast.
Crownvetch	Primarily used for erosion control; does fairly well in pasture mixture with grasses such as orchardgrass.	Tolerant of soil pHs of 8.0 to 5.5 and low fertility; does best on well-drained soil; not affected by cotton root-rot.	Northern two-thirds of U.S.
Cicer milkvetch	A nonbloating pasture legume; reproduces by underground stems.	More tolerant of soil acidity and alkalinity than alfalfa.	Great Plains and western states. Resistant to drought, frost, insects, and diseases.

(Continued)

Table 13.2 (Continued)

Name	Primary Use	Soil Adaptation	Where Best Adapted
Hairy vetch	Soil improvement; early-season pasture and hay (if cut early before it becomes stemmy).	Adapted to both sandy and clay soils.	Can be grown in most of United States but used primarily in southern states for soil improvement during winter.
Red clover	Used for hay, pasture, and soil improvement; crop fits well into rotations.	Practically any well-drained and non-acid soil.	Eastern Dakotas, Nebraska, Kansas, Oklahoma, and Texas to Atlantic Coast; humid northwestern U.S.
White and ladino clovers	Provide good early spring and fall grazing; used with bermudagrass and other grasses; very high in palatability as pasture.	Practically any soil; prefer fertile, well-drained, moist soils.	Where moisture is adequate; ladino clover may winterkill in northern states. Selected white clover strains are very cold-tolerant.
Annual lespedeza	A reseeding annual which provides mid- to late-summer grazing in combination with bluegrass.	Practically any well-drained soil.	Iowa southward to the Gulf and eastward to the Atlantic.
Sericea lespedeza	Hay and pastures; new varieties such as "serala" show excellent promise; a perennial.	Does well on both droughty and acid soils.	Southern U.S.
Sainfoin	A nonbloating, deep-rooted, forage legume for hay or pasture.	Calcareous soils, dryland or irrigated.	Northern Rocky Mountain states. Subject to several insects and diseases.

[1] Primary Source: *Forage Handbook*. Amoco Oil Company.

Fig. 13.4 — Regions of the United States in which various forage legumes are well-adapted and are of primary importance. (Courtesy, Amoco Oil Company)

Fig. 13.5 — Red clover is a desirable legume in the Midwest and the Northwest for pasture and for making hay. (Courtesy, USDA - Agricultural Stabilization and Conservation Service)

13.4 indicates the regions of the United States in which various forage legumes are well-adapted and of primary importance.

13:4 □ PREPARING THE SOIL FOR SEEDING

A firm, weed-free seedbed is of primary importance in the successful establishment of a pasture. The soil should have just enough loose surface soil for uniform, shallow seed coverage. Any tillage method that accomplishes this is satisfactory. Tillage also provides a means of incorporating slow-moving materials such as lime and phosphorus into the root zone.

A disc or spring-tooth harrow can be used and may often be preferred to plowing because of the tendency to leave a protective surface mulch. The surface mulch protects the young germinating plants from desiccation (drying).

After discing or plowing, firming the seedbed with a cultipacker is sometimes necessary. Cultipackers firm the soil more uniformly and leave desirable small ridges which aid in increasing water absorption and seed germination.

On soils of low fertility, the best stand insurance is to use band seeding with a press (packer) wheel (Figure 13.6). The seed should be placed no more than ¼ to ½ inch (0.6 – 1.3 cm) in depth. Band seeding places a band of fertilizer below the seed with 1 to 2 inches (2.5 – 5.1 cm) of soil separating the fertilizer and the seed. It places the seed evenly at a uniform depth and in firm contact with moist soil.

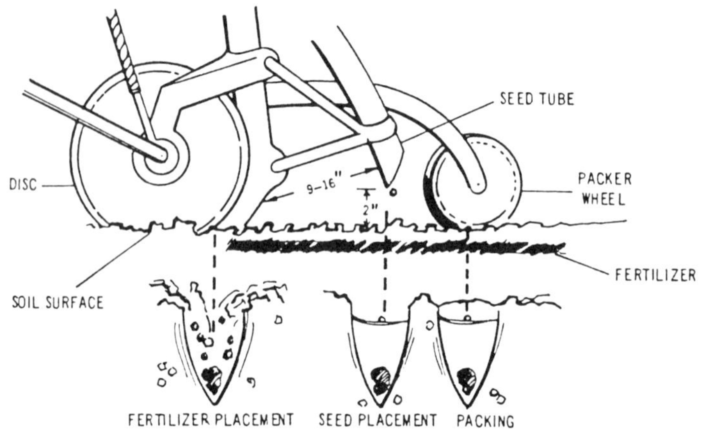

Band Seeding — Press Wheel

Fig. 13.6 — Band seeding with press wheel provides additional stand insurance, especially for August seeding. (Courtesy, The Ohio State University)

Grain drills equipped with forage seed boxes can be converted for band seeding by attaching flexible tubing from the seed box to a point 12 inches (30 cm) behind the openers that place the fertilizer. This distance is needed to allow soil to flow back into the furrow, with soil separating fertilizer from seed (Figure 13.7).

The selection of the proper pasture grass or seeding mixture will depend upon the kind of livestock, the climate, and the soil. Dairy cattle respond best to highly productive and intensively managed legume-grass mixtures. Beef cattle, sheep, and goats are able to utilize the less productive and more extensive grass pastures. When offered the best pastures, however, all classes of livestock do better. It is best to contact the county agent, agricultural experiment station, agricultural education teacher, district conservationist, or local seed dealer for the latest recommendations of seeding mixtures and seeding rates for the local environment.

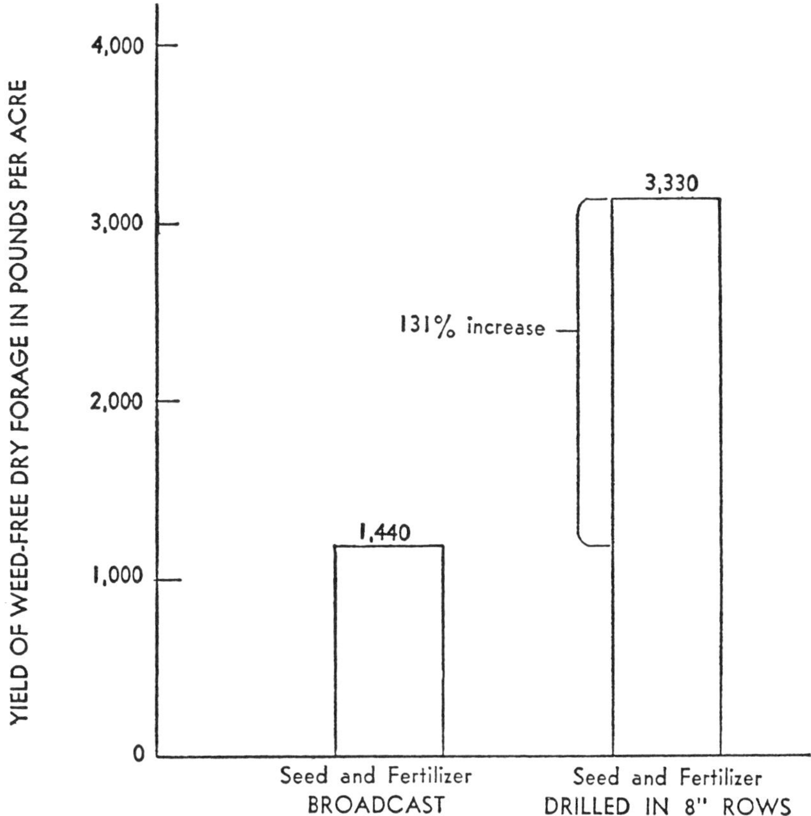

Fig. 13.7 — Forage was increased 131 percent by *drilling* seed and fertilizer in 8-inch (20-cm) rows, as compared with *broadcasting* the same amount of seed and fertilizer. Drilling is more efficient on soils of low fertility. (Maryland) (Courtesy, USDA)

Note: Pounds per acre × 1.12 = kg/ha.

13:5 □ FERTILITY MANAGEMENT

Marked improvement of forage yield and quality have been achieved by fertilizer and lime. The improved nutrition of such forages has in turn improved animal health and productivity. The primary attributes of liming and fertilization are (1) to obtain high economic yields, (2) to improve the seasonal growth pattern, (3) to maintain desirable plants, and (4) to improve mineral content of forages grown in mineral-deficient soils (Figures 13.8 and 13.9).

The following principles may help to analyze more clearly the problems of pasture fertility management:

1. Lime supplies calcium or both calcium and magnesium. Since it is usually needed "by the ton," adequate amounts can be applied about once every three to five years or at each time of reseeding. Lime is also applied satisfactorily as a top-dressing on established pastures, but when

Fig. 13.8 — A cooperator with the Soil Conservation Service is shown checking the seed in his drill seed box prior to planting a mixture of western wheatgrass, intermediate wheatgrass, and slender wheatgrass. (Montana) (Courtesy, USDA - Soil Conservation Service)

Fig. 13.9 — This cow is suffering from calcium and phosphorus deficiency because even tnougn there is plenty of grass here, the soil and grass are deficient in calcium and phosphorus. (Courtesy, Texas A&M University)

lime is applied in this way the plant roots are slow to get what they need. (See Chapter 3.)

2. Nitrogen fertilization results in sharp increases in forage and animal products (Figure 13.10). Nitrogen fertilizers are highly mobile; therefore, they are capable of moving downward for quick plant-root absorption. This means that nitrogen can be used successfully as a top-

Fig. 13.10 — *Top:* Native pasture not fertilized; yield — 1.5 tons per acre (3.4 mt/ha); beef production, 58 pounds per acre (65 kg/ha) per year. *Bottom:* Native pasture seeded and fertilized each year with 80 pounds of N per acre (90 kg/ha) — a nongrazed (protected) area in the pasture; yield — 4.5 tons per acre (10.1 mt/ha); beef production 143 pounds per acre — (160 kg/ha) per year. Fertilization thus increased yield of grass by 573 percent; yield of beef production by 147 percent. (Courtesy, USDA - Agricultural Research Service)

dressing. In a desirable legume-grass mixture, nitrogen favors grasses more than it does legumes. Also, legumes are capable of manufacturing some of their own nitrogen from the air. When legumes receive nitrogen fertilizer, bacteria on their roots become "lazy" and do not manufacture so much from the air. With these facts in mind, on established stands it is best to use high-nitrogen fertilizers only where grasses are the main pasture plants.

3. Phosphate is usually most efficient when applied in fairly large amounts at one time. For this reason, as much as is called for by a soil test should be used at or just before seeding time. When more is needed, however, phosphate can be applied with good results as a top-dressing on young seedlings. Movement of phosphate downward in the soil is very slow, but apparently the leaves absorb some of it, and shallow roots are also able to utilize some of the surface-applied phosphate.
4. Potassium is required in humid regions in fairly large quantities by most forages. In a legume-grass pasture mixture, grasses rob legumes of available potassium. To add to the confusion, legumes actually require more potassium than do grasses. The answer to the riddle appears to be to apply potassium fertilizers each year early in the spring and again after each cutting or grazing period in a rotational grazing system.
5. Minor elements such as boron are needed on many humid-region pasture soils, especially where alfalfa is in the seeding mixture. Where needed, approximately 30 pounds of borax per acre (34 kg/ha) should be applied at the time of seeding (borax is 11 percent boron). Fertilizers containing borax should also be used when top-dressing pastures which contain alfalfa. (See Chapter 4.)
6. Barnyard and poultry manure are highly recommended as top-dressings for pastures. Annual applications of 10 tons or less of manure per acre (22 mt/ha) are more efficient than heavier rates. (See Chapter 5.)
7. Grass tetany is becoming more of a problem in many states. Grass tetany is a magnesium (Mg) deficiency in the blood of cattle. Grass tetany is caused by insufficient or unavailable magnesium in the ration or pasture forage and causes irritability, leg muscle spasms, and convulsions. Treatment consists of intravenous administration of calcium and magnesium. To prevent grass tetany, cattle should be provided with a mineral supplement high in magnesium. Good pasture fertilization will also help prevent grass tetany. Balance the fertility program with additions of magnesium along with nitrogen and phosphorus fertilizer. Figure 13.11 illustrates the relationship between blood serum Mg and herbage Mg. Note also the importance of a balanced nitrogen and potassium fertilizer program. When soils in pastures are high in potassium and ammonium nitrogen (in addition to being low in Mg), the danger of

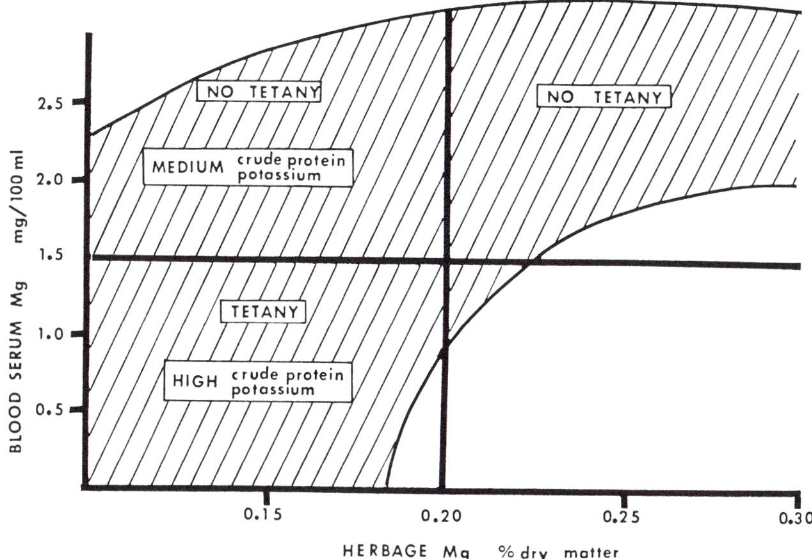

Fig. 13.11 — Low magnesium (Mg) in the blood serum of cattle is related to Mg content of the herbage (forage). This results in a disease called *grass tetany.* Low magnesium, high nitrogen, and high potassium forage are causes of grass tetany. (Courtesy, Duval Sale Corporation, Houston, Texas)

grass tetany is even higher. High application of broiler manure on forage may also cause grass tetany.

8. High levels of nitrogen fertilization may result in the accumulation of high nitrates in certain forages such as oats, corn, wheat, sudangrass, orchardgrass, and tall fescuegrass. High nitrates in plants are also caused by low light intensity and drought.

   Nitrate toxicity may develop when livestock eat large quantities of such forage high in nitrates (NO_3^-). This condition is called **methemoglobinemia** and results in the nitrate being converted to nitrite (NO_2^-). The nitrite reduces the ability of the blood to carry oxygen throughout the animal's body. If not treated, the animal may die for lack of sufficient oxygen, a disease known as **anoxia**.

9. Hydrocyanic acid (HCN) (also called prussic acid) may develop to toxic levels for livestock when they are under conditions of plant stress, such as drought, frost, and high levels of nitrogen fertilization. Plants accumulating high HCN concentrations include sudangrass, sorghums, and sudangrass – sorghum hybrids.

   Since both nitrate toxicity and HCN toxicity may be caused by excessive nitrogen fertilization, livestock symptoms of the two may at times be confused.

10. Smooth bromegrass may be deficient in calcium, magnesium phosphorus, copper, zinc, and sulfur for adequate dairy cow nutrition unless limed and fertilized to supply these nutrients.

13:6 □ GRAZING MANAGEMENT

There are three principal kinds of grazing management systems: (1) continuous, (2) rotational, and (3) deferred. **Continuous grazing** means the grazing of a specific pasture throughout a season or a year. **Rotational grazing** is a system of heavy stocking followed by periods of no grazing to allow the pasture forage to recover and grow again. **Deferred grazing** means the start of grazing is delayed to allow the pasture forage to bear mature seed for reproduction, new forage plants to become better established, and/or for restoration of plant vigor (Figure 13.12). A fourth kind of deferred grazing is practiced under specific conditions. In subhumid and semiarid regions, many of the tall grasses can be deferred until after frost and then grazed as standing dry "hay." This system is also called **accumulated grazing** or **stockpiling of forage**.

For all species of grasses and legumes, the best general rule is to *graze half and leave half* for reserve and residue. There is a scientific reason for this. When more than half of the grass and legume tops are grazed, root growth is slowed. And when 80 percent or more of the tops are grazed, all roots *stop growing* for several days.

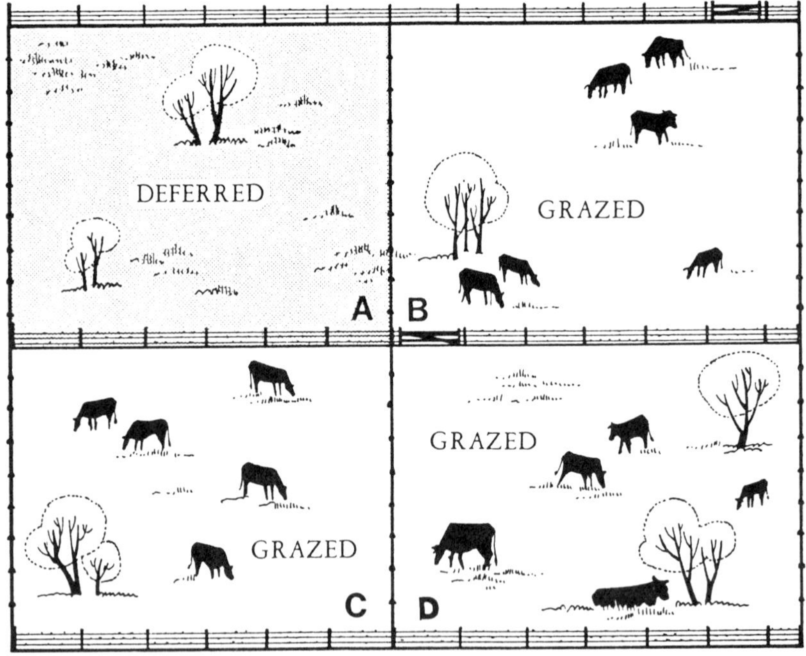

Fig. 13.12 — A suggested four-pasture *deferred* grazing management system. For example, pasture "A" is *not* grazed for about three months, followed by pasture "B" not grazed for three months, then similar deferments for "C" and "D." (Courtesy, Instructional Materials Services, Texas A&M University)

Note: Time not grazed will depend on the growing season.

With *low- to medium-grazing* pressure, the following *low-growing species* of pasture plants are well-adapted to *continuous grazing:* Kentucky bluegrass, common white clover, perennial ryegrass, bermudagrass, and dallisgrass. However, with *high-grazing* pressure, higher yields of milk or meat per acre are obtained by *rotational grazing* because the grass and legume roots need time to start growing again. Grazing should start in the spring when low-growing pasture plants are about 4 inches (10 cm) in height. With high-grazing pressure, livestock should be removed when the plants have been grazed to about 2 inches (5 cm) in height.

Under most kinds of grazing pressure, *tall species* such as alfalfa, bromegrass, orchardgrass, tall fescuegrass, and ladino clover are more productive when rotational grazing is practiced. Grazing should start in the spring when these plants are about 10 inches (25 cm) in height and stop grazing when they have been grazed to a height of about 5 inches (13 cm). This is an example of *graze half and leave half.*

Both maximum *yield* of pasture forage dry matter and maximum *quality* of pasture forage are always objectives of the livestock manager. However, these two objectives cannot be achieved at the same time. For example, maximum dry-matter yield of alfalfa occurs when alfalfa is harvested at the half-bloom stage, whereas maximum digestible protein occurs at the bud stage (before blooming).

In summary: Achieving maximum dry matter yields in any one season depresses quality and quantity of forage over a period of years.

A suggested layout for deferred grazing is seen in Figure 13.12. Since this book is concerned primarily with soils, a closer study of soil compaction caused by grazing animals may be justified.

In New York, a grazed woods soil was packed one-third tighter than the soil in comparable ungrazed woods. This means that a cubic foot of soil in the ungrazed woods weighed 56 pounds (0.9 gm/cc) while the grazed soil weighed 75 pounds (1.2 gm/cc). The percentage of organic matter was reduced one-third by grazing.

13:7 ▫ GRAZING CALENDARS

One of the best ways to reduce the damage to pastures by overgrazing is to prepare a grazing calendar for each farm and to manage the pastures and fields to provide the grazing for each month as indicated by the calendar.

For example, a grazing calendar for North Carolina is given in Figure 13.13. The height of the diagrams indicates the relative amount of available forage for the respective months. This figure indicates that alfalfa can be expected to supply forage beginning in March, reaching a peak in May, and ending in October. Small grains can supply forage from January to May and from October to December. Thus, on any one farm with a sufficient acreage of alfalfa in certain fields and small grains in others, 12 months of green grazing is possible during favorable

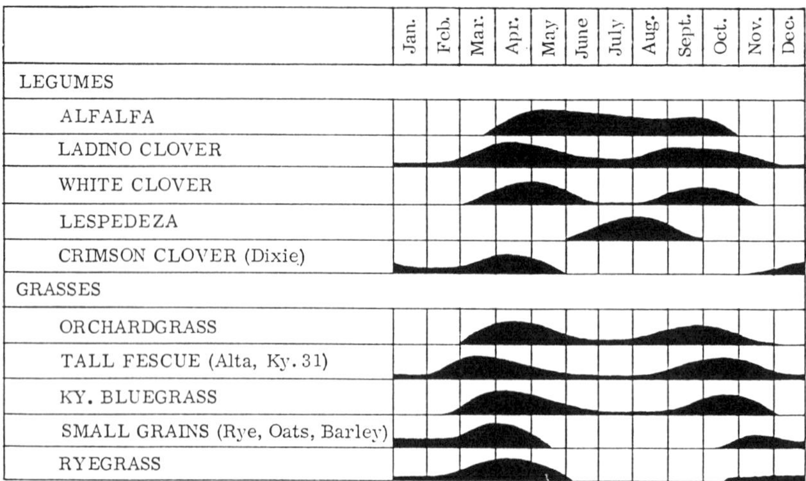

Fig. 13.13 — A grazing calendar for North Carolina. With proper grazing management, at least 10 and sometimes 12 months of green grazing may be obtained. (Courtesy, University of North Carolina)

seasons in North Carolina. Other cropping systems can also supply adequate green forage for most of the year. Year-long grazing cannot be expected in more northerly states, but a longer grazing season can be attained at any location by careful planning and management.

13:8 □ RENOVATING PASTURES

Pasture renovation means chiseling, discing, and/or applying herbicide to an old pasture sod, applying the recommended amounts of lime and fertilizers, then reseeding with a modern pasture mixture, including an adapted legume such as red clover or ladino clover.

Pasture renovation is needed more often on land which is not suitable for a cultivated crop. The land may be too stony, too steep, too stumpy, or too brushy for use in growing an inter-tilled crop such as corn (Figure 13.14). Depending upon the region, the soil, and the farmer's personal choice, pastures may be renovated either in the spring or in the fall.

Many pastures are established in the spring. Because more moisture is present in the soil, it is usually necessary to plow shallow or to use a recommended herbicide to kill the old sod completely. If only a disc or harrow is used, the old grasses and weeds may recover rapidly and offer serious competition to the new seedlings. If the old growth is weak and contains very little Kentucky bluegrass or orchardgrass, only spring discing may be effective.

When new seedlings are made in the spring, it is customary to plant oats at the rate of about 1 bushel per acre (36 kg/ha) as a companion crop to help keep down weed competition. Spring-seeded pastures may be ready for some light grazing by late summer but will not give full production until the next season.

Pastures renovated in the late summer provide a full season of grazing the following year. Late summer renovation usually results in fewer weeds, but often soil moisture will not be adequate. The soil may be prepared by discing, harrowing, or using a combination of the two. A cutaway disc or a bush-and-bog harrow may be more effective on stony, brushy, or rough land.

The first discing may be started by July 1, with the area being disced in two directions. Discing loosens the sod pieces from the soil so that hot, dry weather will kill the old sod. Later discings to finish killing the sod can best be made after a light rain has softened the soil surface. Small areas in the field may have to be disced or harrowed more often to kill the old stand completely. Discing or harrowing will leave dead plant residues to hold moisture and aid the establishment of the young seedlings, especially if dry weather follows seeding.

The no-tillage (**sod-seeding**) approach to pasture renovation is used in some states. The availability of certain herbicides has made possible the suppression of existing vegetation without tillage. A heavy, no-tillage drill with press wheels is used to introduce seeds of a more productive forage species into the areas that are unproductive. The no-tillage system eliminates the standard tillage operations and reduces the potential loss of soil and water.

Sod-seeding can be established in the spring with a vigorous warm-season annual crop such as millet, sudangrass, or a sorghum-sudan hybrid. Fall sod-seeding can be with wheat, or in the lower South, with crimson clover. In either

Fig. 13.14 — This steep and stony pasture in Kentucky is in need of renovation. (Courtesy, USDA – Soil Conservation Service)

season, the technique is to overgraze, use a recommended contact herbicide and apply fertilizer and seed into mineral soil through the existing sod with a heavy drill-and-fertilizer distributor.

13:9 □ IRRIGATING PASTURES

Grasses and legumes used for irrigated pastures are those responding best to intensive management. The principal ones include these cool- and warm-season grasses and legumes.

- □ ***Cool-season*** — Smooth bromegrass, meadow bromegrass, intermediate wheatgrass, orchardgrass, Russian wildrye, tall wheatgrass, creeping foxtail, tall fescuegrass, and reed canarygrass.
- □ ***Warm-season*** — Switchgrass, sand lovegrass, big bluestem, little bluestem, bermudagrass, and coastal bermudagrass.
- □ ***Legumes*** — Alfalfa, clovers, and birdsfoot trefoil.

The roots of well-established grass can absorb water from depths of up to 4 feet (1.2 m); legumes can reach somewhat deeper. It must be remembered however that most of the water the grass uses comes from the top foot of soil and that plants grow best when the soil air space is about half full of water. To determine more accurately when it is time to irrigate and when to stop, tensiometers can be buried in the soil at a depth of about 1 foot (30 cm). However, the lead wires from the tensiometers must be fenced against damage by livestock. (See Chapter 9.) Each season, in the mid-United States, about 35 inches (90 cm) of rainfall plus irrigation water are required for pastures.

In general, the highest yields of the best quality forage are obtained through a system of sprinkler irrigation, rotation grazing, clipping of mature plants and weeds at the end of each grazing cycle, and fertilization and irrigation to promote quick recovery. This program will stimulate growth and maintain relatively similar forage maturity and palatability among individual plants.

Research results in Nebraska on eight irrigated pasture grasses concluded that the highest gain per acre of beef cattle and the most efficient use of irrigation water were made by these five grasses: orchardgrass, smooth bromegrass, meadow bromegrass, Garrison creeping foxtail, and intermediate wheatgrass. Least productive and least water-efficient were tall fescuegrass, reed canarygrass, and Russian wildrye.

13:10 □ REFERENCES

Casler, M. D., Michael Collins, and J. M. Reich. "Location, Year, Maturity, and Alfalfa Competition Effects on Mineral Element Concentration in Smooth Bromegrass." *Agronomy Journal*, Vol. 79, 1987, pp. 774 – 778.

Kilgore, Gary L., Frank K. Brazle, and Marvin R. Fausett. "Tall Fescue: Production and Utilization." Kansas State University, Manhattan, C-622, 1980.

Klebesadel, Leslie J. "Natural Selection May Modify Introduced White Clover Toward Superior Winterhardiness." University of Alaska, Fairbanks, *Agroborealis*, Vol. 18, No. 1, July 1986.

Kroth, Earl M., and Louis Meinke. "Topdressing Nitrogen, Phosphorus, and Potassium on Cool-Season Grasses for Pasture Production." University of Missouri - Columbia, Res. Bul. 1039, 1981.

Marten, G. C., F. R. Ehle, and E. A. Ristau. "Performance and Photosensitization of Cattle Related to Forage Quality of Four Legumes." *Crop Science*, Vol. 27, 1987, pp. 138 - 145.

Rehm, George W. "Fertilizing Pastures." *Soil Science News*, Vol. 3, No. 5, 1981. University of Nebraska - Lincoln.

Roberts, C. A., F. E. Barton, and K. J. Moore. "Estimation of *Acremonium coenophialum* Mycelium in Infected Tall Fescue." *Agronomy Journal*, Vol. 80, pp. 737 - 740, 1988.

U.S. Dept. of Agriculture. *Animal Health — Livestock and Pets. The 1984 Yearbook of Agriculture*. Washington, D.C., 1984.

CHAPTER 14

Field Crops: Soil, Water, and Fertility Management

"We have the means, we have the capacity to eliminate hunger from the face of the earth in our lifetime. We need only the will." — President John F. Kennedy, 1963

OUTLINE

The world food challenge can be met only by family planning (*top*), increased production on the soil (*bottom left*), and research and agricultural education (*bottom right*). (Courtesy, World Bank)

□ □ □

14:1 □ OVERVIEW

Climate determines the area where a crop may be grown, but the soil largely determines the yield. This is assuming that the variety is locally adapted, diseases and insects are controlled, and calamities of nature such as hailstorms do not interfere.

It is a credit to our farmers, our agricultural scientists and educators, and those in agricultural business who serve the farmer that crop yields per acre (ha) of all major crops have been continually on the increase.

Not only has the production per acre (ha) increased, but also the production per farm worker has increased. Two centuries ago, when more than 90 percent of the population lived on farms, one farm worker produced enough food for two other persons. In 1850, one farm worker supplied enough food for 4 other persons; in 1900, 6 others; in 1950, 15 others; and in 1965, 37 others. One farm worker today in the United States produces the food and fiber for 78 other persons.

The tremendous increases in yields can be emphasized by a look at corn production in the United States since 1930. In 1930, the average yield of corn was 20.5 bushels per acre (1,287 kg / ha). In 1985, the average yield was 118.0 bushels per acre (7,408 kg / ha).

14:2 □ THE FOOD CHALLENGE

A generation ago many farmers were self-sufficient; today, off-farm fuel, fertilizer, capital, and technology are essential for their economic survival. A generation ago many nations were self-sufficient; today, the United States and many other exporters provide the margin between life and death for many millions of people.

Thus, food has become the central element of the international economy. A world of energy shortages, rampant inflation, and a weakening trade and monetary system is a world of food shortages as well.

The world has more than enough food to adequately feed every person. The problem lies in lack of purchasing power, poor distribution systems, and the lack of both transportation and storage facilities.

The worst hunger problems are concentrated in the world's poorest nations (Figure 14.1). While extreme cases such as widespread starvation in the wake of warfare and turmoil seem to take place in certain countries and seize the international spotlight, milder but continuous food shortages affect far more people. Se-

Fig. 14.1 — The worst hunger problems are concentrated in the world's poorest nations. (Courtesy, USDA)

vere systematic malnutrition — whereby people are unable, day after day, to obtain the nutrients they need to maintain good health — is common in about 65 developing countries, according to the United Nations Food and Agriculture Organization (FAO). Many developing countries — and even some developed nations — are unable to produce enough food and must depend on food aid, cash purchases, or both.

At the time of Christ, the world population was estimated at 250 million. The first doubling of human population to 500 million occurred in about 1,650 years. The second doubling required only 200 years to arrive at a population of 1 billion in 1850. That was about the time of the discovery of the nature and cause of infectious diseases and the dawn of modern medicine. The third doubling of the population since the time of Christ, to 2 billion, occurred by 1930 — only 80 years after the second doubling. Then, sulfa drugs, antibiotics, and improved vaccines were discovered. World population again doubled in only 45 years to 4 billion people in 1975. At the current world rate of population growth, the population will double again in the next 40 years, reaching 8 billion by 2015 (Figure 14.2). Can they all be fed?

Hunger is a major world problem. Today, we hear statements from people in all walks of life concerning the need for increased food production for the expanding population of the world. Most of the world's population live on only 15 staple crops: rice, wheat, corn, sorghum, millet, rye, barley, cassava, sweet potatoes, potatoes, coconuts, bananas, common beans, soybeans, and peanuts. These crops furnish about 90 percent of the world's food and occupy about 75 percent of the total tilled land on the earth.

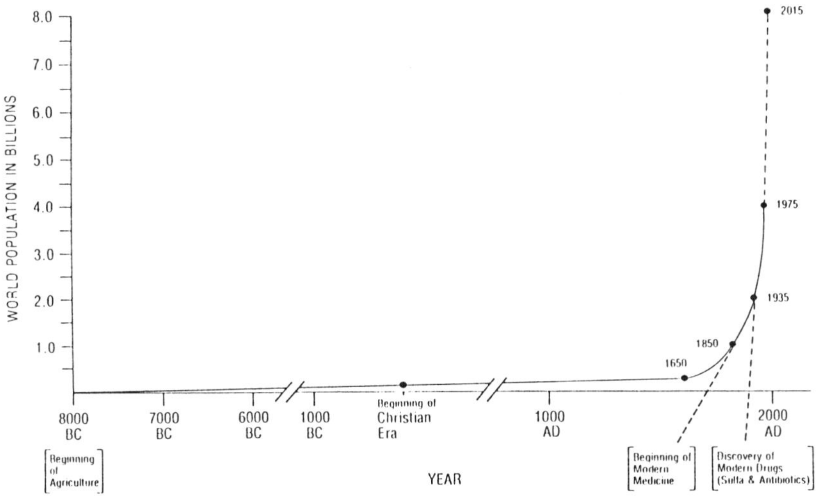

Fig. 14.2 — The world population is now increasing by 200,000 people each day (more than 2 each second). At the current rate of population growth, the world population will reach 8 billion by the year 2015. Can they all be fed?

14:3 ▫ CROP PRODUCTION

Early humans lived on wild game, seeds, berries, leaves, roots, and fruits. As the population increased, the food supply was not always sufficiently stable or plentiful to supply their needs. Crop production began when the domestication of plants became essential to supplement natural supplies. The essential features of crop production have remained almost unchanged since the dawn of history. These features include:

1. Gathering and preserving the seeds of the desired crop plants.
2. Destroying other kinds of competing vegetation growing on the land.
3. Stirring the soil to form a seedbed.
4. Planting the seed when the season and weather are right, as shown by past experience.
5. Destroying weeds.
6. Protecting the crop from natural pests.
7. Gathering, processing, and storing the products.

Improved cultural methods followed observations made by primitive farmers. They found better crops in spots where manure, ashes, or burned limestone had been applied. Observations revealed the most favorable time, place, and manner of planting and cultivating various crops. Observation, the only means of acquiring new knowledge until the nineteenth century, continued to enrich the fund of crop lore. Eventually, the exchange of ideas, observations, and experiences, through agricultural societies and rural papers and magazines, spread the knowledge of successful crop production.

Better crop production follows the application of new discoveries, the adoption of improved machines, and the breeding of improved crop varieties. New developments must be publicized to inform a few enterprising farmers in each community in which such developments might apply. Thereafter, in a free country, sound demonstrated improvements spread with extreme rapidity. This is evident from the almost universal adoption of hybrid corn, hybrid sorghum, disease and insect-resistant grains, **monogerm** (single-seed) sugar beets, and the use of herbicides in the United States. Improvements in crop production from research and invention often arrive only after long, costly, and painstaking effort.

14:4 ▫ AGRONOMIC CLASSIFICATION OF CROP PLANTS

Crop plants may be classified on the basis of a similarity of plant parts. From an agronomic standpoint, they may be classified on the basis of use as follows:

- **Cereal** or **grain crops** — *Cereals* are grasses grown for their edible seeds, the term being applied to either the grain or the plant itself. *Grain* is a collective term that applies to cereal crops and includes wheat, barley, oats, rye, rice, corn, grain sorghum, proso millet, and pearl millet.
- **Legumes** for seed (pulses) — Legume crops include peanuts, soybeans, field beans, field peas, cowpeas, lima beans, mung beans, chickpeas, pigeonpeas, broadbeans, and lentils.
- **Root crops** — Root crops, grown for their enlarged roots, include sugar beets, mangels, turnips, carrots, rutabagas, sweet potatoes, yams, and cassava (manioc).
- **Tuber crops** — Tuber crops include the potato and the Jerusalem artichoke. A tuber is not a true root; it is a short, thickened, underground stem.
- **Forage crops** — Forage crops refer to vegetable matter, fresh or preserved, which is utilized as feed for animals. Forage crops include grasses, legumes, and other crops that are cultivated and used for hay, pasture, fodder, forage, or silage. Grain crops, such as corn and other cereals, are also forage crops. (Livestock producers often use the word **roughage** instead of **forage**.)
- **Sugar crops** — Sugarcane and sugar beets are grown for their sweet juice from which sucrose (common sugar) is extracted and crystallized.
- **Sugar sorghums (sorgos)** — Sugar sorghums as well as sugarcane are grown for syrup production. Dextrose and fructose sweeteners are made commercially from corn.

- **Oil crops** — Soybeans, peanuts, sunflowers, flax, castorbeans, and rapeseeds (canola) are primary members of the oil crops.

- **Fiber crops** — Cotton, flax, hemp, and ramie are some of the more important crops grown for their fiber. Broomcorn (a sorghum type) is grown for its brush fiber for the manufacture of brooms.

- **Drug crops** — Drug crops include tobacco, coffee, tea, peppermint, mustard, and hops.

- **Rubber crops** — The only field crop that has been used for rubber in the United States is guayule.

14:5 ▫ WHEAT

Wheat is the most important world cereal grain crop. The USSR has the largest acreage and total production. Wheat has a wide climatic and soil adaptation range. It is an important crop in both humid and subhumid areas. In the United States, most of the wheat is produced in regions having an annual rainfall of less than 30 inches (76 cm). In 1985, the states ranking highest in harvested acreage in decreasing order were Kansas, North Dakota, Texas, Oklahoma, and Montana.

Wheat is grouped into five principal classes:

- ***Hard red spring*** — Hard red spring wheat is grown principally in the northern Great Plains states where there exist a fairly deep, black soil and hot, dry summers. These environmental characteristics are believed important for the production of a high grade of spring wheat suitable for milling into bread flour.

- ***Durum*** — The center of durum wheat production has continued to be in the Dakotas, Montana, and Minnesota.

- ***Hard red winter*** — The center of production of hard red winter wheat is in Kansas and the adjoining states of Nebraska, Oklahoma, Colorado, and the Panhandle of Texas. The annual precipitation for this area averages less than 25 inches (63 cm), with frequent dry periods and sub-zero winter temperatures. Hard red winter wheat is the most important class of wheat in the United States on the basis of the number of acres.

- ***Soft red winter*** — The production of soft red winter wheat centers in Ohio, Indiana, and Illinois.

- ***White*** — White wheat may be either spring or winter. It is produced mostly in the Pacific Coast states, with some acreage in southern Michigan and the western part of New York.

Water Needs

The best time to sow winter wheat is usually near the date of the average first frost in the fall (Figure 14.3). At this time there is a good chance of getting a satisfactory yield of wheat if the soil is moist to a depth of at least 2 or 3 feet (61 or 91 cm). If there is not this much moisture at that time, it is best not to plant wheat but to wait until spring before planting anything. In the spring, grain sorghum may be grown with a good chance for a successful crop if the soil is moist to a depth of 1 to 2 feet (30 – 61 cm) at the time for planting. If the soil is still too dry for grain sorghum, millet is often planted at a later time if the soil is moist to a depth of approximately 1 foot (30 cm). If there is not enough moisture for millet, the field should be fallowed for the summer and plans should be made to sow wheat in the following fall if there is sufficient soil moisture.

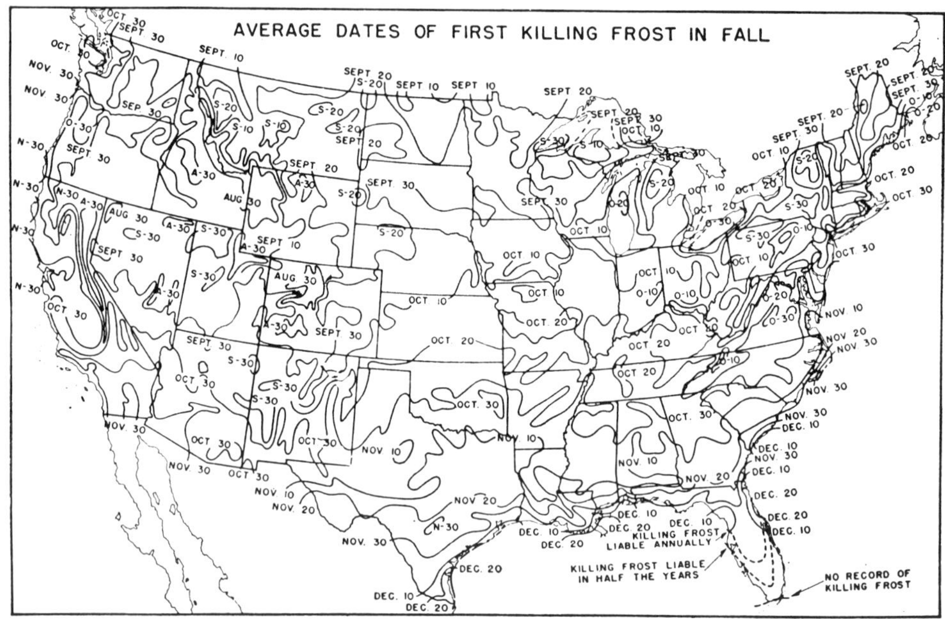

Fig. 14.3 — The best time to sow winter wheat is usually at the time of the first killing frost in the fall. What is this date where you live? (Courtesy, U.S. Department of Commerce)

A comprehensive study was made of soil moisture in the soils of the Great Plains in relation to wheat yields. The data were summarized and are portrayed by a bar graph in Figure 14.4. At the time of seeding wheat, when the soil was wet to a depth of 0 to 1 foot (0 – 30 cm), wheat yielded only 1.5 bushels per acre (101 kg/ha) when the growing season rainfall was not favorable, and 9.0 bushels per acre (605 kg/ha) when the rainfall was favorable. Such yields must be called crop failures. In other words, the soil moisture must be deeper than 1 foot (30 cm) at seeding time, regardless of the rainfall received during the growing season, if a successful crop is to be expected.

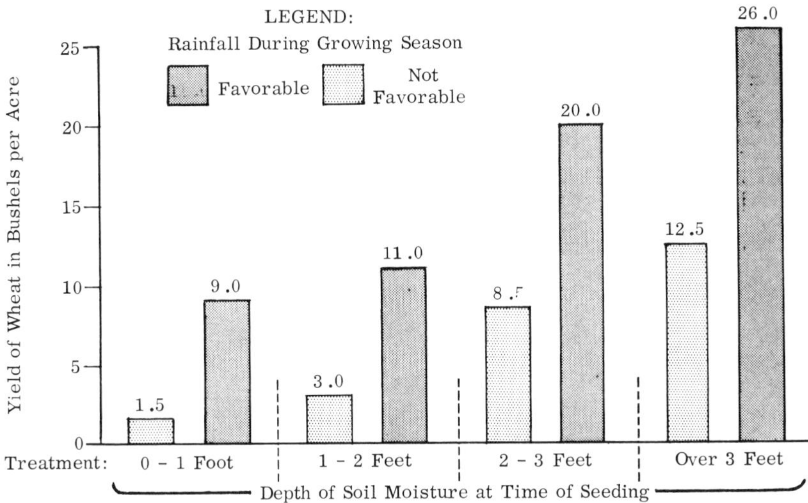

Fig. 14.4 — The highest yields of wheat are obtained only when at seeding time the soil is wet for over 3 feet (91 cm) and the rainfall during the growing season is favorable. (Courtesy, USDA)

Note: Values are averages of 1,107 determinations.

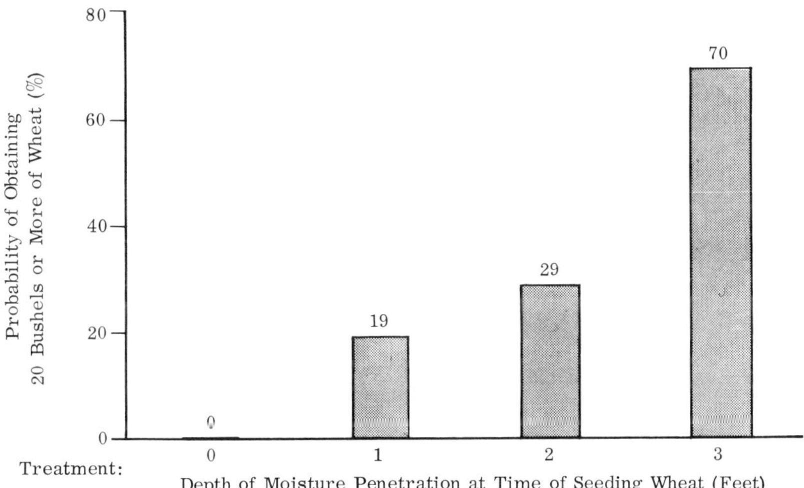

Fig. 14.5 — The probability of obtaining 20 bushels of wheat per acre (1,340 kg / ha) when the soil is moist to depths of 0, 1, 2, or 3 feet (0, 30, 61, or 91 cm) at seeding time. (Kansas) (Courtesy, Kansas State University)

Note: Average annual precipitation, 18 inches (46 cm); average annual evaporation, 57 inches (145 cm).

When the soil was moist to a depth of over 3 feet (91 cm), unfavorable growing season rainfall resulted in a wheat yield of 12.5 bushels per acre (838 kg/ha), and favorable rainfall gave 26 bushels per acre (1,742 kg/ha).

It can be concluded from this study that it is not recommended to seed wheat unless there is at least 2 feet (61 cm) of soil moisture at the time of seeding.

Another study of the relationship between soil moisture and wheat yields was made in Kansas (Figure 14.5). When there was no moisture in the soil at the time of seeding wheat, there was never a likelihood of getting as much as 20 bushels of wheat per acre (1,340 kg/ha). However, with 3 feet (91 cm) of soil moisture at seeding time, a yield of 20 bushels per acre (1,340 kg/ha) can be expected about 70 percent of the time.

Use of Fertilizers

Wheat is not usually considered a soil exhausting crop; nevertheless, when moisture is not the first limiting factor, fertilizers of some kind usually will increase yields.

In recent years experimental plots in Kansas, Colorado, and Nebraska have often shown yield increases of 10 to 20 bushels per acre (670 – 1,340 kg/ha) from applied nitrogen.

The importance of available soil nitrogen to the wheat crop is now being recognized in many wheat-growing areas. Several states now offer an available nitrogen test. This test measures the residual nitrogen that is in the nitrate form or the nitrate plus ammonium form. The available nitrogen test results are used to adjust the basic nitrogen recommendations if residual nitrogen is found to have accumulated in the soil. Accumulations of residual nitrogen may occur under summer fallow conditions or in fields receiving high nitrogen additions from fertilizers, manure, or sewage sludge.

In the central Corn Belt and eastward, most of the wheat is fertilized at seeding time with a combination drill. Nitrogen is applied as a top-dressing in the spring for a large amount of the wheat acreage. In the eastern part of the Great Plains increased yields are obtained from nitrogen or phosphorus or both. In the West, increases have been almost entirely from nitrogen when moisture was available. Irrigated wheat will very often respond to fertilizer.

Wheat is known to respond well to applications of phosphorus on soils testing low in available phosphorus. Wheat plants do not tiller adequately with severe phosphorus deficiency and often winterkill in low-phosphorus soils.

In the eastern states a complete fertilizer is used, and the application will include nitrogen, phosphorus, and potassium. Soil tests should be used to predict the need for phosphorus, potassium, or other nutrients on wheat.

At Rothamsted, England, wheat has been grown *continuously* since 1843 (Figure 14.6). One plot received no manure and no fertilizer, another manure only, and a third plot fertilizer only. Wheat with no manure and no fertilizer produced, on an average, 13 bushels per acre (874 kg/ha); manure only, 32 bushels (2,144 kg/ha); and fertilizer only, 34 bushels per acre (2,278 kg/ha).

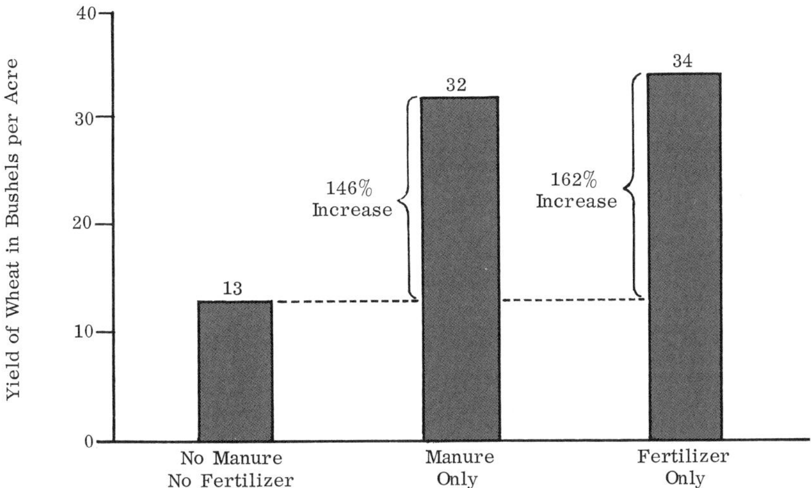

Fig. 14.6 — Continuous wheat yields in England increased greatly with manure only or fertilizer only during a period of more than 100 years. The soil is a clay loam high in organic matter and derived from calcareous sediments.

Note: Bushels (of wheat) per acre × 67.2 = kg / ha.

14:6 □ RICE

Although wheat is the world's number one grain for both livestock and humans, rice is number one for direct consumption by humans. Six states produce all of the rice that is grown in the United States; in 1985, this amounted to 2 million acres (0.8 million ha). The six states, in decreasing order of harvested acreage of rice, are Arkansas, Texas, Louisiana, Mississippi, Missouri, and California.

Rice is adapted to loams and clay soils that are 2 to 3 feet (61 – 91 cm) deep, and that have a pH from 6.0 to 7.5. The soil must be level or capable of being leveled so that, when the field is flooded, the water will be uniform in depth. An adequate amount of good-quality irrigation water must also be available at a reasonable cost. Rice is classified as having a medium degree of tolerance to salt concentrations in the soil, ranking with wheat, grain sorghum, corn, alfalfa, and Irish potatoes in this respect.

Summer temperatures for the best rice production should average about 82° F (28° C), but they sometimes reach 100° F (38° C). Spring and fall temperatures should average 70° F (21° C). The relative humidity must always be high to reduce losses of water by evaporation.

A common method of growing rice is to seed and fertilize it when the soil is dry, just as wheat is sown. The land is usually flooded to hasten the germination of rice. The level of the water is raised to a maximum of about 6 inches (15 cm) as the rice plants emerge and grow. At no time is the rice plant completely under water. Sometimes the water level is lowered and then raised to kill pond weeds. About two weeks before the rice is ready for harvest, all of the water is drained and the soil is allowed to dry. This permits the use of a combine for harvesting, in a manner similar to the harvesting of wheat.

Cropping Systems

In China, Pakistan, Japan, and India, where most of the world's rice is grown, continuous crops of rice have been grown for centuries. By contrast, in the United States, even three successive crops of rice will result in lowered yields. This difference in rice culture makes one wonder why conditions in Asia and the United States are so different.

One contrast in cultural conditions that the authors have observed is that in Asia the rice soils are plowed and cultivated thoroughly by "puddling" them while water stands on the surface. Between crops of rice the soils are allowed to dry thoroughly. Tillage when the soils are saturated with water and thorough drying seem to aerate the soils properly. In the United States, no puddling is done when the soils are saturated, and the soils do not aerate by drying as much as in Asia. At any rate, crop rotations are necessary in the United States in order to maintain a high yield per acre (ha) of rice.

Sometimes the rice land is in a rice – fallow – rice cropping system, but this is not the most desirable system for making the best use of the land. The best cropping systems are those that permit the land to remain out of rice for at least two years.

In Louisiana and Texas, the common two-year rotations are rice-lespedeza, rice-cotton, rice-wheat, rice – winter oats, and rice – grain sorghum. A popular cropping system in Arkansas is a four-year rotation of rice-cotton (winter legume)-soybeans (winter oats)-lespedeza. Sometimes rice is grown in rotation with the raising of fish. Where more forage is needed for livestock, a six-year rotation system is often followed: rice-rice-pasture-pasture-pasture-corn (or soybeans or grain sorghum). The pasture is usually a mixture of white clover, alsike clover, redtopgrass, and ryegrass.

One rotation system in the rice-growing areas of California is rice – wheat (or barley or grain sorghum or beans) – ladino clover. Another system is rice-beans-wheat-beans.

Unless a good cropping system is followed, the soil organic matter will decrease rapidly and crop yields will decline. A study was made of the organic matter and nitrogen level of a soil in southern Louisiana that had been in a rice cropping system for 40 years. The supplies of soil nitrogen and soil organic matter had decreased 60 percent during the period (Figure 14.7).

Use of Fertilizers

Nearly all rice soils are deficient in nitrogen. The prairie and terrace areas in Texas and Louisiana are also low in available phosphorus. Rice on the clays and

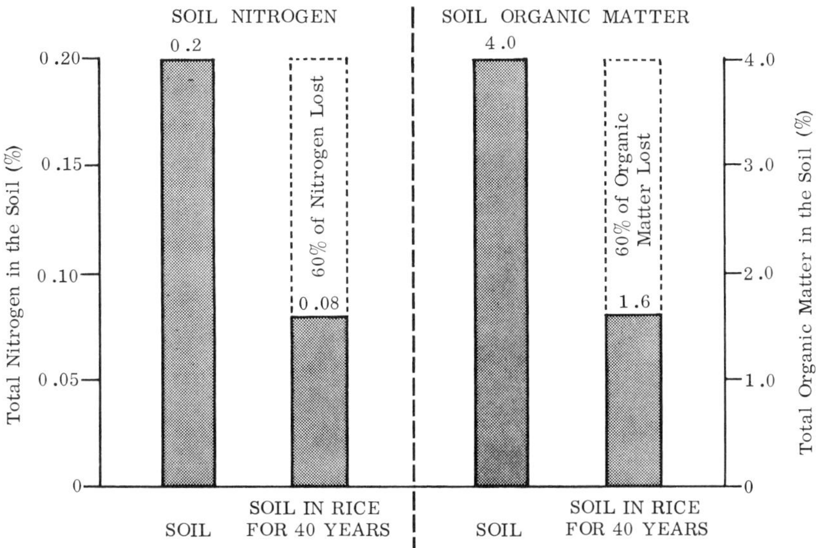

Fig. 14.7 — Sixty percent of original nitrogen and organic matter is lost after cropping to rice for 40 years. (Southern Louisiana) (Courtesy, USDA)

Note: The soil is a Crowley slit loam.

clay loams of these areas does not usually respond to applications of potassium. Relatively few areas with **alluvial soils** (soils deposited by flowing streams) are deficient in either phosphorus or potassium. Phosphorus is more likely to be deficient than potassium. Nitrogen is commonly the only nutrient element applied to rice on recent alluvial soils.

If soil tests indicate a need for phosphorus and potassium, these fertilizer elements should be applied pre-plant and worked into the soil 2 to 4 inches (5 – 10 cm) deep. If the phosphorus and potassium cannot be applied pre-plant, they should be applied no later than the first flood.

If the soil is kept flooded, or if it is drained briefly but kept saturated, all of the nitrogen may be applied pre-plant. If the soil is drained for a period of 2 to 3 weeks, which permits the soil to dry, half of the nitrogen should be applied just before the first flood and the remainder at the time of first jointing of the rice plant.

Ammonia (NH_3) or the ammonium (NH_4) forms of nitrogen drilled into the soil to a depth of 3 to 4 inches (8 – 10 cm) at or just prior to seeding will be held by the soil in the more available ammonium form within easy reach of the rice roots.

The application of top-dressing by broadcasting the fertilizer into floodwater is not as effective as applying the fertilizer after the fields have been drained. Some ammonia may be lost by volatilization if the floodwater is alkaline and warm.

A rather common practice in the **terrace** (old alluvial soil) areas of Texas, Louisiana, and Arkansas is to apply 300 pounds an acre (336 kg/ha) of 15-15-15 fertilizer or its equivalent by drilling or discing at or just before seeding, and then top-dressing not later than eight weeks after emergence with 30 to 50 pounds per acre (34 – 56 kg / ha) of nitrogen as ammonium sulfate or urea. No *nitrate* fertilizers should be used as a basal application on rice because of rapid loss of nitrogen as a gas by **denitrification** (loss of nitrogen into the atmosphere as a gas.)

The best time to apply fertilizer in relation to the growth periods of rice is seen in Figure 14.8. It is not advisable to apply fertilizers after about 60 days from seeding rice.

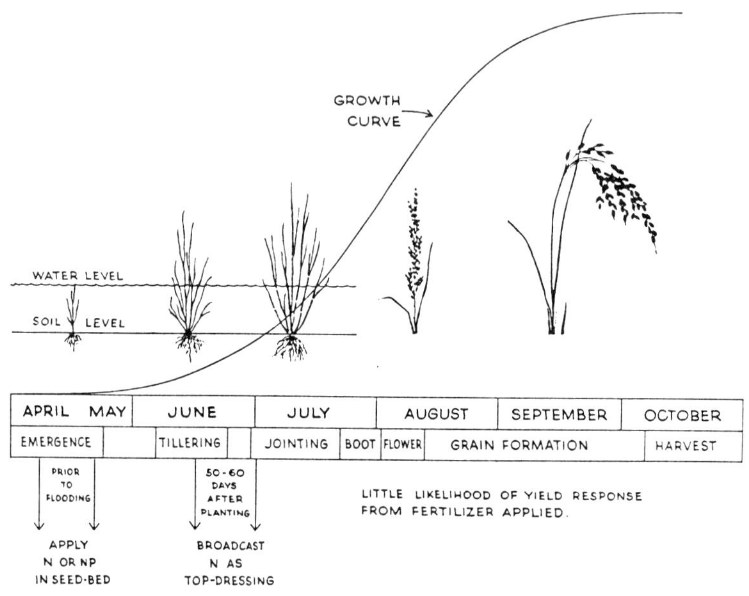

Fig. 14.8 — Growth of rice in relation to fertilizer needs. (Courtesy, University of California)

14:7 ▫ POTATOES

Potatoes are grown in all 50 states, but 4 states produced over 50 percent of the total acreage in 1985. These were Idaho, 345,000 acres (139,000 ha); North Dakota, 139,000 acres (56,000 ha); Washington, 126,000 acres (51,000 ha); and Maine, 97,000 acres (39,000 ha).

Potatoes are a cool-season crop and production is largely limited to the climates that are fairly cool and moist (Figure 14.9). Potatoes grow best where the mean growing temperature is from 60° to 70° F (16° – 21° C). Potatoes are grown in muck soils and in fine sands, sandy loams, loams, and silt loam soils. Clay soils are not very suitable because they produce dirty potatoes that are not attractive to buyers. All soils must be well-drained.

Potatoes are tolerant of soil acidity. In the eastern states they make a satisfactory growth and yield in a soil with a pH range of 4.8 to 5.5. At a soil pH of 5.5 to 6.5, potato scab makes the potatoes dirty in appearance and therefore not very

Fig. 14.9 — Aroostook County, Maine, is the number one potato county in the United States. The principal soil is a Caribou silt loam. Note that the rows are on the contour to conserve soil and water. (Courtesy, University of Maine)

saleable. However, yields at a higher pH are satisfactory. In the West, potatoes are commonly grown at a pH of 7.0 or greater and there is very little damage from scab.

Use of Fertilizers

Potatoes are a short-season crop with a very high demand for plant nutrients, especially during certain periods. For example, only 8 percent of all plant nutrients absorbed for the season are taken in during the first 50 days after planting. During the next 30 days, however, 70 percent of the season's requirements are absorbed. In other words, the most rapid absorption of plant nutrients takes place between the fiftieth and the eightieth day after planting. During this period, most of the nutrients are within 5 inches (13 cm) of the surface, although the plant roots extend to 2 feet (61 cm) in depth. Nearly everywhere potato yields are increased severalfold by the application of large quantities of fertilizer. Applications of 1 ton of complete fertilizer per acre (2.2 mt / ha) are now fairly common. In some areas potatoes are irrigated.

14:8 □ CORN

Corn has been fertilized in the United States longer than any other crop. In the early 1600's the American Indians taught the first English settlers how to fertilize corn by burying a fish beside each hill of corn. As the corn grew, the fish rotted and fertilized the corn. Except for the extra cost and the odor, this practice of fertilizing corn could be used to produce bumper crops even today (Figure 14.10). (Corn is known as *maize* in most countries of the world.)

Fig. 14.10 — The American Indians taught the Pilgrims how to fertilize corn by burying a fish beside each hill. (Courtesy, The Fertilizer Institute)

Most ancestors of corn hybrids trace through open-pollinated corn to two main lines: dent corns, probably originating in Mexico, and flint corns, similar to some in Guatemala. Plant scientists are looking at other types of corn which might improve the nutritional value of U.S. commercial hybrids (Figure 14.11).

Corn requires a very fertile soil, a large amount of water, and a warm temperature. The proper planting date is about 10 days after the average date of the last killing frost in the spring. Usually this period coincides with a mean daily temperature of 60° F (16° C). Below 50° F (10° C) corn will not germinate, and it will grow twice as fast at 86° F (30° C) as it does at 68° F (20° C). Corn is therefore a warm-season crop that can withstand almost any summer temperature, provided there is enough available soil water (Figure 14.12).

Plant breeders have recently developed corn **cultivars** (cultivated varieties) that sprout at a lower than normal soil temperature. This permits earlier planting.

In 1985, there were 75 million acres of corn (30.4 million ha) harvested for grain in the United States, with an average yield of 118.0 bushels per acre (7,408 kg / ha). States ranking highest in harvested acreage of corn in decreasing order are Iowa, Illinois, Nebraska, Minnesota, Indiana, and Ohio. However, some corn is raised in all 50 states.

Fig. 14.11 — The primitive Coroico corn at left may be used to improve the nutritional value of U.S. commercial hybrids such as the other ear. Coroico corn grows on the eastern slopes of the Andes Mountains and is higher in B vitamins and certain amino acids than the U.S. hybrid. (Courtesy, USDA – Agricultural Research Service)

Water Needs

The root system of a corn plant in a friable soil that is well-drained may have a lateral spread of 8 feet (244 cm) and penetrate to a depth of more than 6 feet (183 cm). This means that corn should be irrigated to a depth of at least 3 feet (91 cm) or perhaps deeper for the greatest efficiency in the use of water. The deeper the irrigation water is permitted to penetrate into the root zone, the less the percentage of water that is lost by evaporation.

In the Corn Belt a high-yield crop of corn will use from 16 to 25 inches (41 – 64 cm) of water per year. Only about 10 percent of this is used during the first 45 days of the corn plant's life. Within the next 60 days the crop uses 70 to 75 percent of its total water needs. Timely, adequate irrigations must be made, since the crop is vulnerable to lack of moisture during the pre-tassel stage. Any injury to the crop during this stage, due to insufficient moisture, cannot be compensated for by adequate irrigation at either an earlier or a later date.

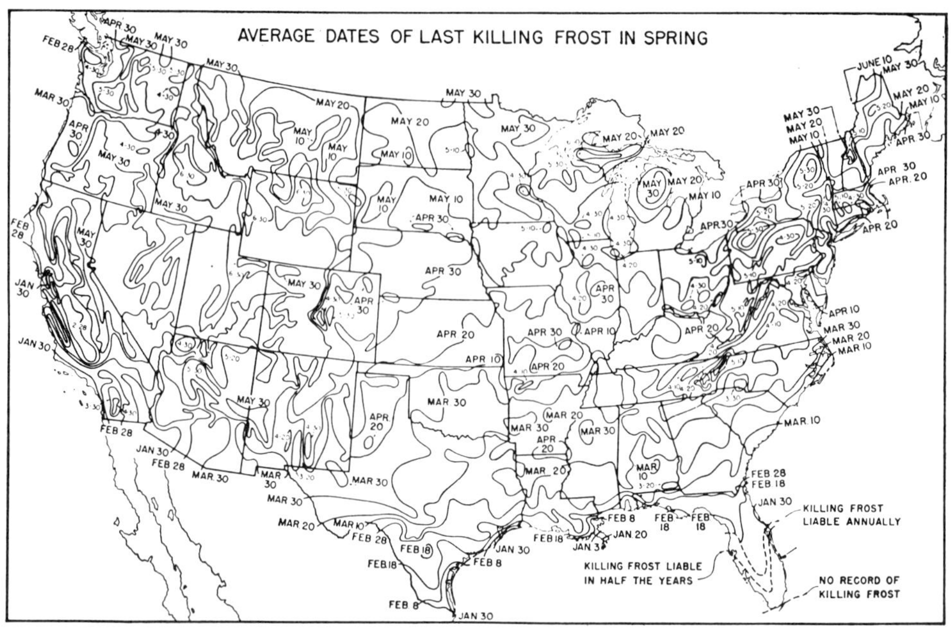

Fig. 14.12 — Corn should be planted approximately 10 days after the last killing frost in the spring. What is the proper time to plant corn where you live? (Courtesy, U.S. Department of Commerce)

Use of Fertilizers

More fertilizer is used on corn than on any other crop grown in the United States. Even with good management few mineral soils will grow corn crops of 100 or more bushels per acre (6,300 kg/ha) without the addition of chemical nutrients.

Nitrogen is the single most needed nutrient for growing corn. Nearly all acres of corn will need some nitrogen fertilizer. **Yield goal** is one of the major considerations to use in determining the optimum rate of nitrogen application for corn. The yield goal is a reasonable expectation of what an individual grower can produce on a given field. Approximately 2 pounds per acre (2 kg/ha) of actual nitrogen (N) fertilizer should be applied for each 1 bushel per acre (63 kg/ha) increase in corn yield expected.

Phosphorus and potassium are the next most likely needed nutrients for corn and such need is best determined by a reliable soil test. The proper use of fertilizer on corn increases crop yields; it also increases the efficiency of the use of water (Figure 14.13). For instance, in Missouri, adequate fertilization increased corn yields from 18 to 79 bushels per acre (1,134 – 4,977 kg/ha) but decreased the gallons of water required to produce *each bushel* of grain per acre, from 21,000 to 5,600 (79,485 – 21,196 liters). However, higher yields require more total water per acre (ha).

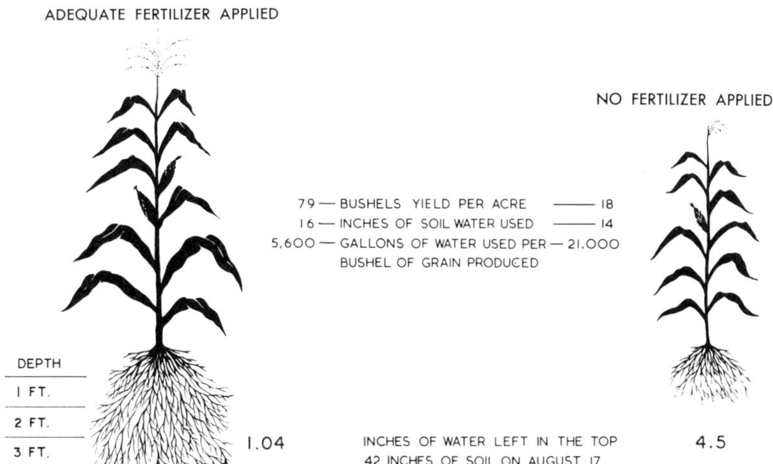

Fig. 14.13 — Fertilizer applied in timely and adequate amounts tripled corn yield per acre and resulted in only one-third as much water required *per bushel of grain.* (Courtesy, University of Missouri)

14:9 ▫ BARLEY

Barley is one of the more extensively grown of the cereals, with worldwide distribution and adaptation. It appears to have been grown in prehistoric times, long before some of the other cereals were known. Barley is grown near the Arctic Circle and also within a few degrees of the equator. The crop is better suited to hot, dry conditions than to hot, humid climates. Barley is also more tolerant of high soluble salts in the soil than any other cereal crop. A combination of high temperature and high humidity favors diseases to which barley is susceptible, such as rust, mildew, and scab.

In 1985, there were 11.5 million acres (4.6 million ha) of barley harvested for grain in the United States, with an average yield of 51.0 bushels per acre (2,744 kg / ha). States ranking highest in harvested acreage of barley in decreasing order are North Dakota, Montana, Minnesota, California, and Idaho. Barley is used in the United States primarily as a livestock feed, but it is also used as a food crop and is important in the malting industry for making beer.

The barleys are divided into *spring* barleys and *winter* barleys. The important areas for spring barley production are the Dakotas, Montana, and Minnesota. The winter type of barley is grown from the southern part of Pennsylvania to northern Texas.

Another division for barley is on the basis of the number of rows of kernels: *six* rows and *two* rows. In addition, an important designation has been made according to whether the beards are *smooth* or *barbed.* Through the years, improved varieties have been developed which are entirely free of barbs on the beards.

Barley is more tolerant of sandy soils but less tolerant of low pH (best pH is 7.0 to 8.0) than either wheat or oats. It performs satisfactorily on soils with surface texture ranging from sandy loam to clay. Barley is also one of the field crops most tolerant to soil salinity. (See Chapter 10.)

The water and fertilizer requirements for barley are similar to those for the other small grains such as wheat, oats, and rye. The fertilizer needs are determined by evaluating the current fertility level of the soil (soil testing) in combination with the particular nutrient needs of the crop to be grown and the expected crop yield. Malting barley should not be overfertilized with nitrogen, because excess nitrogen will increase grain protein content to a level not acceptable for making beer.

14:10 □ SUGAR CROPS

About 46 percent of the sugar production in the United States is from sugar beets. The rest is from sugarcane. Sugar from both sources is identical (sucrose, $C_{12}H_{22}O_{11}$). The continental production of sugarcane is limited to Florida, Louisiana, and Texas. The sugar from Puerto Rico and Hawaii is derived entirely from sugarcane. The bulk of the sugar beets is produced in 12 Western and North Central states, with Minnesota, California, Idaho, North Dakota, Michigan, and Nebraska growing the larger acreages.

Sugar Beets

The sugar beet is particularly well-adapted to irrigated agriculture in the temperate zone. In 1985, the sugar beet produced about 27 percent of the world's supply of sugar. It is a dual purpose crop which provides both a cash crop of sugar and a feed crop in tops and pulp for livestock. The sugar beet far outranks all other temperate zone plants in its ability to **photosynthesize** (manufacture food) and accumulate energy on an acre basis.

The sugar beet is a relatively cold-hardy plant, which in 5½ to 7 months is capable of producing a large tonnage of roots with high sucrose content. Warm days and fairly cool nights favor rapid growth. The sugar beet requires a plentiful and well-distributed supply of moisture during the growing season. A large part of the Intermountain and West Coast sugar beet acreage is grown under irrigation. The crop exhibits a good tolerance to soil salinity, which is present in large areas of land in the arid regions of the western states.

Optimum nitrogen fertility is that which supplies adequate nitrogen early in the development of the beet to promote rapid top and root growth. The tops need a large surface area for maximum carbohydrate manufacture, which in turn increases root size and sucrose storage. If the nitrogen supply is too high, top growth continues into the fall and the sucrose content is still low at harvest. The best sequence of nutrients and environment for sugar beets involves early nitrogen feeding at warm summer temperatures for maximum growth followed by progressively cooler nights, exhaustion of available nitrogen, and a decreased moisture supply to slow up vegetative growth and accelerate sugar storage.

Sugarcane

Sugarcane is a giant, grass-like tropical plant usually requiring 12 to 24 months to reach maturity. It requires an abundant supply of moisture and a fertile, well-drained soil for successful growth.

Louisiana State University recommends 80 to 120 pounds of actual nitrogen per acre (90 – 135 kg / ha). If the soil tests are low in phosphorus and potassium, a recommendation of 40 pounds per acre (45 kg / ha) of P_2O_5 and 80 pounds per acre (90 kg / ha) of K_2O is suggested for sugarcane.

14:11 ▫ SORGHUMS

The acreage of sorghum harvested for grain in 1985 was 16.7 million acres (6.8 million ha), with an average yield per acre of 66.7 bushels (4,187 kg / ha). States ranking highest in acreage in decreasing order are Texas, Kansas, and Nebraska. A total of 20 states report some acreage of sorghum harvested for grain.

Droughts that occurred in the 1950's throughout many states encouraged the acreage of sorghum, especially in Missouri, Iowa, Colorado, South Dakota, Arizona, and California. In these and other states grain sorghum has been replacing the acreage of corn because grain sorghum is more drought-resistant. Another factor favoring the popularity of grain sorghum is the development of dwarf varieties that can be harvested with a combine.

Grain sorghum is a warm-season crop that gives higher yields when planted early. Too early planting, however, results in poor seed germination. A good guide in judging when to plant grain sorghum is to plant it according to one of the following environmental indicators: (1) about two weeks after the average date of the last killing frost in the spring, (2) as soon as the air temperature during the day averages 65° F (18° C), or (3) when the soil in the spring at a depth of 2 inches (5 cm) reaches a temperature of 65° F (18° C). (See Figure 14.12.)

In areas of 15 to 25 inches (38 – 64 cm) of average annual rainfall, sorghums, especially grain sorghums, are well-adapted. In order to permit more water to be stored in the soil, it is sometimes a practice to grow a crop of sorghum one year, no crop (fallow) the next year, and a crop of sorghum the third year.

Although grain sorghum is more drought-resistant than corn, yields of grain sorghum are increased by timely rains or when irrigation water is applied scientifically. Like most crops, grain sorghum roots obtain most of their water from the top 2 to 3 feet of soil (61 – 91 cm), but withdrawal of water to a depth below 5 feet (152 cm) is possible if the soil permits root extension to this depth.

In irrigating grain sorghum, it is usually practical to wet the soil to a depth of 5 feet (152 cm) *before* the crop is planted. This moisture alone may be enough to mature a good crop, especially on loam or clay soils that have a high water-holding capacity. On most soils, however, irrigating the growing crop is necessary. In this connection, the irrigation water should be applied in adequate amounts to meet the water requirements of the crop at its different stages of growth. Figure 14.14 portrays this information. It may be pointed out that the maximum water requirement is at the boot stage of growth (head of grain enclosed by sheath) and is in excess of 0.3 inch (0.8 cm) of water a day.

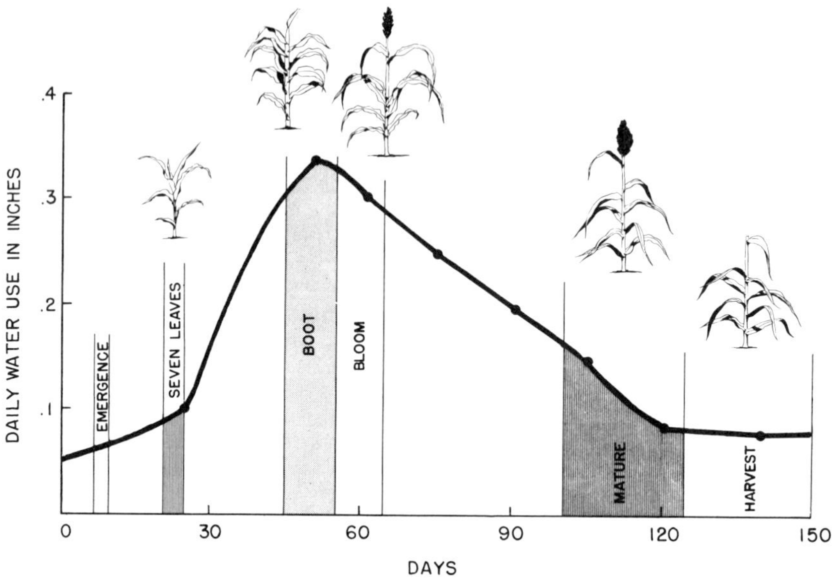

Fig. 14.14 — The daily water requirement of grain sorghum, by stages of growth. Maximum yields are obtained only when these amounts of water are available to the crop at each stage. (Courtesy, Texas A&M University)

Use of Fertilizers

Sorghums have almost the same fertility requirements as corn, but since sorghums are more often grown in semiarid regions, less fertilizer is used. In recent years grain sorghum has been planted in increasing amounts in humid areas of the southern and central regions. In these locations of higher rainfall, grain sorghum can be fertilized the same as corn.

More care must be used in placing fertilizer at the proper location with respect to the seed, because grain sorghum seedlings are more easily injured than those of corn by fertilizer "burn" (salt injury, **exosmosis**).

In the 25-inch (64-cm) rainfall belt of northwest Texas, a five-year nitrogen fertilizer test on grain sorghums resulted in a doubling of the yields of grain. With no nitrogen fertilizer the yields averaged 16 bushels per acre (1,004 kg/ha). With an application of 30 pounds of nitrogen per acre (34 kg/ha), the yields averaged 32 bushels (2,009 kg/ha), double the average yield of the no-nitrogen plots. Doubling the nitrogen fertilizer increased the yield only 1 more bushel, to 33 bushels per acre (2,072 kg/ha). Under these conditions, 30 pounds per acre (34 kg/ha) of nitrogen appears to be the recommended rate. Under irrigated conditions or in higher rainfall areas, grain sorghum may be expected to respond to much higher rates of nitrogen (Figure 14.15).

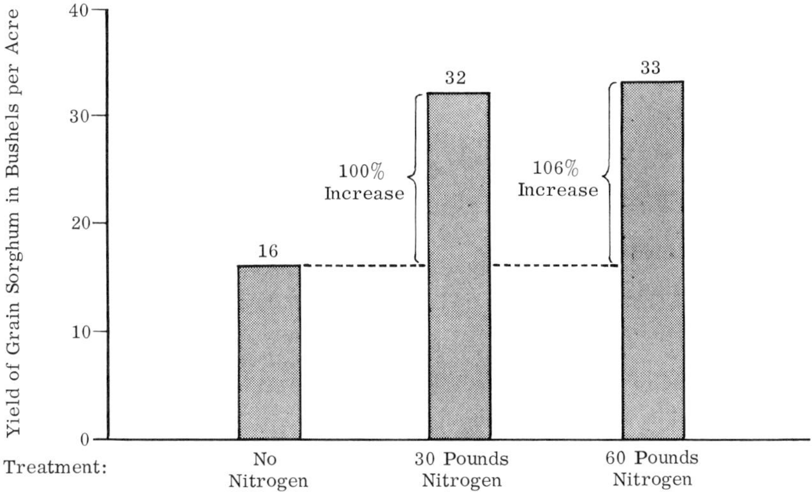

Fig. 14.15 — Grain sorghum yields are doubled by 30 pounds / A (34 kg / ha) of nitrogen fertilizer. (Texas) (Courtesy, USDA)

Note: (1) Average annual rainfall is 25 inches (64 cm). (2) Soil is a Miles fine sandy loam. (3) Dryland conditions (not irrigated).

14:12 □ RYE

There are approximately 2.5 million acres (1.0 million ha) of rye seeded annually in the United States. Slightly over 700,000 acres (283,000 ha) of the total are harvested for grain, with the leading states being North Dakota, South Dakota, Georgia, Minnesota, and Nebraska. In the United States about 50 percent of the grain crop is used in bread and in the distilling industry. The remainder is fed to livestock. Rye is the best suited of the cereals for general use as a cool-season pasture crop. It is also a good winter cover and green-manure crop, especially on sandy soils. (See Chapter 5.)

The procedures for growing winter rye are similar to those common in the production of winter wheat. It is generally not advisable to grow winter rye for grain on farms where winter wheat is grown, because it is almost impossible to keep the rye seed from becoming mixed with the wheat seed. It is almost certain that volunteer plants will appear in the following crops.

Rye is usually not fertilized, though it will respond to fertilizers on the less fertile soils.

14:13 □ OATS

There were approximately 8.1 million acres (3.6 million ha) of oats harvested in the United States for the year 1985. The leading states in oat production are Minnesota, South Dakota, North Dakota, Wisconsin, and Iowa. In the U.S.

Corn Belt, oats rank second only to corn as a feed crop with most of the crop consumed on the farm where it is produced. About 4 percent of the oat crop in the United States is used in making breakfast foods such as rolled oats and oat bran.

Oats perform well on soils within the range of fine sandy loam to clay texture, if drainage is adequate. As a general rule, oats are grown in a rotation. Therefore, it has been more profitable to apply commercial fertilizer to other crops in the rotation than to oats. In areas where legumes are established by seeding in oats, an application of phosphorus is generally recommended for oats.

14:14 □ ALFALFA

Alfalfa is grown in 43 states on a total acreage of 25.7 million (10.4 million ha), and 10 states each harvested more than 1 million acres (0.4 million ha) in 1985. States ranking highest in total acres, in decreasing order, are Wisconsin, South Dakota, Minnesota, Iowa, and North Dakota — all contiguous states. Alfalfa is the best of all forage crops when harvested as hay.

Only very fertile soils that are abundantly supplied with lime will produce alfalfa successfully. The soil must be deep, well-drained, and moist. Alfalfa is moderately tolerant to salt, ranking along with sweetclover, wheat, corn, and grain sorghum in this respect (Figure 14.16). (See Tables 10.3 and 10.4.)

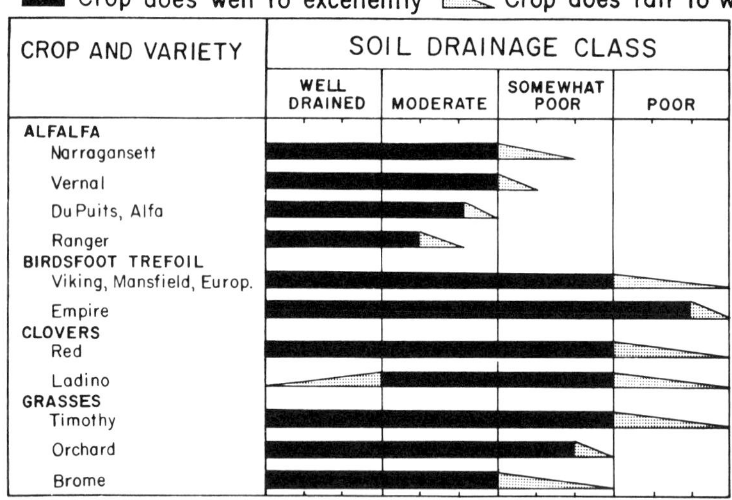

Fig. 14.16 — Choosing a crop variety (legume or grass) to fit the soil drainage class in your field is of prime importance to assure successful crop production. Information on the drainage class in each field may be obtained from the Soil Conservation Service or the extension service in each state. (Courtesy, Department of Agronomy, Cornell University)

Note: (1) Drainage means *internal* soil damage. (2) All soil survey reports indicate the drainage class of each soil map unit.

Alfalfa is well-adapted to irrigated conditions in the West. Under dryland conditions in areas of approximately 20 inches (51 cm) of average annual precipitation, alfalfa may exhaust most of the subsoil moisture.

Like other crops, alfalfa has a characteristic water use pattern. This pattern may vary in magnitude and length from season to season, but the same general pattern persists. The rate of water use by the crop is low immediately following harvest. As the crop begins to grow, the rate increases sharply and reaches a peak at the pre-bud stage. From this stage until maturity of the plant, the water use rate decreases sharply and reaches a low again after the next harvest (Figure 14.17).

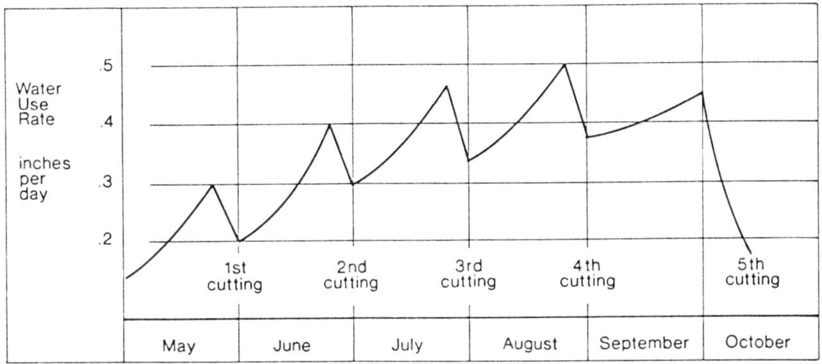

Fig. 14.17 — Characteristic water use pattern for alfalfa. (Courtesy, Kansas State University)

For successful long-lived stands of alfalfa, The Ohio State University recommends the following 10 steps to maximum alfalfa yields and forage quality:

1. Select a deep, well-drained soil.
2. Determine soil nutrient levels by soil tests and apply these amounts.
3. Lime soil to pH 7.
4. Raise soil P level to 90 pounds per acre (101 kg / ha) and K level to 300 pounds per acre (336 kg / ha).
5. Use certified seed of an adapted variety.
6. Band seed with press wheels in April or August. Press wheels used with band seeding provide additional stand insurance. (See Figure 13.6.)
7. Reduce weed competition and give alfalfa seedlings a chance.
8. Control alfalfa insects.
9. Harvest on time (when about 1/10 of full bloom).
10. Fertilize by top-dressing the crop annually, based on a soil test.

14:15 ▫ SOYBEANS

Soybeans are grown in 29 states on a total harvested acreage of 61.5 million

(24.8 million ha). States ranking highest in total acres in decreasing order are Illinois, Iowa, Missouri, Minnesota, Indiana, and Ohio.

Soybeans are grown satisfactorily under a wide variety of climatic and soil conditions. Soybeans give relatively better yields under soil conditions unfavorable for corn or cotton because they utilize soil nutrients more efficiently. In general, however, for best production of soybeans, the soil and climatic conditions approximate those best for corn.

The characteristic growth and water use pattern for soybeans is shown in Figure 9.6. It is noted in Figure 9.6 that water use rate is very low during the germination and seedling stages; increases slowly during the rapid growth stage; reaches a peak during the reproductive stage; and decreases rapidly during the latter portion of the maturity stage. For this total cycle of development, the total annual water use by the crop (evaporation plus transpiration) will be within the range of 20 to 24 inches (51 – 61 cm).

Under normal conditions, properly inoculated soybeans are assumed to need no nitrogen fertilizer, as the nodule bacterium *Rhizobium japonicum* will fix sufficient nitrogen for optimum growth. To ensure nodulation, inoculation of the soybean seed is recommended even if soybeans have been grown on the land previously.

The soybean plant is a relatively heavy user of phosphorus and potassium. Therefore, phosphorus and potassium should be maintained at a medium or preferably high level, as determined by a soil test. The soil should be kept at a soil pH of at least 6.0 by a good liming program.

14:16 □ COTTON

Cotton is grown in 17 states on a total harvested acreage of about 10.1 million (4.1 million ha). States ranking highest in total acreage of cotton, in decreasing order, are Texas, California, Mississippi, Arkansas, and Oklahoma. The yield of lint harvested by all states in 1985 averaged 628 pounds per acre (789 kg / ha).

Cotton in the past was produced only with the expenditure of large amounts of labor. A shortage of low-priced labor led cotton producers to look for labor-saving devices (Figure 14.18). Large acreages of cotton are now produced under humid conditions in the South and Southeast and under irrigated subhumid and low-rainfall conditions in the Southwest and West.

The average summer temperature along the northern edge of the Cotton Belt in the United States is 77° F (25° C). This temperature appears to be the lower limit at which cotton production becomes profitable. The length of the growing season along the northern edge of the Cotton Belt is from 180 to 200 days.

Cotton can be grown on a variety of soils but does best on loamy ones that are deep, well-drained, and moderately fertile. Fine clay soils tend to delay cotton maturity and increase vegetative growth. Cotton has a high tolerance to saline soils. (See Table 10.4.)

Fig. 14.18 — Mechanical harvester in a cotton field in Texas. (Fort Stockton, Texas) (Courtesy, Texas Highway Department)

14:17 ▫ REFERENCES

Agricultural Statistics. U.S. Dept. of Agriculture, 1986, 551 pp. (updated annually).

Anderson, Frank N., and Gary A. Peterson. "Effect of Incrementing Nitrogen Application on Sucrose Yield of Sugarbeet." *Agronomy Journal.* Vol. 80, pp. 709 – 712, 1988.

Brick, M. A., R. H. Follett, and D. H. Smith. "Alfalfa Forage in Colorado." Colorado State University, Co-op. Ext. Serv., in Action No. 703, 1984.

Donahue, Roy L., Raymond W. Miller, and John C. Shickluna. *Soils: An Introduction to Soils and Plant Growth,* 5th ed. Englewood Cliffs, New Jersey: Prentice-Hall, Inc., 1983. (See especially Chapter 11, "Plant Diagnosis and Fertilizer Recommendations.")

Eck, Harold V. "Winter Wheat Response to Nitrogen and Irrigation." *Agronomy Journal,* Vol. 80, pp. 902 – 908, 1988.

Elmore, C. Dennis, and Larry G. Heatherly. "Planting System Weed Control Effects on Soybean Grown on Clay Soil." *Agronomy Journal.* Vol. 80, pp. 818 – 821, 1988.

"Grain Sorghum Handbook." Kansas State University, Co-op. Ext. Serv., C-494, 1980, 27 pp.

Hofman, W. C., D. L. Kittock, and M. Alemayehu. "Planting Seed Density in Relation to Cotton Emergence and Yield." *Agronomy Journal.* Vol. 80, pp. 834 – 836, 1988.

Jones, J. H., and F. J. Olsen. "Alfalfa Establishment and Production on Soils with Different Drainage Characteristics." *Agronomy Journal,* Vol. 79, 1987, pp. 152 – 154.

Metcalf, Darrel S., and Donald M. Elkins. *Crop Production: Principles and Practices.* New York: Macmillan Publishing Co., Inc., 1980, 774 pp.

Plucknett, D. L., and Nigel J. H. Smith. "Sustaining Agricultural Yields." *BioScience*, Vol. 36, 1986, pp. 40 – 45.

Quarberg, D. M., and F. J. Wooding. "Successful Barley Practices." Alaska Commercial Agr. Ser., University of Alaska, A-00245, 1986.

"Soybean Handbook." Kansas State University, Co-op. Ext. Serv., C-449, 1980, 26 pp.

Stoskopf, Neal C. *Understanding Crop Production.* Reston, Virginia: Reston Publishing Co., Inc., 1981, 433 pp.

"Wheat Production Handbook." Kansas State University, Co-op. Ext. Serv., C-529, 1986, 31 pp.

Will There Be Enough Food? The Yearbook of Agriculture, 1981. U.S. Dept. of Agriculture, 302 pp.

CHAPTER 15

Vegetable Gardens: Soil, Water, and Fertility Management

"People who have wild ideas about how to run the earth ought to start with a small garden." — Lou Erickson in *Atlanta Journal*

OUTLINE

Soil, water, and fertility management of vegetable gardens varies from climate to climate, soil to soil, and gardener to gardener. Here tensiometers are being used at two root depths (at arrows) to determine when to start and when to stop adding irrigation water to this commercial field of tomatoes on a saline clay loam. (Courtesy, Irrometer Company, Inc., Riverside, California)

□ □ □

15:1 □ OVERVIEW

Currently, about 34 million families enjoy the outdoor work of raising vegetables for the home table. The main reasons for gardening are for relaxation and for more healthful foods.

The average garden size is about 300 square feet (28 square m). This is about the same area as that of two rooms in an average house. By comparison, in the 1970's the average garden size was twice as large. The primary reason for smaller gardens is that more members of the family are now working for wages outside the home.

A well-managed, two-room – size garden can furnish as much stress-relieving activity and garden-fresh vegetables as an area much larger. However, very little can be grown for canning or freezing unless the garden size is at least 900 square feet (83.7 sq m), three times the average size.

Successful management of soil, water, and fertility for a vegetable garden varies from climate to climate, soil to soil, and person to person. There are, however, compatible sets of principles and practices that are nearly universal.

Sometimes there is no choice of a vegetable garden site. In these instances more attention must be paid to closely following recommended practices with greater precision and timeliness. When a choice of site does exist, a stone-free, well-drained loam or sandy loam should be selected that is at least 6 feet (1.8 m) deep to any root-restrictive layer such as bedrock. The site should be as far away from trees as possible to avoid root competition and shade. If the slope is not gentle enough to control erosion, the site should be terraced and the rows established parallel to the terraces. On slopes less steep, contoured rows without terraces should control erosion. The soil organic matter level of the soil should be as high as possible.

Before the gardener adds any lime or fertilizer, a chemical test should be made. Vegetable seed should be bought from a reliable seed dealer that has the best local reputation. Planting should be done at a time depending on the temperature-tolerance of the plant in relation to the frost (freeze) dates.

There are successful vegetable gardeners that use *only* an organic mulch and no chemical fertilizer. Equally successful gardeners use *both* an organic mulch and a commercial fertilizer; others use *only* a chemical fertilizer and no mulch. A fourth group of gardeners use *neither* organic mulch nor fertilizer. Either system may be successful if the supplying capacity of the soil for *all* essential plant nutrients is satisfactory. This can be determined only by a soil test. (See Chapter 2.)

Even in humid regions, all vegetable gardens must have some means of adding supplemental water at critical times. These critical times are at seedling and plant establishment and during dry periods between rains. The irrigation system must be suited to the local climate and infiltration capacity of the soil.

15:2 □ SELECTING A GARDEN SITE

Many people have very little choice in selecting a site for a garden because of the small size of most homesite lots. The back of the lot is the traditional location. On large lots, farms, and on land set aside for community garden sites, there can be some choice, and this freedom should be used to select the best of the available sites. Selection of sites for commercial vegetable production usually offers wider alternatives. The ideal soil is classified as a well-drained, stone-free loam or sandy loam, at least 6 feet (1.8 m) deep to a restricting layer such as compacted clay or bedrock. For producing vegetables for the earliest market, a south or west slope is preferred; although on these aspects the hazard of frost heaving is greater. Air drainage should be good; that is, between the garden and the nearest valley there should be no buildings or dense stands of trees or shrubs (Figure 15.1).

Even though large trees offer restful shade to the gardener, the garden site selected should be as far from a tree as possible. All trees cast shade on the garden, and this will reduce crop yields. Some trees such as elm, sassafras, white poplar, black walnut, and willow have very vigorous and extensive root systems that will grow into the garden and compete seriously with the vegetables for soil moisture and nutrients.

Fig. 15.1 — A well-selected garden site. Air drainage is not hindered, there is a gentle slope, the soil is a sandy loam which is easily tilled, and the size of the garden is adequate to supply vegetables for a large family, with some surplus to preserve and sell. Trees are far enough away to not compete with the vegetables for light, moisture, and nutrients or to act as a barrier to air drainage. (Courtesy, USDA)

Even loose sand can be made into a productive garden with adequate organic matter, fertilizer, and water. Fine-textured clay soils can also be very productive for gardens by the liberal use of organic matter. Sometimes clay may be added to a sandy garden soil and sand added to a clay soil to increase productivity.

The slope of the garden soil is important for three reasons:

1. Steep slopes are more likely to erode than gentle slopes. A slope of 1 or 2 percent is most suitable.
2. Long and uniform slopes that end in a stream valley are the most desirable because air drainage will be good, thus reducing frost (freeze) hazards. Slopes that end in a blind valley with no water drainage outlet are more susceptible to frost.
3. Long slopes that face north or east are cooler, more moist, and more desirable than hot, dry slopes that face south or west. Earlier vegetables, however, can be produced on south and west slopes.

In some locations, a windbreak should be built on the side of the garden toward the cold winds, usually the northwest. A porous windbreak such as a lath fence or live trees and shrubs is preferred over a solid, nonporous wall.

Regardless of the amount of annual rainfall, every garden needs a source of irrigation water. Especially at the time of transplanting, irrigation is essential for successful establishment.

15:3 □ TILLING

Traditionally, garden soils are kept clean-tilled. The rotary tiller is the most common tillage machine used in medium-sized and large vegetable gardens. In small gardens, perhaps the wheel hoe or common garden hand hoe is used to kill weeds and stir the soil. A bare soil, however, gets hotter and colder than one that is mulched with organic matter. A bare soil also is beat into flowing mud by falling raindrops that strike the soil at a speed of 20 miles (32 km) per hour. The result is splash erosion, sheet erosion, muddy water to seal the cracks and insect holes, and a surface crust upon drying. A bare soil often deteriorates into a barren soil. (See Chapter 7.)

The alternative to a bare garden soil is an organic mulch such as hay or straw. Other organic materials are also usually suitable, such as sawdust, peanut hulls, woodchips, and cotton gin trash. Organic material such as grain straw is perhaps more common because of its availability and because it is usually freer of weed seeds than hay is. Organic gardeners keep 2 to 4 inches (5 – 10 cm) of straw or hay on the soil at all times and practice no-tillage. When they get ready to plant seeds, they rake an area clean, scratch the line where seeds are to be planted, plant the seeds, and gradually pull the organic mulch around the new seedlings as they grow. The mulch smothers out most of the weeds; what few come up through the mulch are easily pulled out because the soil is always soft (Figure 15.2). (See Chapter 5.)

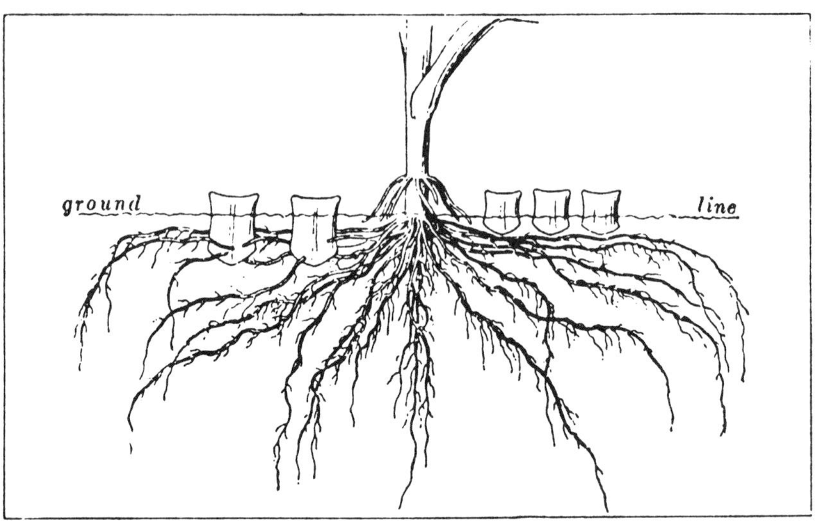

Fig. 15.2 — *Top*: With any kind of clean cultivation and mechanical tillage system of gardening, it is almost impossible to avoid cutting vegetable plant roots. *Bottom*: With an organic mulch system of gardening and biological tillage by the earthworm, root growth is enhanced. (*Top*: Courtesy, Department of Crop and Soil Sciences, Michigan State University. *Bottom*: Courtesy, Julian P. Donahue)

Another system of gardening is to spade only the rows and keep an organic mulch between the rows. This is similar to the "strip tillage" system used for field crops. (See Chapter 6.)

15:4 ◻ SOIL TESTING

Every gardener should take a soil sample and send it to the state laboratory for testing to determine the lime and fertilizer needs. In arid lands, the soil may be too alkaline, too salty, or too sodic; these conditions can be determined by a soil test.

The county extension director should be contacted for detailed information on how to collect the sample, how to fill out the necessary forms, where to send the soil sample for testing, the cost, and how to interpret the results. The extension director is the official representative of the state soil testing laboratory. (See Chapter 4.)

The hazards of *not* having the garden soil tested are that an excess of lime may cause **chlorosis** (yellowing) of leaves and a reduction in yield due to iron deficiency; or other micronutrients may be deficient. If the soil is strongly acid and no lime is added, all yields will be low and some vegetable crops may be failures. The soil may be rich in potassium, and if more potassium is applied, the calcium and magnesium uptake by plants will be less and yields may decline. Too much phosphorus added to a high-phosphorus soil may induce a zinc deficiency in corn or beans; and an excess of nitrogen will produce luxuriant lettuce and spinach but delay maturity and reduce the yield of tomatoes and sweet corn. Too much nitrogen also may increase damage by insects and diseases; whereas, too little nitrogen will stunt growth and cause the leaves to turn yellow. Finally, without a soil test, most vegetable gardeners apply too much fertilizer and thereby waste money.

15:5 ◻ USING ADAPTED SEED

The best soil, water, and fertility management practices in the world cannot substitute for the use of adapted seed.

To be most certain of buying adapted seed:

1. Buy only from a reputable seed house. (Ask acquaintances who garden successfully where they buy their seed.)
2. Select seed that has been grown under ecological conditions similar to those in your garden. (See Note 15.1.)
3. If you have a choice of seeds grown north or south of your garden, choose those that have been grown north of you.
4. If you have a choice of seeds produced at elevations higher or lower than your garden, choose seed grown at higher elevations.

Note 15.1 — Hopkins Bioclimatic Law

The **Hopkins bioclimatic law** states that on land areas in the world:

- For each degree of latitude north or south of the Equator, and
- For each 400 feet (122 m) increase in altitude, the date of flowering of plants of the same species is *retarded* four days, and
- For each five degrees of longitude from east to west, the date of flowering is *advanced* four days.

5. Plan the garden, then order or purchase locally all of the seed that you anticipate needing for an entire year in order to be certain of obtaining what is best adapted. This forethought will reduce the hazard of "impulse buying" of seed at the local drug store or grocery store the day before you are ready to plant. Such seed may not be adapted.

15:6 □ WHEN TO PLANT

Some gardeners plant according to phases of the moon: root crops (such as potatoes) during the dark of the moon and fruit crops (e.g., tomatoes) during the light of the moon. Scientists believe this to be a myth.

Scientists and experienced gardeners plant cool-season crops outdoors about two weeks *before*, and warm-season crops about two weeks *after*, the average last freeze in spring (Figure 15.3). Examples of temperature adaptation of common vegetables and flowers are listed in Table 15.1.

Planting in rows running north and south will permit more sunlight to strike more plants (less shading) and result in higher yields, because the pathway of the sun in our northern hemisphere is always in the southern half of the sky.

Table 15.1 — Temperature Adaptation of Principal Vegetables and Flowers

Cool-Season Crops		Warm-Season Crops	
Vegetables	**Flowers**	**Vegetables**	**Flowers**
Beets	Aster	Beans	Coleus
Cabbage	Carnation	Cucumbers	Balsam
Carrots	Chrysanthemum	Melons	California poppy
Lettuce	Larkspur	Peppers	Marigold
Mustard	Nasturtium	Pumpkins	Morning-glory
Onions	Pansy	Squash	Sunflower
Peas	Sweet pea	Sweet corn	Zinnia
Potatoes		Sweet potatoes	
Radishes		Tomatoes	

Fig. 15.3 — Cool-season plants should be planted about two weeks before the mean date of last 32° F (0° C) temperature in spring. Warm-season plants should be planted about two weeks after this date. (Courtesy, U.S. Department of Commerce)

15:7 ▫ MULCHING

Mulching materials commonly used in gardens include:

1. Clear polyethylene plastic mulches to hasten germination and growth during early spring or in areas with a short growing season, such as Alaska.
2. Black polyethylene mulches to smother difficult-to-kill weeds such as bermudagrass and quackgrass.
3. Organic mulches such as straw, hay, sawdust, pine needles, woodchips, and hardwood leaves (Figure 15.4).

Fig. 15.4 — Three kinds of mulches are used in this home garden: (1) Black polyethylene mulches at arrow (held in place by white bricks) to smother bermudagrass; (2) wheat straw between the rows to control weeds, conserve moisture, and supply micronutrients; and (3) weathered sawdust under straw to help smother weeds. (Courtesy, Roy L. Donahue)

Other surface mulches sometimes used are:

1. Animal manures (fresh is more desirable if kept away from plants).
2. Peat moss (often blows or floats away).
3. Waste paper (blows away and is ugly).
4. Tree bark (floats away).
5. Crushed rock, gravel, and volcanic rock (dulls the hoe and interferes with weeding).
6. Perlite and vermiculite (blow away and float away).
7. Grass clippings (avoid bermudagrass, quackgrass, and other difficult-to-eradicate plants).

Note: Do not use sewage sludge or septage as a garden mulch even if it is composted, heat-treated, or irradiated. (See Chapter 5.)

15:8 ▫ LIMING

If the soil test calls for the application of lime, the lime can be applied at any time that it can be tilled or spaded into the soil. Uniform spreading is essential. Uniformity can be obtained with the use of a hopper-type fertilizer spreader, either tractor-operated or hand-operated. Fairly uniform applications can be achieved by hand-spreading if these suggestions are followed:

1. Do not spread lime on a windy day.
2. Divide the total amount of lime to be spread into two equal parts, then spread one half while moving back and forth in a north and south direction, and spread the other half while moving in an east and west direction.

The usual frequency of lime applications is about once every three to five years, but another soil test during this period will give a more accurate determination.

Recent research in Virginia has indicated that lime can also be effectively applied on fields on which no-tillage is practiced. This means that lime can also be applied efficiently by being spread on top of an organic mulch on the garden. (See Chapter 3.)

15:9 ▫ FERTILIZING

There are many ways to fertilize a garden, from top-dressing by broadcasting to row application only. Even a row application has variations by positions in placement. Under large field conditions and with the use of a machine to sow seed and fertilizer at the same time, the most common recommendation is the placement of the fertilizer in a continuous band 2 inches (5 cm) from one or both sides of the seed and 2 inches (5 cm) below the seed level. This position of the fertilizer reduces the hazard of "burning" (**plasmolysis**) of the seedlings (Figure 15.5). In small gardens, this position can be approximated by making a deep (4-inch) (10-cm) trench, applying the fertilizer in a continuous band along the bottom of the trench, then making a shallower (2-inch) (5-cm) trench in which to plant the seed about 2 inches (5 cm) to one side of the first trench.

Another modified method of row application is to use a four-tine scratcher or narrow (4-inch) (10-cm) hoe to make a flat-bottomed trench about 4 inches (10 cm) deep, apply the right amount of fertilizer in the trench, then scratching or hoeing it to mix the fertilizer with the soil. This should be done a week or more in advance of planting and should be soaked with rain or a garden hose. The "salty" part of the fertilizer will react with the soil and thus reduce the "burning" hazard to earthworms and to seedlings.

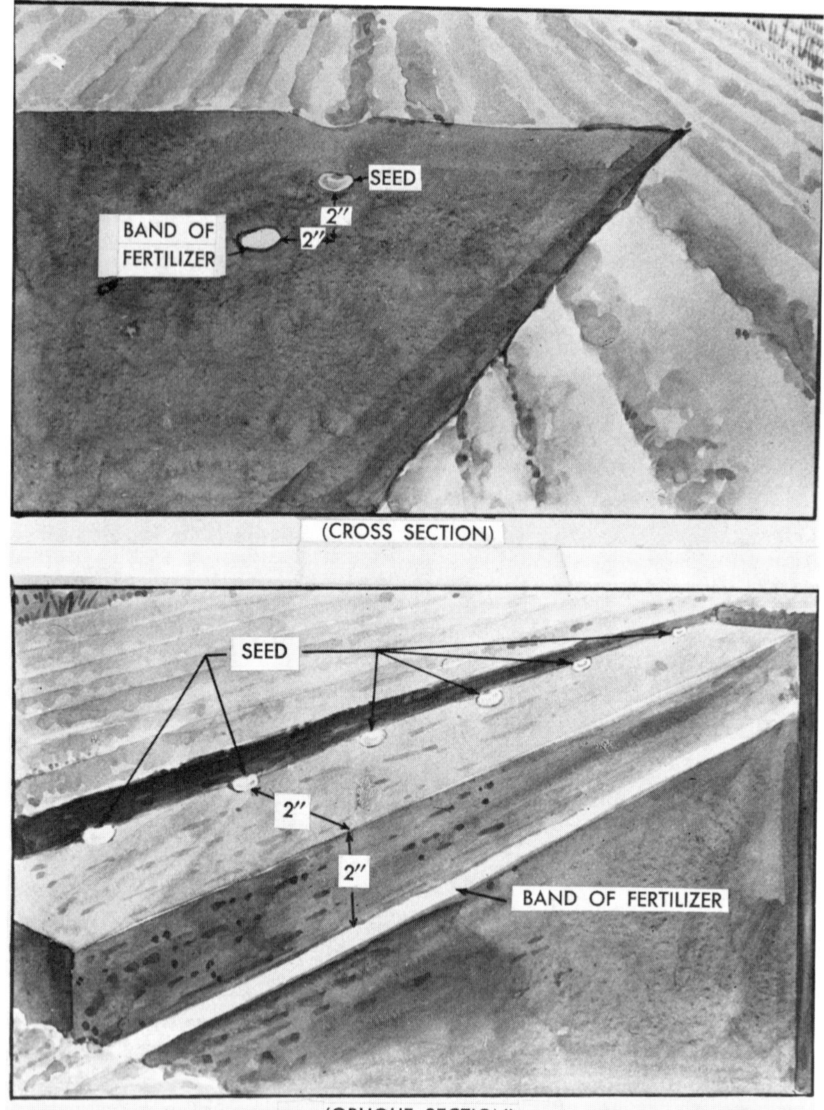

Fig. 15.5 — Fertilizers are most efficiently used when applied approximately 2 inches (5 cm) to one side of the seed and 2 inches (5 cm) below seed level. (Courtesy, U.S. Agency for International Development)

Organic gardeners often apply cottonseed meal or other oil meals, rather than chemical fertilizers. Even these organic fertilizers should not be in direct contact with seeds because, although they will not "burn," they do encourage the damping-off fungus disease that is lethal to seedlings. (See Chapter 4.)

Kinds and amounts of fertilizer to apply must be determined by a soil test. However, a soil test is seldom made for nitrogen or for all micronutrients; furthermore, some tests do not accurately diagnose the deficient nutrients because of the very small concentrations in the soil. Decomposing organic matter supplies some of all essential nutrients. (See Chapter 5.)

15:10 □ PLANTING

Most of the common vegetable seeds that are normally grown in pots and later transplanted in the field require light for germination. This dictates very shallow planting and thin covering with peat, vermiculite, or clear plastic. If plastic is used, it must be removed immediately after germination.

A list of vegetable seeds that must have light for germination includes:

cabbage	muskmelon
cauliflower	onion
celery	pepper
cucumber	summer squash
eggplant	tomato
lettuce	watermelon

Poorly drained soils should be bedded (ridged) to provide better drainage for plant roots on top of the beds. The beds can be made wide enough for one row (about 8 inches) (20 cm) or for two rows of crops such as lettuce or one row of beans or staked tomatoes, perhaps 24 inches (61 cm) wide. The height of the bed can vary from only a few inches to about 6 inches (15 cm). The shape of the top of the beds should be almost flat, or with only a slight crown.

The same or similar system of bedding (ridging) is recommended on well-drained soils in areas of very high spring rainfall, or in arid regions where furrow irrigation is practiced. After the garden is planted, irrigation water is released in the furrow between every two ridges and is moved down each row by gravity and sideways by capillarity to wet the soil around the seeds.

On well-drained soils where spring rainfall is low, it is sometimes recommended to construct similar narrow ridges but wider and flat-bottomed furrows, then plant the garden seed in the furrows. With this system of furrow planting, it is essential that each furrow drain readily following a heavy rain because even a few hours of standing water will kill most vegetable seedlings.

Most gardens on well-drained soils in humid regions can be planted "flat"; that is, with no ridges and no furrows. But, even flat planting needs modification to assure uniform and adequate distribution of water. The flat planting of a tomato plant is illustrated in Figure 15.6. This black, stony, well-granulated soil is satisfactory for flat planting because rain or irrigation water will move into it rapidly. On most soils there should be a depression around each plant to hold about a quart (0.9 liter) of water until it soaks into the soil around the roots.

Fig. 15.6 — These tomato plants are being planted "flat," that is, with no trench or depression to hold the water until it soaks around the roots. This system of planting is satisfactory only on permeable, stony clays, as shown here, and on sands. (Courtesy, Tennessee Valley Authority)

15:11 □ WATERING

From the day that garden seeds are planted until the plants are well-established, the soil around the root zone should be kept moist. Most gardeners do this with a garden hose. But anyone who has tried to get uniform and adequate water distribution around the seeds and seedlings when the garden is planted on the *flat* knows how impossible this is on soils not freely permeable. For this reason, it is highly recommended to use a four- or five-tine scratcher and make a very shallow furrow about 1 inch (2.5 cm) deep and 4 inches (10 cm) wide on otherwise flat land, and plant the seed in the minifurrow. Then by watering with a hose at a water flow rate never fast enough to overflow the furrow, the seeds and seedlings can be kept moist, while the paths between the rows remain dry enough at all times to walk on. The use of the minifurrow system makes it easier to maintain uniform and adequate water around the seeds and plants and saves water. Land that is sloping should have the ridges and furrows established on the contour (Figure 15.7).

Successful vegetable gardening in arid regions requires the availability of an adequate supply of high-quality water for irrigation and a satisfactory delivery system.

The most common anti-quality hazards of water in arid regions are high salt content and disease organisms. In humid regions, water with a high salt content may come from deep wells in any location or from shallow wells near the seashore. In arid regions, water high in salt may come from surface as well as underground sources. Water-carrying disease organisms may come from municipal sewage treatment plants, septic tank effluents, or livestock feed lots.

The methods most commonly used for applying water are the sprinkler, the furrow, the trickle, and the subsurface systems. (See Chapter 9.)

Fig. 15.7 — To reduce erosion on sloping lands 2 to 3 percent or greater, the ridges and furrows should follow the contours with a grade of 0.5 percent along the water furrows. Here, an Irish potato field is being irrigated with a gated pipe system, which releases water uniformly down each furrow. (Idaho) (Courtesy, USDA - Soil Conservation Service)

1. The sprinkler system for the garden may vary from a $5 handmoved sprayer at the end of a hose (like those used for sprinkling lawns) to overhead rigid rotating pipes fixed to ground-line sprinkler heads.
2. The furrow method of irrigation will be successful only on medium- or fine-textured soils. Because of rapid losses of water by deep percolation, the furrow method would not be successful on deep sand soils. Other factors determining success are the establishment of the gradient along the furrows and a system of drainage at the ends of the rows to dispose of or to reuse tailwaters.
3. Trickle irrigation consists of a system for the slow release of water on individual plants or hills of plants over several hours at a time. Hill plants such as cucumbers can be watered efficiently by this method.

4. Subsurface irrigation consists of a series of buried pipes with small openings that regulate the amount of water moving into the soil so that the surface is seldom wet by capillary rise. Once the garden crops are up, weeds are kept at a minimum, thus the greatest efficiency is achieved by the least amount of water.

Regardless of the irrigation system used, it is very difficult to determine when and how much to irrigate the vegetable garden. Four indications of water need are (1) the wilting of plants, (2) the feel of the soil, (3) the use of a homemade evaporation can, and (4) the use of commercial tensiometers.

1. When plants on the most droughty soils (usually sands) *start* to wilt at midday, it is time to irrigate.
2. Determining when a soil *feels* dry enough to start irrigating depends on texture (Table 15.2).
3. A good evaporation can is easily made from a clean 5-quart oil can with one end cut out and the can placed in or near the garden and away from the drip of branches. On top of a fence post is an ideal location. After a soaking rain or irrigation, the oil can should be filled to within 2 inches (5 cm) of the top. Water will evaporate from the oil can, and water from rainfall will be added to the can. If rainfall just equals evaporation, the water level will remain at about 2 inches (5 cm) from the top and no irrigation will be needed. But if the water level drops 0.7 inch (1.8 cm) and the garden soil is a sand, it is time to irrigate by adding 0.7 inch (1.8 cm) of water. Fine-textured soils will not require irrigation until more water is evaporated (Table 15.3).
4. A commercial tensiometer can be used to determine with greater precision when to start and also when to stop irrigating. At the bottom of the tensiometer is a porous tube, mounted in a closed system with a cap that can be removed for adding water for maintaining the water level and a gage that is calibrated to read: "Time to irrigate." Tensiometers are buried at various levels to record the soil moisture at the point where roots are absorbing water and growing most vigorously. Sometimes one tensiometer is buried at the 4-inch (10-cm) level, another at 8 inches (20 cm), one at 12 inches (30 cm), and a fourth at 16 inches (41 cm). Such a system is especially desirable for use on deep-rooted crops such as muskmelons (Figure 15.8).

15:12 □ WEEDING

A system of soil, water, and fertility management practices to control weeds includes the avoidance of applying water on the soil *between* the rows (except in the furrow irrigation system), the use of a mulch to smother the weeds, and the proper placement of fertilizers to hasten growth of the vegetables so they will outgrow the weeds.

Table 15.2 — Guide for Determining by Soil Texture When to Start Irrigating[1]

Relative Moisture Condition	Percent of Available Moisture Remaining in Soil	Soil Texture		
		Sands — Sandy Loams	Loams — Silt Loams	Clay Loams — Clays
Dry	Nearly 0% (Wilting Point)	Dry, loose, flows through fingers.	Powdery, sometimes slightly crusted but easily broken down into powdery condition.	Hard, baked, cracked; difficult to break down into powdery condition.
Low	50% or less	Loose, doesn't feel moist.	Will form a weak ball when squeezed, but won't stick to tools.	Pliable, but not slick, will ball under pressure, sticks to tools.
Time to Irrigate Here				
Fair	50 to 75%	Tends to ball under pressure but seldom will hold together when bounced in the hand.	Forms a ball somewhat plastic, will stick slightly with pressure, doesn't stick to tools.	Forms a ball, will ribbon out between thumb and forefinger, has a slick feeling.
Good	75 to 100%	Forms a weak ball, breaks easily when bounced in the hand, feels moist.	Forms a ball, very pliable, sticks readily, clings slightly to tools.	Easily ribbons out between thumb and forefinger, has a slick feeling, very sticky.
Ideal	100% (Field Capacity	Soil mass will cling together. Upon being squeezed, outline of ball is left on hand.	Wet outline of ball is left on hand when soil is squeezed, sticks to tools.	Wet outline of ball is left on hand when soil is squeezed. Sticky enough to cling to fingers.

[1] Mark Peterson, William J. Murphy, and John Falloon. "When and How Much to Irrigate." Science and Technology Guide No. 1650, University of Missouri, Columbia, 1971.

Table 15.3 — The Relationship Between Soil Texture and Time to Irrigate When a Specific Amount of Water Has Evaporated from an Evaporation Can[1]

Soil Texture	Irrigate When Water Level in Evaporation Can Drops These Amounts
	(*in*)
Sand soils	0.7
Sandy loams	1.3
Loams	1.8
Silt loams	2.2
Clay loams	1.9
Clays	1.8

[1]"Schedule Irrigations with Evaporation Cans." Science and Technology Guide No. 1656. Department of Agricultural Engineering, College of Agriculture, University of Missouri - Columbia, 1969.

Note: Inches × 2.5 = cm.

Fig. 15.8 — *Left*: A tensiometer showing a porous base, a hollow tube, a watertight cap, and a vacuum gage for reading "Time to irrigate." *Right*: Muskmelons with four tensiometers mounted at four levels in the root zone. (Courtesy, Irrometer Company, Inc., Riverside, California)

The most feasible method of watering garden seeds and plants without hastening the germination of weed seeds between the rows is to use the minifurrow method previously explained in Section 15:10. Careful manipulation of an organic mulch can very effectively smother weeds, as explained earlier in this chapter; and fertilization applied with precision can hasten seedling growth and thus help the garden plants to grow faster than the weeds.

The use of herbicides in the garden to control weeds is feasible only under restricted conditions. A pre-emergent herbicide may be effective and safe, but the authors of this book have never used one except around borders of gardens and on field crops. Herbicides to kill weeds in one part of the garden are a hazard to nontarget vegetables because of drift. The authors believe that the safe uses for herbicides are to kill all vegetation before establishing a garden, and to kill all vegetation around the borders. Border sterilization can be accomplished cheaply, safely, and effectively by the application of used crankcase oil, which will be effective for about two years. Spraying weeds with diesel fuel oil is also safe and effective.

There are many herbicides on the market today, and new ones are developed each year. Each herbicide bears a label specifying the amounts to use on specific crops according to the Federal Environmental Pesticide Act of 1972, as amended. Any county extension service can furnish the latest information.

The traditional methods of hoeing and pulling weeds are still the only 100 percent safe and certain methods, except when the gardener gets tired. In large gardens it is more feasible to use a rotary tiller to control weeds; however, weeds in the row must be hoed out or pulled by hand (Figure 15.9).

Fig. 15.9 — In large gardens, weeds are usually controlled with a rototiller. However, excess tillage hastens the decomposition of organic matter, reduces desirable soil structure, destroys some plant roots, and induces surface crust formation. (Courtesy, Ariens Company)

15:13 □ USING PESTICIDES

The uses of pesticides are related to soil management practices in several indirect ways. The liberal use of organic mulches encourages the various species of lady beetles, the praying mantis, and several species of predatory wasps. Overuse of many pesticides will kill these insect killers. Toads are also more numerous when organic mulches are used to smother weeds, and so are most birds; toads and most birds such as robins and bluebirds eat many insects. Piece-row planting, such as the planting of a few feet of any one vegetable in several different places at the same time, confuses the insects, since insects use their sense of smell to locate the plants. Organic gardeners claim that the following plants repel insects and, therefore, should be interplanted in the vegetable garden: nasturtiums, chrysanthemums, marigolds, asters, cosmos, onions, garlic, and coreopsis.

The use of organic mulches also makes the garden environment more favorable for insects that kill other insects. During the past 90 to 95 years, the U.S. Department of Agriculture has brought into this country about 520 species of **parasitic** (living on or in) and **predatory** (capturing and eating) **insects** to control harmful insects. Approximately 115 of these have been reproducing, but only about 20 are successful parasites and/or predators. Some of these and the insects that they destroy are:

1. Two species of lady beetles, the *black* that destroys orchard mites and scale insects and the *convergent* that feeds on many species of harmful insects.
2. Praying mantis and robber fly that prey on many species of insect pests.
3. Three species of tachinid fly that are parasitic to (a) the European corn borer, (b) many species of caterpillars, and (c) orchard mites, respectively.
4. Green lacewing that eats aphids and other insects.
5. Four species of wasps that control (a) the Oriental fruit moth, (b) the European corn borer, (c) grain aphids, and (d) many species of insect larvae.

Even though the authors of this book believe in the use of biological control methods, they also believe that there will be times when insect populations build up to such great numbers that chemical insecticides should be used. During the past several years, the senior author has used rotenone, a pesticide that has a low hazard to humans, and as a last resort, Sevin, an insecticide that is toxic to humans. *Last resort* means that when biological control is not effective, rotenone is used; and when rotenone does not control the harmful insects, Sevin is used. Sevin is also effective in discouraging rabbits and other wild animals from eating vegetables. The Sevin insecticide is applied with a crank duster. It should be used only in the early morning when there is no wind and when there is a heavy dew on the vegetables to hold the dust in place. With more difficulty for the small gardener, Sevin can also be applied as a spray. Methoxychlor and malathion are also good garden insecticides; they can be alternated with Sevin to overcome insect resistance. Sometimes a fungicide is necessary. However, labels on most pesticides, by law, restrict their use. All directions on pesticide labels should be read and followed.

15:14 □ REFERENCES

Colt, W. Michael, *et al.* "Vegetable Gardening: Planning and Preparing the Vegetable Garden Site." University of Idaho, Current Inf. Ser. 755, December 1984.

Encyclopedia of Organic Gardening. Emmaus, Pennsylvania: Rodale Press, 1978, 1,236 pp.

Fletcher, Robert F., and Peter A. Ferretti. "Soil Testing for the Organic Gardener." The Pennsylvania State University, Hort. Ser. 1, undated, 8 pp.

Follett, R. Hunter, and B. R. Sabey. "Land Application of Municipal Sludge." Colorado State University, Ext. Pub. 547, 1981.

Gomez, Ricardo E. "Tips on Energy Saving for the Home Gardener." U.S. Dept. of Agriculture. *Cutting Energy Costs. The 1980 Yearbook of Agriculture,* pp. 46 – 48.

Hayes, Jack, ed. "Animal Health." U.S. Dept. of Agriculture. *The 1984 Yearbook of Agriculture,* pp. 372 – 416.

Hull, Richard. "Gardening for Those Who Cannot or Don't Like to Toil with Soil." *In Touch,* Vol. 4, No. 2, May – June 1981. University of Rhode Island, Kingston.

Kowal, Norman E. "Health Effects of Land Application of Municipal Sludge." U.S. Environmental Protection Agency, EPA-600/S-1-85-015, 1986.

Rothenberger, R. R., *et al.* "Garden and Home Grounds Weed Control." "Grounds for Gardening," No. 6952. University of Missouri – Columbia, 1981.

Soil Improvement Committeee, California Fertilizer Association. *Western Fertilizer Handbook,* 7th ed. Danville, Illinois: The Interstate Printers & Publishers, Inc., 1985, 288 pp.

Soule, James. *Glossary for Horticultural Crops.* New York: John Wiley & Sons, Inc., 1985, 898 pp. (See especially pp. 7 and 8.)

Sutton, A. L., D. W. Nelson, and D. D. Jones. "Utilization of Animal Manure as Fertilizer." Purdue University, Ext. Cir. ID-101, 1983.

Note: The serious vegetable gardener may want to write for one or more of these garden catalogs: (1) Gurney's, Yankton, South Dakota 57079; (2) Henry Field's, Shenandoah, Iowa 51602; (3) Jung, Randolph, Wisconsin 53957; (4) Savage Farms, P.O. Box 125, SF 12, McMinnville, Tennessee 37110; (5) R. H. Shumway's, P.O. Box 1, Graniteville, South Carolina 29829; (6) Wayside Gardens, 1 Garden Lane, Hodges, South Carolina 29695.

CHAPTER 16

Turf and Ornamental Plants: Soil, Water, and Fertility Management

"And God said, let the earth bring forth grass . . ."
— Genesis 1:11

OUTLINE

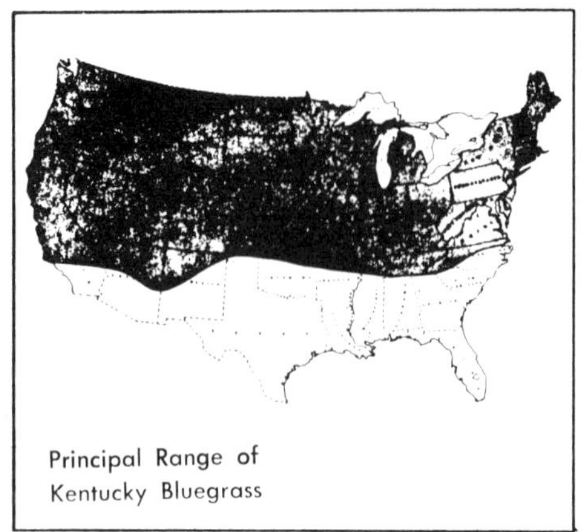

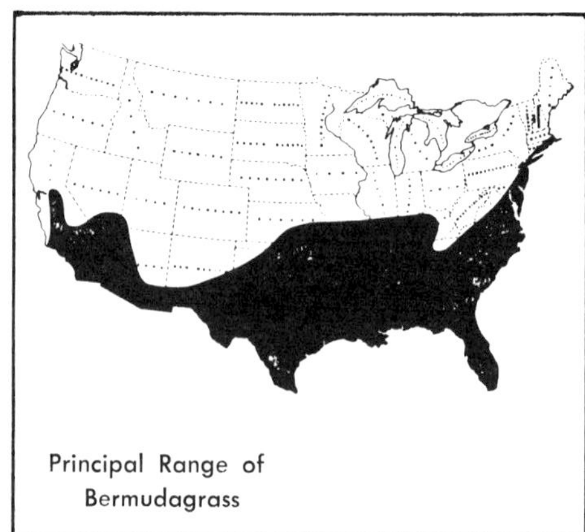

The range of the two most extensively used turfgrasses, a homesite that has been beautified by proper selection and maintenance of turf and ornamental plants, and a leaf and fruit of American holly. (*Power mower*: Courtesy, Ariens Company; *maps* and *leaf insert*: Courtesy, USDA)

□ □ □

Turfgrasses

16:1 □ OVERVIEW

Turfgrasses and ornamental plants enhance the environment by providing color; moderating temperatures; attracting birds; increasing humidity and oxygen; reducing unpleasant odors, dust, glare, and noise; and controlling erosion. Plants can be pleasing to all five senses:

- ***Sight*** — "A thing of beauty is a joy forever." Nothing is more beautiful than green grass, ornamental trees and shrubs, and colorful flowers.
- ***Smell*** — "It smells fresh" is a statement that can be made about an area that has been scientifically and artfully landscaped.
- ***Touch*** — The softness of blades of grass, the velvety feel of the pussy willow, and the roughness of oak tree bark can appeal to the sense of touch.
- ***Taste*** — "Things taste better when you pick them yourself" is a statement that includes the thrill of discovery coupled with tree-, shrub-, and vine-ripened fruits, nuts, and berries. Wild juneberries, blackberries, wild grapes, hawthorn, hickory nuts, and mulberries are examples of edible fruits, nuts, and berries that are often used in ornamental plantings.
- ***Hearing*** — The whispering sounds of the wind through pine trees, the rustling of hardwood leaves by gentle breezes, and the quietness without echo of all sounds "confused" by grasses, shrubs, and trees are appealing to the human ear to "unconfuse" the mind.

Modern urban living in the United States is often ugly, dirty, smelly, harsh, either too hot or too cold to the touch, almost tasteless, and too loud. To compensate for this dehumanizing environment, everyone needs a home with beautiful grass, shrubs, trees, and flowers; the fragrance of chlorophyll-cleansed air; soft and cool sod; flavorful fruits; and the echoless sounds of everything.

In the early days, nature provided the trees and shrubs and grasslands for human beings to enjoy. Now they must use what is left of the natural, rearrange it to suit their desires, and improve it by methods of selection and plant breeding. Now available in the United States are grasses and other ornamentals that have been collected from around the world and improved upon for greater utility and beauty. Whereas the choice of plants to grow and techniques to grow them are increasing each year, the principles of growth of turf and ornamental plants remain the same and will be discussed in this chapter.

16:2 □ SOILS AND TURFGRASSES

Over two-thirds of the soils of the United States have been mapped, and many reports are available that describe the soils and interpret the information. One of these interpretive tables is entitled "Estimated Degree and Kinds of Limitations for Town and Country Planning . . . Golf Fairways."

Characteristics of each soil-mapping unit that are used in rating the suitability of soils for grassed fairways include the following:

Drainage (general)
Seasonal wetness
Texture
Depth to water table
Depth to bedrock
Stability
Stoniness

The foregoing soil characteristics of each soil-mapping unit are considered in making judgments on their suitability for establishing grass on golf fairways and many other turf areas. Typical entries in the suitability table are (see Chapter 1):

Severe limitations — stoniness
Severe limitations — high water table
Moderate limitations — droughtiness
Moderate limitations — occasional flooding
Slight limitations — suitable

All of the natural soil characteristics that are important to the selection of the species and cultivars (species improved by breeding) of grasses are shown in the soil survey maps and reports. Human-induced soil characteristics that are important to grass establishment must be determined by inspection of each parcel of land. For example, some areas will have severe limitations for grass establishment because of tillage pans, foot paths, old roads, livestock trails, oil spills, surface scalping, refuse dumping, livestock salt licks, or residues from highway salting. Subsoil from basements is not suitable for turf grasses.

Rooting depths of turfgrasses depend upon their growth characteristics as well as upon the physical, chemical, and biological conditions of the soil. In soils without root-inhibiting layers to a depth of 5 feet (1.5 m), the approximate maximum rooting depth of the following five grasses has been reported:

Grass	*Maximum Rooting Depth*	
	(inches)	*(cm)*
Red fescuegrass	8 - 10	20 - 25
Bentgrasses	12	30
Kentucky bluegrass	30	76
Tall fescuegrass	42	107
Bermudagrass	60	152

16:3 □ ADAPTATION OF PRINCIPAL TURFGRASSES

Fifteen grass cultivars are characterized in Table 16.1 according to their adaptation to temperature, sun / shade, internal soil drainage (wetness), relative water requirement, general soil fertility level, and soil acidity. Figure 16.1 shows the regions in the United States where these 15 grasses are adapted.

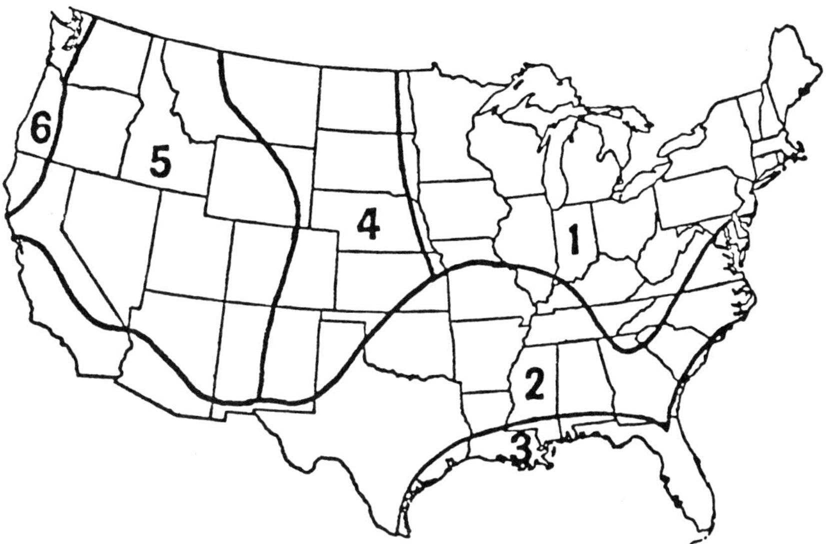

Fig. 16.1 — Regions of grass adaptations in the United States. See text for grasses adapted to these regions. (Courtesy, USDA)

- □ ***Bentgrasses (Region 1)*** — There are three principal species of bentgrass and many vegetative varieties: colonial, creeping, and velvet are the main species. All species are adapted to cool seasons, poorly drained soils, low fertility, acid soils, and sunny places, and all have a high water requirement. The principal exception is that velvet bentgrass will grow well in full sun as well as in partial shade. Creeping bentgrass is the major grass species used on golf greens in the cool, humid regions of the world (Figure 16.2).

- □ ***Bermudagrass (Regions 2 and 3)*** — There are many improved bermudagrasses, but for use as turf, common bermudagrass is the most widespread. This is a warm-season grass that grows best in the Gulf Coast and Southeastern states. More winter-hardy selections now grow in Kansas, Missouri, Kentucky, Illinois, and Virginia: Bermudagrass grows best in full sunlight on well-drained soils that are highly fertile and slightly acid (Figure 16.3). It has a relatively low water requirement and is very tolerant of

Table 16.1 — Principal Turfgrasses and Their Ecological Adaptation[1]

Grass	Temperature		Sun/Shade		Soil Drainage		Water Requirement		Soil Fertility		Soil Acidity	
	Cool-Season	Warm-Season	Sun	Shade	Well	Poorly	High	Low	High	Low	pH 6 – 8	pH 5 – 7
Bentgrass												
Colonial	*		*		*	*	*		*			*
Creeping	*		*			*	*		*			*
Velvet	*		*			*	*		*			*
Bermudagrass		*	*		*			*	*			*
Blue gramagrass		*	*		*			*		*	*	
Buffalograss		*	*		*			*		*	*	
Carpetgrass		*		*		*	*			*		*
Centipedegrass		*		*	*			*		*		*
Crested wheatgrass	*		*		*			*		*	*	
Kentucky bluegrass	*		*		*		*		*		*	
Fine fescuegrass	*		*	*	*			*		*		*
Perennial ryegrass	*		*		*		*			*		*
St. Augustinegrass		*		*	*		*		*			*
Tall fescuegrass	*		*	*	*			*		*		*
Zoysiagrass		*		*		*		*		*		*

[1]Adapted from the U.S. Dept. of Agriculture.

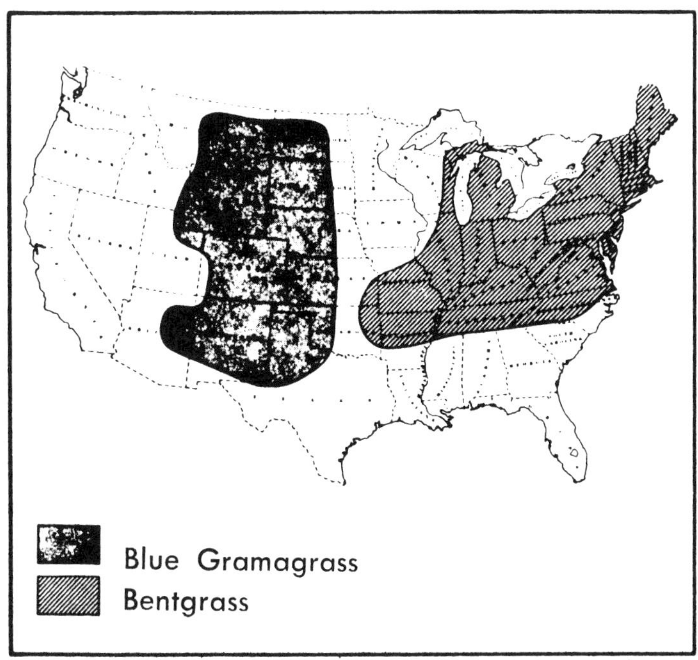

Fig. 16.2 — All of the bentgrasses have approximately the same range in humid central and northeastern United States, but in contrast, blue gramagrass is best adapted to the northern and southern semiarid Great Plains. (Courtesy, USDA)

Fig. 16.3 — This sunny exposure indicates that sun-adapted grass species such as bermudagrass should be used. (See Table 16.1.) (Courtesy, Toro Mfg. Corp., Minneapolis)

salt. Common bermudagrass can be seeded; however, almost all improved bermudagrass varieties must be vegetatively established. Bermudagrass is the major species used for golf greens in the tropical regions of the world. It is excellent as a summer turf but not suitable as a winter turf because of its straw color after a heavy frost. In the Los Angeles area, bermudagrass has been observed to be damaged by smog. (The range of bermudagrass is shown opposite the first text page of this chapter.)

- ***Blue gramagrass (Region 4)*** — A native of the Great Plains, blue gramagrass makes a fair turf in arid locations where irrigation water is costly or scarce. It is a warm-season grass adapted to the sunny locations on well-drained soils of low fertility with a high pH. Its principal use is as a range grass, and for this purpose it ranks very high (Figure 16.2).

- ***Buffalograss (Region 4)*** — A native of the United States, buffalograss is adapted to the dry clay plains from Wyoming and South Dakota to New Mexico. It grows best in the warm season in full sunlight on well-drained, fine-textured soils. It has a low water, low fertility requirement, and it tolerates sodic soils. As nonirrigated turf on lawns in the Great Plains, probably no grass excels buffalograss (Figure 16.4).

- ***Carpetgrass (Region 3 and southern half of Region 2)*** — Although it originated in Central America and the West Indies, carpetgrass is now well-adapted to the poorly drained sandy loam soils of the South where it grows wild as if it were a native. Carpetgrass makes a good warm-season lawn with minimum care. It will tolerate moderate shade and grows best on a poorly drained soil, but it does not tolerate flooding. A seepage area with a moving water table and an acid soil with low fertility appear to be its ecological adaptation. Although it grows wild on poorly drained soils, when grown in lawns, it does not have a high irrigation water requirement (Figure 16.4).

- ***Centipedegrass (Region 3 and southern third of Region 2)*** — Southern United States is the area of ecological adaptation of this warm-season, shade-tolerant, low-maintenance – cost lawn grass. Centipedegrass has all of the adaptability of carpetgrass except that centipedegrass grows best on well-drained soils (Figure 16.5).

- ***Crested wheatgrass (northern three-fourths of Regions 4 and 5)*** — Crested wheatgrass is a cool-season, coarse-textured, perennial bunchgrass that grows best in the northern Great Plains and northern Intermountain Region at an annual range in rainfall of 10 to 15 inches (25 – 38 cm). It grows best in sunny places and on well-drained soils. A low water requirement, low fertility, and high pH further characterize the ecological adaptation of crested wheatgrass. It has fairly low value as a turf because of its coarseness and bunchiness, but it persists in arid areas where most other grasses would die for lack of water. It is an excellent grass for seeding ranges (Figure 16.5).

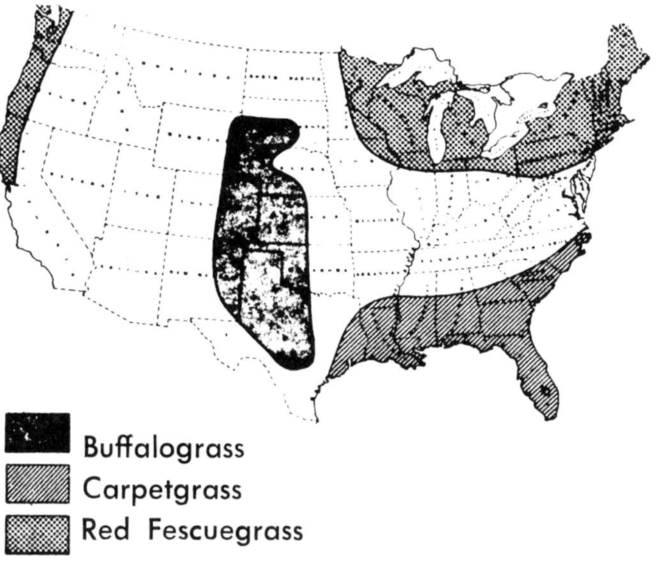

Fig. 16.4 — Buffalograss is adapted to the dry clay soil regions of the northern and southern Great Plains; carpetgrass to the lower South; and red fescuegrass to cool and humid areas of western Washington, Oregon, and California and to the Great Lakes States and the Northeast. (Courtesy, USDA)

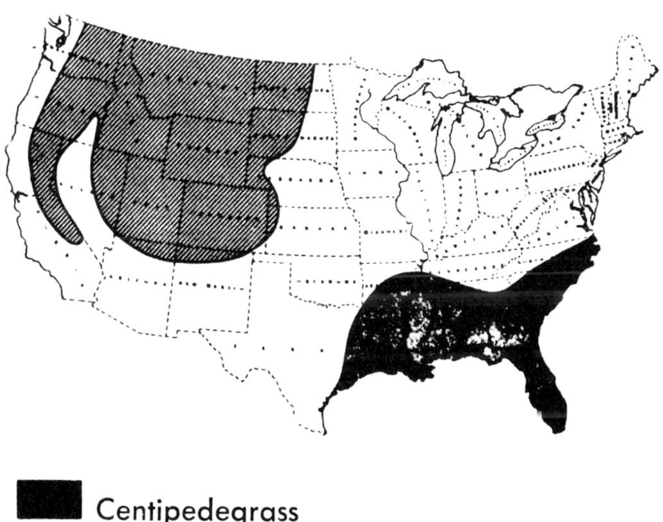

Fig. 16.5 — The adapted range of centipedegrass is the southeastern states where annual precipitation averages 50 inches (127 cm); by strong contrast, the range of crested wheatgrass is the northwestern semiarid to arid region where the annual precipitation is only 10 to 15 inches (25 - 38 cm). (Courtesy, USDA)

- ***Kentucky bluegrass (all regions, but not abundant in southern and southwestern states)*** — Kentucky bluegrass occurs in all 50 states and in Canada, but it grows best in the northern two-thirds of the United States and in southern Canada. Scores of improved Kentucky bluegrass cultivars have been developed. Kentucky bluegrass is a fine-textured, cool-season, sod-forming grass that grows best in sunny places and on well-drained soils; it has a high water requirement, a high fertility requirement, and a soil acidity pH range adaptation between 6 and 8. The adaptation of Kentucky bluegrass to annual precipitation variations is from 20 to 50 inches (51 – 127 cm); and adaptation to extremes of temperature is from 105° F (41° C) to −30° F (−34° C). (See map of adaptation opposite the first text page of this chapter.)

- ***Fescuegrasses (Regions 1 and 2)*** — Fine-leafed fescuegrasses, including creeping red, chewings, and hard fescues, are adapted to Region 1 and the northern third of Region 2. These three species are similar in texture and adaptation. Fine-leafed fescuegrasses are cool-season, fine-textured, sod-forming grasses that are adapted to cool, humid areas. They tolerate shade (but also grow in full sun), have low soil fertility, high acidity, and a relatively low water requirement. They do, however, need a well-drained soil. They are used extensively in turf mixtures, especially with Kentucky bluegrass; more than a dozen varieties of fine-leafed fescues are being sold commercially (Figure 16.6).

Fig. 16.6 — The large amount of shade indicates that only shade-tolerant grasses such as the fescuegrasses should be used to seed the turf. An estimated 20 percent of all turf is grown in shade. (See Table 16.1.) (Courtesy, Toro Mfg. Corp., Minneapolis)

Tall fescuegrass is the major turfgrass in the "transition zone" around and south of the northern boundary of Region 2. For example, nearly 90 percent of lawns in Atlanta are tall fescuegrass. Tall fescuegrasses are similar to fine-leafed fescues in adaptation except that they are better adapted to full sun and higher temperatures. At least a dozen varieties are available, and the new varieties are somewhat finer in texture than the original variety Kentucky 31. Tall fescues are not as fine, however, as the fine-leafed fescues.

- ***Perennial ryegrass* (northern, humid parts of Regions 1, 5, and 6)** — Perennial ryegrass is similar in texture, adaptation, and growth to Kentucky bluegrass, including poor drought tolerance. The bunch type is very wear-tolerant.

- ***St. Augustinegrass (Region 3)*** — A warm-season grass of tropical origin, St. Augustinegrass has found its place among the turfgrasses along the Gulf and Atlantic states at low elevations, and in southern California. It is also an important grass for pasture on well-drained organic soils of southern Florida. It is reproduced by above-ground **stolons** and by underground **rhizomes.** The important ecological adaptations are shade, a well-drained soil, a high water requirement, high fertility, and a pH of 6 to 7. At least three improved varieties have been released. Sometimes St. Augustinegrass is susceptible to damage by chinch bugs and certain fungus diseases.

- ***Zoysiagrass (Regions 2 and 3)*** — Several improved varieties of zoysiagrass are sold commercially, and all are reproduced primarily by above-ground stolons and below-ground rhizomes. Its climatic adaptation is the low elevations along the Atlantic and Gulf coasts and in the irrigated Southwest and southern California. Zoysiagrass resembles bermudagrass in appearance and in adaptation, except that zoysiagrass is more shade-tolerant, will grow on soils more poorly drained, and at a lower fertility level. This is one grass that is being advertised extensively in magazines as a "grow-anywhere" grass; for this reason it is sometimes planted too far north and at too-high elevations where it winterkills. Its main problems are a long winter dormancy and slow establishment (Figure 16.7).

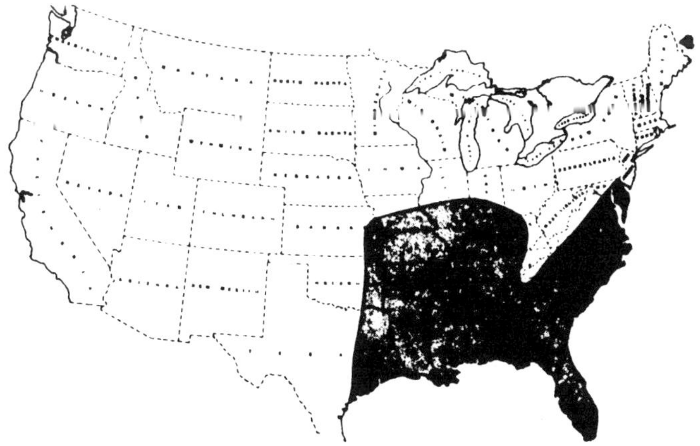

Fig. 16.7 — Zoysiagrass is so highly advertised as the "grow-anywhere" grass that care must be taken to plant it within its range, shown here. (Courtesy, USDA)

16:4 □ WATER MANAGEMENT

Turfgrasses contain from 70 to 90 percent water. Although grasses can absorb some water through their leaves, most of the water is absorbed by their roots from the soil. Therefore, the physical conditions of the soil must be suitable to hold adequate water, oxygen, and all essential elements simultaneously. Suitable physical conditions are very difficult to describe and much more difficult to achieve. (See Chapter 2.)

A soil in ideal physical condition for the growth of turfgrasses should have between 40 and 50 percent of its volume occupied by pore spaces and 50 to 60 percent occupied by sand, silt, clay, and organic matter. This is easier said than done. A loam soil that has been growing satisfactory vegetables for a few years should have a composition that is ideal for the growth of turfgrasses. How do you build a productive garden soil every time you want to establish a good turf? A look at football fields with natural turf may help to understand what constitutes an ideal physical soil medium for excellent water management for turfgrasses and for hard use.

About 2 feet (61 cm) of dune sand, classified mostly as a fine sand, with particle diameters between 0.25 and 0.10 mm, appears physically ideal for the growth and intensive care of turfgrasses on heavy traffic areas. To reduce compaction and to improve air and water relations for grass roots, some kind of organic material should be intermixed into the surface 6 inches (15 cm). Suitable organic materials are moss peat, animal manures, seed hulls, and weathered hardwood sawdust. With these soils, irrigation is necessary for turf survival.

When the layers of fine sand and organic material are added to a clay base, a thorough mixing of the materials is essential if water movement is to be satisfactory. For example, a fine sand over a clay, with no intermixing, will create saturation in the fine sand before water will move into the clay. This is conventional wisdom. What is not conventional wisdom is the fact that the reverse in textural sequence also produces the same result. In other words, a clay over sand or gravel will also become saturated before water will move into the sand or gravel below. Soil in the entire root zone should be intermixed.

The most clever manager of turf observes that a slight moisture deficiency stress results in a darker green color and that a moderate moisture stress will result in a bluish-green color. The astute turf manager also can read signs of excess soil water by a yellow-green color and a "sappy" feel to the grass. Although these are scientific observations, some turf managers may use a tensiometer to determine "when" and "how much" irrigation water should be added. The tensiometers are usually buried at perhaps two levels of 4 inches (10 cm) and 8 inches (20 cm) (Figure 16.8). (See Chapters 9 and 15.)

16:5 □ FERTILITY MANAGEMENT

Turfgrasses require the same 16 elements essential for all higher plants. (See

Fig. 16.8 — Intensively managed turf must be watered at the right time with the right amount of water. The most scientific as well as practical way to determine when to start irrigating and when to stop is to use tensiometers such as these. Their bulbs are set at different root depths. (Courtesy, Irrometer Company, Inc., Riverside, California)

Chapters 2 and 4.) Unlike most crop plants, however, grasses are useful primarily when they are continuously vegetative and dark green. Warm-season grasses go dormant and therefore are not dark green the entire year. The dormant leaves and stems maintain many of the physical characteristics of green-growing vegetation. Because the weather is cool or cold when the grass is dormant, the leaves and stems decay very slowly.

The fertilizer element most conducive to a delay in maturity and to a dark green color is nitrogen. Some amateurs conclude, therefore, that *only* nitrogen is needed and that the more nitrogen fertilizer is applied the better. The fallacy of this reasoning can be observed when an excess of nitrogen is applied and the possible results include:

1. A shallow root system which will be injured more by drought.
2. Higher water use requirement.
3. More insect damage.
4. More disease damage.
5. More winterkilling.
6. A turf more easily injured by people and by machine traffic.
7. Nitrate pollution of the waters.

Soils are so variable, dynamic, and changeable from season to season that they keep the very best soil scientists guessing. Moreover, the turfgrasses are also so variable that they baffle the **agrostologists** (grass scientists). A meeting of the minds of the soil scientists, the agrostologists, the ultimate manager of the turf (such as a golf course manager), and the supplier of fertilizers requires that they must temper their human judgments with the ecology of the soils and grasses.

A *balanced* fertilizing schedule at a specified time and for a particular tract of turf is, therefore, a blend of the science of soils with the science, art, and judgment of the grass manager and the interpreter of the soil and tissue tests. (Actually, tissue tests are of relatively little help.) As opposed to crop production, in turf one is more interested in reducing growth rate. The growth and appearance of the grass remaining is much more important than the growth (or yield) of that which is removed by the mower. Every few months a new assessment must be made to determine the need for a change in the balanced fertilizing schedule.

All 50 states have soil testing laboratories, and many also analyze plant tissues as a diagnostic technique for determining the kind, amount, and timing of fertilizer and its application. The teacher of vocational agriculture and the county extension director can guide anyone who wants reputable service on soil and tissue testing.

The best soil and tissue testing available and their interpretation cannot relieve the responsibility of the turfgrass manager to know how to apply the recommendations and how to judge a new turfgrass problem.

For assistance in interpreting new problems as they arise (and they are *always* arising on turf with intensive care), consult Table 16.2. Twelve of the 13 essential elements supplied from the soil to plants are listed in the table under the three headings: "Deficiency Symptoms Indicated on Leaves," "Ecological Conditions Inducing Deficiency," and "Techniques for Correcting Deficiency." Only chlorine deficiency is not discussed because its occurrence is so rare. Many specialty turf fertilizers can be purchased. These often contain many of the needed micronutrients.

□ □ □

Ornamental Plants

16:6 □ OVERVIEW

There are thousands of plants that are grown for their beauty. Even poison ivy fits this category because of the reddening of the leaves in the early fall. Poison ivy may not be planted intentionally as an ornament, but birds carry the seeds, and volunteer plants in many cities are not killed because of their brilliant red color at about the time of the first frost.

Many ornamental plants may be grouped according to the following ecological adaptations:

1. Plants for each of the 10 temperature hardiness zones (Figure 16.9).
2. Plants for each of the 32 plant growth regions. (See Figure 16.10 and Table 16.3.)
3. Plants for sunny places.
4. Plants for partial shade.

Table 16.2 — Deficiency Symptoms of Turfgrasses, Ecological Conditions Inducing Them, and Techniques for Their Correction[1]

Nutrient	Deficiency Symptoms Indicated on Leaves	Ecological Conditions Inducing Deficiency	Techniques for Correcting Deficiency
Nitrogen	Light green to yellow-green	High rainfall / heavy irrigation	N fertilizer to soil or leaves, add decomposed organic matter
Phosphorus	Blue-green to purple	Organic, calcareous, cold soils	P fertilizer, adjust organic soils to pH 5.5 and mineral soils to pH 6.5
Potassium	Brown on tips and margins	Organic, sandy, leached soils	K fertilizer
Calcium	Reddish-brown between veins	Sandy soils, high rainfall, sodic soils	Lime acid soils, gypsum on sodic soils
Magnesium	Yellow-green stripes	Organic, sandy soils, high rainfall	Dolomitic limestone
Sulfur	Yellow all over	Soils low in organic matter	Elemental S, gypsum, organic matter
Copper	Yellow growing tips	Organic, calcareous soils, excess N	Cu fertilizer to soil or leaves, reduce N fertilizer
Molybdenum	Pale yellow all over	Sandy and ironstone soils	Mo fertilizer to soil or leaves, lime acid and ironstone soils
Iron	Yellow-green between veins	Calcareous soils, excess P and Mn	Fe fertilizer to soil or leaves, sulfur to lower pH, reduce P and Mn fertilizer
Zinc	Small, yellow all over	Alkaline, cold soils, excess P	Zn fertilizer to soil or leaves, reduce P fertilizers
Manganese	Young leaves yellow-green	Calcareous soils	Mn fertilizer to leaves or soil
Boron	Pale green tips, bronzing	Sandy, organic, acid, calcareous soils	B fertilizer on soil or leaves, lime acid soils, sulfur on calcareous soils

[1]Adapted from R. R. Davis in Ohio, H. D. Chapman in California, and from many other sources.

Note: The table lists 12 of the 13 essential elements supplied to plants from the soil; the thirteenth element is chlorine. However, as far as known, chlorine deficiency is almost nonexistent.

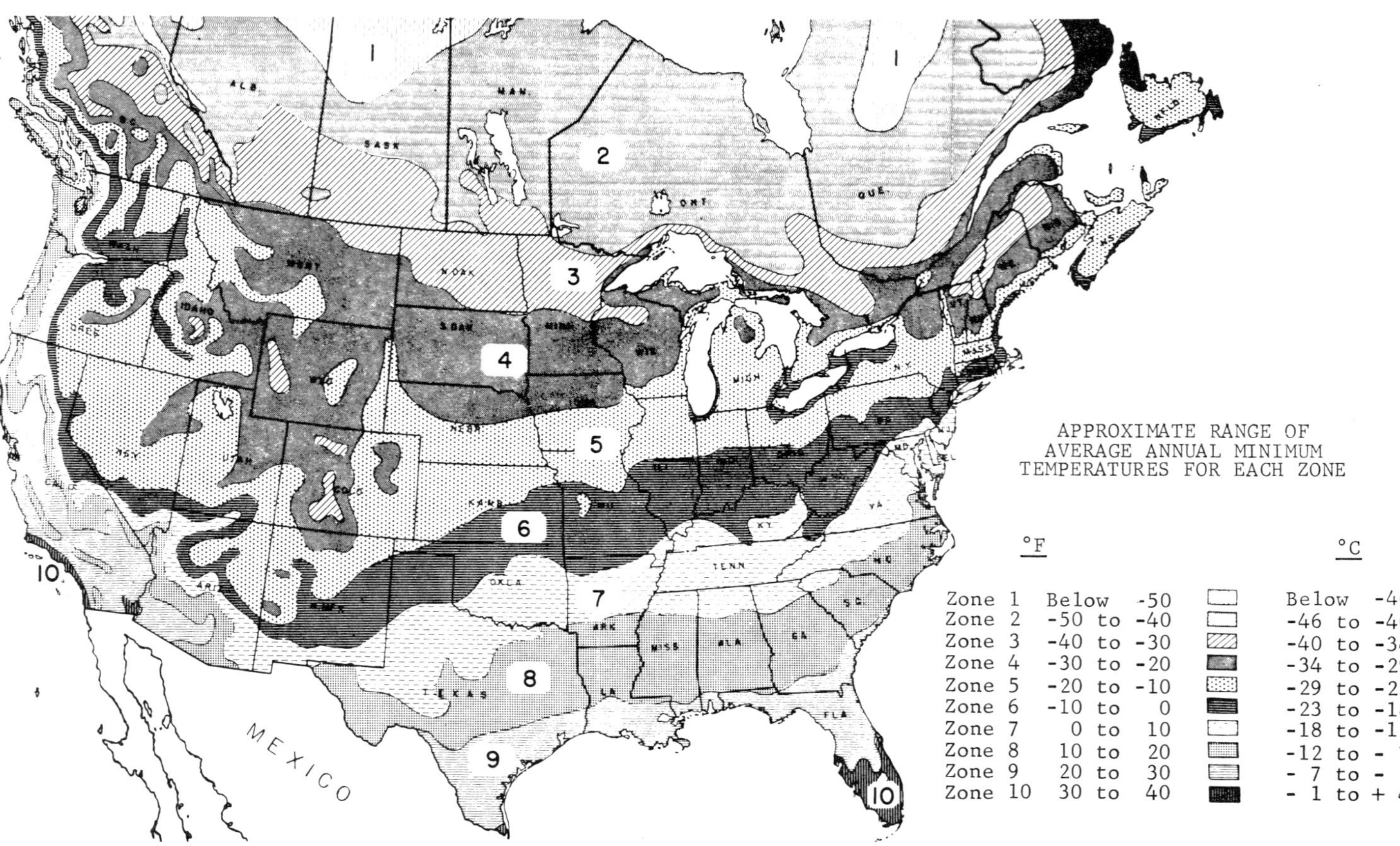

Fig. 16.9 — Less winterkilling of plants ordered from nurseries will result if this map of plant hardiness (temperature-tolerant) zones is used as a guide. Most reliable nurseries now indicate in which hardiness zone each plant offered for sale will thrive. (Source: USDA – Agricultural Research Service, Misc. Pub. 814)

Table 16.3—Fifteen Native Trees and Shrubs, Regions Where They Can Be Grown, and Notes on Their Beauty / Ecology (Use with Figure 16.10)[1]

Tree / Shrub	Plant Growth Region	Notes on Beauty / Ecology
American holly	20, 25, 27 – 30	Evergreen leaves, red berries in fall; English and Japanese holly usually preferred.
Blackgum	20, 22, 24, 25 – 30	Leaves brilliant red in fall.
Cranberry bush	1, 2, 4, 12, 13, 15, 18, 21, 27	White flower clusters, brilliant red fruit.
Creosote bush	9, 10, 11, 16, 17	Yellow flowers, evergreen leaves, for desert garden.
Flame azalea	1, 2, 3, 20, 22, 24 – 29	Flowers yellow, orange, or red, brilliant, evergreen.
Flowering dogwood	1, 2, 20, 22 – 25, 27 – 30	White or pink flowers, very conspicuous.
Magnolia	28 – 30	Very conspicuous white flowers.
Ohio buckeye	22, 24, 25, 27 – 29	Flowers yellowish, very conspicuous.
Pacific dogwood	1 – 5, 12	Giant tree, giant white to pink flowers, red to orange fruit (shrubs from Region 12 will grow in Regions 27 and 28).
Red maple	[illegible]9 – 31	Leaves red in fall.
Redwood		Will not grow outside Region 1.
Sagebrush	[illegible], 4 – 13, 15, 27, 28	Silvery leaves, requires alkaline soil.
Sequoia	[illegible]	Will not grow outside Region 4.
Sugar maple	[illegible], 15, 18, 21 – 29	Leaves yellow, orange, and red in fall.
Sweetgum	[illegible], 20, 22, 25, 27 – 30	Leaves yellow to red in fall.

[1]William R. Van Dersal. "Adventures with Native Plants That Passed Their Tests." *Landscape for Living. The 1972 Yearbook of Agriculture,* U.S. Dept. of Agriculture, pp. 176 – 184.

5. Plants for full shade.
6. Plants for wet places.
7. Plants for dry places.
8. Plants for poor soils.
9. Plants for rich soils.
10. Plants for rock gardens.
11. Plants for beginners.
12. Plants for fragrance.
13. Plants to provide shade.
14. Plants for birds, butterflies, and other wildlife.

Many books have been written on each of the foregoing groups of plants, and details here may be superfluous. The authors have decided to present some principles and interpretations to guide the reader toward more scientific decisions, regardless of the specific kinds of ornamental plants that are grown.

Scientific considerations include:

1. Whether it is an indoor plant or an outdoor plant, check on the light requirements of the ornamental plant to be grown.
2. Is the temperature suitable? For an indoor plant, check this with a thermometer; for an outdoor ornamental plant, look at Figure 16.9 to determine the temperature zone where you live.
3. Consider the soil. Most outdoor ornamental plants will grow well in any soil that supports many earthworms. Potting soil can be obtained from the same earthworm soil or it can be purchased from nurseries, greenhouses, or seed companies. Planters can mix their own by adding one-third sand, one-third vermiculite, and one-third well-decomposed compost or well-decomposed manure.
4. Fertilizing is difficult for many people. The most scientific way to determine the amount and kind of fertilizer to apply is to have the soil tested. (See Chapter 2.) Without a soil test, the authors predict that the "first-timer" will add about 10 times too much fertilizer. As a rule of green thumb, fertilizer should be used no thicker than the salt you put on your meat and potatoes.
5. Watering is nearly always in excess. A potting mixture of one-third each of sand, vermiculite, and compost will need watering every third day in hot and dry weather and about half as frequently in cool and moist weather. If the ornamental plant is grown in a pot, holes in the bottom are necessary to permit free drainage.

16:7 □ PLANT GROWTH REGIONS

If your interest is in *native* ornamental shrubs or trees, here are some suggestions. Figure 16.10 is a map of the United States that has been divided into 32

Fig. 16.10 — This map of plant growth regions of the United States was compiled by Furman L. Mulford and updated by William R. Van Dersal. The criteria used for delineating the areas included rainfall, temperature and soil, tempered by judgments of two superior field scientists. (Use with Table 16.3.) (Courtesy, USDA)

fairly uniform plant growth (ecological) units. It represents the composite judgments of two field scientists of the U.S. Department of Agriculture who have spent their professional lives developing, testing, and revising the map. The map can be interpreted in this way:

1. If a native shrub or tree grows in any part of a region, it will likely grow all over the same region under similar sun-soil-rainfall conditions.
2. Some plants can be transported from one region to another and thrive because of compensating ecological conditions, because of their ability to adapt, and / or because they have been given tender loving care.
3. By trial-and-error it has been proved that plants from one region can adapt to other plant growth regions, as follows:

Plants Now Growing in Region	*Will Grow in Region*
1, 4	2, 29, 30
2	29, 30
5	31, 32
12, 13	24, 25, 27, 28
20	29
27, 28	1, 2, 24, 25, 29

Examples of common native shrubs and trees that have high ornamental value, their plant growth regions of adaptation, and some special comments on their beauty and culture are given in Table 16.3.

16:8 ▫ REFERENCES

Bachelor, Charles. "Sod Isn't Simple." *Cooperative Farmer,* Richmond, Virginia, Vol. 43, No. 4, May – June 1987, pp. 21–23.

Beard, James B. *How to Have a Beautiful Lawn: Easy Steps in Turfgrass Establishment and Care for Aesthetic and Recreational Purposes,* 2nd ed. College Station, Texas: Beard Books, 1979, 114 pp.

Carrow, R. N., B. J. Johnson, and R. E. Burns. "Thatch and Quality of Tifway Bermudagrass Turf in Relation to Fertility and Cultivation." *Agronomy Journal,* Vol. 79, 1987, pp. 524 – 530.

Johnson, B. J., R. N. Carrow, and R. E. Burns. "Bermudagrass Turf Response to Mowing Practices and Fertilizer." *Agronomy Journal,* Vol. 79, 1987, pp. 677 – 680.

Powell, A. J., Jr. "Improving Turf Through Renovation." University of Kentucky, Co-op. Ext. Serv., AGR-51, November 1987, 4 pp.

Powell, A. J., Jr. "Lawn Establishment, in Kentucky." University of Kentucky, Co-op. Ext. Serv., AGR-50, November 1987, 4 pp.

Powell, A. J., Jr. "Lawn Fertilization in Kentucky." University of Kentucky, Co-op. Ext. Serv., AGR-53, November 1987, 4 pp.

Powell, A. J., Jr. "Mowing, Dethatching, Coring, and Rolling Kentucky Lawns." University of Kentucky, Co-op. Ext. Serv., AGR-54, November 1987, 4 pp.

Powell, A. J., Jr. "Selecting the Right Grass for Your Kentucky Lawn." University of Kentucky, Co-op. Ext. Serv., November 1987, 4 pp.

Powell, A. J., Jr., and M. L. Witt. "Home Lawn Irrigation." University of Kentucky, Co-op. Ext. Serv., ID-79, September 1987, 8 pp.

Simson, C. R., E. E. Schulte, and J. R. Love. "Sampling Lawn and Garden Soils." University of Wisconsin, Co-op. Ext. Serv., August 1982, 4 pp.

Tekulsky, Mathew. *The Butterfly Garden.* Boston: The Harvard Common Press, 1985, 144 pp.

Throssell, C. S., R. N. Carrow, and G. A. Milliken. "Canopy Temperature Based Irrigation Scheduling Indices for Kentucky Bluegrass Turf." *Crop Science,* Vol. 27, 1987, pp. 126 – 131.

Titko, Steve, III, John R. Street, and Terry J. Logan. "Volatilization of Ammonia from Granular and Dissolved Urea Applied to Turfgrass." *Agronomy Journal,* Vol. 79, 1987, pp. 535 – 540.

"Trees for Shade and Beauty: Their Selection and Care." U.S. Dept. of Agriculture, 1972, 8 pp.

Tylka, Dave. "Butterfly Gardening and Conservation." Pamphlet No. 2, The Conservation Commission of the State of Missouri, Jefferson City, 18 pp.

Witt, M. L., A. J. Powell, Jr., R. Hartman, W. O. Thom, and W. M. Fountain. "Principles of Home Landscape Fertilization," University of Kentucky, Co-op. Ext. Serv., ID-72, June 1986, 8 pp.

CHAPTER 17

Greenhouses and Nurseries: Soil, Water, and Fertility Management

"Who loves a garden loves a greenhouse too.
Unconscious of a less propitious clime,
There blooms exotic beauty, warm and snug,
While the winds whistle, and snows descend."
— William Cowper (1731 - 1800)

OUTLINE

Greenhouses come in many sizes and designs. The main purpose of some is pleasure, while others produce food or flowers for profit. Greenhouses also provide for year-round plant growth, allowing many types of research to be continued through the winter. (Courtesy, Tennessee Valley Authority)

□ □ □

17:1 □ OVERVIEW

While agricultural production has been increasing in almost every area, increased production in greenhouses and nurseries has been especially spectacular. The high fixed costs of production have forced producers to increase efficiency. The high values of greenhouse and nursery crops have further encouraged producers to raise their yields. Increased population, urban sprawl, and rising land prices have all contributed to the pressure to produce more on less land.

Many factors have been responsible for increased yields, including more research on which to base decisions, better varieties, better propagation techniques, improved pesticides, and more effective growth media. Watering and fertilizing techniques have improved steadily with new technology.

Production costs are high for greenhouse and nursery crops; establishing either on a large scale may be excessively expensive. Most greenhouse and nursery crops are high-risk crops. It is important, then, to consider several points before you establish a greenhouse or nursery.

One alternative to establishing a greenhouse for bedding plant production is the use of coldframes (Figure 17.1). Commercial disposable-type plastic tunnels and inflatable greenhouses and coldframes are available at relatively low cost. Coldframes can be built in the school shop or at home with readily available and relatively inexpensive materials. This allows the growing of plants in a protected environment on a small scale with a limited investment of time and money. Costly heating systems, which will increase the range of plants that can be produced and increase the likelihood of success, can be added later.

Another point to consider is location. Location affects light intensity, temperature, and humidity. Too high a temperature or humidity may cause direct damage to plants or increase diseases and be more difficult to control than too low a temperature or humidity.

The site for a greenhouse or nursery should be carefully selected. Available markets should also be considered. The land should be level or slightly sloping to the south to give good exposure to the sun. The greenhouse should be located in an open area away from woods and other plants that may transmit insects or diseases. The area should have good air and water drainage and should be located so as not to be flooded or consistently wet or where air pollution may be a hazard (Figure 17.2).

Another point to check is good water supply. An ample supply of usable water is necessary for greenhouse and nursery production. Water taken from some sources may contain harmful organisms, excessive salts, iron, chlorine, sodium, or boron, which may damage plants and especially plant roots.

Fig. 17.1 — A coldframe can produce quality bedding plants with a much lower investment than is normally required in a greenhouse. (Courtesy, Dean Knavel, University of Kentucky)

Fig. 17.2 — Even after this flood recedes, it will leave a polluting coating of silt. Floodwaters also carry disease organisms that may damage greenhouse and nursery plants. This is not a good location for a greenhouse or nursery. (Courtesy, U.S. Environmental Protection Agency)

Soil must be selected wisely. While plants will grow on a variety of soils and artificial media, some media are much better than others. The following section should help in the selection of greenhouse and nursery soils.

17:2 ▫ SELECTING GREENHOUSE AND NURSERY SOILS

The normal pattern in production, whether it be field crops, pasture and range, forests, or orchards, is that a certain kind of soil exists and the variety of crop must be chosen to fit the soil with only rather minor adaptations. In the case of greenhouses, and to a lesser extent nurseries, the best soils can be selected or prepared.

There are four important functions that soil and other plant growth media must perform to do a good job under normal circumstances; they are: (1) physical support for the plants, (2) available water, (3) an exchange of gases, and (4) available nutrients in balanced proportions. A serious deficiency in any one of these may cause severe crop damage. Normally the finer-textured the media, the better they will support plants; however, they will also be harder to handle and move around. The exchange of gases is more important in most modern media than water-holding capacity because with automatic watering systems (which are mentioned later in this chapter), watering is usually a minor problem. Likewise, nutrients can be added rather easily, but the gas exchange capacity is not always so easy to change; therefore, media with good ability to exchange gases should be selected originally. When media are to be used for an extended period of time, it is important that they be material that will remain relatively unchanged over the period of time they are to be used.

Several factors should be considered in the selection of soils for greenhouses and nurseries. Important selection factors are pore space, structure, texture, organic matter content, available nutrients, disease organisms, cost, and availability. (See Chapters 1 and 2.)

▫ ***Pore space*** — Pore space in a soil is the space not taken up by organic or mineral solids. Pore space is normally filled with either air or water. When water is added the amount of air decreases. When water evaporates, is drained away, or is used by plants, it is replaced by air. Pore space is important because the water it contains is used by plant roots. Plant roots also require oxygen which is stored in soil pore spaces if the soil is not overly wet. Soil air contains up to 100 times as much carbon dioxide as the atmosphere, so the soil must get rid of excess carbon dioxide, which must be replaced by oxygen from the air. Small soil pore spaces tend to be filled with water, while larger ones, if not saturated with water, usually contain more air. So the size of pore spaces as well as the quantity of pore space is important in good plant growth. (See Chapters 6 and 8.)

- ***Soil structure*** — *Soil structure* is a term used to describe how soil particles are grouped together into units or aggregates. *Peds* are natural aggregates held together by nature's cements. Nature's cements come from earthworm slime, decaying bacterial bodies, fungus roots, clay, and plant roots, especially grass roots. *Clods* are aggregates caused by the soil's having been manipulated when it was too wet.

  Soil scientists classify structural aggregates into three main groups based upon their shape, size, and strength. The main shapes which occur in surface soils are **crumb, granular, block-like,** and **plate-like.** Sizes are *coarse, medium,* and *fine. Strength* refers to the ability of the structural unit or aggregate to resist crushing or breakdown from the impact of a raindrop. Descriptive terms of strength are **weak, friable,** and **firm** (Figure 17.3).

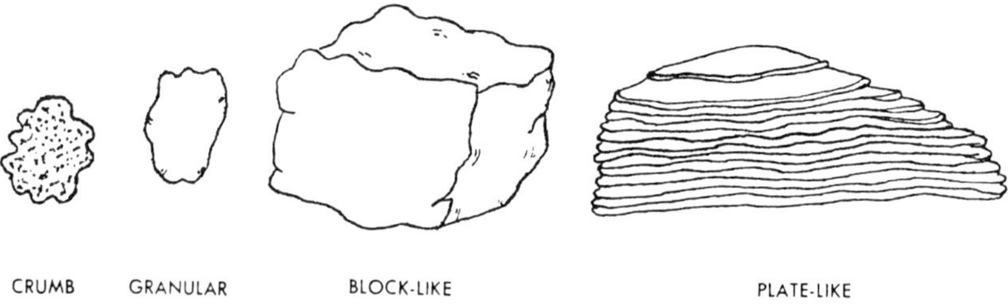

Fig. 17.3 — The four main types of soil structure are crumb, granular, block-like, and plate-like.

Good soil structure is necessary so that plant roots may "eat," "drink," and "breathe" at the same time. Without sufficient water, plants will not absorb enough water or plant nutrients for normal growth. Without plenty of air at the same time, plants are not able to absorb plant nutrients or to carry on their necessary exchange of gases (respiration).

The structure of a virgin prairie soil, considered ideal for plant growth, would be described as a friable, medium crumb. Such a soil would be excellent for a greenhouse or nursery.

While it is true that tillage is necessary for making a desirable seedbed, the long-time effect of tillage is to destroy desirable soil structure, therefore making conditions less desirable for a seedbed. The problem of soil structure deterioration can be solved by using a minimum amount of tillage, by growing sod-forming crops, and by using as much barnyard manure as possible (Figure 17.4).

- ***Soil texture*** — The texture of a soil is determined by the percentage of sand, silt, and clay it contains. The largest particles are sand and the smallest are clay. Clay particles are especially important in soils. Half of the names in the soil triangle contain the word *clay.* (See the soil triangle in Figure 1.2 and the definitions of sand, silt, and clay in Chapter 11.) Most of the chemical properties of the soil are dependent on the negatively

Fig. 17.4 — The long-time effect of tillage is to destroy soil structure that is best for plant growth. Desirable soil structure is built and maintained under grass, as shown on the left. Continuous clean-tillage cropping with corn for many years results in destruction of desirable soil structure, as seen on the right. (Courtesy, USDA - Soil Conservation Service)

charged clay particles to which positively charged plant nutrient ions adhere. The adsorbed ions are held rather tightly and are not normally removed by watering and leaching. Some of these ions can be freed for plant use with the addition of lime or fertilizers. It takes more lime to change pH in clay soils than in sand because of the clay's ability to attract a large number of positive hydrogen ions. Therefore, clays are more highly buffered.

Silt, which is the middle-sized particle, has less ability to attract nutrient ions than clay has. Silt particles also do not tend to clump like clay does.

Sand particles have less surface area than silt and clay particles; therefore, they have less chemical activity. Sand grains act somewhat as a **skeletal framework** for the soil.

Texture is important because it largely determines the water-holding and nutrient-holding capacity of the soil. Texture also influences consistence, structure, permeability, and erodibility of the soil.

- ***Organic matter*** — Greenhouse and nursery growers use large quantities of organic materials. Organic materials help hold water and nutrients which can be used by plants and help improve soil structure. Whether worked into the soil or used as a mulch, organic materials have good biological, physical, and chemical effects on soils and should be used according to recommendations of the state agricultural experiment station. In most soil mixtures one or more of the following organic materials is used to make up 25 to 33 percent of the mixture: manure, compost, peanut hulls, peat, muck, sawdust, humus, corncobs, and leaf mold.

- ***Available nutrients*** — Because of the high costs involved in producing greenhouse crops, investment in soil testing can bring very good returns. Many state soil testing laboratories now have special soil tests to determine not only available fertilizer nutrients and pH but also nitrate and nitrite nitrogen, total soluble salts, calcium, magnesium, and micronutrients for greenhouse soils.

- ***Disease organisms*** — While it is important to have soils for greenhouse and nursery production which are free of disease organisms, this is not necessarily an important factor in soil selection. This is because most greenhouse soils and some nursery soils are sterilized to kill disease organisms.

In selecting greenhouse and nursery soils, it is important to look for a deep, well-drained, fertile soil. A sandy loam or silt loam is preferred, and a soil with a hardpan present should be avoided. While organic matter and additives can be used, a good natural soil is advantageous (Figure 17.5). Soil from areas where harmful chemical residues are present, such as corn fields which have been heavily sprayed with atrazine, should be avoided. If good natural soil is not available, the following suggestions for preparing greenhouse and nursery soils and artificial media may be helpful.

Fig. 17.5 — A good natural soil is the starting point for economical production in a greenhouse or nursery. Notice the dark-colored organic matter present and the crumb structure in this soil. (Courtesy, Dean Knavel, University of Kentucky)

17:3 ▫ PREPARING GREENHOUSE AND NURSERY SOILS AND ARTIFICIAL GROWING MEDIA

Growing mixtures may be made up of one or more of three broad classes of material: soil, organic matter, and artificial materials. Regardless of what they consist of, mixtures for greenhouses and nurseries should be well-drained, but at the same time they must not dry out too rapidly and must be easy to manage (Figure 17.6).

Fig. 17.6 — This soil contains so much shrink-swell clay that it would be difficult to manage and would be undesirable for use in a greenhouse or nursery. (Courtesy, USDA)

Soils

Soils are made up of clay, silt, or sand in varying proportions. (See Chapters 1 and 11.) Clay alone, which is the smallest soil particle, makes a tight or compact soil not favorable to root growth. Where clay predominates, the soils are wet, poorly drained, and tend to crust or crack severely. The primary reason that clay is important in a soil mixture is because it has a high cation-exchange capacity. In other words, it adsorbs large quantities of plant nutrients which are held for later plant use. Clay soils can be improved as a plant medium by the addition of large quantities of organic materials or coarse sand. Additions of sand without organic matter will, when dry, likely result in a material resembling concrete.

Silt consists of larger particles that do not have high cation-exchange capacities like clay does. Silty soils are better drained, but sand, organic matter, or other material is required to make them manageable.

Sand is used in many plant-growing mixtures and is easy to work. It loses both water and nutrients rapidly, but the addition of peat moss and / or muck can create a very desirable plant-growing mixture. A major drawback to the use of sand is that it is very heavy — about 100 pounds per cubic foot (1.6 gm / cc).

Any combination of these natural soil particles is likely to contain weed seeds, soil insects, nematodes, and / or harmful fungi. This means that any soils used in greenhouse production must be sterilized. Sterilization can be done by using steam, aerated steam, hot water, methyl bromide, or **metham sodium** (Vapam®). (Caution: Methyl bromide must not be inhaled, and metham sodium is irritating to the skin.)

Using steam is considered by many growers to be the most satisfactory method of soil sterilization (Figure 17.7). Each batch of steamed soil should be checked to see that the coolest part of the soil is kept at 180° F (82° C) for 30 minutes. Excessive time or temperature may cause release of soluble manganese, which will injure many crops. Soil should not be allowed to become too moist or

Fig. 17.7 — A barrel with a cover can be used to sterilize small lots of soil. Steam is injected through the vertical pipe. (Photo by Rodney W. Tulloch)

Table 17.1 — Amount of Steam Heat Required to Sterilize 1 Cubic Foot (28.32 Liters) of Various Soil / Peat Mixtures with a Temperature Increase from 70° F (21° C) to 150° F (66° C)

Soil Mixture	BTU's	Kilogram-Calories
½ loam plus ½ peat	1,054	266
½ sand plus ½ peat	1,374	346
⅔ sand plus ⅓ peat	1,699	428

piled too deep for storage because nitrites (NO_2) may be formed, small quantities of which are toxic to plants. This process may also reduce the phosphorus, zinc, and copper availability, thus causing deficiencies of these elements. Another problem is the rather large amount of heat needed to sterilize the soil (Table 17.1).

Another method of soil sterilization is with the use of aerated steam. Using aerated steam at 180° F (82° C) to 212° F (100° C) often causes undesirable side effects. These side effects include compaction of the soil, toxic accumulation of ammonia or manganese, extreme kill of microflora which allows the rapid growth of many pathogens introduced into the treated mixture, increased soluble salts, and water soluble toxins from organic matter.

The cited side effects can largely be eliminated by using temperatures of 140° F (60° C) to 160° F (71° C) and forcing air with the steam through the soil mixture. When treated for 30 minutes, most weed seeds, pathogens, and soil insects will die.

Hot water treatment is generally not acceptable because of the large quantity of hot water required and the resultant wet soil condition.

Another method of soil sterilization is through the use of chemicals. Where steam is not available, using chemicals may be less expensive than having steam installed. The type of soil and its condition both affect the results of the treatment; therefore, the directions on the label for the use of the chemical should be followed very carefully.

There are several disadvantages to using chemicals, some of which follow.

1. Some chemicals are highly toxic to humans.
2. Some chemicals are harmful to plants and may damage those nearby.
3. Following chemical sterilization, complete aeration of the soil is necessary; an operation that involves up to two weeks to accomplish. In a greenhouse "time is money," and it is costly to leave beds vacant for any length of time.

Methyl bromide, a liquid formed from compressed gas and used to fumigate soils, is very toxic to humans and to some crops such as certain flowers. Because it is compressed it vaporizes at atmospheric pressure and enters the soil. A tight plastic cover is kept over the soil at the time of fumigation and for three days after a methyl bromide application. Temperature should be at least 60° F (16° C) for best results. The soil should then be exposed to the air for several weeks before it is used. Some plants do not grow or germinate well on soils treated with methyl bromide; therefore, the instructions for its use should be carefully read and followed.

Vapam®, on the other hand, is drenched on the soil and washed in. A cover is not needed, but its use will result in a better job on the top inch of soil. Vapam® kills fungi, nematodes, and weed seeds but is not effective against growing weeds.

Many other chemicals are available, and new ones are continually being developed. The county agent, vo-ag teacher, or extension specialist should be consulted for the latest available recommendations.

After a soil mixture has been sterilized, be careful to avoid reinfestation. Because many organisms that control diseases have been eliminated, special care needs to be taken to prevent reinfestation. Avoid walking over the soil. Sterilize tools and containers in 10 parts water / 1 part liquid bleach. Keep the bleach mixture in a sealed container, and mix only what can be used in a 24-hour period.

Organic Matter

Organic matter is another material used in preparing growing mixtures. In general, the addition of organic matter improves moisture-holding capacity, the relationship of air to water, and the ability to physically manage the mixture. (See Chapter 5.) Five commonly used organic materials in greenhouse and nursery soils are manure, bark, sawdust, muck, and sphagnum peat moss.

Manure is difficult to obtain in some areas and varies greatly in physical properties, nutrients, and state of decomposition. Because fresh manure is high in ammonia and salt, it can "burn" the roots of plants. Well-rotted manure has less ammonia and is quite safe. The addition of manure to sandy soils increases their water-holding capacity. Added to clay soils, it improves the air-water relationships.

Redwood, fir, pine, and a number of other barks are used in soil mixes. Bark is less expensive than most other materials used in soil mixtures. Some bark contains compounds injurious to plants. Bark needs to be composted (partially decomposed) to lower problems such as tying up nitrogen in the decomposition process.

Sawdust is similar to bark and should be composted. Sawdust will compost quite rapidly. Commercially composted manure, bark, and sawdust are also available and are suitable for use as greenhouse and nursery growing media.

Muck is black, partly decomposed peat. Because it is nonfibrous in the sense that the fibers are not readily visible, it is easy to work with. Deposits of muck are located in several states where they are dug and sold in bulk or bags, since the material will not hold together in bales. Muck has a high adsorptive quality. Some mucks are strongly acid, while others are about neutral. It is important to check the pH of muck to determine its suitability for the plants to be grown.

Sphagnum peat moss is a brown moss peat which is often mixed with other growth media such as vermiculite (Figure 17.8).

Fig. 17.8 — The tomato plants behind this researcher are growing in a "soil-less" medium of peat and vermiculite. Vermiculite is shown in his hand. (Courtesy, Ontario Ministry of Agriculture and Food, reproduced through the courtesy of Sperry - New Holland Corporation)

It is important to get the horticultural grade to use in transplanting, rather than the coarser grades which are often used for poultry bedding. Sphagnum peat moss commonly comes in 6-cubic foot (0.17-cu m) bales, which when spread out will equal about 12 bushels (0.422 cu m). Sphagnum peat moss has a pH of 3.5 to 4.3, and when used for many plants it must be neutralized by the addition of about 5 pounds (2.3 kg) of hydrated lime or 8 pounds (3.6 kg) of ground limestone per each 6-cubic foot (0.17-cu m) bale. If hydrated lime is used, it will "burn" plants if sufficient time is not allowed for it to react with air and water. However, acid-loving plants such as rhododendrons and azaleas need an acid medium.

Artificial Materials

Artificial materials commonly used as growth media for greenhouses and nurseries are vermiculite, perlite, calcined clay, and polystyrene foam.

- **Vermiculite** — Vermiculite is a mica that has been heated to 1,400° F (760° C), is sterile, and weighs 6 pounds per cubic foot (96 kg / cu m) (0.1 gm / cc). The unique platelike structure of vermiculite enables it to hold and release large quantities of water and minerals for plant growth. Vermiculite has a relatively high cation-exchange capacity. It does not change pH rapidly, and somewhat higher levels of fertility can be used without plant damage (Figure 17.9).

  Vermiculite contains enough magnesium and potassium for growth of most crops. Additional calcium is needed and must be supplied, usually in

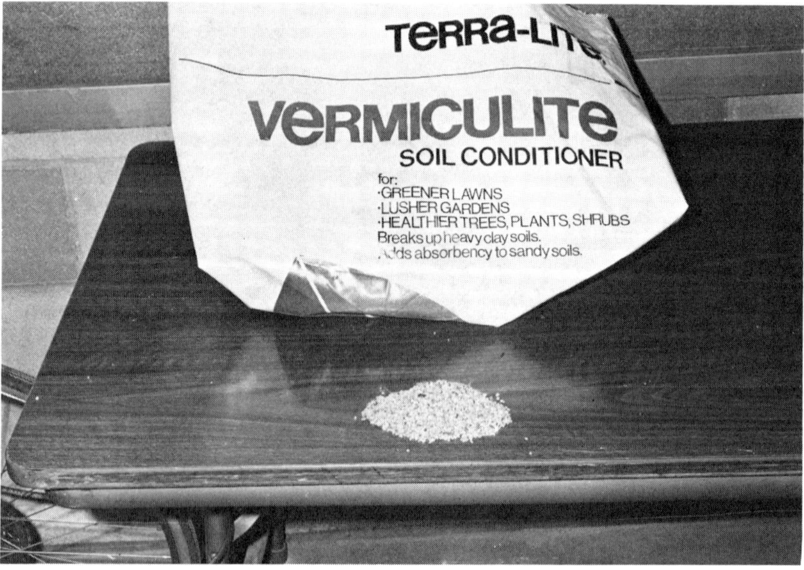

Fig. 17.9 — Vermiculite is one of the most widely used artificial materials for growing plants. (Photo by Rodney W. Tulloch)

the form of limestone, to bring the pH of the mixture into the proper range. Only horticultural grade vermiculite should be used for plant growing. The quality control for insulation grades is less rigorous. Vermiculite can lose many of its good characteristics through rough handling which causes the particles to be broken into smaller bits.

- **Perlite** — is a form of volcanic rock that has expanded by being heated to 1,800° F (982° C). It is sterile, has a pH of 7.0 to 7.5, and weighs 6 to 9 pounds per cubic foot (96 to 144 kg / cu m) (0.1 to 0.14 gm / cc). This material contains no mineral nutrients and, unlike vermiculite, has no cation-exchange capacity or buffering capacity. It does not decay or deteriorate. Perlite holds water on its irregular surface areas similar to the way sand does.

- **Calcined clay** — This is montmorillonite clay (high shrink-swell clay), which when baked at high temperature, improves the air-water relationship when it is added to a soil mixture. After being baked, the clay does not shrink or swell. Since it is porous, it can absorb water when soil moisture is high and then hold water for plant use as the soil dries out. It has no direct fertilizer value but is very resistant to breakdown from handling or tillage. Calcined clay is lighter in weight than sand.

- **Polystyrene foam** — Better known by trade names such as Styrofoam® and Styromul®, it is a very light-weight synthetic, plastic, inert material. Polystyrene foam increases aeration in a mixture but does not absorb water or nutrients. Therefore, it must be mixed with something that does hold water and nutrients. Polystyrene foam does not decay or deteriorate, which makes it useful for beds or other places which are used over and over. The foam comes in flakes or beads for use as plant growth media.

Artificial materials have several advantages that favor their use.

1. They are generally very light in weight.
2. They have a known fertility level, making it much easier to fertilize plants properly.
3. They contain fewer disease organisms than do natural soils or organic matter.
4. They are more uniform from batch to batch than soil is.
5. They drain well and supply outstanding aeration for plant roots.

Mixtures

Thorough mixing of the components is very important, whether soils, organic matter, artificial material, or a combination of these is being used. If fertilizers, and especially micronutrients, are added to the mix, it is important that they be uniformly distributed throughout the mixture (Figure 17.10). Several companies sell various mixtures ready for growing specific kinds of plants.

Fig. 17.10 — This greenhouse medium has been fertilized and run through a concrete mixer and is ready for use as a potting soil. (Photo by Rodney W. Tulloch)

The kind of mixture to make depends partly on the purpose for which the mixture will be used. A quality such as how long the mixture will hold up with little change is obviously more important in seed germination media that are reused than in germination media or bedding plant mixes that will not be used again.

Any mixture using organic matter will break down over time. A mixture using soil is more likely to develop diseases and weeds. It is hard to find uniform supplies of soils, and some of them contain herbicides, such as atrazine, that can destroy an entire crop. These and other factors have led many growers to use artificial materials, sometimes mixed with soils and organic materials.

Mixtures used as seed germination media should be well-drained and well-aerated. They should also have high water-holding capacity, although modern watering methods and the placing of plastic over the germinating seeds have made water holding less important than it once was.

A good mixture for seeding can be obtained by mixing top soil with high proportions of perlite or vermiculite and peat. Each year more and more growers are using a vermiculite or perlite-and-peat mixture, usually using about 50 percent of each. Some growers have even used 100 percent pure sand, 100 percent perlite or vermiculite, or 100 percent peat, all with good germinating success.

Potted plants require mixtures which are coarse and porous and which will drain readily. Materials such as coarse peat moss, chopped straw, or peanut hulls have been used successfully. The amount to use varies with the soil and the plant being grown.

Foliage plants generally grow best in high organic soils. Usually half or more of the mixture should be either leaf mold or acid peat moss. The finer the soil texture, the higher the percentage of organic materials necessary for ideal plant growth. Most foliage plants require a pH of 5.5 if the soil mixture is mostly organic and pH 6.5 if it is mostly a mineral soil. (See Chapter 3.)

Cut-flower crops, such as roses and carnations, require a coarse and porous mixture which has been well-prepared. Organic matter, such as coarse peat, chopped straw, or rotted manure, can be added to improve the soil structure. One-fourth or more of the volume of the mixture should be organic matter. If clay soils are used, the volume of organic matter should be increased to one-third.

Common potting mixtures for nursery use are one or two parts sand, one part loam soil, and one part peat moss, shredded bark, or leaf mold. For general container-grown nursery stock, one part sand, two parts loam soil, and one part peat moss, shredded bark, or leaf mold is used. Many producers are using more artificial mixtures for container stock such as fine sand mixed with redwood sawdust, shredded fir bark, or rice hulls. For germinating and growing mixes, peat moss is mixed with vermiculite or perlite. Both of these artificial materials are low in nutrients, and based on a soil test, fertilizer must be added.

17:4 □ WATERING GREENHOUSE AND NURSERY PLANTS

There are three types of soil water: gravitational, capillary, and hygroscopic. Gravitational water is of little use to plants because it passes through the soil rapidly, thus giving plants little chance to use it. Capillary water is held in capillary pores against the force of gravity. Capillary water is most important in plant growth since plants can utilize it. Hygroscopic water is the water left in air-dry soil and is not available for plant use.

When soil is put in a bench or pot, the resulting capillary columns are shorter than in normal field conditions, since the soil is shallower. The result is wetter soils in the bench or pot. If clay pots are used, half the water applied may move through the porous pot and be evaporated. If moist soil or water surrounds all or part of the pot, water can be taken into the pot and be absorbed by the pot and then by the plant roots.

For best growth, plants must have a constant supply of water. To take up the nutrients necessary for plant growth, plants take in considerably more water than is really used. The extra water is released through very small openings in the leaves called **stomata** and is evaporated into the air. This process is called **transpiration.**

Water serves as a solvent for many nutrients. Once these nutrients are dissolved they move into plants. Within the plant, nutrients are moved from one place to another via the medium of water. Water is therefore essential in the transportation process.

Water is also involved in many of the important chemical reactions that take place within a plant. Water vapor in the air may collect many soluble materials that can then be carried into a plant through the leaves.

The amount of water required varies greatly with the type of plant, humidity, temperature, stage of growth, presence of nutrients, and other conditions under which the plant is being grown. For example, a potted plant may use as much as a quart (0.9 liter) in one summer day. A rose plant growing in a bench may also consume a quart (0.9 liter) of water on a summer day. Many greenhouse plants may need as little as ⅛ quart (0.1 liter) of water on a cloudy winter day.

While plant roots need a constant water supply for best growth and production, they need air at the same time. If too much water and not enough air is present, plants may not take up as much water as they require. This may result in wilting, even though the soil is very moist.

The soil mixture used for growing plants is very important in providing an appropriate air-water ratio for good root growth. Many soil mixtures used for greenhouses and nursery potting contain ¼ to ⅓ peat, ¼ to ⅓ medium sand, and ⅓ to ½ loam soil. Peat will hold up to 15 times its dry weight in water. Medium sand allows the water to drain out and leaves space for air.

Use of good artificial soil mixtures has improved the grower's ability to apply the correct amount of water at the right interval of time. Much of the watering can be done automatically or semiautomatically with various kinds of equipment.

Cuttings and germinating seeds are often watered with a "mist propagation" system (Figure 17.11). Pipes rising about 2 feet (61 cm) above the bench and 4 feet (1.2 m) apart, each having a mist nozzle on top, are located down the middle of a propagation bench. This system is usually set to automatically throw a 6-foot (1.8-m) circle of mist from each nozzle for about two seconds every two to five minutes. The system is operated only during the sunlight hours when high evaporation takes place. The "mist evaporation" is excellent for keeping cuttings or seeds moist until roots have developed or until germination has taken place.

Another system which is widely used for potted plants is the "spaghetti" system. Small plastic tubes are run to each pot so that when the water is turned on each plant receives the same amount of water (Figure 17.12). With this system the grower normally turns the water on once or twice a day depending on the media used and the needs of the plants. It can be set to shut itself off after each plant is given a specific amount of water (Figure 17.13).

Fig. 17.11 — These are common nozzles used for adding a mist for plant propagation. The areas being mist propagated are often surrounded with plastic or other material to keep the moisture in the desired area. (Photos by Rodney W. Tulloch)

Fig. 17.12 — The "spaghetti" watering system for container-grown plants is used in many greenhouses. (Photo by Rodney W. Tulloch)

Fig. 17.13 — This automatic control connected to the electrically operated valve allows automatic watering as much and as often as desired. (Photo by Rodney W. Tulloch)

Some benches and pots can be watered from the bottom. A solid bench can be temporarily flooded. Watering should generally be done in the morning to let the plants dry before evening. This can help prevent disease.

A grower may test the moisture in selected pots to determine when watering is to take place. (See Chapter 9 for a description of how to determine the amount of water in a soil.) When the medium is cool to the touch, it is normally wet enough. While severe wilting may cause permanent damage or death, slight wilting often causes little or no permanent damage. Watering by the spaghetti or bottom watering system may take less than 1 percent of the time required to water potted plants by hand.

Cut-flower crops, such as chrysanthemums, roses, and carnations, are often watered with perforated plastic tubes or flat spray nozzles. A single valve may be used to water from one to several benches. Approximately a quart (0.9 liter) of water is used for each square foot (0.1 sq m) of media each time water is added.

When watering by hand, use a **water breaker** (Figure 17.14) to apply larger volumes of water at slow velocities without damaging the plants or washing or compacting the media. Water breakers are often used in high school greenhouses because they are relatively inexpensive and are easy to use.

Since most of these methods do the job effectively, it is most important to know how much water to apply and when to apply it. The amount of water needed depends on factors such as the kind of plant, the size of plant, the stage of growth, the season of the year, and the weather. We can generalize and say that each time water is applied the entire soil volume should become wet to the field capacity. (See Figure 9.4.) We can be reasonably sure we have reached this point when a uniform application of water begins to drain from the bottom of a pot or bench.

A little extra water above the field capacity can be beneficial in that it can leach out excessive soluble salts which can inhibit plant growth. (See Chapter 10.) These salts come from fertilizers and, in some cases, from water applied to the plants. If a soil test shows that a high salt content is present, the soils must be thoroughly leached. This is done by applying four times the water normally used. The application may have to be repeated to remove most of the salts. Since leaching also removes nitrates, additional nitrogen fertilizers will often need to be added.

17:5 □ NUTRIENT LEVEL AND BALANCE

A thorough discussion of fertilizers can be found in Chapter 4, and of plant nutrition in Chapter 2. However, a few of the major points will be reemphasized in this section along with information which is specific to greenhouse or nursery crops. The 16 essential plant nutrients are as important to the growth of greenhouse and nursery plants as to any other kinds of plants. Because of the high costs of greenhouse and nursery production, it is essential that nutrients be applied with greater timeliness and precision for the best possible production and profit.

Fig. 17.14 — This water breaker allows hand application of large volumes of water at slow velocities without damage to plants. (Photo by Rodney W. Tulloch)

Many growers are using somewhat more fertilizer than in the past on both greenhouse and nursery crops. There are two main reasons for this practice. First, using more fertilizer increases growth, thereby cutting the time necessary to produce plants. This cuts the cost, especially high heating bills. Second, added fertilizer often increases the quality of the plants produced, thus bringing a better price and making them easier to sell.

Care must be taken to avoid the overuse of some nutrients. For example, many vegetable plants, such as tomatoes, are raised for their fruit. Too much nitrogen can drastically cut the amount of fruit set on such plants and can also contribute to the production of "leggy" plants.

When plants are raised at cool temperatures to make them short and compact, little fertilizer is needed since there is little growth. As the weather warms and more water is needed, fertilizer application must be increased as well.

Many greenhouse growers and some nursery growers add fertilizer as they water their plants. Fertilizer injectors are used to put a measured amount of liquid or dissolved fertilizers in the water. While this is an excellent method of adding fertilizer, problems can arise. If water is not applied heavy enough for pots or beds to drip, the buildup of soluble salts will be very rapid. This buildup can damage or kill plants in a short time. Injectors can have many kinds of difficulties and should be checked and calibrated at frequent intervals. If the injectors malfunction, they may not add enough fertilizer, thus causing a nutrient deficiency.

Long-lasting (slow-acting) **fertilizers,** pellets, and even **fertilizer "spikes"** (stick-like, slow-acting fertilizers) are now available. Each one makes smaller amounts of nutrients available over a longer period of time than traditional fertilizers. The long-lasting fertilizers may be added to the surface of the soil or worked into the soil in the bin. Then for 2½ to 4 months it is necessary to add only water. When mixing these long-lasting materials, the grower must mix them thoroughly, or some plants will get too much fertilizer while others will get too little. It may be necessary to use a concrete mixer or other mechanical device to do the job well. Mixing the fertilizer into the soil with a shovel is tedious, and it is nearly impossible to get a uniform mixture. Fertilizer spikes that can be driven into the ground may be used on large shrubs and trees. Smaller fertilizer spikes are available for driving into the soil around small plants.

Two long-lasting fertilizers have the trade names of **MagAmp®** and **Osmocote®.** MagAmp® is a pelletized material with a grade of 7-40-6. It will last three or four months in the soil, which is long enough to raise most bedding plants. Osmocote® is a resin-coated material with a grade of 18-9-13 or 14-14-14. It will last 2½ to 3 months in the soil. Steam sterilization will break down the resin and release the nutrients. It is not advisable to use much, if any, long-lasting fertilizer on early plants as it may become impossible to control their growth. For the first 2½ months new plants need only ½ the normal recommended dosages of fertilizer. Tables 17.2 and 17.3 indicate the kinds and forms of long-lasting fertilizers available and recommended application of fertilizers for greenhouse plants.

Table 17.2 — Kinds and Forms of Long-lasting Fertilizers Available[1]

Brand Name	Fertilizer Grade ($N-P_2O_5, K_2O$)	Trace Elements Listed (number)	Frequency of Application	How Applied
Capsules, Spikes, Sprays, and Timed-Release				
Dramm Mini-Pill	14-4-6	—	4 months	On top of soil
House in Bloom	7-7-7	—	3 - 4 weeks	Spray onto soil
Jobe's Houseplant Spikes	12-6-6	—	2 months	Bury in soil
Ortho General Purpose Plant Food	12-6-6	—	3 - 4 months	Work into top of soil
Osmocote	14-14-14	—	3 - 4 months	Mix into potting soil
Plantabbs	11-15-20	—	1½ - 2 months	Dissolve in water
Precise Plant Food	12-6-6	—	3 - 4 months	Work into top of soil
Liquid Concentrates				
Agro Fish 'n Six	14-6-8	6	2 - 3 weeks	Dilute in water
Cole's 100% Organic (from seaweed)	0.28-0.02-0.07		—	Foliar spray
Dr. Sher's Gourmet Food for Plants Foliage & Leaf Diet	15-5-5	8	every watering	Dilute in water
Eco-Gro Fish & Seaweed Fertilizer	8-4-4	—	3 - 4 weeks	Dilute in water
Liquinox Bloom	0-10-10	—	3 - 4 weeks	Dilute in water
Liquinox Grow	10-10-5	—	2 weeks	Dilute in water
Ortho House Plant Food	5-10-5	3	weekly	To soil

(Continued)

Table 17.2 (Continued)

Brand Name	Fertilizer Grade (N-P_2O_5, K_2O)	Trace Elements Listed (number)	Frequency of Application	How Applied
Dry Mixes				
Agro Aqua Feed	20-20-20	—	—	Foliar or in water
Eco Grow	10-8-14	12	1 - 3 weeks	Dissolve in water
Hyponex House Plant Food	7-6-19	—	weekly	Dissolve in water
Miller's Booster House Plant Formula	5-10-14	9	7 - 10 days	Dissolve in water
Nulife Soluble Plant Food	20-20-20	—	weekly	Foliar or in water
Restore Foliage Plant Formula	3-4-4	—	every watering	Dissolve in water

[1]Source: Instructional Materials Services, Texas A&M University.

If a soil test indicates low P and K, a fertilizer containing phosphorus and potassium may be added to the soil mixture as needed. Nitrogen is better added after plants have begun to grow since the nitrate and urea forms are very soluble and leach readily.

17:6 □ CONTROLLING WEEDS AND MULCHING

Weeds in the greenhouse are generally controlled by the selection of weed-free soil or media or by sterilization. Mulches are not widely used in greenhouse-crop production except with special crops such as orchids.

Weeds are a problem in nursery production just as they are in many field crops. Weeds compete for light, moisture, and nutrients and often succeed in getting what they want while choking out or slowing the growth of nursery plants. Weeds serve as hosts to insects and diseases that are detrimental to nursery plants.

Methods used to control weeds when growing nursery plants include cultivation, soil fumigation, application of herbicides, and black plastic mulches (Figure 17.15). Major factors in selecting weed control methods are labor required, cost, effectiveness, and side effects.

Mechanical cultivation has been the predominant method of weed control in the nursery (Figure 17.16). Mechanical tilling of the soil kills small and sometimes larger weeds, but at the same time brings more weed seeds to the surface to germinate. Cultivation may thus become an almost endless task. It still may be cheaper than applying herbicides, especially to weed species which keep coming back. Cultivation has the additional advantage of breaking up any surface crust and loosening the soil to allow more air and water to enter plant roots. It may still be necessary to apply a herbicide in the row to control the weeds and/or reduce hand labor.

Table 17.3 — Recommended Application of Fertilizers for Greenhouse Plants[1]

Fertilizer	Bench- or Bed-grown Plants		Pot-grown Plants		
	Dry (lbs. / 100 sq. ft.)	Liquid (oz. / gal. of water)	Dry (lbs.)	Liquid (Apply weekly or every other week)	
				Young Plants	Established Plants
Nitrogen					
Urea	½	1 to 7	1 to 3 for each 2½ bu. (1 wheelbarrow)	20 to 50 ppm	75 to 90 ppm
Ammonium nitrate	½	1 to 5	"	"	"
Ammonium sulfate	1	1 to 2	"	"	"
Sodium nitrate	1	1 to 2	"	"	"
Calcium nitrate	1	1 to 2	"	"	"
Potassium nitrate	1	1 to 2	"	"	"
Phosphorus					
Superphosphate	5	1 to 5	"	"	"
Monocalcium phosphate	1	1 to 2	"	"	"
Potassium					
Potassium nitrate	1	1 to 2	"	"	"
Potassium chloride	1	1 to 2	"	"	"
Fritted	3	not recommended	"	"	"

[1]Source: Instructional Materials Services, Texas A&M University.

Fig. 17.15 — The black plastic mulch on these marigolds at the University of Kentucky Landscape Center is effectively controlling the weeds. (Photo by Rodney W. Tulloch)

Fig. 17.16 — This arborvitae nursery stock grown on the contour is practically weed free through the use of mechanical cultivation. (Courtesy, USDA - Soil Conservation Service)

Soil fumigants, such as Vapam® and methyl bromide, can often be used to control weeds in plant beds on an economical basis. These materials kill all plant life and many soil-borne diseases and insects. They can be dangerous to use, and label directions should be followed carefully, the same as when any pesticide is being used. Soil temperatures must be about 60° F (16° C) or higher for fumigants to be effective. It takes 10 to 14 days for the fumigant to dissipate and before the crop can be planted. The high cost of fumigation prohibits its use except on very high-value crops. It is usually used only on plant beds at nurseries and not around the nursery in general.

Selective herbicides that can be used around nurseries come in granular and liquid forms. Careful selection and application of the correct amount of herbicide is necessary to kill the weeds without damage to nursery plants. Herbicides are usually applied after the plants have been planted. Special precautions are necessary to assure that the material does not come into contact with nursery plant leaves, as severe injury could occur. Most herbicides work best when tilled into freshly cultivated soils. Weeds are then killed as they germinate. The timing and amount of herbicide to apply vary with type of nursery plants, weeds to be controlled, the stage of growth of weeds, the kind of herbicide being used, texture and organic matter content of soil, and temperature at time of application. If herbicides can be applied while nursery plants are not actively growing, there is less chance of damage to the plants.

A heavy organic mulch can be used effectively to control weeds. Materials such as hay that may contain large numbers of weed seeds should be avoided. (See Chapter 15.)

Besides controlling weeds, organic mulches also are useful for increasing water infiltration, promoting more uniform germination and faster growth, conserving water, and improving soil tilth. Organic mulches also make it easier to walk through a nursery and to harvest crops without compacting the soil or getting muddy.

Black polyethylene plastic may be used to control weeds, and it is particularly good for smothering hard-to-kill weeds such as bermudagrass, quackgrass, and johnsongrass. Machines can be used to cover entire fields in black plastic.

Weed blocker mulch (a woven material) lets moisture enter but stops light and thus weeds. Weed blocker mulches are quite expensive but are popular in container nurseries as they can be used for up to three years.

17:7 □ REFERENCES

Ball, Vic, ed. *Ball Red Book: Greenhouse Growing,* 14th ed. Reston, Virginia: Reston Publishing Co., Inc., 1985, 720 pp.

Blessington, Thomas M., Connee C. Luttrell, and N. Curtis Peterson. "Growing Media Selection & Fertilizer Programs for Optimum Growth and Quality of Geraniums." *Research Report,* Vol. 10, No. 6, Mississippi State University: Mississippi Agricultural & Forestry Experiment Station, April 1985.

Boodley, James W., and Raymond Sheldrake, Jr. "Cornell Peat-Lite Mixes for Commercial Plant Growing." *Information Bulletin 43.* Ithaca: New York State College of Agriculture and Life Sciences, Cornell University, July 1982.

Cook, Ray L., and Boyd G. Ellis. *Soil Management.* New York: John Wiley & Sons, 1987, 414 pp. (See especially Chapter 17, "Soil Management for Gardens, Greenhouses, and Lawns.")

Donahue, Roy L., Raymond W. Miller, and John C. Shickluna. *Soils: An Introduction to Soils and Plant Growth,* 5th ed. Englewood Cliffs, New Jersey: Prentice-Hall, Inc., 1983.

Duncan, George A., and C. R. Roberts. "Hotbeds for Transplant Production." University of Kentucky, Co-op. Ext. Serv., 1976, 9 pp.

Follett, Roy H., Larry S. Murphy, and Roy L. Donahue. *Fertilizers and Soil Amendments.* Englewood Cliffs, New Jersey: Prentice-Hall, Inc., 1981, 558 pp.

Link, Conrad, and David Ross. "Year-Round Gardening with a Greenhouse." *Living on a Few Acres. The 1978 Yearbook of Agriculture,* U.S. Dept. of Agriculture, Washington, D.C.

Nelson, Paul V. *Greenhouse Operation and Management,* 3rd ed. Reston, Virginia: Reston Publishing Co., Inc., 1985.

Schneider, R. W., ed. *Suppressive Soils and Plant Disease.* St. Paul: The American Phytopathological Society, 1982, 88 pp.

CHAPTER 18

Orchards: Soil, Water, and Fertility Management

Much labour is required in trees;
Well must the ground be digg'd, and better dress'd.
New soil to make, and meliorate the rest.
— Dryden (1631 - 1700)

OUTLINE

Most orchards are established on sloping land to obtain good air drainage. Sloping lands, however, are subject to severe erosion that robs the orchard of the most productive soil and of water. Contour terraces, sodded middles, and cover crops can be used to control both soil and water losses. In orchards that require irrigation, the water channel should be established above each terrace on grades of 0.5 to 1.5 percent, depending on soil texture and infiltration rate and the trees planted on the terraces. (Courtesy, Irrometer Company, Inc., Riverside, California)

□ □ □

18:1 □ OVERVIEW

Orchards, vineyards, and planted nut trees are most numerous in California, Oregon, Washington, Florida, Mississippi, Texas, Georgia, Maryland, New York, and Michigan. However, they are planted in all 50 states. According to the 1986 Census of Agriculture, the total acreage was in excess of 3.5 million (1.4 million ha) (Figure 18.1).

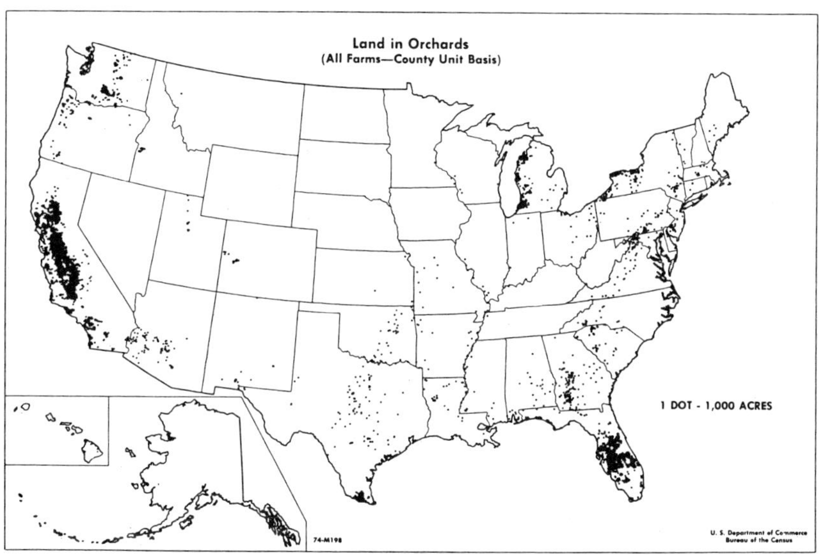

Fig. 18.1 — There are more than 3.5 million acres (1.4 million ha) of orchards, vineyards, and planted nut trees in the United States, indicated by dots on this map. (Courtesy, Bureau of the Census, U.S. Department of Commerce)

Citrus fruit crops, which account for 27 percent of the acreage, are produced mainly in Florida, California, Texas, and Arizona. Apples and grapes also rank high in acreage.

The major orange-producing states are Florida, California, Texas, and Arizona. Florida has the largest acreage, accounting for 71 percent of the total crop. These four states account for almost 100 percent of the entire acreage of oranges.

Although the value per acre of fruit and nut crops is high, they are also among the most sensitive crops in their adaptation to climate, slope, soil, water, and fertility. This chapter will discuss some of the soil, water, and fertility management practices most important to successful site selection and orchard management.

18:2 □ ECOLOGICAL REQUIREMENTS

The ideal soil for most kinds of orchards is a slightly acid, well-drained loam that permits roots to proliferate to a depth of at least 5 feet (1.5 m). Unrestricted root proliferation requires the absence of compacted soil layers, bedrock, water table, very strongly acid horizon, or salt layer. Good air drainage is just as important as good soil drainage. Frost pockets are to be avoided. Good air drainage exists on slopes that have no obstructions to the valleys below. Flowing waters or quiet waters in the valley also cause better air drainage. A band of trees, shrubs, or buildings across a slope below an orchard can prevent cold air from moving down the slope to the valley, causing fruit buds to freeze in the spring. A lower slope that receives cold air from the upper slope is to be avoided for an orchard site. Since prevailing winter winds in the United States are from the west, a large lake or ocean west of the orchard site will moderate the temperature of the wind and improve the chances of success of an orchard that is located east of such a body of water. The states of Washington, Oregon, California, New York, and Michigan are examples of states that receive frost protection from large bodies of water located on their western side.

The direction of slope (**aspect**) of an orchard is of some importance. Most fruit growers prefer a north-facing slope because growth begins later in the spring; therefore, risk of frost injury is less. The second choice of most orchardists is a northeast slope, and the third choice is an east slope (Figure 18.2).

South-facing slopes are warmer than north-facing slopes because south slopes receive more direct rays of the sun. However, warmer temperatures on south-facing slopes encourage fruit buds to start growing earlier, and therefore, late frosts do more damage.

Fig. 18.2 — For most orchards a north slope is preferred because it warms more slowly in the spring. This delays the premature opening of fruit buds and thus reduces damage to the fruit crop by a late spring freeze. (Courtesy, University of New Hampshire)

Research to confirm the statement of conventional wisdom that valleys are frost pockets and that slopes should be chosen for establishing orchards is cited from near Athens, Georgia, in the Piedmont Plateau. Here, four sandy loam sites were selected for a detailed study of climatic factors relating to topographic position and slope exposure (aspect). All four sites were within 2,000 feet (610 m) of each other. Average growing season (freeze-free days) for a five-year period were:

Soil	*Topographic Position*	*Freeze-Free Days per Year*
Congaree sandy loam	Level river bottom	179
Cecil sandy loam	Northwestern slope	187
Cecil sandy loam	Level hilltop	198
Cecil sandy loam	Southeastern slope	199

The data from the Piedmont Region in Georgia indicate that a river bottom site may have a freeze-free period 20 days shorter, and a northerly slope 12 days shorter, than a southeastern slope. The level hilltop had a freeze-free period only one day shorter than the southeastern slope.

The probable rating (from most desirable to least desirable) of these sites for establishing a peach orchard would be:

1. Northwestern slope (because late frosts do less damage).
2. Level hilltop (because frost damage to fruit buds is intermediate between northwestern and southeastern slopes).
3. Southeastern slope (because late frosts kill fruit buds).
4. Level river bottom (because it is a frost pocket).

Under practical field conditions, when an orchard site has been chosen in relation to slope direction and air drainage, there is seldom much choice left in selecting the right kind of soil. But soil is no less important. Probably the most important soil factor to consider in selecting a good orchard site is the internal drainage of the soil. Seepage places, wet spots, and shallow soils should be avoided. To be more certain of selecting the proper soil, dig several holes to a depth of 5 feet (1.5 m) in order to see the soil. Are there any compact layers that may hinder root growth? Is the soil uniform in color or is it mottled with yellow, brown, and gray? A uniform gray color indicates very poor drainage, and mottling indicates poor drainage during a part of the year (Figure 18.3). Did you strike bedrock? How deep do grass and weed roots grow? The chances are that orchard trees will root to approximately the same depth. Is the soil moist (but not wet) all the way down? Is the soil dark, indicating a high organic matter content? (For more details on how to judge desirable soils, see Chapter 11.)

Modern technology is available to overcome or compensate for certain unfavorable soil and site factors. Open-ditch or tile drainage can overcome unfavorable soil factors associated with a soil that is too wet for the ideal orchard. (See Chapter 8.) A bulldozer can be used to eliminate a frost pocket, if the cost is not prohibitive. Sprinkler irrigation and wind machines can be used during periods of

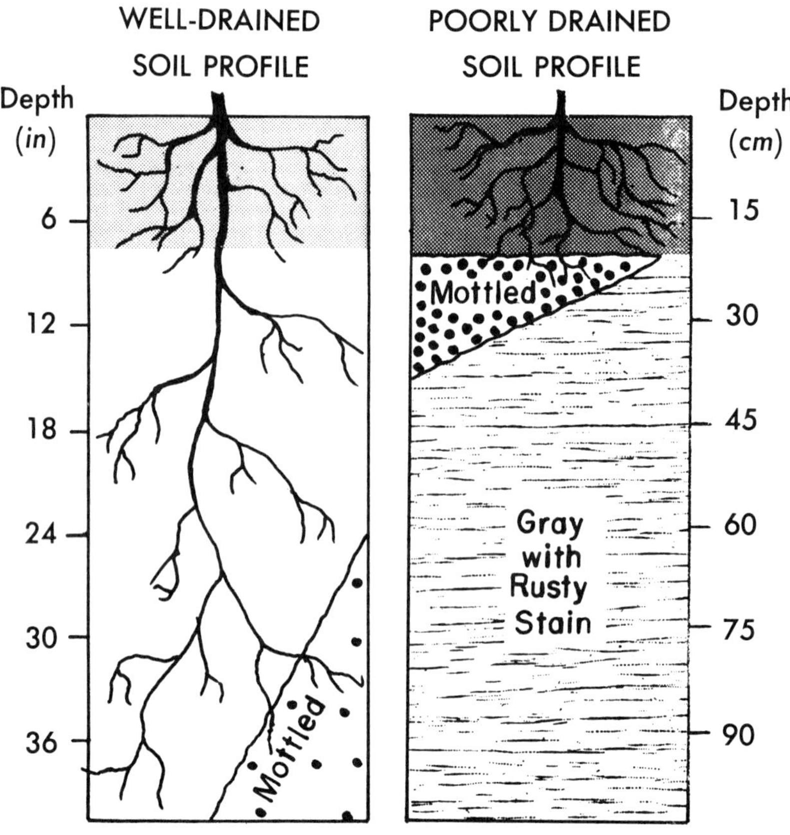

Fig. 18.3 — Mottling in a soil indicates intermittently poor internal drainage, and a solid gray with rusty stains indicates year-round poor drainage. Soils for orchards should be well-drained with no mottling within the surface 36 inches (91 cm). (Courtesy, Department of Agronomy, Cornell University)

potential frost damage to fruit buds. The cost of fuel for orchard heaters has made their use prohibitive in many cases. A site where wind erodes soil or causes "wind-whip" of fruit trees can be protected by a windbreak planting of trees and shrubs. (See Chapter 19.)

18:3 □ SOIL TESTS AND FOLIAR ANALYSES

Assuming that the proper soil and slope have been selected and the orchard has been established with the recommended cultivars, how does the manager determine what fertilizers should be applied to maximize yields?

Soil tests use a chemical extractant that approximates the ability of plant roots to absorb nutrients from the soil. All scientists know that different soils vary in releasing nutrients and that different plant roots vary in absorbing them. Maximum tree root extension may be 3 feet (91 cm) in some soils and 6 feet (183 cm) in others. With woody perennial crops such as orchard trees, a further complication is the storage of nutrients in roots and stems during one year for use in subsequent years. Another variable is the rate of growth in response to seasonal variations in temperature and moisture. Within any one species, fast-growing leaves have a lower concentration of nutrients than slow-growing ones. Additionally, the concentration of any one nutrient in a leaf is determined by the relative concentration of other nutrients.

Soil tests should be used in conjunction with foliar (leaf) analysis. A foliar analysis will tell the grower if trees are becoming deficient in nutrients, such as boron, before symptoms are present on the leaves. Foliar analysis can also be used to evaluate a fertilization program. For example, a soil test may show that the soil phosphorus level is low, but the foliar analysis indicates that the tree is obtaining enough phosphorus and that additional soil applications of phosphorus are not needed. Frequently, foliar analysis shows that foliage nutrient levels are low, and the soil test indicates a low soil pH. In this case the situation is often corrected by liming to raise the soil pH. This makes many soil nutrients more available to the tree, and specific nutrients may not need to be applied. (See Chapters 2 and 3.) When taking a foliar sample for analysis, follow these instructions:

1. Obtain a mailing kit and a submission form from your local county extension office for each foliar sample that you want analyzed.
2. Avoid dust- or soil-contaminated tissues. If all the tissue available is dusty, wash it gently in flowing clean water for about a minute. Further washing may leach some of the nutrients.
3. Avoid taking diseased, insect-infested, or mechanically damaged plant tissue for sampling.
4. When sampling nutrient-element deficient plants, take two samples if possible, one from a normal plant and one from an abnormal plant. Use individual mailing containers and submission forms. Reference to the other can be made on each submission form.
5. Both the *plant part* collected and the *time* it is collected are important. When sampling, be sure to collect the proper plant part at the recommended time, as specified in Table 18.1.
6. If specific sampling instructions are not given for the crop you wish to have analyzed, a good rule of thumb is to sample mature leaves which are

Table 18.1 — How to Take Foliar Samples from Fruit Trees

Crop	Time	Method of Sampling	No. of Trees to Sample
Peach	12 to 14 weeks after bloom	Select mature leaves from mid-portion or near base of current season's terminal growth, taking four to eight leaves per tree.	25
Apple	8 to 10 weeks after bloom	Select leaves from spurs or near base of current season's growth, taking four to eight leaves per tree.	25
Pear	Midsummer	Select mature leaves from spurs, taking four to eight leaves per tree.	25

representative of the current season's growth during the mid-period of growth cycle or just prior to seed or fruit set.

Allow the plant sample to air dry for at least a day, then place the plant tissue sample in a clean envelope or paper bag. Avoid using a plastic or polyethylene bag since this may cause the sample to spoil.

7. Complete a submission form in its entirety for each sample. Use a pencil to fill out the form(s). A pen may smear if the sample is moist. Missing information will result in an incomplete interpretation and recommendation. Place the completed form with the leaf sample in the mailing kit and mail to the laboratory.

After reviewing the foliar sampling instructions and Table 18.1, contact your local extension service for further assistance. While visual observation can detect deficiencies, serious damage may have already been done to the current crop (Figure 18.4).

Fig. 18.4 — A normal apple leaf is on the left and a magnesium-deficient leaf is on the right. This deficiency can be cured by applying Epsom salt ($MgSO_4$) or dolomitic limestone. Both a soil test and a tissue analysis can accurately predict magnesium deficiency *before* visual symptoms appear. (Courtesy, University of New Hampshire)

18:4 ▫ SOILS AND SITES FOR TART CHERRY ORCHARDS

Moderation of westerly wintry winds by more than 100 miles (161 km) of the deep waters of Lake Michigan makes cherry production possible on the sandy slopes near the lake in the state of Michigan. Not only is cherry production possible, but Michigan accounts for more than 80 percent of all tart cherry production in the United States.

In the heart of the tart cherry orchards in Michigan, a cooperative soil and site study was made of Grand Traverse County in the northwestern part of the Lower Peninsula of the state by personnel of the Soil Conservation Service, Michigan State University, and the National Weather Service. On a series of 37 aerial maps for the county, four categories of potential red tart cherry orchard sites were identified based on these factors: soil, physiography, biotic influences, and climate.

- ***Soil*** — The ideal soil for establishing a cherry orchard is a coarse loamy sand that is well-drained, high in natural fertility, moderately high in available water capacity, moderately high in permeability, and at least 4 feet (122 cm) deep to a compacted soil layer or bedrock.
- ***Physiography*** — Ideal physiographic factors favorable for a cherry orchard include:
    - A slope of 2 to 12 percent.
    - Unrestricted air-flowage – ways on the slope. This implies the crest of a hill or top of the slope so that cold air from above cannot move on to the site. It also means no obstructions along the slope to hinder the ready movement of air down-slope to the valley below. Tall grass, shrubs, or trees will act as a dam to hold cold air. Undulations in the slopes may also retard the movement of cold air down hill.
    - A lake at the base of the slope to heat the mass of air that comes down-slope; the cold air thus heated will rise and thereby make space for more cold air to descend. ("Heating" of down-slope cold air means increasing the temperature by about 2° F [about 1° C], which is enough to expand the air, lighten the air mass, and cause it to rise.)
    - Location of the orchard site above the "spring freeze line." The spring freeze line is a locally determined elevation caused by cold air masses in valleys. Above the top of the normal cold air mass, freezing of the cherry crop in the spring may occur only 2 years in 10, which is considered satisfactory for establishing a cherry orchard in Grand Traverse County, Michigan.
- ***Biotic influences*** — Although not mentioned in the reference cited, biotic factors should also be considered in selecting a cherry orchard site. The most important biotic factors include the relative hazard of damage by birds, wild animals, people, insects, and diseases.

- ***Climate*** — Ideal climatic factors for a cherry orchard include:
  - A minimum winter temperature of −20° F (−29° C).
  - During pollination of cherry blossoms, daytime temperatures should be above 50° F (10° C), and nighttime temperatures no lower than 28° F (−2° C).

18:5 ▫ GROUND COVER AND FERTILITY MANAGEMENT FOR APPLE ORCHARDS

In establishing an orchard, it is very important to adjust the soil pH to the recommended level and to work in needed nutrients (usually phosphorus, potassium, magnesium, and calcium). Nutrients such as phosphorus are very difficult to get down into the root zone after the orchard is planted.

Following planting, many orchards require only an annual application of nitrogen early in the spring before growth starts. Most growers adjust the amount of nitrogen that they apply based on annual shoot growth, leaf color, and foliar analysis.

In the apple-growing regions near Cashmere, Washington, the Agricultural Research Service reports an experiment with nitrogen fertilizer. The highest nitrogen application of 1.5 pounds (680 grams) of N per tree gave not only the highest yield of apples but also the greatest percentage of off-color (green) Golden Delicious apples. High nitrogen resulted in more leaves and darker green leaves that shaded the apples from the sun (that would have brought out their normal color) and thus produced more green-colored apples. The green apples are discounted in price because they do not keep in storage as well as yellow apples and because the consumer prefers a yellow-colored apple to a green-colored one (Figure 18.5).

Fig. 18.5 — The orchard manager should try to obtain not only high yields but also a color that the consumer wants. Will these apples "color-up" by picking time or has too much nitrogen fertilizer been used? (Courtesy, Tennessee Valley Authority)

The conclusion drawn was that the lowest level of nitrogen used, 0.75 pound (340 grams) of nitrogen (N) per tree, resulted in the highest percentage of yellow apples and was, therefore, the recommended amount of nitrogen fertilizer to apply each year. There was a visual correlation between nitrogen treatment and intensity of green in the leaves. A standard color chart is available for determining the shade of green at which it is recommended to apply nitrogen fertilizer.

Cumulative yields of three varieties of apples from a 16-year study of nine ground cover – fertility management systems in Delaware is summarized in Figure 18.6.

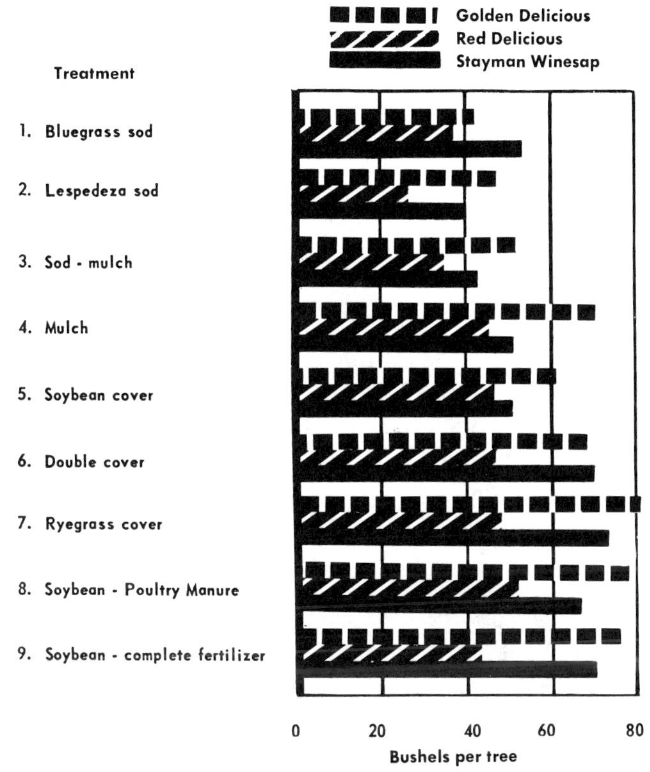

Fig. 18.6 — Cumulative yields of apples per tree to the age of 16 years for three varieties and nine ground cover – fertility management systems. (Source: University of Delaware)

Note: A bushel of apples weighs 48 pounds (21.8 kg).

In addition to the treatments shown in Figure 18.6, each spring all plots received 300 pounds per acre (336 kg / ha) of 0-12-12 fertilizer. Plots 1 through 7 also received a nitrogen fertilizer at the rate of ¼ pound (113 gm) of N per tree per each year of age. This means, for example, at 4 years of age each tree received 1 pound (454 gm) of N, and at 16 years of age each tree received 4 pounds (1.8 kg) of N.

Variations in yield occurred among both treatments and varieties. The highest yields of all three varieties were obtained by Golden Delicious apples with perennial ryegrass ground cover, followed by soybean – poultry manure and soybean – complete fertilizer. Yields of Stayman Winesap were next highest where the ground cover – fertility treatment was perennial ryegrass cover, then a crop of soybeans plus complete fertilizer, and third highest was a double cover of soybeans plus a mixture of rye and vetch. Yields of Red Delicious were highest on soybeans plus poultry manure, perennial ryegrass cover, and a double cover of soybeans plus mixed rye and vetch, respectively. Looking at the cover – fertility treatments for the three apple varieties, the two best for all varieties were perennial ryegrass and soybeans plus poultry manure (Figure 18.7).

Fig. 18.7 — A fertilized perennial ryegrass cover crop for apple orchards in the northeast rates among the best for most varieties of apples and for protection against erosion. (Courtesy, USDA)

Most commercial orchards in the United States are grown using sod middles with herbicide strips down the rows for weed control. Erosion problems and the problem of getting spray machinery into clean, cultivated orchards when they are wet has reduced the number of clean, cultivated orchards in the United States. Exceptions are young orchards and peach orchards, which are greatly benefited by cultivation. The limited availability and cost of organic residue mulches has just about eliminated the use of mulch in commercial orchards. Mulches can also increase rodent populations and damage.

Neither alfalfa nor sweetclover should be used as a cover crop on orchards because both transpire large amounts of water which is obtained at great depths and thus use soil moisture that the fruit trees need. In an apple orchard in Nebraska, alfalfa depleted soil moisture to a depth of more than 15 feet (4.6 m) and sweetclover to a depth of 13 feet (4 m) two years after seeding.

18:6 ▫ WATER MANAGEMENT

Water management of orchards and vineyards means draining wet spots and wet soils by an appropriate system, as discussed in Chapter 8. However, no amount of artificial drainage can change a wet soil into a first-class orchard soil. Only naturally well-drained soils should be selected for establishing orchards (Figure 18.8).

For mature orchards, maximum annual loss of water from the soil by evaporation from the soil surface and transpiration through the leaves may be as high as 36 acre-inches an acre (38 ha-cm). This amount of water is equal to rain or irrigation water standing to a depth of 36 inches (91 cm) over an acre of land (38 cm over a hectare), assuming no runoff and no infiltration. *Average* water use, however, ranges from 18 to 24 acre-inches per acre (18.5 – 24.7 ha-cm) per year. On a daily basis during the growing season, mature orchards may use, in rainfall

Fig. 18.8 — A peach orchard on a well-drained soil is being sprayed for both insect and disease control. A lake is at the base of the slope, making an ideal "sink" for cold air moving down the slope. (Kentucky) (Courtesy, USDA–Soil Conservation Service)

equivalents, from 0.1 to 0.4 inch (0.25 – 1 cm) of water depending mostly on the temperature, relative humidity, kinds of fruit trees, and stage of growth.

Since good air drainage dictates sloping orchard soils, managing water is more difficult than on level land. To reduce erosion and at the same time increase infiltration of both rainfall and irrigation water, terracing is sometimes practiced. The orchard trees are usually planted on the terrace, and above each terrace a water channel is constructed to carry irrigation water and to control rainwater. The grade along the water channel can vary between 0.5 and 1.5 percent, on fine-textured and coarse-textured soils, respectively. The grade of the channels should be established steep enough to get uniform absorption of water all along the full run of the channel, yet not fast enough to cause erosion. The length of runs must also be adjusted with the grade of the channel for the same purpose. Runs may be 300 feet (92 m) on sandy soils, 600 feet (183 m) on loam soils, and longer on fine clay soils.

Instead of terraces with water channels, check basins are sometimes constructed. These basins are constructed as rectangles on level land and along the contour on sloping soils, and in each basin there may be 10 or more trees; to irrigate the trees the basin is flooded. Each basin has retaining walls of soil around it to hold the floodwater (Figure 18.9).

Sprinkler irrigation is adapted to orchards on both level and sloping land. Many orchards use movable under-tree sprinkler irrigation systems; however, many of the newer orchards, particularly on the West Coast, are using solid set

Fig. 18.9 — A pecan orchard is being irrigated by the check-basin method. Each basin is flooded by the use of siphon tubes that deliver water from the irrigation canal in the foreground (New Mexico). (Courtesy, Bureau of Reclamation, U.S. Department of the Interior)

over-the-tree sprinkler irrigation. The over-the-tree sprinkler irrigation systems are being used for both frost control and irrigation. Trickle irrigation is a newer method well-suited to orchards. (See Chapter 9.)

18:7 ▫ REFERENCES

Childers, Norman F., *Modern Fruit Science,* 9th ed. Gainesville, Florida: Horticultural Publications, 1983, 583 pp.

"Fruit and Nut Varieties for Home Plantings." *Agricultural Guide.* Bul. No. 6005, University of Missouri - Columbia, 1987.

Johnson, James W., Hall Shaffer, and Einar Palm. "Home Fruit Spray Schedule." "Grounds for Gardening" No. 6010, University of Missouri - Columbia, 1988.

Petrie, Steven E., W. Michael Colt, and Walter Kochan. "Idaho Fertilizer Guide: Orchards." University of Idaho, Current Inf. Ser. No. 655, undated, 4 pp.

"Red Tart Cherry Site Inventory for Grand Traverse County, Michigan." Soil Conservation Service in cooperation with Michigan State University and the National Weather Service, 1971, 12 pp. and 37 maps.

Soule, James. *Glossary for Horticultural Crops.* New York: John Wiley & Sons, Inc., 1985, 898 pp.

Stang, E. J., L. A. Peterson, and E. E. Schulte. "Fertilizing Small Fruits in the Home Garden." University of Wisconsin, Co-op. Ext., 1981, 4 pp.

Stiegler, Jim, and Glen Taylor. "Soils for Fruit Trees." Oklahoma State University, Ext. Facts No. 6216, undated.

Strang, J. G., M. L. Witt, R. T. Jones, and G. R. Brown. "Growing Fruit at Home in Kentucky." University of Kentucky, HO-64, 1987, 30 pp.

That We May Eat. The 1975 Yearbook of Agriculture. U.S. Dept. of Agriculture, Washington, D.C.

Thom, William O. "Can Plant Analysis Help You?" *The Kentucky Farmer,* September 1981.

U.S. Dept. of Agriculture. *Agricultural Statistics, 1986.* Washington, D.C., 1986, 552 pp. (see especially, Chapter 5, pp. 184 - 240).

Van Den Brink, C., N. D. Strommen, and A. L. Kenworthy. "Growing Degree Days in Michigan." Michigan State University, Res. Rpt. 131, 1971, 48 pp.

Will There Be Enough Food? The 1981 Yearbook of Agriculture. U.S. Dept. of Agriculture, Washington, D.C.

CHAPTER 19

Forests: Soil, Water, and Fertility Management

"The forest industry must as a whole look beyond the cutting of God-given forest stands to future man-generated forests. The soil that becomes suspended sediment in streams needs to remain on site to maintain productivity of the land for tree growth." — Dr. William H. Lawrence, Weyerhaeuser Co.

OUTLINE

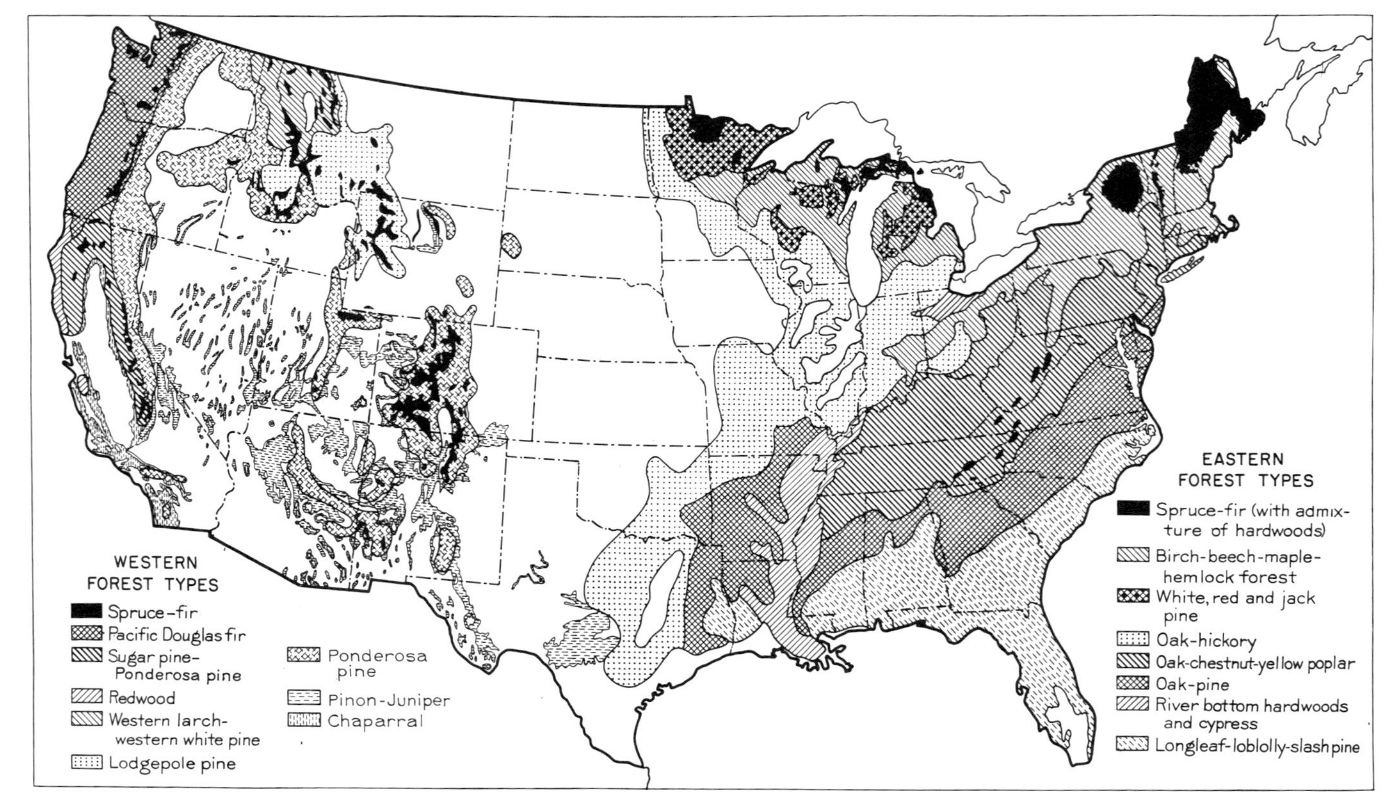

Forestlands comprise about one-third of the area of the United States and commercial forests more than one-fifth. Forests are the best protectors of soil, water, and fertility. (Courtesy, U.S. Forest Service)

□ □ □

19:1 □ OVERVIEW

A tree is often defined as "a woody plant at least 20 feet (6.1 m) high and supported by a single stem." Scientists have identified more than 1,000 tree species in the United States. Each tree species has unique soil, water, and fertility requirements. Furthermore, trees are very useful as lumber for buildings, wood for heating, and windbreaks and shelterbreaks to protect the soil and provide a better environment for livestock and people (Figures 19.1 and 19.2).

An example of available information on soil suitability for forest species to plant or favor in management in a forest can be found in all soil survey reports made and published by the National Cooperative Soil Survey. All soil survey reports of forested counties that have been published after 1965 have detailed tables that interpret the following information on each soil series:

1. Site index (predicted height of trees at a specified age, usually at 10, 25, or 50 years), based on soil, water, and fertility.
2. By groups of soil series, the tree species preferred for planting or management as a woodland or windbreak.
3. Erosion hazard in establishing a forest plantation or a windbreak.

Fig. 19.1 — The owner of these trees has a right to be emotional about them because they provide shade, add beauty, and moderate the winds. This is a typical oak-hickory forest with wild rhododendrons as an understory. (Appalachian Mountains) (Courtesy, Tennessee Valley Authority)

Fig. 19.2 — This is a commercial forest of Douglas-fir, the number one lumber tree in the United States, being clear-cut in patches. The clear-cutting is necessary to encourage natural reseeding. (Gifford Pinchot National Forest in the state of Washington) In the background, on the left, is Mt. St. Helens that erupted in 1980, and on the right is Mt. Adams. (Courtesy, USDA - Forest Service)

4. Predicted percentage of plant trees that will survive, based on soil, water, and fertility.
5. Relative ease of use of machinery to plant trees, based on topography and stoniness.
6. Anticipated native plant competition for planted hardwood or coniferous trees.
7. Cultural practices, such as weeding.
8. Hazard of **windthrow** (uprooting of trees by wind), based on soil depth.

(See Section 19:5 for an example of using soil survey information to predict forest site quality.)

19:2 ▫ TOLERANT AND RESISTANT TREE SPECIES

Everyone wants a tree species that is beautiful; grows rapidly; can be planted in any climate on all kinds of soils; is tolerant of wet soils, flooding, dry soils, acid soils, alkaline soils, and sea spray; and is resistant to attack by insects, diseases, and strong winds. No *one* species is that "tough," but there is a significant difference in tolerance and resistance among species.

Trees adapted to flooded soils are (in order of tolerance, high to low):

Baldcypress
Willow
Silver maple
Green ash
Cottonwood
Sycamore
Honeylocust
Pecan
Black walnut
Box elder
Bur oak
Black ash

The following species are well-adapted to wet soils:

Tupelo gum
Arborvitae
Red maple
Swamp white oak
Hemlock
Sweetbay
Sweetgum
White pine

Dry-soil adaptation of trees that are especially suited for use as shade trees:

Black oak
California black oak
Golden-raintree
Japanese pagodatree
Sassafras
Sawtooth oak
Serviceberry
Velvet ash

Trees resistant to salt injury from sea spray and deicing salt on roadways are (from high to low):

Live oak
Palm (many species)
Austrian pine
Japanese black pine
Scotch pine
Arborvitae
Slash pine
Longleaf pine
Loblolly pine
Redcedar
Baldcypress
Sweetgum
Water oak
Sycamore
Southern red oak
Hickory (all species)
Pecan
Red maple
Dogwood
Red oak

Shade trees with some resistance to insects and diseases include:

American hornbeam (blue beech)
American yellowwood
Golden-raintree
Hophornbeam (ironwood)
Japanese pagodatree
Maidenhair tree (ginkgo)
Sweetgum
White oak

Among trees that are windfirm and resist uprooting (from more windfirm to less) are:

Live oak
Palm (many species)
Baldcypress
Redcedar
Slash pine
Loblolly pine
Water oak
Red maple

Sweetgum	Dogwood
Sycamore	Hickory (all species)
Longleaf pine	Pecan
Southern red oak	Black locust

Tree species resistant to very low soil temperatures are willows (many species), hemlock, sugar maple, flame maple, ponderosa pine, red pine, Scotch pine, white spruce, white birch, and yellow birch.

Note: (1) Tree-of-heaven and Austrian pine are resistant to pollution in inner cities. (2) Nearly all forest trees in the United States are resistant to the present levels of acid precipitation received.

The acidity and alkalinity tolerance of common tree species is listed in Table 19.1.

The most acid–soil-tolerant tree species are eastern redcedar, cypress, holly, magnolia, pine, spruces, tamarack, and western hemlock. Species most alkaline–soil-tolerant are American elm, junipers, pines, and sycamore. The pines have the widest range of tolerance, from pH 3.5 to 7.5.

19:3 □ FORESTS AND WATERSHEDS

Forests are known to be the best protector of the soil — better even than grasses. However, it is true that a given watershed will yield more runoff water if planted to grass than if planted to trees. Stated in another way, planting trees on a grassed watershed may reduce streamflow as much as 20 percent. There is even a difference of water yield as much as 20 percent, depending on whether pines or hardwoods are growing on the watersheds (more water yield from hardwoods).

One reason why trees are so effective in holding the soil in place on watersheds is that the roots of many species grow extensively, both horizontally and vertically (Figure 19.3). Another reason is that the leaves and branches intercept from 15 to 25 percent of the total precipitation, and this amount never strikes the soil. A third reason is that some species of trees are able to grow on very infertile soils.

What happens to annual rainfall received on fully stocked stands of loblolly pine (the principal pine in the South)? Research results are as follows:

Mean annual rainfall	53 in.	(134.6 cm)	
Evaporation loss from the soil plus transpiration loss through leaves (needles)	25 in.	(63.5 cm)	(47%)
Runoff	18 in.	(45.7 cm)	(34%)
Interception by leaves (needles)	10 in.	(25.4 cm)	(19%)
Total	53 in.	(134.6 cm)	(100%)

Table 19.1 — Acidity and Alkalinity Tolerance of Common Tree Species[1]

	3.5–4.0	4.0–4.5	4.5–5.0	5.0–5.5	5.5–6.0	6.0–6.5	6.5–7.0	7.0–7.5	7.5–8.0
	Very Acid		Acid		Slightly Acid		Neutral		Alkaline
American elm				x	xxxxxxxxxxxxx	xxxxxxxxxxxxx	xxxxxxxxxxxxx	xxxxxxxxxxxxx	xxxxxxxxxxxxx
Ash		xxxxxxxxxxxxx	xxxxxxxxxxxxx	xxxxxxxxxxxxx	xxxxxxxxxxxxx				
Aspen			xxxxxxx	xxxxxxxxxxxxx	xxxxxxxxxxxxx				
Beech	xx	xxxxxxxxxxxxx	xxxxxxxxxxxxx	xxxxxxxxxxxxx	xxxxxxxxxxxxx				
Cypress	xxxxxxxxxxxxx	xxxxxxxxxxxxx							
Dogwood						xxxxxxxxxxxxx	xxxxxxxxxxxxx	x	
Douglas-fir				xxxxxxxxxxxxx					
Eastern redcedar	xxxxxxxxxxxxx	xxxxxxxxxxxxx	xxxxxxxxxxxxx	xxxxxxxxxxxxx	xx				
Fir		xxxxxxxxxxxx	xxxxxxxxxxxxx	xxxxxxxxxxxxx	xxxxxxxxxx				
Holly	xxxxxxxxxxxxx	xxxxxxxxxxxxx	x						
Juniper			xxxxxxxxxxxxx	xxxxxxxxxxxxx	xxxxxxxxxxxxx	xxxxxxxxxxxxx	xxxxxxxxxxxxx	xxxxxxxxxxxxx	
Larch				x	xxxxxxxxxxxxx	xxxxxxxxxxxxx	xxxxxxxx		
Locust			xxxxxxxxxxxxx	xxxxxxxxxxxxx	xxxxxxxxxxxxx	xxxxxxxxxxxxx	xxxxxxxxxxxxx		
Madrone			xxxxxxxxxxxxx	xxxxxxxxxxxxx	xxxxxxxxxxxxx	xxxxxxxxxxxxx	xxxxxxxxxxxxx	xx	
Magnolia	xxxxxxxxxxxxx	xxxxxxxxxxxxx	xxxxxxxxxxxxx	xxxxxxxxxxxxx					
Maple				xxxxxxxxxxxxx	xxxxxxxxxxxxx	xxxxxxxxxxxxx	xxxxxxxxxxxxx	xxx	
Pine	xxxxxxxxxxxxx	xxxxxxxxxxxxx	xxxxxxxxxxxxx	xxxxxxxxxxxxx	xxxxxxxxxxxxx	xxxxxxxxxxxxx	xxxxxxxxxxxxx	xxxxxxxxxxxxx	x
Spruce	xxxxxxxxxxxxx	xxxxxxxxxxxxx	xxxxxxxxxxxxx	xxxxxxxxxxxxx	xxxxxxxxxxxxx	xxxxxxxxxxxxx			
Sycamore			xxxxxxxxxxxxx	xxxxxxxxxxxxx	xxxxxxxxxxxxx	xxxxxxxxxxxxx	xxxxxxxxxxxxx	xxxxxxxxxxxxx	x
Tamarack	xxxxxxxxxxxxx	xxxxxxxxxxxxx	xxxxxxxxxxxxx	xxxxxxxxxxxxx					
Western hemlock	xxxxxxxxxxxxx	xxxxxxxxxxxxx	xxxxxxxxxxxxx	xxxxxxxx					
Western redcedar				xxxxxxxxxxxxx	xxxxxxxxxxxxx	xxxxxxxxxxxxx	xxxxxxxx		

[1]Source: John M. Harkin and John W. Rowe. "Bark and Its Possible Uses." USDA – Forest Service, Res. Note FPL-091, 1971.

Fig. 19.3 — A partially excavated western redcedar tree showing the extensive root system to a depth of 10 feet (3 m) that is capable of holding the soil in place. (Northern Idaho) (Courtesy, USDA - Forest Service)

When commercial forests are harvested, what happens to the soil and the water? The answer depends on the number of trees removed, the soil characteristics, and the amount and intensity of the precipitation. In the northern hardwood forest, precipitation is usually gentle and adequate each month in the year, and forested soils are generally sandy loams that are not very fertile. Under these conditions, the removal of an increasing number of trees during harvest results in the following changes in the watershed: (1) it increases the total runoff water, (2) it increases the amount of erosion sediment, (3) it increases the amount of nutrients lost, and (4) it increases the temperature of soil and surface waters.

Watershed studies have been conducted in the southern Appalachian Mountains of southern North Carolina for many years. On one watershed, the second-growth oak-hickory forest was cut on a 40-acre (16-ha) watershed and was planted to white pine. After 16 years, the monthly runoff was measured for a year and compared with predicted runoff values, assuming that the same watershed was still in hardwoods (Figure 19.4). Figure 19.5 graphically displays the fact that for every month in the year the runoff from the pine forest is *less* than if the same watershed were in hardwoods. On an annual basis the runoff from the white pine forest is 20 percent *less* than the predicted runoff from the original oak-hickory forest.

Clear-cutting (cutting all trees) in an oak-hickory forest and planting it to white pine resulted in an annual *reduction* in streamflow of about 10 million gallons per acre (93.4 million liters / ha). If the management objective is to reduce floods, this conversion from hardwoods to pine is desirable; however, if the objective is to obtain maximum water yield for domestic use, agriculture, and industry, the conversion to white pine is not favorable.

It can be assumed that other conifers such as southern pines and northern spruces would also reduce water yield on watersheds as compared with hardwood species on similar sites. The explanation is that conifers intercept 20 percent more precipitation than hardwoods which evaporates into the atmosphere.

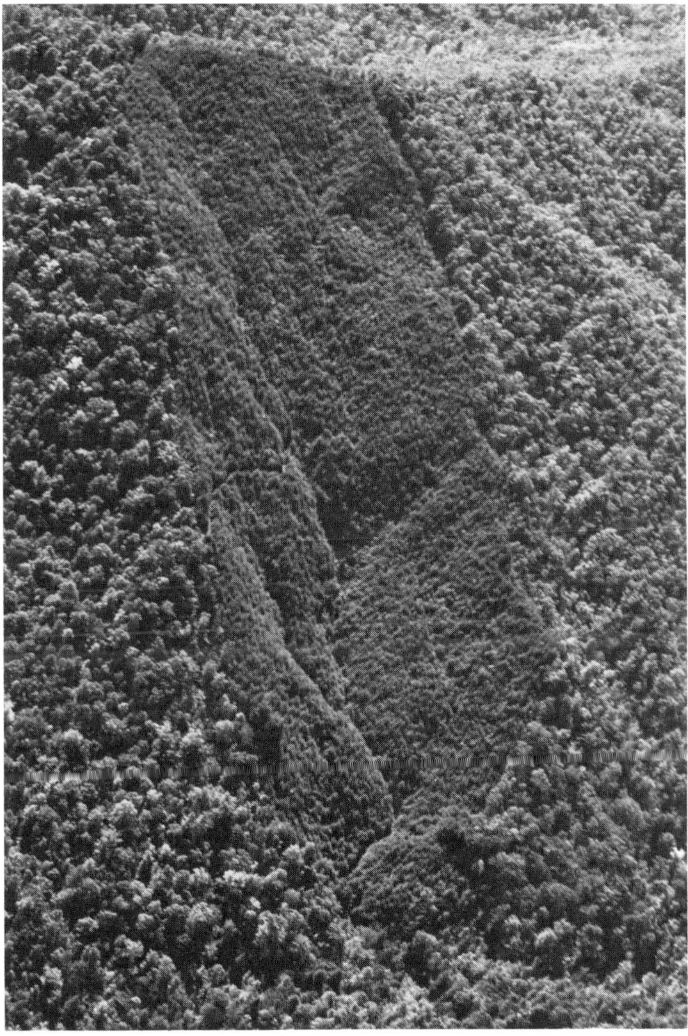

Fig. 19.4 — An aerial view of the 40-acre (16-ha) watershed covered with 16-year-old white pine that was planted where a stand of oak-hickory had been clear-cut. (See graphic data in Figure 19.5.)

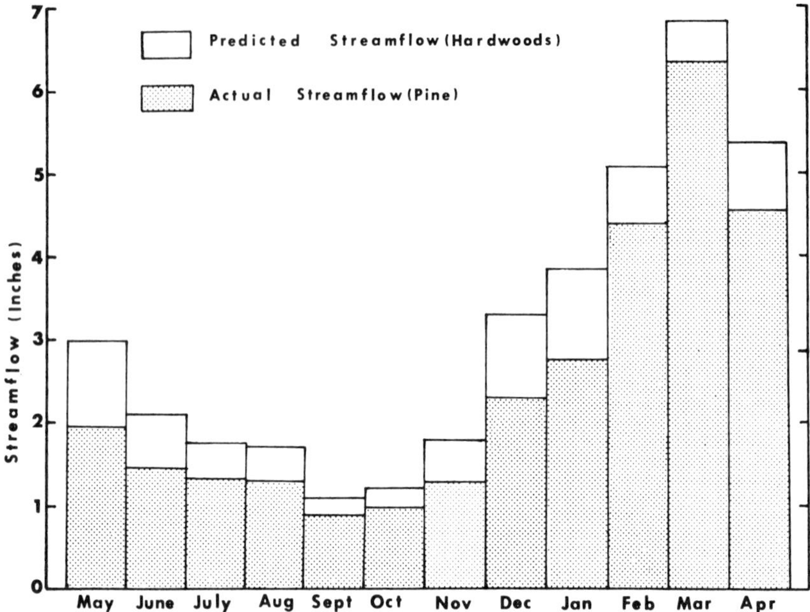

Fig. 19.5 — In the southern Appalachian Mountains of southern North Carolina, with an annual rainfall variation of 98 inches (250 cm) on upper slopes and 75 inches (190 cm) at lower elevations, a white pine watershed yielded 20 percent less runoff water in a year than a comparable hardwood watershed. (See aerial view of watershed in Figure 19.4.) (Courtesy, Southeastern Forest Experiment Station, USDA - Forest Service)

Note: To convert inches to centimeters, multiply inches by 2.54.

19:4 □ TREES AS SHELTERBELTS AND WINDBREAKS

During the years 1935 – 1942, prolonged drought in the Great Plains of the central United States prompted a massive tree-planting campaign known as the Prairie States Forestry Project. The trees were planted in narrow bands (windbreaks) around farmsteads and fields to moderate the winds and to reduce wind erosion. During an eight-year period, 217 million trees and shrubs were planted that totaled 18,600 miles (29,927 km) of **windbreaks** and **shelterbelts** (extensive windbreaks) extending from western Texas to North Dakota. Financial and technical assistance to farmers and ranchers for tree planting has continued.

Although many of the original windbreaks and shelterbelts no longer exist, each year more than 2,000 miles (3,218 km) of *new* plantings are made. This is more than enough to offset those destroyed by people and by natural deterioration. However, as of February 1986, 3.3 million acres (1.3 million ha) in the 10 Great Plains states were in need of more protection from wind erosion.

In the 18- to 20-inch (46- to 51-cm) mean annual precipitation belt in eastern North Dakota, a study of wheat yields in relation to a single-row planting of Siberian elm was made. The conclusions from this study were:

1. Because of tree root competition, wheat yields were lowest near the windbreak for a distance equal to the height of the trees.
2. Beyond the area of root competition, on the leeward side wheat yields were increased for a distance of five times the height of the trees in the windbreak.
3. On average for the entire wheat field, there was no *net* increase or decrease of wheat yields as the result of the windbreaks.
4. The principal values of the windbreaks are for moderating the winds for humans and livestock, reducing wind erosion, adding beauty, and increasing wildlife (Note 19.1 and Table 19.2).

Note 19.1 — Trees Modify Temperature and Save Energy

Well-placed trees around a home, barn, or feedlot make winter temperatures warmer by 9 percent and summer temperatures cooler by as much as 38 percent (detailed in Table 19.2). Furthermore, wind speeds are reduced, relative humidity is increased, sunlight is filtered, and the esthetic environment is enhanced.

Table 19.2 — Percentage Home Energy Saved by Well-placed Trees[1]

Source	During Winter	During Summer
	(%)	(%)
Shading	-3	15
Reduced air temperature	0	15
Reduced wind speed	12	3
Reduced lawn care	0	5
Total	9	38

[1]Source: John M. Norman, "Home Energy Savings from Urban Trees in Nebraska." In David L. Hintz, "International Symposium on Windbreak Technology — Proceedings." Lincoln, Nebraska, June 23 - 27, 1986, pp. 199, 200.

Another study of windbreaks in North Dakota revealed that weeds were the principal enemy.

All recent soil survey reports from counties in the U.S. Great Plains include a section entitled "Management of Windbreaks." This section indicates the most desirable species of forest trees or shrubs to plant on each soil mapping unit. All soil mapping units are placed in "windbreak groups." For each group the recommended forest or shrub species for planting is given, together with a relative suitability from excellent to poor.

Guidelines for establishing and managing windbreaks and shelterbelts include:

1. All windbreaks and shelterbelts should be oriented at right angles to the prevailing erosive winds. In the U.S. Great Plains this means planting along the northern and western sides of areas to be protected.
2. Forest tree and shrub species recommended for the U.S. Great Plains include redcedar, hackberry, black locust, green ash, Siberian peashrub, Russian olive, autumn olive, crabapple, hawthorn, wild plum, chokecherry, and fragrant sumac.
3. Species offering the least competition to adjoining field crops are green ash, Siberian peashrub, and Russian olive.
4. Temporary windbreaks can be established by planting sudangrass or one of the small grain crops: wheat, barley, or rye.

19:5 □ SOILS AND TREE GROWTH

Many studies have used the standard National Cooperative Soil Survey information as mapped, described, and interpreted by soil series as a basis for predicting forest site quality.

In the Green Mountains of central Vermont, the soils are predominantly sandy loams, acid, low in essential elements, and since 1965 classified in the soil order of Spodosols. (See Chapter 1.) Elevations range from 500 feet (152 m) to more than 4,000 feet (1,220 m), and annual precipitation varies from 36 inches (91 cm) to 50 inches (127 cm).

Seventy-seven field plots were established to study height growth in relation to age of the principal commercial hardwood forest species: sugar maple, white ash, yellow birch, and white birch. The soil and site characteristics that accounted for most of the variations in height growth were:

1. Soil drainage class (the best was moderately well-drained).
2. Elevation (the lower elevations were best).
3. Latitude (the more southerly latitudes were best).
4. Depth to bedrock (the best sites were more than 24 inches [61 cm] in depth).
5. Aspect (exposure) (the best were north- and east-facing slopes) (Figure 19.6).

Detailed field studies in Maine compared the chemical composition of red spruce foliage (green needles) with that of the soil in which the spruce were growing.

The soils under study were glacially deposited, acid, infertile, sandy, and classified as Spodosols.

Fig. 19.6 — A soil profile typical of the soils in Vermont where the tree growth studies were conducted. This is a Spodosol, moderately well-drained, and more than 36 inches (91 cm) to bedrock. At the lower left is a large granite boulder. (The heavy marks on the yardstick are at 6-inch (15-cm) intervals.) (Photo by Roy L. Donahue)

Note: To convert inches to centimeters, multiply inches by 2.54.

It was concluded that the best growth of red spruce occurred on soils high in calcium, magnesium, and manganese and low in potassium. Under these soil conditions, the relative concentration of the nutrients in red spruce needles corresponded to nutrients in the soil.

Among the southern pines, shortleaf pine is the most extensive. It grows on upland soils in 22 states in the southeastern United States. The ideal soil for shortleaf pine is a loam or sandy loam at least 20 inches (51 cm) thick to bedrock or dense clay. Northern or eastern slopes are more suitable because soil moisture is more adequate.

19:6 ▫ FOREST FIRES AND SOIL FERTILITY

Contrary to popular opinion, there are *good* forest fires. Good forest fires are those that kill "weed" species of competing trees and shrubs and thus permit seedlings of the more valuable tree species to become established in mineral soil. Examples of tree species that need a fire to become established in nature are the four commercial southern pines (loblolly, shortleaf, longleaf, and slash), jack pine and quaking aspen in the Great Lakes states, lodgepole pine in the Rocky Mountains, and Douglas-fir in the Pacific Northwest.

Fires burn the organic debris, prepare seedbeds for seed germination and growth, release plant nutrients to fertilize the tree seedlings, and encourage wildlife. Over a 43-year period of study, low-intensity ("cool") fires had no lasting effect on physical and chemical properties of the forest soil.

Although organic matter and nitrogen are destroyed by fire, in about two years the amount of both will have recovered to about their former levels. Also forest fires result in more soil nitrogen due to symbiotic atmospheric nitrogen fixation by invading legumes, nonsymbiotic fixation bacteria such as *Azotobacter,* and fixation by blue-green algae. Some studies show that over a period of years, controlled burning *increases* available soil nitrogen, calcium, phosphorus, and soil pH. Controlled burning is also desirable because it reduces excessive vegetation in which a wildfire would be difficult to control.

19:7 ▫ FERTILIZING TREES

Although trees require fewer nutrients than most farm crops, trees almost always are growing on soils of very low fertility. In addition, in the United States erosion depletes essential tree nutrients on 23 million acres disturbed by grazing. Tree fertilization in commercial forests started about 1967, and the practice has increased rapidly.

In contrast to the technique of fertilizing farm crops, fertilizer to supply adequate nutrients to forest trees are in these ranges (see Chapter 4):

1. Nitrogen (N) — 100 to 400 pounds per acre (112 – 448 kg / ha) every five years.

2. Phosphorus (P) — (Soluble and / or available) 20 to 100 pounds per acre (28 – 112 kg / ha) every five years *or* three times this amount of rock phosphate every 20 to 30 years, or once each **cutting cycle** (each time a forest tract is cut).
3. Potassium (K) — 100 pounds per acre (112 kg / ha) every 20 years, when exchangeable potassium (K) in the surface soil is less than 20 pounds per acre (22.4 kg / ha), as determined by a soil test.

The most common fertilizer applied to forest trees is urea (46 percent nitrogen). This is true for Douglas-fir in the Pacific Northwest as well as for the pines in the Southeast. In addition, on poorly drained soils in the Southeast, phosphorus fertilizer gives increases in pine tree growth equal to artificial drainage. The explanation is that under low soil oxygen, phosphorus is not readily available to most plants (Figure 19.7).

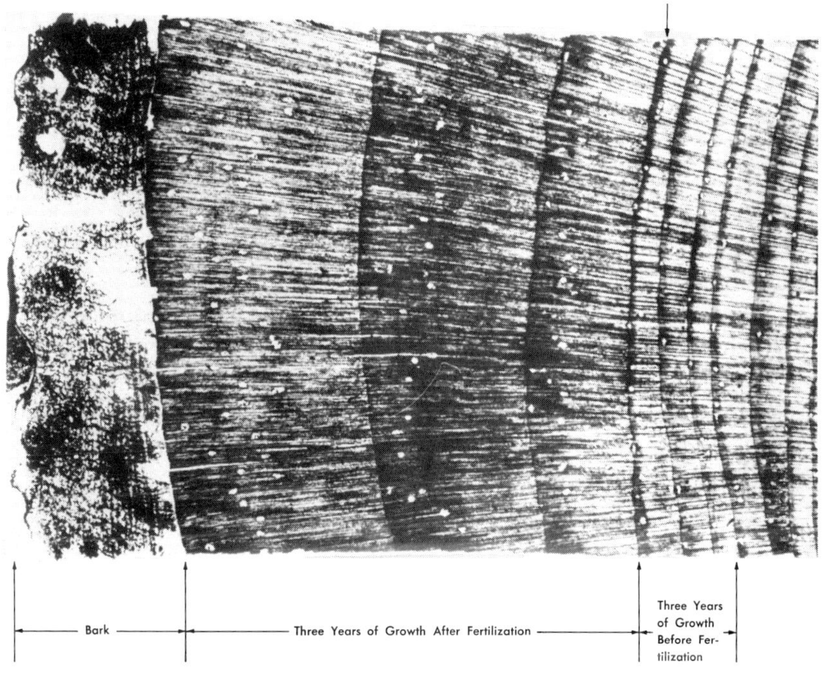

Fig. 19.7 — Cross-section of a Scotch (Scot) pine tree showing annual rings before and after fertilization. (Courtesy, Pulp and Paper Research Institute of Canada)

Many forested soils, especially in the southern United States, are shallow to bedrock, strongly acid, and low in natural fertility. Such soils erode rapidly.

An experiment to increase protective vegetation by fertilization was reported in 1986. A severely eroded forest soil in Louisiana was fertilized with 600 pounds per acre (672 kg / ha) of 16-30-13. It was applied by aircraft. Protective litter was increased in one year by 337 percent. Because of the mild and humid climate in the South, pine trees can absorb fertilizers every month in the year.

When applied with timeliness and precision, sewage sludge and sewage wastewaters are excellent fertilizers / soil amendments.

Furthermore, the hazard of **pathogens** (disease-producing organisms) and of heavy metals in sewage sludge is less of a problem when applied to forestlands than when applied at some point in the human food chain, as on gardens.

19:8 □ BENEFICIAL FUNGI

When a tree species is planted in soils where this species never grew before, it may not thrive if the proper beneficial fungus is not present. Such a fungus increases the plant's tolerance of droughty and infertile soils. The beneficial fungus helps roots of trees and shrubs to absorb phosphorus. One of the fungi species most beneficial to the southern pines is a mycorrhiza, meaning "fungus root." Its scientific name is *Pisolithus tinctorius.* Many mycorrhizae can be added to seedlings before transplanting.

19:9 □ AIR POLLUTION AND FORESTS

Air pollutants include large amounts of oxides of nitrogen and sulfur. Both are essential for plant growth but are acid-forming (nitric and sulfuric acids). Pollutants also contain smaller amounts of essential plant nutrients: phosphorus, potassium, calcium, magnesium, copper, and zinc. With such "free" fertilizer who can complain?

Soils-supporting forests are in humid regions where rainfall leaches the basic ions: calcium, magnesium, sodium, and potassium. The result is acid soils. Acid pollutants plus *acid soils* equals *very acid soils.* Even with the existing acid concentration of rain in the United States, very little injury to forest trees has been reported. However, acid precipitation (known as *acid rain*) has been very harmful to aquatic life in the northeastern United States. (See Chapter 3.)

19:10 □ REFERENCES

"Acidic Deposition and Forests." Position statement of the Society of American Foresters. Bethesda, Maryland, 1984.

Agricultural Statistics, 1986. U.S. Dept. of Agriculture, Washington, D.C., pp. 472, 473.

Berry, Charles R. "Use of Municipal Sewage Sludge for Improvement of Forest Sites in the Southeast." Res. Paper SE-266, Southeastern Forest Exp. Sta., Asheville, North Carolina, U.S. Dept. of Agriculture – Forest Service.

Fernandez, Ivan J. "Potential Effects of Atmospheric Deposition on Forest Soils." Main Agr. Exp. Sta. Pub. 1015. Proceedings of a symposium "Air Pollutants Effects on Forest Ecosystems." St. Paul, Minnesota, May 8, 9, 1985, pp. 237 – 250.

Fernandez, Ivan J., and R. A. Struchtemeyer. "Correlation Between Element Concentration in Spruce Foliage and Forest Soils." *Comm. in Soil Science, Plant Analysis,* Vol. 15, No. 10, 1984, pp. 1243 – 1255.

Garrett, Lawrence D. "Forests and Woodlands — Stored Energy for Our Use." In *Cutting Energy Costs. The 1980 Yearbook of Agriculture.* U.S. Dept. of Agriculture, Washington, D.C., pp. 101 - 108.

Graney, David L. "Site Quality Relationships for Shortleaf Pine." Proceedings of a symposium on the shortleaf pine ecosystem. Little Rock: University of Arkansas, Ext. Serv., March 1986.

Hatchell, Glyndon E., and Donald H. Marx. "Response of Longleaf, Sand, and Loblolly Pine to *Pisolithus* Ectomycorrhizae and Fertilizer on a Sandhills Site in South Carolina." *Forest Science.* Society of American Foresters, Bethesda, Maryland, Vol. 33, No. 2, pp. 301 - 315.

Hintz, David L., and James R. Brandle, eds. "International Symposium on Windbreak Technology." Great Plains Agricultural Council, Pub. No. 117, Lincoln, Nebraska, June 23 - 27, 1986, 309 pp.

Hocker, Harold W., Jr. *Introduction to Forest Biology.* New York: John Wiley & Sons, Inc., 1979.

Kraemer, James F., and Richard K. Hermann. "Broadcast Burning: 25-Year Effects on Forest Soils in the Western Flanks of the Cascade Mountains." *Forest Science,* Vol. 25, No. 3, 1979, pp. 427 - 439.

Krug, E. C., and C. R. Fink. "Acid Rain on Acid Soil: A New Perspective." *Science,* Vol. 221, 1983, pp. 520 - 525.

McNab, W. Henry. "Hardwoods and Site Quality." Southeastern Forest Exp. Sta., Asheville, North Carolina, U.S. Dept. of Agriculture - Forest Service, Workshop Proceedings, Morgantown, West Virginia, May 24 - 26, 1988.

Pope, P. E. "Some Soil Fungi Are Beneficial to Tree Seedling Growth." Purdue University, FNR-104, August 1980, 7 pp.

Shoulders, Eugene, and Allan E. Tiarks. "Slash Pine Shows Mixed Response to Potassium Fertilizer." Southern Silvicultural Research Conference, Atlanta, Nov. 4 - 6, 1986.

"Silvics of Forest Trees of the United States." *Agriculture Handbook No. 271.* Washington, D.C.: U.S. Dept. of Agriculture - Forest Service, 1975.

Silvicultural Systems for the Major Forest Types of the United States. Washington, D.C.: U.S. Dept. of Agriculture - Forest Service, 1979.

Thill, Ronald E., and John C. Bellemore. "Understory Responses to Fertilization of Eroded Kisatchie Soil in Louisiana." U.S. Dept. of Agriculture - Forest Service, Res. Note SO-330, December 1986.

Tomlinson, G. "Forest Vulnerability and the Cumulative Effects of Acid Deposition." Semmeville, Quebec, Canada: Domtar, Inc., 1984.

U.S. Environmental Protection Agency. "The Acidic Deposition and Its Effects." Critical Assessment Review Papers, Vol. I, "Atmospheric Science." EPA-600/8-83-016A, Washington, D.C., 1983.

U.S. Environmental Protection Agency. "The Acidic Deposition and Its Effects." *Critical Assessment Review Papers,* Vol. II. "Effects Sciences." EPA-600/8-83-0116B, Washington, D.C., 1983.

Waldrop, Thomas A., David H. Van Lear, F. Thomas Lloyd, and William R. Harms. "Long-Term Studies of Prescribed Burning in Loblolly Pine Forests of the Southeastern Coastal Plains." Gen. Tech. Rept. SE-45, Southeastern Forest Exp. Sta., Asheville, North Carolina, U.S. Dept. of Agriculture - Forest Service, 1987.

CHAPTER 20

Vegetating Disturbed Areas: Soil, Water, and Fertility Management

"The laws of a country ought to bear reference to its physical character, to the climate, whether warm, cold, or temperate: to the quality of the soil, to its situation, to its size, to the kind of life led by the people, whether farmers, hunters, or laborers." — C. L. Montesquieu, 1748

OUTLINE

Top: Surface mining has drastically disturbed these soils. *Bottom:* Three years after planting, the soils have been stabilized by black alder, black locust, and Russian olive. (Kentucky) (Courtesy, USDA – Soil Conservation Service)

□ □ □

20:1 □ OVERVIEW

Each year in the United States, soils that have been drastically disturbed and are lying bare erode at a rate of about 75 tons per acre (168 mt / ha). After these same disturbed soils have been stabilized by vegetation, the annual rate drops to about 3 tons per acre (7 mt / ha), the approximate geologic rate of erosion.

The principal kinds of soil-disturbing activities include construction, urbanization, and surface mining. The *annual* disturbance in a recent year was about 400,000 acres (160,000 ha) and is increasing.

When soil-disturbing activities have ended, the bare soil that must be stabilized by vegetation may vary from boulders to gravel, sand, loam, clay, or a mixture of these. Water may be deficient (droughty) or in surplus (ponded). Acidity may vary from pH 3 (very strongly acid) to 10 (very strongly alkaline), and fertility may vary from adequate to deficient. Total soluble salts may be low to excessive. Regardless of the physical and chemical characteristics of the surface soil material, usually some vegetation can be established.

The first step in establishing vegetation is to have the soil tested. Sampling a soil with such contrasting properties requires care in obtaining a sample from each area that appears different. Based on the soil test, the recommended amount of lime, fertilizer, manure, or sewage sludge can be applied. In addition, adapted plants can be selected and established by seeding or planting. The last step is to spread a protective organic mulch of straw, hay, or woodchips or a synthetic material over the seeding or planting until vegetation can protect the soil.

20:2 □ SEQUENCE OF ESTABLISHING VEGETATION

20:2.1 □ Introduction

Whether in humid or arid regions, the following sequence is necessary for successfully establishing vegetation on soils disturbed by mining, urbanization, or construction activities.

20:2.2 □ Soil Testing

As soon as feasible after soil disturbance has been completed, professional help should be obtained in taking soil samples for chemical analysis in a standard laboratory. Humid-region soils should be tested for pH, phosphorus, and potassium. If toxic quantities of elements are suspected, such as in mine or mill spoils from a copper, lead, or zinc mine, these elements should also be determined. In arid-region soils, tests should also be made for total soluble salts and exchangeable sodium percentage. Irrigation water should also be tested.

20:2.3 □ Fertilizers and Soil Amendments

The recommended amounts of lime, fertilizers, manures, and / or sewage sludge should be applied and mixed with the surface soil by a suitable tillage technique. (See Chapters 2 and 6.)

20:2.4 □ Tillage

The preparation of a seed / plant bed to a 6-inch (15-cm) depth by tillage is essential. The spoils from mining or construction may be clay that was manipulated and compacted by heavy machinery when wet. Or a **cut slope** (upper slope) during highway construction may be naturally compacted clay subsoils. Regardless of the cause, all dense soil layers must be tilled until the air / water / solid soil particles are in proper proportion to support plant growth. (See Chapter 1.)

20:2.5 □ Selection of Species

In humid regions there are many species of grasses, legumes, trees, shrubs, vines, and brambles adapted for planting to stabilize disturbed soils (Table 20.1). The selection is primarily based on soil-acidity tolerance and adaptability to internal soil drainage. Only a few grasses and one shrub are recommended for such purposes in arid regions. Here the limiting factors are tolerance of total soluble salts and of low soil moisture.

The most **acid-tolerant** grasses (pH 3.8 to 4.5) are deertongue, redtopgrass, bahiagrass, and bermudagrass. Acid-tolerant (to pH 5.0) legumes are alsike clover, flatpea, common and sericea lespedeza, and birdsfoot trefoil (Table 20.1). Black locust, European black alder, and bristly locust tolerate a pH as low as 4.0 (Table 20.2).

Plants that are tolerant of high (8 to 12 mmhos / cm) total soluble salts are used in stabilizing mine spoils, urban areas, and construction sites and include: alkaligrass, alkali sacaton, barley (cereal), fourwing saltbrush (a shrub), bermudagrass, fescuegrass, perennial ryegrass, saltgrass, sweetclover, switchgrass, wheatgrass, and wild ryegrass (Figure 20.1).

20:2.6 □ Proper Time of Planting

The ultimate objective is to establish long-lived perennials on the drastically disturbed soils. However, it is most important to establish protective vegetation as quickly as possible. For example, in humid regions, if the soil disturbance is completed in late summer or early fall, only a cool-season perennial such as Kentucky bluegrass or a cool-season annual such as rye should be planted. If cereal rye or another cool-season annual is established, the following spring a warm-

Table 20.1 — Humid Region: Grass and Legume Species Used for Stabilizing Soils Disturbed by Mining and Construction Activities[1]

	Longevity		Season of Maximum Growth		Soil Suitability				
Grasses / Legumes	Annual	Perennial	Cool	Warm	pH Range	Well-Drained	Moderately Well-Drained	Somewhat Poorly Drained	Poorly Drained
Grasses									
Bahiagrass		x		x	4.5 – 7.5	x	x		
Barley	x		x		5.5 – 7.8	x	x		
Bermudagrass		x		x	4.5 – 7.5	x	x	x	
Bluegrass, Kentucky		x	x		5.5 – 7.0	x	x	x	
Bromegrass		x	x		5.5 – 8.0	x	x	x	
Buffalograss		x		x	6.5 – 8.0		x	x	
Canarygrass, reed		x	x		5.0 – 7.5	x	x	x	x
Deertongue		x		x	3.8 – 5.0	x	x	x	x
Fescuegrass, tall		x	x		5.0 – 8.0	x	x	x	
Gramagrass, blue		x		x	6.0 – 8.5	x	x	x	
Lovegrass, sand		x		x	6.0 – 7.5	x			
Lovegrass, weeping		x		x	4.5 – 8.0	x	x	x	
Redtopgrass		x	x		4.0 – 7.5	x	x	x	x
Rye, cereal	x		x		5.5 – 7.5	x	x		
Ryegrass, annual	x		x		5.5 – 7.5	x	x	x	

(Continued)

Table 20.1 (Continued)

Grasses / Legumes	Longevity		Season of Maximum Growth		Soil Suitability				
	Annual	Perennial	Cool	Warm	pH Range	Well-Drained	Moderately Well-Drained	Somewhat Poorly Drained	Poorly Drained
Grasses (cont.)									
Sudangrass	x			x	5.5 – 7.5	x	x	x	
Switchgrass		x		x	5.0 – 7.5	x	x	x	
Timothy		x	x		4.5 – 8.0	x	x	x	x
Wheat, cereal	x		x		5.0 – 7.0	x	x	x	
Legumes									
Alfalfa		x	x		6.5 – 7.5	x	x		
Clover, alsike		x	x		5.0 – 7.5	x	x	x	x
Clover, white		x	x		6.0 – 7.0	x	x	x	
Crownvetch		x	x		5.5 – 7.5	x	x		
Flatpea		x	x		5.0 – 6.0	x	x	x	
Lespedeza, common	x			x	5.0 – 7.0	x	x		
Lespedea, sericea		x		x	5.0 – 7.0	x	x	x	
Sweetclovers	Biennial			x	6.0 – 8.0	x	x		
Trefoil, birdsfoot		x	x		5.0 – 7.5	x	x	x	

[1]Environmental Protection Agency. "Erosion and Sediment Control: Surface Mining in the United States, No. 1 Planning." EPA-625 / 3-76-006, October 1976, pp. 47, 48, 50.

Table 20.2 — Humid Region: Tree, Shrub, Vine, and Bramble Species Used for Stabilizing Soils Disturbed by Surface Mining and Construction Activities[1]

	Tolerance to Lowest pH Range		
Tree Species	**4.0**	**4.0 - 5.5**	**5.6 - 6.0**
Black locust	X		
European black alder	X		
Green ash		X	
Loblolly pine		X	
Northern red oak		X	
Norway spruce		X	
Red pine		X	
Scotch pine		X	
Shortleaf pine		X	
Virginia pine		X	
White pine		X	
Shrub, Vine, and Bramble Species			
American cranberrybush		X	
Amur privet		X	
Autumn-olive		X	
Bristly locust	X		
Crabapple, Japanese, Siberian, tea, toringo		X	
Honeysuckle, amur, tartarian			X
Indigobush		X	
Kudzu			X
Lespedeza, bicolor, Japanese		X	
Rose, memorial, multiflora[2]		X	
Russett buffaloberry		X	
Sumac, fragrant, shining		X	

[1] Soil Conservation Service. "Plant Performance on Surface Coal Mine Spoil in Eastern United States." USDA - SCS-TP-155, 1978.
[2] Multiflora rose is not recommended because it soon becomes a thorny pest in wild areas.

season perennial grass such as bermudagrass should be planted in the rye residue.

In arid regions the proper time to plant an adapted grass or shrub is immediately prior to the average date of the beginning of precipitation.

20:2.7 ▫ Seeding and Planting

If seed is to be sown, it must be drilled uniformly and at a rate about 50 percent heavier than recommended for fields on a farm. The heavier rate is because of the harsh environment of disturbed soils. If broadcast, the seed should be mixed thoroughly with the top layer of the soil previously tilled (Figure 20.2).

If tree seedlings or sprigs of grass are to be planted, they should be about 50 percent more closely spaced than the rate recommended for undisturbed soils.

20:2.8 ▫ Mulching

After seeding or planting, 2 to 4 inches (5 – 10 cm) of an organic mulch

Fig. 20.1 — Grasses vary in their tolerance to soil acidity, as confirmed by this research information. *Top:* Weeping lovegrass will tolerate soil pH as low as 3.7. *Center:* Switchgrass as low as pH 4.1. *Bottom:* Fescuegrass as low as 4.5. (Courtesy, USDA - Forest Service)

Fig. 20.2 — This surface-mined site is a hostile environment for establishing *any* vegetation. The slopes must be returned to their original contour to meet regulations. Topsoil that has been stockpiled (saved) must then be returned to the surface before adequate vegetation will grow to stabilize the mine spoils. (Courtesy, USDA - Agricultural Research Service)

should be spread uniformly over the entire area. Usual mulching materials are straw, old hay, woodchips, excelsior, bark, pine needles, **bagasse** (sugarcane wastes), wood fiber, manure, sewage sludge, ground paper, or sawdust. Inorganic (mineral) materials such as fiber glass and gravel are sometimes used. Liquid emulsion soil binders and mulching blankets can be used to limit erosion and help establish new growth.

20.3 ▫ VEGETATING AREAS DISTURBED BY SURFACE MINING

20:3.1 ▫ Introduction

Mining for coal accounts for about half of the land areas drastically disturbed by all kinds of mining; about 40 percent of the coal mined comes from a deep underground shaft and 60 percent from surface mining. Because of the lower cost, a larger percentage of future coal is expected to come from surface mining. For each ton of coal produced, surface mining disturbs far more soil than shaft mining does (Figures 20.3 and 20.4).

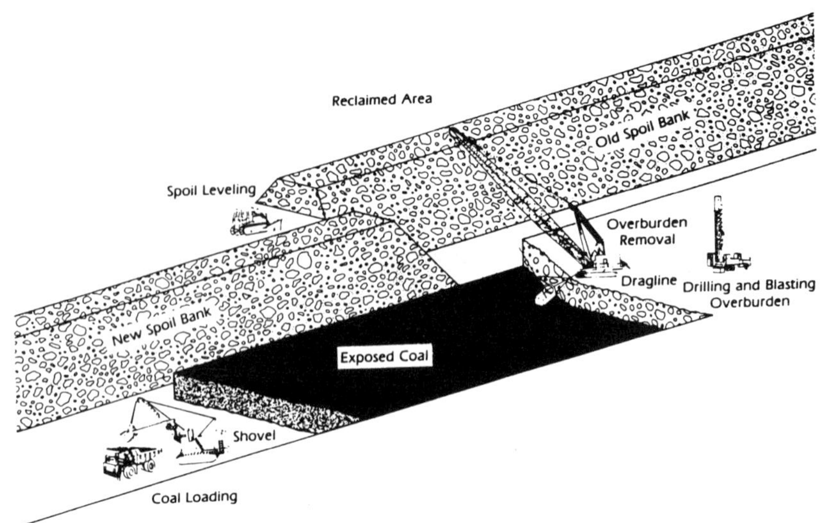

Fig. 20.3 — In area mining, long strips are excavated to uncover the coal. The overburden from the strip being mined is deposited in the strip previously mined. (Courtesy, U.S. Energy Information Administration)

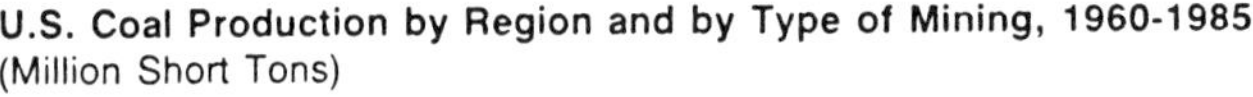

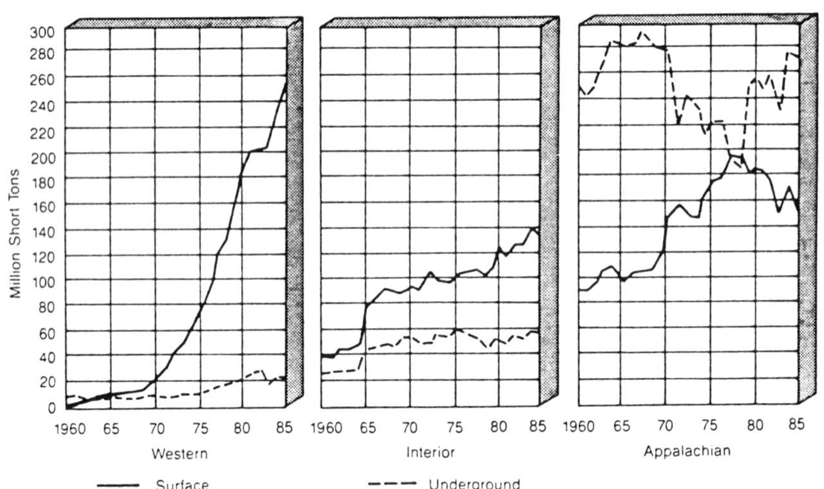

Fig. 20.4 — Surface-mined coal production in the West has increased markedly since 1969, largely because of a rapid rise in the coal output from the Powder River Basin in northeastern Wyoming and southeastern Montana. Underground coal production in the Appalachian Region has increased significantly since 1978, reversing the downward trend of earlier years. (Courtesy, U.S. Energy Information Administration)

Coal-bearing rocks underlie 458,600 square miles (1,187,768 sq km) of the United States, about 13 percent of the total land area. Coal is present in 38 states and was mined in 27 states in 1985 (Figure 20.5).

Coal is mined from about 350 beds, but more than half of the annual production comes from only about 20 beds. The thickness of the coalbeds mined ranges from less than 2 feet to about 100 feet (0.6 to 30.5 m). The average thickness is a little more than 4 feet (1.2 m) in the Appalachian states, about 6 feet (1.8 m) in the interior coal fields, and more than 30 feet (9.2 m) in the West. Historically, underground mines have provided most of the production. Since the early 1970's, however, surface mines have accounted for more than half of the total coal produced.

Coal was produced in 1985 by both underground and surface mining in 16 of the 27 coal-producing states. In 10 of the states, coal was produced only by surface mining. One state, Utah, had only underground coal production. The mining method used, whether underground or surface, depends on the depth of the coalbed from the surface.

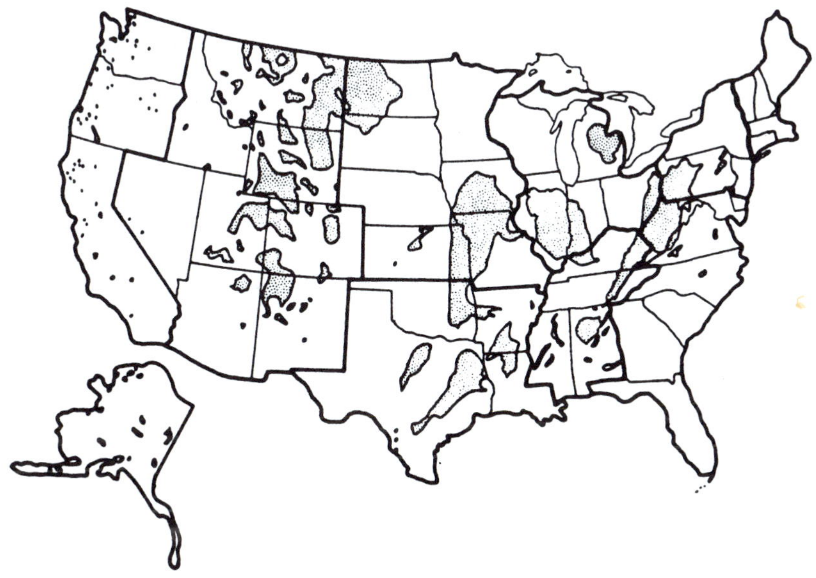

Fig. 20.5 — Coal fields of the United States. These cover much of the same area as coal reserves. (Courtesy, U.S. Geological Survey)

Surface mining requires a variety of heavy equipment, such as dragline excavators, power shovels, bulldozers, front-end loaders, and carryall scrapers. Coalbeds exposed along a hillside are sometimes mined with large augers that can drill more than 100 feet (30.5 m) into the bed. The coal recovery rate at surface mines can exceed 90 percent. At some surface mines, mainly those in the Appalachian coal fields, two or more coalbeds are mined during the same mining operation. In the anthracite area of northeastern Pennsylvania, **culm banks** and **silt banks** (waste accumulations of coal and rock from earlier mining operations) are mined and used as fuel. In some parts of the Appalachians, dredges recover coal that was dumped into rivers from early preparation plants or was eroded into the waterways from coal stockpiles.

20:3.2 ◻ Surface Mining Legislation

There are 29 states with major reserves of surface minable coal, and there are laws regulating this industry. In 1977, the U.S. Congress passed a law which is enforced by each of the 29 states. Mandates of the federal law relating to the stabilization of the residual spoils by vegetation include the following:

1. Restore all mined land to its original contour.
2. Restore all mined land to its original use or to an approved more intensive use.
3. Revegetate all abandoned and active surface mines.
4. Sponsor research and demonstrations on improved surface mining techniques such as replacing topsoil and establishing perennial vegetation.

The topsoil is to be removed and put in a separate pile, away from contamination by strongly acid or other toxic spoils, protected against water and wind erosion by seeding fast-growing plants on it, and after the final grade of the spoil materials has been achieved, the topsoil is to be spread uniformly over all the surface. If sufficient topsoil is not available, subsoil can be used.

A soil surveyor can readily identify **topsoil** as specified in this legislation. It includes all of the A-horizons, most B-horizons, and a few C-horizons that will support plant growth. (See Chapter 1.)

20:3.3 ◻ Regional Variations

More coal has been mined by surface mining than underground mining in recent years. Figure 20.6 shows U.S. coal production from 1970 to 1985. Experts predict that in the eastern United States, less coal will be surface-mined and more underground-mined in the next few years.

In the eastern United States most areas receive more than 40 inches (102 cm) of annual precipitation; therefore, water is usually not a limiting factor in establishing vegetation.

In the arid West, however, the problem is more difficult. The largest reserves of coal in the West are in Montana and Wyoming, where the annual precipitation is 15 inches (38 cm); and in the states of New Mexico, Arizona, Utah, Nevada, and Colorado, where precipitation is 5 inches (13 cm) or less. Water in arid and semiarid regions is the principal limiting factor in establishing vegetation on mine spoils, but research to date has been only moderately successful in areas of 10 inches (25 cm) or more of annual precipitation.

In the Appalachian Region, rainfall is adequate for establishing vegetation, but the limiting soil factors are steep slopes, stoniness, strong acidity, low phos-

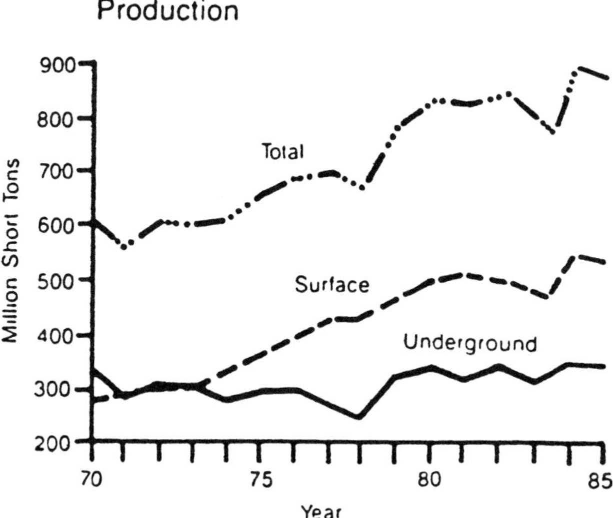

Fig. 20.6 — U.S. coal production trends, 1970 - 1985. (Courtesy, U.S. Energy Information Administration)

phorus, low nitrogen, toxicity of aluminum and manganese, low organic matter, high soluble salts, high surface soil temperatures, frost heaving, very low legume bacteria, lack of nonlegume nitrogen-fixing bacteria, and a scarcity of symbiotic mycorrhizal fungi. (See Chapter 19.)

20:4 ▫ VEGETATING AREAS DISTURBED BY ROAD CONSTRUCTION AND URBANIZATION

20:4.1 ▫ Introduction

In the United States there are about 5 million miles (8 million km) of roadways. These include interstate highways, state highways, county roads, farm roads, and forest and range truck trails. All grading and all cuts and fills to make a suitable roadbed expose bare soil that must be protected from erosion by roadbed surfacing or by perennial vegetation. The roadbed and shoulders are compacted to stabilize the road by reducing the amount of water which infiltrates. **Cut slopes** (upper road banks) are usually steep, shallow, dry, infertile, and naturally too compact to support luxuriant vegetation. **Fill slopes** (lower road banks) are often steep, loose, deep, cloddy, stony, and infertile. Fill slopes are more moist because more rain will infiltrate them and because runoff water from the impervious roadway usually is allowed to infiltrate. Both cut and fill slopes that face south or west are warmer and drier than those facing north or east. More drought-resistant vegetation should be selected for south- and west-facing slopes.

Each year *new* housing developments and *new* shopping centers drastically disturb almost 1 million acres (405,000 ha). Technology to vegetate such urbanizing areas is essentially the same as that for cuts and fills for new roadways (Figure 20.7).

Criteria for selection of species to vegetate roadsides and urbanizing areas are about the same as indicated in Section 20:2.5, except for these variations:

1. Vegetation selected must be low-growing so that it will not obstruct vision for motorists using the roads.
2. Vegetation must be beautiful.
3. Vegetation must be long-lasting and easy to maintain.

Fig. 20.7 — In urbanizing areas, without regulation or proper technical assistance, soil erosion losses have been reported as high as 150 tons per acre (336 kg / ha) per year. Soil loss is only one of the costs. When will the $200,000 house fall into the gully? (Alabama) (Courtesy, USDA — Soil Conservation Service)

Examples of techniques used successfully to vegetate roadsides in Alaska, in the mountains of California and Nevada, in the arid Southwest (New Mexico), on the Atlantic Coast (Virginia), and in the Northeast (Massachusetts) are given here.

20:4.2 □ Alaska

Astride the Arctic Circle and along the 789-mile (1,270-km) Alaskan oil pipeline, soils disturbed by construction are usually loamy, very high in organic matter in the surface, always moist to wet, cold, and unstable primarily because of naturally occurring **ice lenses (permafrost).** This is a word coined to mean *per-*

manent frost. Once the highly organic surface layer is removed during construction or fire, the soil warms to greater depths, buried ice lenses melt, and water may move downward *vertically,* causing sink hole erosion even on level areas. On slopes, traditional sheet and gully erosion also occurs but at a rate faster than in more temperate climates. Until they melt, ice lenses are impermeable and therefore cause more surface runoff and more erosion.

Because they are so fragile, soils disturbed by construction in Alaska must be vegetated as quickly as possible to stabilize them. However, with a freeze-free period of only 65 to 115 days, adapted vegetation is scarce. Recommended seeding mixtures of species and varieties include:

1. Creeping foxtail (Garrison) + Kentucky bluegrass (Nugget or Merion) + white clover or alsike clover
2. Timothy (Engmo) + Kentucky bluegrass (Nugget or Merion) + white clover or alsike clover
3. Smooth bromegrass (Manchar or Polar) + white clover or alsike clover
4. Hard fescuegrass
5. Meadow foxtail

Seed should be sown from May 15 to June 15 to take advantage of the very short growing season (Figure 20.8).

Fig. 20.8 — Grasses for forage production and soil stabilization are on trial in central Alaska along the service roads of the Alaska pipeline. *Right:* Hard fescuegrass. *Left:* Meadow foxtail. Both are four years old. Other grasses that are successful include species and selections of bluegrasses, other fescuegrasses, bromegrasses, canarygrasses, wheatgrasses, and foxtail grasses. (Courtesy, William W. Mitchell, Institute of Agricultural Sciences, University of Alaska)

20:4.3 ▫ California

In the Sierra Nevadas of central eastern California, a cut slope was successfully vegetated by burying bundles of willow cuttings in trenches established on the contour (Figure 20.9). Also orchardgrass, fairway crested wheatgrass, and bromegrass were established successfully. Tree species that were used with success are ponderosa pine, lodgepole pine, and quaking aspen.

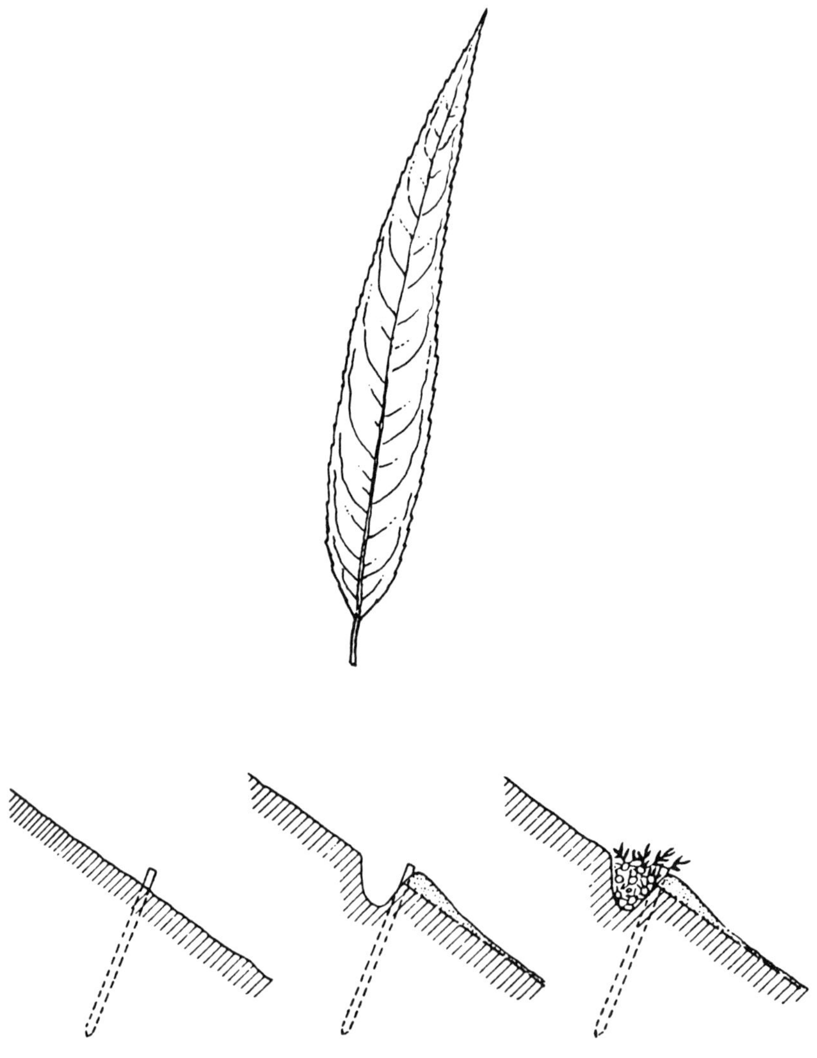

Fig. 20.9 — *Top:* Leaf of black willow, one of 300 or more species of willow worldwide. *Bottom:* For stabilizing steep slopes in humid regions, individual willow cuttings can be pushed into the soil, or bundles of willow cuttings can be buried in trenches established on the contour. (Courtesy, USDA)

20:4.4 □ New Mexico

In Hidalgo County, in southwestern New Mexico, annual rainfall averages only 10 inches (25 cm). Half of the annual rainfall is received during June, July, and August. Grasses such as Lehmann lovegrass, sideoats gramagrass, blue gramagrass, black gramagrass, and alkali sacaton have successfully stabilized slopes along a highway when they were seeded in early June. Also one shrub, fourwing saltbush, was successfully established. All of these plants are native species. Tall native grasses were used for mulching; grain straw blew away, thus proving to be unsatisfactory as a mulch. One or two irrigations were given at the time of establishment.

20:4.5 □ Virginia

Soils in the Piedmont Plateau in the southeastern United States are old, low in fertility, and highly erosive when disturbed by construction. Rainfall averages about 50 inches (127 cm) and is well-distributed. Calcium, phosphorus, potassium, and nitrogen are very low in the soil, and toxic exchangeable aluminum is very high.

Soil tests to obtain lime and fertilizer recommendations for establishing protective vegetation for stabilizing cut and fill slopes along roadways have averaged about as follows:

- ***Lime*** — 2 tons per acre (4.5 mt / ha) tilled into the soil before fertilizing.
- ***Fertilizer*** — 1,000 pounds per acre (1,120 kg / ha) of 10-20-10 tilled into the soil before seeding. Six months after seeding, soil should be top-dressed with 500 pounds per acre (560 kg / ha) of 10-20-10, and a year later top-dressed with a third application of 150 pounds per acre (168 kg / ha) of the same fertilizer grade.

The recommended seeding mixture depends on the season as follows: (*Note:* All plants are perennials unless indicated otherwise.)

For Spring and Fall Seeding

Kentucky-31 fescuegrass
Creeping red fescuegrass
Annual ryegrass
Sericea lespedeza or crownvetch

For Summer Seeding

Kentucky-31 fescuegrass
Creeping red fescuegrass
Weeping lovegrass
German millet (an annual)
Sericea lespedeza or crownvetch

For Winter Seeding

Kentucky-31 fescuegrass
Creeping red fescuegrass
Cereal rye (an annual)

After seeding, a heavy mulch, usually of straw or woodchips, is used. On steep slopes, straw or other recommended materials need to be held in place with stakes.

20:4.6 □ Massachusetts

Precipitation in Massachusetts varies from 40 to 50 inches (102 – 127 cm), with 3 to 4 inches (8 – 10 cm) being received each month. The relative humidity is usually high. Unfavorable climatic factors related to establishing vegetation are glaze ice formation and **frost heaving** (ice crystals pushing plants out of the ground).

Soils are predominantly sandy, acid, and low in fertility. Lime and fertilizer are to be applied according to a soil test. More lime and fertilizer are required to maintain grasses than are required to maintain trees and shrubs. Legumes usually require more lime and fertilizer than grasses do.

If mowing is planned, creeping red fescuegrass, fine fescuegrass, durar hard fescuegrass, tall fescuegrass, annual ryegrass, weeping lovegrass, ladino clover, and / or alsike clover should be seeded. When no mowing is to be done, perennial ryegrass, creeping red fescuegrass, and / or crownvetch is recommended.

In locations where woody perennials will not obstruct visibility, the use of trees, shrubs, and vines such as autumn olive, bayberry, bearberry, bittersweet (oriental), indigo-bush, inkberry, larch (Japanese), locust (bristly), plum (beach), redcedar, and sweetfern is recommended.

Woodchips are the recommended organic mulch, to be applied immediately after seeding or planting to a depth of about 4 inches (10 cm).

20:5 □ REFERENCES

Barnett, James P. "Containerized Pine Seedlings for Revegetating Difficult Sites." *Journal of Soil and Water Conservation,* November – December 1983, pp. 462 – 464.

Coal Data: A Reference. Energy Information Administration, U.S. Department of Energy, DOE / EIA-0064, 1985, 95 pp.

Donahue, Roy L., Raymond W. Miller, and John C. Shickluna. *Soils: An Introduction to Soils and Plant Growth,* 5th ed. Englewood Cliffs, New Jersey: Prentice-Hall, Inc., 1983.

Ettinger, William S. "Impacts of a Chemical Dust Suppressant / Soil Stabilizer on the Physical and Biological Characteristics of a Stream." *Journal of Soil and Water Conservation,* March – April 1987, pp. 111 – 114.

Gaffney, F. B., and J. A. Dickerson. "Species Selection for Revegetating Sand and Gravel Mines in the Northeast." *Journal of Soil and Water Conservation,* September – October 1987, pp. 358 – 361.

"Grow Where No Grass Has Grown Before: Put a Mulching Blanket on Those 'Impossible' Areas." *Cooperative Farmer*, April 1987, p. 15.

Hossner, L. R., ed. "Reclamation of Surface-mined Lignite Soil in Texas." Texas A&M University, RM-10, 1980.

Joost. R. E., F. J. Olsen, and J. H. Jones. "Revegetation and Mineral Development of Coal Refuse Amended with Sewage Sludge and Limestone." *Journal of Environmental Quality*, Vol. 16, No. 1, 1987, pp. 65 - 68.

Reclamation of Drastically Disturbed Lands. Madison, Wisconsin: American Society of Agronomy, Crop Science Society of America, and Soil Science Society of America, 1978, 742 pp.

Second RCA Appraisal: Soil, Water, and Related Resources on Nonfederal Land in the United States — Analysis of Condition and Trends, The Review Draft. U.S. Dept. of Agriculture, August 1987, 486 pp.

Stein, O. R., *et al.* "Runoff and Soil Loss as Influenced by Tillage and Residue Cover." *Soil Science Society of America Journal*, Vol. 50, pp. 1527 - 1531, 1986.

Steinhardt, Gary C., Donald P. Franzmeier, and Jerry V. Mannering. "Soil Erosion in Indiana — An Overview." *Agronomy Guide.* West Lafayette, Indiana: Purdue University, February 1981, 4 pp.

Troeh, Frederick R., J. Arthur Hobbs, and Roy L. Donahue. *Soil and Water Conservation for Productivity and Environmental Protection.* Englewood Cliffs, New Jersey: Prentice-Hall, Inc., 1980, pp. 360 - 419.

CHAPTER 21

Careers in Soil and Crop Management and Allied Subjects

"Career [French, *carrière*] . . . one's progress through life or in a particular vocation . . . a profession or occupation which one trains for and pursues as a life work." — *Webster's New World Dictionary of the American Language.* Second College Edition, 1984.

OUTLINE

□ □ □

21:1 □ OVERVIEW

With a continuing decrease in the *number* of farms and ranches, a casual conclusion may be that the opportunity for employment in agriculturally related business is also shrinking. Just the reverse is true. Numbers of farms and ranches are fewer, but their size, complexity, and diversity are far greater, requiring more management skill, science, and technology. The result is a greater need for persons trained and educated in all phases of agriculture, but especially soil and crop management (see Figure 21.1).

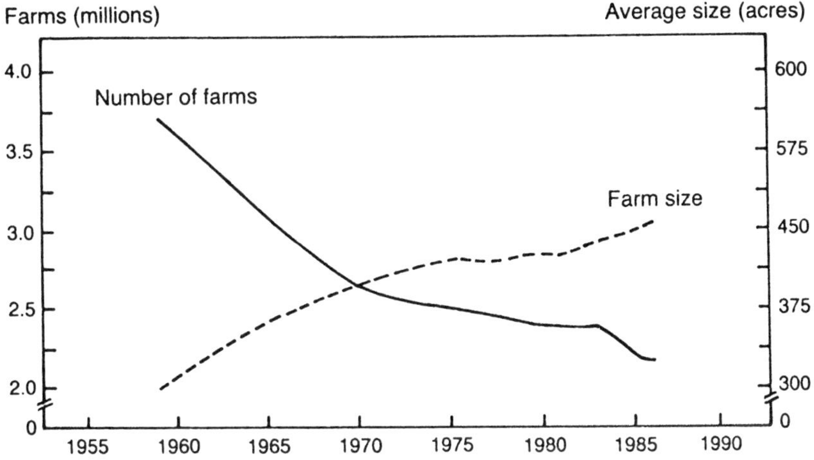

Fig. 21.1 — Farms in the United States are decreasing in numbers but increasing in size. Large farms require more knowledge and skills to operate, including knowing more about efficient soil and water management techniques. (Courtesy, USDA)

21:2 □ THE CHOICES OF CAREERS

From agricultural administrator to weed control specialist, the authors have selected 49 career titles. In Table 21.1 each of these career titles has been keyed to relevant chapters in this sixth edition of *Our Soils and Their Management*. This gives a person some notion of the knowledge and skills required for a particular career.

Most of the 49 careers relating to soil and crop management require more information than that found in this text. Please consider the facts here to be only exploratory in nature.

Table 21.1 — Career Choice Information in Text on Soil and Crop Management and Allied Subjects[1]

Career	Chapters Giving More Information on Each Career
Agricultural administrator	1–21
Agricultural climatologist	1
Agricultural consultant	1–21
Agricultural education teacher	1–21
Agronomist	1–14, 20
Conservationist	1–21
County agent	1–21
Crop ecologist	1–21
Crop marketing specialist	14–19
Crop physiologist	2–5
Crop production specialist	1–21
Crop quality specialist	1–5, 7–11
Crop scientist	1–21
Crop specialist	1–21
District conservationist	1–21
Edaphologist	1–21
Environmental scientist	1–21
Erosion and sediment control specialist	1–21
Farmer	1–21
Farm manager	1–21
Farm operator	1–21
Farm worker	1–21
Fertilizer technologist	1–21
Fertilizer use specialist	1–21
Field scout for plant diseases and insects	1–21
Forest soil scientist	1–21, esp. 19
Forester	1–21, esp. 19
Forest technician	1–21, esp. 19
Forest worker	1–21, esp. 19
Gardener	1–21, esp. 14–17
Hazardous waste management specialist	1–21
Irrigation specialist	1–21, esp. 9
Land management specialist	1–21
Lawn service worker	1–21, esp. 16
Park management specialist (see Figure 21.2)	1–21
Pedologist	1–21
Pest management specialist	1–21, esp. 13–19
Plant biochemist	1–21

(*Continued*)

Table 21.1 (Continued)

Career	Chapters Giving More Information on Each Career
Plant breeder	1–21
Plant care worker	1–21, esp. 13–17
Plant ecologist	1–21
Plant physiologist	1–21
Soil-water-plant specialist	1–21
Surface mine reclamation specialist (see Figure 21.6)	1–21, esp. 20
Tree farmer	1–21, esp. 19
Tree planter	1–21, esp. 19
Turfgrass specialist	1–21, esp. 16
Waste management specialist	1–21, esp. 11
Weed control specialist	1–21, esp. 13–18

[1] *Note:* For further information on preparing yourself for the career(s) of your interest, write to the university or college of your choice, preferably the one(s) in your state.

Fig. 21.2 — Park management specialists educate people about the proper care of national forests and national parks. (California) (Courtesy, USDA–Forest Service)

Fig. 21.3 — A range management specialist and a rancher are examining the range grasses and other vegetation that grows when protected from grazing. Such a specialist requires at least a Bachelor of Science degree in Range Science at an accredited agricultural college or university. (Colorado) (Courtesy, Colorado State University)

Fig. 21.4 — Soil judging by FFA students is an excellent activity to help you in deciding if 1 of the 17 soil careers is the profession to follow. (Kansas) (Courtesy, Roy Hunter Follett)

Fig. 21.5 — A soil surveyor is mapping soils in a subdivision in Madisonville, Kentucky. Careers such as this require at least a Bachelor of Science degree in Soil Science at an accredited agricultural college or university. (Courtesy, USDA-Soil Conservation Service)

Fig. 21.6 — A surface mine reclamation specialist is needed here to select adapted vegetation to stabilize this soil disturbed by surface mining in Ohio. (Courtesy, USDA-Agricultural Research Service)

21:3 ▫ THE CAREER OF FARM MANAGER

Suppose for example that you believe you want to consider a career as a farm manager. Table 21.1 indicates all 21 chapters in this text as essential information.

By asking questions of your agricultural education teacher, agriscience instructor, the county extension agent, and farmers and by reading farm magazines you will learn that:

1. Farm managers are involved in the world's largest and most productive agricultural sector. For this reason, at least a Bachelor of Science degree in Agriculture is useful and may be required.
2. Farm managers may be managers of farms owned by someone else, farm owners and operators, or tenants.
3. Farm managers may be managing a crop farm that grows grain, fiber, fruit, tobacco, and / or vegetables.
4. Farm managers are responsible for planning, tilling, planting, fertilizing, cultivating, spraying, irrigating, harvesting, and selling the produce.
5. Farm managers may operate a livestock farm with dairy cattle, beef cattle, sheep, horses, poultry, game animals, goats, and / or swine. Daily jobs include feeding, breeding stock, checking on the health of animals, record keeping, and marketing. Many livestock farms also raise supplementary crops and hay / pasture / range.
6. Farm managers hire, fire, train, supervise, evaluate, and pay employees. Their decisions are influenced by weather, disease, insects, and market fluctuations. Often what happens world-wide is more influential in determining on-farm decisions than are domestic occurrences. A good farm manager therefore must be an internationalist.
7. The preparation to be a successful farm manager requires:
   a. Skill in selecting a suitable farm, including productive soil.
   b. Knowledge of crops suitable to the soil and the climate.
   c. Judgment and knowledge of livestock.
   d. Mechanical ability.
   e. Leadership ability.
   f. Skill in operating microcomputers and in using software to help keep records, solve management problems, and make other decisions.
   g. Technical knowledge that is most readily available from an agricultural college or university. ***Note:*** When writing to the college or university of your choice, indicate the career in which you are interested. Also ask the cost of tuition, room, and meals. If your high school grades are excellent, you may be able to receive a scholarship. Ambitious college students can always get part-time jobs.

The best of luck in choosing a life-long career as a farm manager.

Instead of becoming a farm manager, suppose you decided to explore the possibility of a career in soil and crop management. What guidance is available?

21.4 □ CAREERS IN SOIL AND CROP MANAGEMENT

Ask your agricultural education teacher to write for the publication "Exploring Careers in Agronomy, Crops, and Soils." (See "References.")

Entry careers are usually available in soil and crop management. Some of these are soil surveyor, soil instructor, conservation technician, and fertilizer applicator. Often you can get a start on these careers through summer employment during the years you are studying for a Bachelor of Science degree in a four-year college. Students many times, however, find greater satisfaction and a life-long career if they complete a Master of Science degree or a Doctor of Philosophy degree in Soil Science or a service-oriented field such as Agricultural Education. (***Note:*** Two of the authors of this book have Ph.D. degrees in Soil Science. The third has a Ph.D. degree in Agricultural Education.)

The following 132 colleges and universities offer special courses and degrees in soil and crop management as well as in other agricultural subjects. For additional information on agricultural careers of your choice, write to or visit the colleges, schools, or divisions of agriculture in your state.

Alabama

Alabama A&M University
Normal, AL 35762

Auburn University
Auburn University, AL 36849

Tuskegee Institute
Tuskegee Institute, AL 36088

Alaska

University of Alaska
Fairbanks, AK 99701

Arizona

The University of Arizona
Tucson, AZ 85721

Arizona State University
Tempe, AZ 85281

Arkansas

Arkansas State University
State University, AR 72467

Southern Arkansas University
Magnolia, AR 71753

University of Arkansas
Fayetteville, AR 72701

University of Arkansas
Monticello, AR 71655

University of Arkansas
Pine Bluff, AR 71601

California

California Polytechnic State University
San Luis Obispo, CA 93407

California State Polytechnic University
Pomona, CA 91768

California State University
Chico, CA 95926

California State University
Fresno, CA 93740

Humboldt State University
Arcata, CA 95521

University of California
Davis, CA 95616

California (cont.)

University of California
Riverside, CA 92521

Colorado

Colorado State University
Fort Collins, CO 80523

Fort Lewis College
Durango, CO 81301

Connecticut

The University of Connecticut
Storrs, CT 06268

Delaware

Delaware State College
Dover, DE 19901

University of Delaware
Newark, DE 19711

Florida

Florida Southern College
Lakeland, FL 33802

University of Florida
Gainesville, FL 32611

Georgia

Abraham Baldwin Agricultural College
Tifton, GA 31794

Berry College
Mount Berry, GA 30149

Fort Valley State College
Fort Valley, GA 31030

University of Georgia
Athens, GA 30602

Hawaii

University of Hawaii
Honolulu, HI 96822

Idaho

College of South Idaho
Twin Falls, ID 83301

Ricks College
Rexburg, ID 83440

University of Idaho
Moscow, ID 83843

Illinois

Illinois State University
Normal, IL 61761

Southern Illinois University
Carbondale, IL 62901

University of Illinois
Urbana-Champaign, IL 61801

Western Illinois University
Macomb, IL 61455

Indiana

Purdue University
West Lafayette, IN 47907

Iowa

Iowa State University
Ames, IA 50011

Kansas

Fort Hays State University
Hays, KS 67601

Kansas State University
Manhattan, KS 66506

McPherson College
McPherson, KS 67460

Kentucky

Eastern Kentucky University
Richmond, KY 40475

Morehead State University
Morehead, KY 40351

Kentucky (cont.)

Murray State University
Murray, KY 42071

University of Kentucky
Lexington, KY 40506

Western Kentucky University
Bowling Green, KY 42101

Louisiana

Louisiana State University
Baton Rouge, LA 70803

Louisiana Tech University
Ruston, LA 71270

McNeese State University
Lake Charles, LA 70609

Nicholls State University
Thibodaux, LA 70302

Northeast Louisiana University
Monroe, LA 71209

Northwestern State University
Natchitoches, LA 71497

Southeastern Louisiana University
Hammond, LA 70402

Southern University
Baton Rouge, LA 70813

University of Southwestern Louisiana
Lafayette, LA 70504

Maine

University of Maine
Orono, ME 04469

Maryland

University of Maryland
College Park, MD 20742

University of Maryland
Princess Anne, MD 21853

Massachusetts

University of Massachusetts
Amherst, MA 01003

Michigan

Michigan State University
East Lansing, MI 48824

Michigan Technological University
Hancock, MI 49930

Northern Michigan University
Marquette, MI 49855

Minnesota

University of Minnesota
St. Paul, MN 55108

University of Minnesota Technical College
Crookston, MN 56716

University of Minnesota Technical College
Waseca, MN 56093

Mississippi

Alcorn State University
Lorman, MS 39096

Mississippi State University
Mississippi State, MS 39762

Missouri

Central Missouri State University
Warrensburg, MO 64093

Lincoln University
Jefferson City, MO 65101

Missouri Western State College
St. Joseph, MO 64507

Northeast Missouri State University
Kirksville, MO 63501

Northwest Missouri State University
Maryville, MO 64468

Missouri (cont.)

Southwest Missouri State University
Springfield, MO 65802

University of Missouri
Columbia, MO 65211

Montana

Montana State University
Bozeman, MT 59717

Northern Montana College
Havre, MT 59501

Nebraska

University of Nebraska
Lincoln, NE 68583

Nevada

University of Nevada
Reno, NV 89557

New Hampshire

University of New Hampshire
Durham, NH 03824

New Jersey

Rutgers University
New Brunswick, NJ 08903

New Mexico

New Mexico State University
Las Cruces, NM 88003

New York

Cornell University
Ithaca, NY 14853

North Carolina

North Carolina A&T State University
Greensboro, NC 27411

North Carolina State University
Raleigh, NC 27650

North Dakota

North Dakota State University
Fargo, ND 58105

Ohio

The Ohio State University
Columbus, OH 43210

Wilmington College
Wilmington, OH 45177

Oklahoma

Cameron University
Lawton, OK 73505

Langston University
Langston, OK 73050

Oklahoma Panhandle State University
Goodwell, OK 73939

Oklahoma State University
Stillwater, OK 74074

Oregon

Oregon State University
Corvallis, OR 97331

Pennsylvania

Delaware Valley College of Science
and Agriculture
Doylestown, PA 18901

The Pennsylvania State University
University Park, PA 16802

Temple University
Ambler, PA 19002

Puerto Rico

University of Puerto Rico
Mayagüez, PR 00708

University of Puerto Rico
Rio Piedras, PR 00928

Rhode Island

University of Rhode Island
Kingston, RI 02881

South Carolina

Clemson University
Clemson, SC 29631

South Dakota

South Dakota State University
Brookings, SD 57007

Tennessee

Austin Peay State University
Clarksville, TN 37040

Middle Tennessee State University
Murfreesboro, TN 37130

Tennessee Technological University
Cookeville, TN 38505

University of Tennessee
Knoxville, TN 37901

University of Tennessee
Martin, TN 38238

Texas

Abilene Christian University
Abilene, TX 79601

East Texas State University
Commerce, TX 75428

Prairie View A&M University
Prairie View, TX 77445

Sam Houston State University
Huntsville, TX 77341

Southwest Texas State University
San Marcos, TX 78666

Stephen F. Austin State University
Nacogdoches, TX 75962

Sul Ross State University
Alpine, TX 79832

Texas A&I University
Kingsville, TX 78363

Texas A&M University
College Station, TX 77843

Texas Tech University
Lubbock, TX 79409

West Texas State University
Canyon, TX 79015

Utah

Brigham Young University
Provo, UT 84602

Utah State University
Logan, UT 84322

Vermont

University of Vermont
Burlington, VT 05405

Virginia

Old Dominion University
Norfolk, VA 23508

Virginia Polytechnic Institute and
State University
Blacksburg, VA 24061

Virginia State College
Petersburg, VA 23806

Washington

University of Washington
Seattle, WA 98195

Washington State University
Pullman, WA 99164

West Virginia

West Virginia University
Morgantown, WV 26506

Wisconsin

University of Wisconsin
Green Bay, WI 54302

Wisconsin (cont.)

University of Wisconsin
Madison, WI 53706

University of Wisconsin
Platteville, WI 53818

University of Wisconsin
River Falls, WI 54022

University of Wisconsin
Stevens Point, WI 54481

Wyoming

University of Wyoming
Laramie, WY 82071

21:5 ▫ A WORLD VIEW

World-wide careers in soil and crop management and allied subjects are available to those persons with proper aptitude and scientific education.

For example, there are 26 international centers conducting basic and applied agricultural research. The commodities researched include rice, wheat, sorghum, millet, barley, cassava (tapioca), Irish potato, food and forage legumes, vegetables, and livestock.

More information about 13 of these centers may be obtained by writing to: Consultative Group on International Agricultural Research, 1818 H Street, N.W., Washington, DC 20433.

21:6 ▫ REFERENCES

"Exploring Careers in Agronomy, Crops, and Soils." American Society of Agronomy, Crop Science Society of America, and Soil Science Society of America, Madison, Wisconsin (undated).

Occupational Outlook Handbook, 1988–1989. U.S. Dept. of Labor–Bureau of Labor Statistics, Bul. 2300, Washington, DC 20213 (revised periodically).

APPENDIX

A:1 □ CONVERSION TABLES

To Convert from (Column A)	To (Column B)	Multiply Column A by:
Acres	Hectares	0.404 686
	Square feet	43 560.000
	Square kilometers	0.004 047
	Square meters	4 046.856 422
	Square miles (statute)	0.001 562
	Square rods	160.000
	Square yards	4 840.000
Acre-feet	Acre-inches	12.000
	Cubic feet	43 560.000
	Cubic meters	1 233.482 0
	Cubic yards	1 613.333
	Gallons (U.S.)	325 851.560
	Hectare-centimeters	12.335
	Hectare-meters	0.123 35
Acre-feet / day	Cubic feet / second	0.504 17
	Cubic meters / second	0.014 28
Acre-inches	Acre-feet	0.083 33
	Cubic feet	3 630.000
	Cubic meters	102.790 15
	Gallons (U.S.)	27 154.286
	Hectare-centimeters	1.028
Ångstroms	Centimeters	1×10^{-8}*
	Inches	0.907 × 10 [illegible]
	Meters	1×10^{-10}
	Millimeters	1×10^{-7}
Atmospheres	Bars	1.013 25
	Centimeters of Hg (0° C)	76.000
	Centimeters of H_2O (4° C)	1 033.260
	Feet of H_2O (4° C)	33.900
	Grams / square centimeter	1 033.230
	Inches of Hg (0° C)	29.921 3
	Millimeters of Hg (0° C)	760.000
	Pascals	101 325.000
	Pounds / square inch	14.696 0

*This is a negative logarithmic number. The conventional number would be written as 0.000 000 01.

(Continued)

A:1 □ CONVERSION TABLES (Continued)

To Convert from (Column A)	To (Column B)	Multiply Column A by:
Bars	Atmospheres	0.986 923
	Centimeters of Hg (0° C)	75.006 2
	Feet of H_2O (15° C)	33.488 3
	Grams / square centimeter	1 019.716
	Inches of Hg (0° C)	29.530 0
	Megapascals	0.100
	Millibars	1 000.000
	Pounds / square inch	14.503 8
Board-feet	Cubic centimeters	2 359.737 2
	Cubic feet	0.083 333
	Cubic inches	144.000
British thermal units (BTU)	Joules	1 054.350 264
	Kilowatt-hours	0.000 293
Bushels (U.S.) (level full, known as stricken or struck bushel. A heaping bushel is 1¼ stricken bushel)	Cubic centimeters	35 239.070
	Cubic feet	1.244 456
	Cubic inches	2 150.420
	Cubic meters	0.035 239
	Cubic yards	7.481
	Gallons (U.S., dry)	8.000
	Gallons (U.S., liquid)	9.309 177
	Liters	35.239 07
	Pecks (U.S.)	4.000
	Pints (U.S., dry)	64.000
	Quarts (U.S., dry)	32.000
	Quarts (U.S., liquid)	37.236 71
Calcium (Ca)	Calcium carbonate ($CaCO_3$)	2.497 255
	Calcium hydroxide ($Ca(OH)_2$)	1.848 802
	Calcium oxide (CaO)	1.399 202
Calcium carbonate ($CaCO_3$)	Calcium (Ca)	0.400 440
	Calcium hydroxide ($Ca(OH)_2$)	0.740 334
	Calcium oxide (CaO)	0.560 296
Calcium hydroxide [$Ca(OH)_2$]	Calcium (Ca)	0.540 891
	Calcium carbonate ($CaCO_3$)	1.350 742
	Calcium oxide (CaO)	0.756 815
Calcium oxide (CaO)	Calcium (Ca)	0.714 693 2
	Calcium carbonate ($CaCO_3$)	1.784 772
	Calcium hydroxide ($Ca(OH)_2$)	1.320 856 6
Calories, gm, mean	British thermal units (BTU)	0.003 974
	Calories, kg, mean	0.001
	Foot-pounds	3.090 40
	Horsepower-hours	$1.560\ 81 \times 10^{-6}$
	Joules	4.190 02
	Kilowatt-hours	$1.163\ 90 \times 10^{-6}$

(Continued)

A:1 □ CONVERSION TABLES (Continued)

To Convert from (Column A)	To (Column B)	Multiply Column A by:
Celsius (° C)	Fahrenheit (° F)	1.8° C + 32
Centimeters	Ångstrom units	1 × 10^8**
	Feet	0.032 808
	Inches	0.393 701
	Meters	0.010
	Microns	10 000.000
	Millimeters	10.000
	Millimicrons	1 × 10^7
Centimeters / second	Feet / minute	1.968 504
	Kilometers / hour	0.036
	Meters / minute	0.600
	Miles / hour	0.022 369
Chains (Gunter's)	Feet	66.000
	Furlongs	0.100
	Links (Gunter's)	100.000
	Meters	20.116 8
	Miles (statute)	0.012 5
	Yards	22.000
Cords (of wood), standard	Cubic feet	128.000
	Cubic meters	3.624 573
Crop yield conversions		
Barley (bushels / acre) (48 lb / bu)	Kilograms / hectare	53.808
Corn (bushels / acre) (56 lb / bu)	Kilograms / hectare	62.776
Cotton, lint (bales / acre) (500 lb / bale)	Kilograms / hectare	560.500
Flax seed (bushels / acre) (56 lb / bu)	Kilograms / hectare	62.776
Millet (bushels / acre) (48 lb / bu)	Kilograms / hectare	53.808
Millet (bushels / acre) (50 lb / bu)	Kilograms / hectare	56.050
Peanuts (bushels / acre)		
Virginia type (17 lb / bu)	Kilograms / hectare	19.057
Spanish type (25 lb / bu)	Kilograms / hectare	28.025
Potatoes, Irish (bushels / acre) (60 lb / bu)	Kilograms / hectare	67.260
Rice, rough (bushels / acre) (45 lb / bu)	Kilograms / hectare	50.445
Rye (bushels / acre) (56 lb / bu)	Kilograms / hectare	62.776

**This is a positive logarithmic number. The conventional number would be written as 100 000 000.

(Continued)

A:1 □ CONVERSION TABLES (Continued)

To Convert from (Column A)	To (Column B)	Multiply Column A by:
Crop yield conversions (cont.)		
Sorghum (bushels / acre) (56 lb / bu)	Kilograms / hectare	62.776
Sorgo seed (bushels / acre) (50 lb / bu)	Kilograms / hectare	56.050
Soybeans (bushels / acre) (60 lb / bu)	Kilograms / hectare	67.260
Sudangrass (bushels / acre) (40 lb / bu)	Kilograms / hectare	44.840
Sweet potatoes (bushels / acre) (55 lb / bu)	Kilograms / hectare	61.655
Vetch (bushels / acre) (60 lb / bu)	Kilograms / hectare	67.260
Wheat (bushels / acre) (60 lb / bu)	Kilograms / hectare	67.260
Cubic centimeters	Bushels (U.S.)	$2.837\ 759 \times 10^{-5}$
	Cubic inches	0.061 024
	Cubic meters	1×10^{-6}
	Cubic yards	$1.307\ 95\ 1 \times 10^{-6}$
	Cups	0.004 227
	Gallons (U.S., liquid)	0.000 264
	Liters	0.001
	Ounces (U.S., fluid)	0.033 814
	Quarts (U.S., liquid)	0.001 057
Cubic feet	Acre-feet	2.296×10^{-5}
	Bushels (U.S.)	0.803 564
	Cords (of wood)	0.007 812
	Cubic centimeters	28 316.847
	Cubic inches	1 728.000
	Cubic meters	0.028 317
	Cubic yards	0.037 037
	Gallons (U.S., liquid)	7.480 520
	Liters	28.316 847
	Ounces (U.S., fluid)	957.506 49
	Pounds of water at 21° C (avoirdupois)	62.366
Cubic feet / second	Acre-feet / day	1.983 47
	Acre-inches / hour	0.991 73
	Cubic centimeters / second	28 316.847
	Cubic meters / second	0.028 317
	Hectare-centimeters / hour	1.019 4
	Liters / minute	1 699.011
	Liters / second	28.316 85
	Million gallons / day	0.646 412

(Continued)

A:1 □ CONVERSION TABLES (Continued)

To Convert from (Column A)	To (Column B)	Multiply Column A by:
Cubic inches	Board-feet	0.006 944
	Bushels (U.S.)	0.000 465
	Cubic centimeters	16.387 064
	Cubic feet	0.000 579
	Cubic meters	$1.638\ 706 \times 10^{-5}$
	Cubic yards	2.143×10^{-5}
	Gallons (U.S., liquid)	0.004 329
	Liters	0.016 387
	Milliliters	16.387 064
	Ounces (U.S., fluid)	0.554 113
	Quarts (U.S., liquid)	0.017 316
Cubic meters	Acre-feet	0.000 811
	Cords (of wood)	0.384
	Cubic centimeters	1×10^{6}
	Cubic feet	35.314 667
	Cubic inches	61 023.740
	Cubic yards	1.307 951
	Gallons (U.S., liquid)	264.172 05
	Liters	1 000.000
Cubic miles (U.S. statute)	Acre-feet	3.379×10^{6}
Cubic yards	Acre-feet	6.190×10^{-4}
	Cubic centimeters	764 554.860
	Cubic feet	27.000
	Cubic inches	46 656.000
	Cubic meters	0.764 555
	Gallons (U.S., liquid)	201.974 03
	Liters	764.554 86
Cups	Cubic centimeters	236.588
	Liters	0.240
	Milliliters	240.000
	Ounces	8.000
Degrees Celsius (° C)	Degrees Fahrenheit (° F)	$1.8 \times {}^\circ C + 32$
Degrees Fahrenheit (° F)	Degrees Celsius (° C)	$({}^\circ F - 32) \times 0.556$
Dynes	Newtons	0.000 01
Ergs	Joules	1×10^{-7}
Feet	Centimeters	30.480 37
	Inches	12.000
	Kilometers	3.048×10^{-4}
	Meters	0.304 8
	Miles (statute)	$1.893\ 93 \times 10^{-4}$
	Yards	0.333 333

(Continued)

A:1 □ CONVERSION TABLES (Continued)

To Convert from (Column A)	To (Column B)	Multiply Column A by:
Furlongs	Chains (Gunter's)	10.000
	Feet	660.000
	Meters	201.168
	Miles (statute)	0.125
	Rods	40.000
	Yards	220.000
Gallons (U.S., liquid)	Acre-feet	3.068 883 $\times 10^{-6}$
	Cubic centimeters	3 785.411 8
	Cubic feet	0.133 681
	Cubic inches	231.000
	Cubic meters	0.003 785
	Cubic yards	0.004 951
	Gallons (British)	0.832 675
	Gallons (U.S., dry)	0.859 367
	Liters	3.785 412
	Ounces (U.S., fluid)	128.000
	Pints (U.S., liquid)	8.000
	Quarts (U.S., liquid)	4.000
Gallons (U.S.) of H_2O (60° F, 15° C)	Pounds of water	8.328 23
Gallons (U.S., liquid) / minute	Acre-feet / day	0.004 419
	Cubic feet / hour	8.020 8
	Cubic feet / second	2.228 $\times 10^{-3}$
	Cubic meters / hour	0.227
	Cubic meters / second	0.631 $\times 10^{-4}$
	Hectare-centimeters / hour	0.002 27
	Liters / second	0.063 1
Gallons / acre	Liters / hectare	9.353
Gallons / day	Liters / second	4.381 $\times 10^{-5}$
Grams	Kilograms	0.001
	Ounces (apothecary or troy)	0.032 151
	Ounces (avoirupois)	0.035 274
	Pounds (apothecary or troy)	0.002 679
	Pounds (avoirdupois)	0.002 205
	Tons (metric) (megagrams)	1 $\times 10^{-6}$
Grams / cubic centimeter	Pounds / cubic foot	62.427 961
	Pounds / cubic inch	0.036 127
	Pounds / gallon (U.S., liquid)	8.345 404
Grams / liter	Parts / million	1 000.000
	Pounds / cubic foot	0.062 426

(Continued)

A:1 □ CONVERSION TABLES (Continued)

To Convert from (Column A)	To (Column B)	Multiply Column A by:
Hectares	Acres	2.471 054
	Square feet	107 639.100
	Square kilometers	0.010
	Square meters	10 000.000
	Square miles	0.003 861
Hectare-centimeters	Acre-feet	0.081 08
	Acre-inches	0.972 76
Hectare-centimeters / hour	Cubic feet / second	0.981
	Gallons (U.S.) / minute	440.300
Hectare-meters	Acre-feet	8.108
	Acre-inches	97.290
Horsepower, mechanical	Foot-pounds / second	550.000
	Horsepower, boiler	0.076 018
	Horsepower, electrical	0.999 598
	Horsepower, metric	1.013 870
	Joules / second	745.700
	Kilowatts	0.745 700
	Tons of refrigeration	0.212 040
	Watts	745.700
Hundredweights (British long)	Kilograms	50.802 345
	Pounds (avoirdupois)	112.000
	Tons (long)	0.050
	Tons (metric)	0.050 817
	Tons (short)	0.056
Hundredweights (U.S. short)	Kilograms	45.359 237
	Pounds (avoirdupois)	100.000
	Tons (long)	0.044 643
	Tons (metric)	0.045 359
	Tons (short)	0.050
Inches	Ångstrom units	2.540×10^{8}
	Centimeters	2.540
	Feet	0.083 333
	Meters	0.025 4
	Yards	0.027 778
Joules	BTU (mean)	9.472×10^{-4}
	Foot-pounds	0.737 684
	Kilowatt-hours	2.778×10^{-7}
	Watt-seconds	1.000

(Continued)

A:1 □ CONVERSION TABLES (Continued)

To Convert from (Column A)	To (Column B)	Multiply Column A by:
Kilograms	Ounces (apothecary or troy)	32.150 747
	Ounces (avoirdupois)	35.273 962
	Pounds (apothecary or troy)	2.679 229
	Pounds (avoirdupois)	2.204 623
	Quintals	0.010
	Tons (long)	9.842×10^{-4}
	Tons (metric)	0.001 000
	Tons (short)	0.001 102
Kilograms / hectare	Pounds / acre	0.892
Kilometers	Centimeters	100 000.000
	Feet	3 280.839 9
	Meters	1 000.000
	Miles (statute)	0.621 371
	Yards	1 093.613 3
Kilometers / hour	Centimeters / second	27.777 778
	Feet / hour	3 280.839 9
	Meters / second	0.277 778
	Miles (statute) / hour	0.621 371
Liters	Bushels (U.S.)	0.028 378
	Cubic centimeters	1 000.000
	Cubic feet	0.035 315
	Cubic inches	61.023 744
	Cubic meters	0.001
	Cubic yards	0.001 308
	Gallons (U.S., liquid)	0.264 172
	Ounces (U.S., fluid)	33.814 023
	Quarts (U.S., liquid)	1.056 688
Meters	Ångstrom units	1×10^{10}
	Centimeters	100.000
	Feet	3.280 840
	Inches	39.370 079
	Kilometers	0.001 000
	Miles (statute)	6.214×10^{-4}
	Millimeters	1 000.000 000
	Rods	0.198 839
	Yards	1.093 613
Microns	Centimeters	0.000 1
	Meters	1×10^{-6}
	Micrometers	1.000
	Millimeters	0.001
	Millimicrons	1 000.000

(Continued)

A:1 □ CONVERSION TABLES (Continued)

To Convert from (Column A)	To (Column B)	Multiply Column A by:
Miles (statute)	Chains (Gunter's)	80.000
	Feet	5 280.000
	Inches	63 360.000
	Kilometers	1.609 344
	Meters	1 609.344
	Rods	320.000
	Yards	1 760.000
Miles / hour	Feet / hour	5 280.000
	Feet / minute	88.000
	Feet / second	1.466 667
	Kilometers / hour	1.609 344
	Meters / minute	26.822 4
	Meters / second	0.477
	Miles / minute	0.016 667
Milligrams	Ounces (apothecary or troy)	3.215×10^{-5}
	Ounces (avoirdupois)	3.527×10^{-5}
	Pounds (apothecary or troy)	2.679×10^{-6}
	Pounds (avoirdupois)	2.205×10^{-6}
Milligrams / liter	Grams / liter	0.001
	Parts / million	1.000
	Pounds / cubic foot	6.243×10^{-5}
Millimeters	Centimeters	0.100
	Feet	0.003 281
	Inches	0.039 37
	Meters	0.001
Millimhos / centimeter	Decisiemens / meter	1.000
Million gallons / day	Acre-inches / day	36.828
	Cubic feet / second	1.547
	Cubic meters / minute	2.629
Nitrogen (N)	Ammonia (NH_3)	1.216 274
	Crude protein	6.250
	Nitrate (NO_3)	4.426 803
Ounces (avoirdupois)	Grams	28.349 523
	Ounces (apothecary or troy)	0.911 458 3
	Pounds (apothecary or troy)	0.075 955
	Pounds (avoirdupois)	0.062 5

(Continued)

A:1 ▫ CONVERSION TABLES (Continued)

To Convert from (Column A)	To (Column B)	Multiply Column A by:
Ounces (U.S., fluid)	Cubic centimeters	29.573 530
	Cubic inches	1.804 688
	Cubic meters	2.957×10^{-5}
	Cups	0.125
	Gallons (U.S., liquid)	7.812×10^{-3}
	Liters	0.029 574
	Quarts (U.S., liquid)	0.031 25
	Tablespoons	2.000
Parts / million	Grams / liter	0.001
	Milligrams / liter	1.000
Phosphorus (P)	Phosphorus (P_2O_5)	2.291 37
Phosphorus (P_2O_5)	Phosphorus (P)	0.436 41
Pints (U.S., dry)	Bushels (U.S.)	0.015 625
	Cubic centimeters	550.610 47
	Cubic inches	33.600 312
	Gallons (U.S., dry)	0.125
	Gallons (U.S., liquid)	0.145 456
	Liters	0.550 610
	Pecks (U.S.)	0.062 5
	Pints (U.S., liquid)	1.163 647
Pints (U.S., liquid)	Cubic centimeters	473.176 47
	Cubic feet	0.016 710
	Cubic inches	28.875
	Cubic yards	0.000 619
	Cups	2.000
	Gallons (U.S., liquid)	0.125
	Liters	0.473 177
	Ounces (U.S., fluid)	16.000
	Pints (U.S., dry)	0.859 367
Potassium (K)	Potassium (K_2O)	1.204 59
Potassium (K_2O)	Potassium (K)	0.830 16
Pounds (avoirdupois)	Grams	453.592 37
	Kilograms	0.453 592
	Ounces (apothecary or troy)	14.583 333
	Ounces (avoirdupois)	16.000
	Pounds (apothecary or troy)	1.215 277
Pounds / acre	Kilograms / hectare	1.121
	Megagrams / hectare	0.001 121
	Metric tons / hectare	0.001 121
	Quintals / hectare	0.011 21

(Continued)

A:1 ▫ CONVERSION TABLES (Continued)

To Convert from (Column A)	To (Column B)	Multiply Column A by:
Pounds / cubic foot	Grams / cubic centimeter	0.016 018
	Kilograms / cubic meter	16.018 463
	Pounds / cubic inch	5.787×10^{-4}
Pounds / cubic inch	Grams / cubic centimeter	27.679 905
	Grams / liter	27 679.905
	Kilograms / cubic meter	27 679.905
Pounds / square foot	Pascals	47.88
Pounds / square inch	Atmospheres	0.068 046
	Bars	0.068 948
	Grams / square centimeter	70.306 958
	Pascals	6 900.0
Quarts (U.S., liquid)	Cubic centimeters	946.352 95
	Cubic feet	0.033 420
	Cubic inches	57.750
	Gallons (U.S., dry)	0.214 842
	Gallons (U.S., liquid)	0.250
	Liters	0.946 353
	Ounces (U.S., liquid)	32.000
	Pints (U.S., liquid)	2.000
Quintals	Kilograms	100.000
	Pounds (avoirdupois)	220.462 26
	Metric tons	10.000
Quintals / hectare	Kilograms / hectare	100.000
	Metric tons / hectare	0.100
	Pounds / acre	89.206 07
Rods	Feet	16.500
	Feet (U.S. survey)	16.499 967
	Furlongs	0.025
	Inches	198.000
	Meters	5.029 2
	Miles (statute)	0.003 125
	Yards	5.500
Square centimeters	Square feet	1.076×10^{-3}
	Square inches	0.155 000
	Square meters	0.000 1
	Square yards	1.196×10^{-4}
Square centimeters / gram	Square meters / kilogram	0.100
Square chains (Gunter's)	Acres	0.100
	Square meters	404.686
	Square miles	0.000 156
	Square rods	16.000
	Square yards	484.000

(Continued)

A:1 ▫ CONVERSION TABLES (Continued)

To Convert from (Column A)	To (Column B)	Multiply Column A by:
Square feet	Acres	2.296 × 10^{-5}
	Hectares	9.290 3 × 10^{-6}
	Square centimeters	929.030 4
	Square inches	144.000
	Square meters	0.092 903
	Square miles	3.587 × 10^{-8}
	Square yards	0.111 111
Square inches	Square centimeters	6.451 6
	Square feet	6.944 × 10^{-3}
	Square meters	6.452 × 10^{-4}
	Square millimeters	645.16
Square kilometers	Acres	247.105 38
	Hectares	100.000
	Square feet	1.076 × 10^{7}
	Square inches	1.550 × 10^{9}
	Square meters	1 × 10^{6}
	Square miles (statute)	0.386 102
	Square yards	1.196 × 10^{6}
Square meters	Acres	2.471 × 10^{-4}
	Hectares	0.000 1
	Square centimeters	10 000.000
	Square feet	10.763 910
	Square inches	1 550.003 1
	Square kilometers	1 × 10^{-6}
	Square miles	3.861 × 10^{-7}
	Square yards	1.195 990
Square miles (statute)	Acres	640.000
	Hectares	258.998 81
	Square kilometers	2.589 988
	Square meters	2 589 988.000
	Square yards	3 097 587.500
Square rods	Acres	0.006 25
	Hectares	0.002 529
	Square centimeters	252 928.526 4
	Square feet	272.250
	Square inches	39 204.000
	Square meters	25.292 853
	Square miles	9.766 × 10^{-6}
	Square yards	30.250
Square yards	Acres	2.066 × 10^{-4}
	Hectares	8.361 × 10^{-5}
	Square centimeters	8 361.273 6
	Square feet	9.000
	Square inches	1 296.000
	Square meters	0.836 127
	Square miles	3.228 × 10^{-7}

(Continued)

A:1 □ CONVERSION TABLES (Continued)

To Convert from (Column A)	To (Column B)	Multiply Column A by:
Tablespoons	Ounces	0.500
	Teaspoons	3.000
Teaspoons	Cubic centimeters	4.928 922
	Cups	0.020 829
	Ounces	0.150 150
	Tablespoons	0.333 333
Temperature (° C + 32)	Temperature (° F)	1.80
Temperature (° F − 32)	Temperature (° C)	0.555
Temperature (° C + 273.15) or (° F + 459.67)	Temperature K***	1.00
Tons (long)	Kilograms	1 016.046 9
	Ounces (avoirdupois)	35 840.000
	Pounds (apothecary or troy)	2 722.220
	Pounds (avoirdupois)	2 240.000
	Tons (metric)	1.016 047
	Tons (short)	1.120
Tons (metric) (tonnes) (megagrams)	Grams	1×10^6
	Kilograms	1 000.000
	Ounces (avoirdupois)	35 273.962
	Pounds (apothecary or troy)	2 679.228 9
	Pounds (avoirdupois)	2 204.622 6
	Quintals	0.100
	Tons (long)	0.984 207
	Tons (short)	1.102 311
Tons (metric) / hectare	Kilograms / hectare	1 000.000
	Pounds (avoirdupois) / acres	892.180
	Quintals / hectare	10.000
	Tons (short) / acre	0.446
Tons (short)	Hundredweights (short)	20.000
	Kilograms	907.184 74
	Ounces (avoirdupois)	32 000.000
	Pounds (apothecary or troy)	2 430.555
	Pounds (avoirdupois)	2 000.000
	Tons (long)	0.892 857
	Tons (metric)	0.907 185
Tons (short) / acre	Megagrams / hectare	2.242
	Tons (metric) / hectare	2.242
Township (survey)	Hectares	9 323.957 0
	Square kilometers	93.239 6
	Square miles	36.000

***Degrees Kelvin (K) (absolute zero) is customarily written without a degree sign (°).

(Continued)

A:1 □ CONVERSION TABLES (Continued)

To Convert from (Column A)	To (Column B)	Multiply Column A by:
Water, weight of	Grams	3 774.653
1 gallon at 20° C	Kilograms	3.774 65
(68° F)	Pounds	8.321 67
Yards	Centimeters	91.440
	Feet	3.000
	Inches	36.000
	Meters	0.914 4

A:2 □ REFERENCES

Agricultural Statistics, 1986. U.S. Dept. of Agriculture, pp. v – ix (revised annually).

Green, Marvin H. *Metric Conversion Handbook.* New York: Chemical Publishing Co., 1978, 225 pp.

Ground Water Manual: A Water Resource Technical Publication, rev. reprint. U.S. Department of the Interior. Water and Power Resources Services, 1981, 480 pp.

Metric Practice Guide. Canadian Fertilizer Institute, Ottawa, Ontario, Canada, 1977, 119 pp.

Page, Chester H., and Paul Vigoureux, eds. "The International System of Units (SI)." National Bureau of Standards, U.S. Department of Commerce, NBS Special Publication, 330, 1974.

"Publications Handbook and Style Manual." 1988, pp. 92. American Society of Agronomy, Crop Science Society of America, and Soil Science Society of America, Madison, Wisconsin.

Weast, Robert C., ed. *Handbook of Chemistry and Physics,* 65th ed. CRC Press, 1984 – 1985, pp. F-296 – F-319 (revised annually).

INDEX OF IMPORTANT TERMS

A

B

C

D

E

F

G

H

I

J

K

L

M

N

O

P

Q

R

V

W

Y

Z

INDEX

T

U

V

W

Y

Z